D0083828

Quaternary Paleoclimatology

TITLES OF RELATED INTEREST

Physical processes of sedimentation
J. R. L. Allen

The formation of soil material
T. R. Paton

Petrology of the sedimentary rocks
J. T. Greensmith

Invertebrate palaeontology and evolution
E. N. K. Clarkson

A dynamic stratigraphy of the British Isles
R. Anderton, P. H. Bridges, M. R. Leeder & B. W. Sellwood

Thresholds in geomorphology
D. R. Coates & J. D. Vitek (eds)

Atmospheric processes
J. D. Hanwell

Microfossils
M. D. Brasier

Fluvial geomorphology
M. Morisawa (ed.)

Glacial geomorphology
D. R. Coates (ed.)

Theories of landform development
W. N. Melhorn & R. C. Flemal (eds)

Geomorphology and engineering
D. R. Coates (ed.)

Geomorphology in arid regions
D. O. Doehring (ed.)

Geomorphological techniques
A. S. Goudie (ed.)

Terrain analysis and remote sensing
J. R. G. Townshend (ed.)

British rivers
J. Lewin (ed.)

Soils and landforms
A. J. Gerrard

Soil survey and land evaluation
D. Dent & A. Young

Environmental archaeology
M. L. Shackley

Adjustments of the fluvial system
D. D. Rhodes & G. P. Williams (eds)

Applied geomorphology
R. G. Craig & J. L. Craft (eds)

Space and time in geomorphology
C. E. Thorn (ed.)

Sedimentology: process and product
M. R. Leeder

Sedimentary structures
J. D. Collinson & D. B. Thompson

Geomorphological field manual
V. Gardiner & R. Dackombe

Historical plant geography
P. Stott

The changing climate
M. J. Ford

Aspects of micropalaeontology
F. T. Banner & A. R. Lord (eds)

Statistical methods in geology
R. F. Cheeney

Groundwater as a geomorphic agent
R. G. LaFleur (ed.)

The climatic scene
M. J. Tooley & G. M. Sheail (eds)

Models in geomorphology
M. Woldenberg (ed.)

Environmental change and tropical geomorphology
I. Douglas & T. Spencer (eds)

Paleopalynology
A. Traverse

Quaternary Paleoclimatology

Methods of Paleoclimatic Reconstruction

R. S. Bradley

University of Massachusetts, Amherst

Boston
ALLEN & UNWIN
London Sydney

Allen & Unwin Inc.,
Fifty Cross Street, Winchester, Mass 01890, USA

George Allen & Unwin (Publishers) Ltd,
40 Museum Street, London WC1a 1LU, UK

George Allen & Unwin (Publishers) Ltd,
Park Lane, Hemel Hempstead, Herts HP2 4TE, UK

George Allen & Unwin Australia Pty Ltd,
8 Napier Street, North Sydney, NSW 2060, Australia

First published in 1985

Library of Congress Cataloging in Publication Data

Bradley, Raymond S., 1948–
Quaternary paleoclimatology.
Bibliography: p.
Includes index.
1. Paleoclimatology. 2. Geology, Stratigraphic—Quaternary.
I. Title.
QC884.B614 1984 551.6 84-9281
ISBN 0-04-551067-9 (alk. paper)
ISBN 0-04-551068-7 (pbk. : alk. paper)

British Library Cataloguing in Publication Data

Bradley, Raymond S.
Quaternary paleoclimatology.
1. Paleoclimatology 2. Geology, Stratigraphic—Quaternary
I. Title
551.6 QC884
ISBN 0-04-551067-9
ISBN 0-04-551068-7 Pbk

Cover Photograph A grindstone, formerly used by neolithic people for crushing grain, is all that remains of an agricultural community that once flourished in what is today one of the most inhospitable regions on Earth. In this part of the Algerian Sahara (the Tassili n'Ajjer), the climate has changed radically during mid- to late-Holocene time – from (at least) seasonally moist, to extremely arid. Agricultural activities, which this *in situ* artifact recalls, would be inconceivable in the region today. Evidence for climatic change in sub-Sahara Africa is discussed in Chapters 7 and 8.

Set in 10 on 12 point Palatino by Paston Press, Norwich
and printed in Great Britain by Butler & Tanner Ltd,
Frome and London

To Jane

Preface

Quaternary paleoclimatology is intended as an introduction to methods used in reconstructing past climates from proxy data series. My objective was to provide a summary of each method for the non-specialist, with enough depth to enable the reader to comprehend current research in each subfield. To this end, the references I have quoted are representative of developments in each subfield, but should not be considered comprehensive by any means. However, they will provide the interested reader with a key to further studies, through which most of the pertinent literature will be revealed. As a byproduct of this methodological approach, the reader will become acquainted with the main characteristics of the paleoclimatic record. Furthermore, by following a methodological path, he or she will become aware of the limitations of each approach as well as their inherent attributes, thereby developing a critical perspective on paleoclimatic reconstructions. It is also my hope that this book will encourage specialists in one or another subfield of paleoclimatology to develop a better understanding of methods used in other subfields, so that interdisciplinary research, the hallmark of Quaternary paleoclimatology, will be fostered and strengthened.

The field of paleoclimatology is huge – so huge, in fact, that one might argue that a book purporting to survey the field would have been better written by a team of specialists. However, in my experience this is rarely satisfactory, often resulting in an uneven text which the non-specialist reader may find bewildering. By writing the text myself, I have, in a sense, acted as a filter between the literature and the reader, providing a uniform perspective at least. I believe this approach has much to commend it, but inevitably the result reflects my own biases, and perhaps misunderstandings. I am therefore very grateful to the many specialist reviewers who took the time to go painstakingly over my early efforts, to correct mistakes and to suggest better approaches. At various stages I have benefited greatly from the advice and suggestions of the following individuals: R. G. Barry, H. J. B. Birks, T. J. Blasing, G. Boulton, J. Cruikshank, F. Duerden, J. Grove, R. Harmon, J. T. Hollin, R. M. Koerner, Henry Lamb, C. C. Langway, Jr, the late Gordon Manley, W. McCoy, G. Miller, A. V. & A. Morgan, H. Nichols, W. Riebsame, G. de Q. Robin, M. L. Salgado-Labouriau, N. J. Shackleton, S. Short, R. N. Starling, R. Switsur, B. Weingarten, C. White, A. Wintle, and the students in my paleoclimatology course who were subjected to various drafts. I am particularly grateful to J. T. Andrews and L. D. Williams who provided very valuable comments on the entire manuscript. To all these people, I can only say that I hope the results meet with their approval, though of course any remaining errors or biases are mine alone.

I was particularly fortunate to spend a sabbatical year at the Scott Polar Research Institute, Cambridge, where I started work on the book. My warm thanks to Drs G. de Q. Robin and T. Armstrong, H. King and his library staff, and to colleagues at Clare Hall for making that year so fruitful and enjoyable. I would also like to thank Marie Litterer, for drafting many of the figures, and my colleagues in the Department of Geology and Geography at the University of Massachusetts who have been very supportive during the period I have spent completing the book. However, it is my wife, Jane, to whom I owe the greatest debt. Not only has she typed the entire manuscript twice, but has encouraged me at every turn and put up with bouts of frustration and depression. There were times when I felt that the job would never be finished and that Leonardo da Vinci was right: "Avoid those studies of which the result dies with the worker"! The fact that I made it is in no small part due to her help.

October 1983 Ray Bradley

Contents

List of tables

1

Paleoclimatic reconstruction

"From the time when written records began, and doubtless long before that, thoughtful men have had two preoccupations: the past and the future. It is this impelling sense of time and process that distinguishes such individuals from the 'practical man', absorbed as he is in the immediate present. Although we still have with us this divergence of attention and emphasis, the birth of modern science has been, in a curious way, the result of their fusion. For man has, at length, learnt that the key to past and future lies in the scrutiny of evidence that is before his eyes, here and now. . . ."

(Sears 1964)

1.1 Introduction

Paleoclimatology is the study of climate prior to the period of instrumental measurements. Instrumental records span only a tiny fraction ($<10^{-7}$) of the Earth's climatic history and so provide an inadequate perspective on climatic variation and the evolution of climate today. A longer perspective on climatic variability can be obtained by the study of natural phenomena which are climate-dependent, and which incorporate into their structure a measure of this dependency. Such phenomena provide a proxy record of climate and it is the study of proxy data that is the foundation of paleoclimatology. As a more detailed and reliable record of past climatic fluctuations is built up, the possibility of identifying causes and mechanisms of climatic variation is increased. Thus, paleoclimatic data provide the basis for testing hypotheses about the causes of climatic change. Only when the causes of *past* climatic fluctuations are understood will it be possible to anticipate or forecast climatic variations in the future (US Committee for GARP 1975).

Studies of past climates must begin with an understanding of the types of proxy data available and the methods used in their analysis. One must be aware of the difficulties associated with each method used and of the assumptions each entails. With such a background, it may then be possible to synthesize different lines of evidence into a comprehensive picture of former climatic fluctuations, and to test hypotheses about the

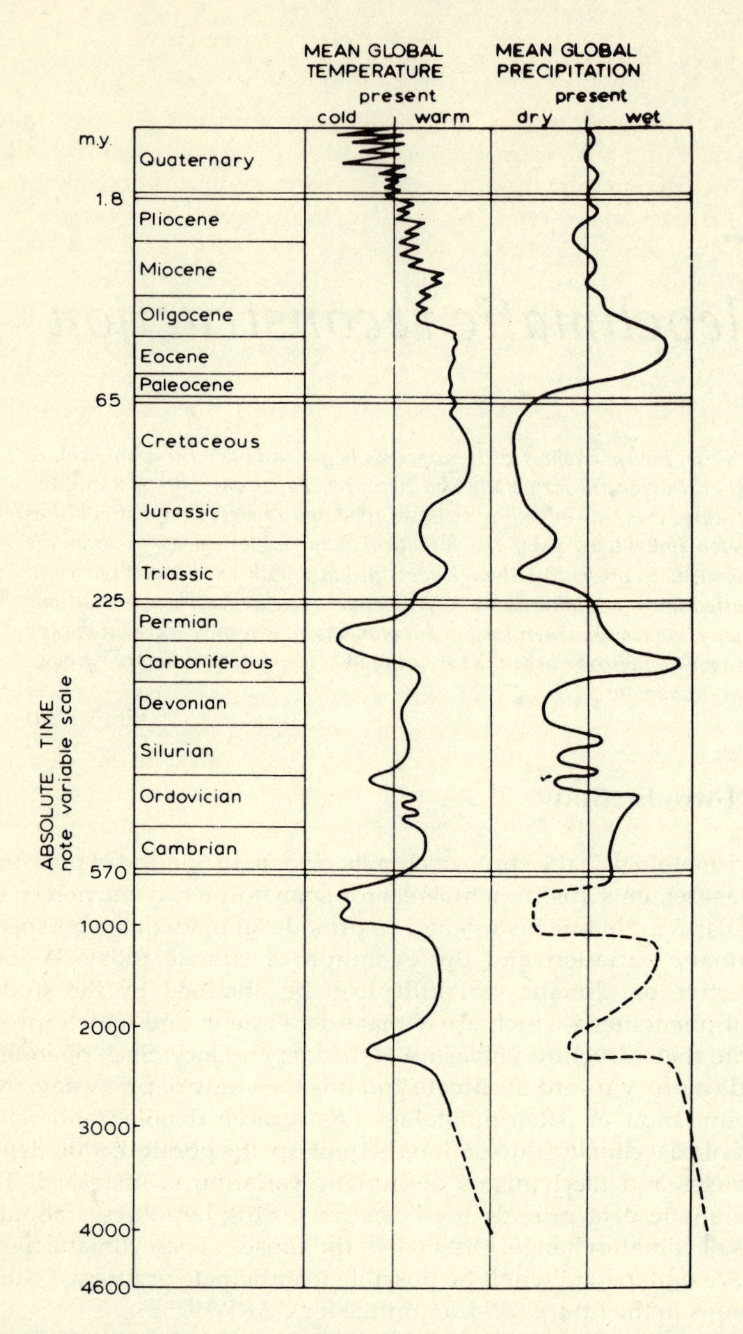

Figure 1.1 Generalized temperature and precipitation history of the Earth. Trends are dashed where data are very sparse. The curves are drawn to represent postulated departures from present global means, but only relative values are indicated (from Frakes 1979). Note that the timescale is progressively expanded in younger time units. Studying the proxy record of paleoclimate is rather like looking through a telescope held the wrong way around; for recent periods there is evidence of short-term climatic variations, but these cannot be resolved in earlier periods.

causes of climatic change. This book deals with the different types of proxy data and how these have been used in paleoclimatic reconstructions. The theme is methodological, but through discussion of examples, selected from major contributions in each field, an overview of the climatic record during the late Quaternary (the last $\sim 0.75 \times 10^6$ years) is also provided. The climate of earlier periods can be studied by some of the methods discussed here (particularly those in Chs 6, 7 & 9) but the further back in time one goes, the greater are the problems of dating, preservation, disturbance, and hence interpretation. For a thorough discussion of climate over a much longer period, the reader is referred to the very comprehensive survey by Frakes (1979).

The Quaternary was a period of major environmental changes, possibly greater than at any other time in the last 60 million years (Fig. 1.1), though our perspective on the past is obviously somewhat myopic. Nevertheless, there is no doubt that an understanding of climatic variation and change during the Quaternary Period is necessary not only to appreciate many features of the natural environment today, but also to fully comprehend present climate. Different components of the climate system (see Sec. 2.2) change and respond to external factors at different rates; in order to understand the role such components play in the evolution of climate it is necessary to have a record considerably longer than the time it takes for them to undergo significant changes. For example, the growth and decay of continental ice sheets may take tens of thousands of years; in order to understand the factors leading up to such events and the effects such events subsequently have on climate, it is necessary to have a record considerably longer than the cryospheric (snow and ice) changes which have taken place. Furthermore, since major periods of global ice build-up and decay appear to have occurred on a quasi-periodic basis during at least the late Quaternary, a much longer record than the mean duration of this period ($\sim 10^5$ years) is necessary to determine the causative factors, and to appreciate how those factors play a role in climate today. A detailed paleoclimatic record, spanning at least the late Quaternary period, is therefore fundamental to comprehension of modern climate, and the causes of climatic variation and change (Kutzbach 1976). Furthermore, unless the *natural* variability of climate is understood, it will be extremely difficult to recognize any man-made effects on climate that may become apparent in the future (Wilson & Matthews 1971).

1.2 Sources of paleoclimatic information

Many natural systems are dependent on climate; where evidence of such systems in the past still exists, it may be possible to derive paleoclimatic information from them. By definition, such proxy records of climate all contain a climatic signal, but that signal may be relatively weak, embedded in a great deal of extraneous "noise" arising from the effects of other (non-climatic) influences. The proxy material has acted as a filter, transforming climatic conditions at a point in time, or over a period, into a more or less permanent record, but the record is complex and incorporates other signals which may be irrelevant to the paleoclimatologist.

To extract the paleoclimatic signal from proxy data, the record must first be calibrated. Calibration involves using modern climatic records and proxy materials to understand how, and to what extent, proxy materials are climate-dependent. It is assumed that the modern relationships observed have operated, unchanged, throughout the period of interest (the principle of uniformitarianism). All paleoclimatic research, therefore, must build on studies of climate dependency in natural phenomena today. Dendroclimatic studies, for example, have benefited from a wealth of research into climate – tree growth relationships, which have enabled dendroclimatic models to be based on sound ecological principles (Ch. 10). Significant advances have also been made in palynological research by improvements in our understanding of the relationships between modern climate and modern pollen rain (Ch. 9). It is apparent, therefore, that an adequate modern data base is an important pre-requisite for reliable paleoclimatic reconstructions. However, not all environmental conditions in the past are represented in the period of modern experience. Obviously, situations existed during glacial and early Postglacial times which defy characterization by modern analogs. One must therefore be aware of the possibility that erroneous paleoclimatic reconstructions may result from the use of modern climate–proxy data relationships when past conditions have no analog in the modern world (Sachs *et al.* 1977). By the use of more than one calibration equation it may be possible to detect such periods and avoid the associated errors (Hutson 1977; see also Sec. 6.4).

Major types of proxy climatic data available are listed in Table 1.1. Each line of evidence differs according to its spatial coverage, the period to which it pertains, and its ability to resolve events accurately in time. For example, ocean sediment cores are potentially available from 70% of the Earth's surface, and may provide continuous proxy records of climate spanning many millions of years. However, these records are difficult to date accurately; commonly there is a dating uncertainty of ±5% of the sample's true age (the absolute magnitude of the uncertainty thus increasing with sample age). Mixing of sediments by marine organisms and generally low sedimentation rates also make it difficult to obtain samples which represent less than 500–1000 year intervals (depending on depth in the core). This large minimum sampling interval means that the value of ocean core studies lies in low-frequency (long-term) paleoclima-

Table 1.1 Principal sources of proxy data for paleoclimatic reconstructions.

(1) GLACIOLOGICAL (ICE CORES)
- (a) oxygen isotopes
- (b) physical properties (e.g. ice fabric)
- (c) trace element and microparticle concentrations

(2) GEOLOGICAL
- (A) *Marine* (ocean sediment cores)
  - (i) *Organic sediments* (planktonic and benthic fossils)
    - (a) oxygen isotopic composition
    - (b) faunal and floral abundance
    - (c) morphological variations
  - (ii) *Inorganic sediments*
    - (a) mineralogical composition and surface texture
    - (b) accumulation rates, and distribution of terrestrial dust and ice-rafted debris
    - (c) geochemistry
- (B) *Terrestrial*
  - (a) glacial deposits and features of glacial erosion
  - (b) periglacial features
  - (c) glacio-eustatic features (shorelines)
  - (d) aeolian deposits (loess and sand dunes)
  - (e) lacustrine deposits and erosional features (lacustrine sediments and shorelines)
  - (f) pedological features (relict soils)
  - (g) speleothems (age and stable isotope composition)

(3) BIOLOGICAL
- (a) tree rings (width, density, stable isotope composition)
- (b) pollen (type, relative abundance and/or absolute concentration)
- (c) plant macrofossils (age and distribution)
- (d) insects (type and assemblage abundance)
- (e) modern population distribution (refuges and relict populations of plants and animals)

(4) HISTORICAL
- (a) written records of environmental indicators (parameteorological phenomena)
- (b) phenological records

tic information (on the order of 10^3–10^4 years; see Ch. 6). By contrast, tree rings are available from much of the continental land mass, can be accurately dated to an individual year, and may provide continuous records of up to several thousand years in duration. With a minimum sampling interval of one year, they provide primarily high-frequency (short-term) paleoclimatic information (Ch. 10). Table 1.2 documents the main characteristics of these and other sources of paleoclimatic data. The value of proxy data to paleoclimatic reconstructions is very dependent on the minimum sampling interval and dating resolution, since it is this which primarily determines the degree of detail available from the record. At the present time, annual and even seasonal resolution of

Table 1.2 Characteristics of major paleoclimatic data sources.†

Data sources	Variable measured	Continuity of evidence	Potential geographical coverage	Period open to study (years)	Minimum sampling interval (years)	Usual dating accuracy (years)	Climate-related inferences
Ocean sediments	Isotopic composition of planktonic and benthic fossils Floral and faunal assemblages Morphological characteristics of fossils Mineralogical composition and abundance	Continuous	Global ocean, except (for carbonate fossils) deepest zones (below $CaCO_3$ compensation depths)				Global ice volume; surface temperature and salinity; bottom temperature and bottom water flux; aridity of adjacent land areas; prevailing wind direction and strength
		Sedimentation rates (cm per 1000 years)					
		<2	Favored areas along continental margins	1 000 000+	1000+	±5%	
		2–5		200 000+	500+	±5%	
		>10		10 000+	50+	±5%	
Ice cores	Oxygen isotope composition Trace chemistry and electrolytic conductivity Fabric	Continuous	Glaciated regions in polar and alpine areas (optimally in dry snow zones)	100 000+	Variable, but optimally 1–10 years for last 10^4 years	Variable, but optimally 0.05% for last 10^3 years	Temperature, accumulation rates, atmospheric composition and turbidity, ice thickness (height), solar output variations

Mountain glaciers	Terminal positions Glaciation levels and equilibrium line altitudes	Episodic	45°S to 70°N	50 000	—	±5–10%	Temperature, precipitation (net accumulation)
Closed basin lakes	Lake level	Episodic	Low to mid latitudes (arid and semi-arid environments)	50 000		±5%	Moisture availability ("effective precipitation")
Bog or lake sediments	Insect assemblage composition Pollen type concentration, geochemical and sedimentological composition	Continuous	All continents	10 000+ (common) 150 000 (rare)	~50	±5%	Temperature, precipitation, soil moisture, air mass frequencies
(varved sediments)			Mid to high latitudes	10 000+	1–10	+1–10	
Tree rings	Ring width anomaly, density, isotopic composition	Continuous	Mid- and high-latitude continents	1000 (common) 8000 (rare)	1	1	Temperature, runoff precipitation, soil moisture, pressure (circulation modes)
Written records	Phenology, weather logs, sailing logs, etc.	Episodic or continuous	Global	1000+	1	1	Varied

† Based on Table 1 of Kutzbach, 1975, but considerably modified.

climatic fluctuations in the timescale 10^1–10^3 years is provided by ice-core and tree-ring studies (Chs 5 & 10). Detailed analyses of pollen in varves may provide annual data, but it is likely that the pollen itself is an integrated measure of the pollen rain over a number of prior years (Jacobson & Bradshaw 1981). On the longer timescale (10^5–10^6 years) ocean cores provide the best records at present, though resolution probably decreases to $> \pm 10^4$ years beyond 1 million years ago. Historical records have the potential of providing annual (or intra-annual) data for a thousand years or more in some areas, but so far this potential has rarely been realized, even for the last few centuries (Ch. 11).

Not all paleoclimatic records are sensitive indicators of abrupt changes in climate; the climate-dependent phenomenon may lag behind the climatic perturbation so that abrupt changes appear as gradual transitions in the paleoclimatic record. Different proxy systems have different levels of inertia with respect to climate, such that some systems vary essentially in phase with climatic variations whereas others lag behind by as much as several centuries (Bryson & Wendland 1967). This is not simply a question of dating accuracy but a fundamental attribute of the proxy system in question. Pollen, for example, derives from vegetation which may take hundreds of years to adjust to an abrupt change in climate. Even with interannual resolution in the pollen record, sharp changes in climate are unlikely to be reflected in pollen assemblages, as the vegetation affected may take many centuries to adjust to a new climatic state (though interannual values of total pollen influx may provide clues to rapid shifts in circulation patterns). By contrast, the record of fossil insects may point to short-term changes of climate (because insect populations are often highly sensitive to temperature fluctuations) which are not resolvable using pollen analysis alone (see Sec. 8.4). Thus, not all proxy data are readily comparable, because of differences in response time to climatic variations.

In terms of the resolution provided by proxy data, it is also worth noting that all data sources provide a continuous record. Certain phenomena provide discontinuous or episodic information; glacier advances, for example, may leave geomorphological evidence of their former extent (moraines, trim-lines, etc.) but these represent discrete events in time, resulting from the integration of climatic conditions prior to the ice advance (Sec. 7.4). Furthermore, major ice advances may obliterate evidence of previous, smaller advances, so the geomorphological record is likely to be not only discontinuous but also incomplete. Studies of continuous paleoclimatic records can help to place such episodic information in perspective and, for this reason, some have advocated that continuous marine sedimentary records be used as a chronological and paleoclimatic reference frame for long-term climatic fluctuations (e.g. Kukla 1977). This should not imply, however, that glacial events are globally synchronous; indeed there is much evidence that this is not the case (Andrews & Barry 1978, Grove 1979).

1.3 Levels of paleoclimatic analysis

Paleoclimatic reconstruction may be considered to proceed through a number of stages or levels of analysis. The first stage is that of data collection, generally involving fieldwork, followed by initial laboratory analyses and measurements. This results in primary or level 1 data (cf. Hecht *et al.* 1979, Peterson *et al.* 1979). Measurements of tree-ring widths or the isotopic content of marine foraminifera from an ocean core are examples of primary data. At the next stage, the level 1 data are calibrated and converted to estimates of paleoclimate. The calibration may be entirely qualitative, involving a subjective assessment of what the primary data represent (e.g. "warmer," "wetter," "cooler" conditions, etc.) or may involve an explicit, reproducible procedure that provides quantitative estimates of paleoclimate. These derived or level 2 data provide a record of climatic variation through time at a particular location. For example, tree-ring widths from a site near the alpine or arctic treeline may be transformed into a paleotemperature record for that location, using a calibration equation derived from the relationship between modern climatic data and modern tree-ring widths (see Ch. 10).

Different level 2 data may also be mapped to provide a regional synthesis of paleoclimate at a particular time, the synthesis providing greater insight into former circulation patterns than any of the individual level 2 data sets could provide alone (e.g. Nicholson & Flohn 1980). In some cases, three-dimensional arrays of level 2 data (i.e. spatial patterns of paleoclimatic estimates through time) have been transformed into objectively derived statistical summaries. For example, spatial patterns of drought in China over the last 500 years (based on level 1 historical data) have been converted into a small number of principal components (eigenvectors) which account for most of the variance in the level 2 data set (Wang & Zhao 1981). The eigenvectors show that there are a small number of modes, or patterns of drought, which characterize the data. The statistics derived from such analyses constitute a third level of paleoclimatic data (level 3 data).

At present, most paleoclimatic research involves level 1 and level 2 data at individual sites. There are few studies of the spatial dimensions of climate at particular periods in the past. Notable exceptions are the CLIMAP reconstructions of marine and continental conditions at 18 000 years BP (CLIMAP Project Members 1976, Peterson *et al.* 1979). On the continents, there is on average only one well dated paleoclimatic record per 10^6 km^2 for this time period, with the best network densities in

Australia and Oceania (3.5 data points per 10^6 km^2) and the worst networks in Africa, South America, and Asia (Peterson *et al.* 1979). In North America, there are 2.7 sites per 10^6 km^2, equivalent to ~60 records scattered (unfortunately non-randomly) throughout the continent.† By comparison, the modern cooperative climate station network in the USA averages ~1100 stations per 10^6 km^2. These statistics dramatize how superficial is our current understanding of climatic conditions during the last glacial maximum. Similar studies are needed for other time periods to shed light on key areas of contradicting evidence and to focus attention on areas of the world where additional data would be most valuable.

†A mean site density of 2.5 sites per 10^6 km^2 implies a characteristic distance between sites of ~600 km and hence the smallest resolvable feature would be ~1200 km (Kutzbach & Guetter 1980).

2

Climate and climatic variation

2.1 The nature of climate and climatic variation

Climate is the statistical expression of daily weather events; more simply, climate is the expected weather. Naturally, for a particular location, certain weather events will be common (or highly probable); these will lie close to the central tendency or mean of the distribution of weather events. Other types of weather will be more extreme and less frequent; the more extreme the event, the lower the probability of recurrence. Such events would appear at the margins of a distribution of weather events characterizing a particular climate. The overall distribution of climatic parameters defines the climatic variability of the place. If we were to measure temperature in the same location for a finite period of time, the statistical distribution of measured values would reflect the geographical situation of the site (in relation to solar radiation receipts and degree of continentality) as well as the relative frequency of synoptic weather patterns and the associated airflow over the region. Given a long enough period of observations, it would be possible to characterize the temperature of the site in terms of mean and variance. Similarly, observations of other meteorological parameters, such as precipitation, relative humidity, solar radiation, cloudiness, wind speed, and direction, would enable a more comprehensive understanding of the climate of the site to be obtained. However, implicit in such statistics is the element of time. For how long should observations be taken to obtain a reliable picture of the climate at a particular place? The World Meteorological Organization has recommended the adoption of standardized 30 year periods to characterize climate; initially, the period 1931–60 was chosen (Mitchell *et al.* 1966, Jagannathan *et al.* 1967). Adoption of a standard reference period is considered to be necessary because the statistics which define climate in one area may vary over time so that climate, strictly speaking, should always be defined with reference to the period used in its calculation.

Climate may vary in different ways. Some examples of climatic variation are shown in Figure 2.1. Variations may be periodic (and hence predictable), quasi-periodic (predictable only in the very broadest terms), or non-periodic. Central tendencies (mean values) may remain

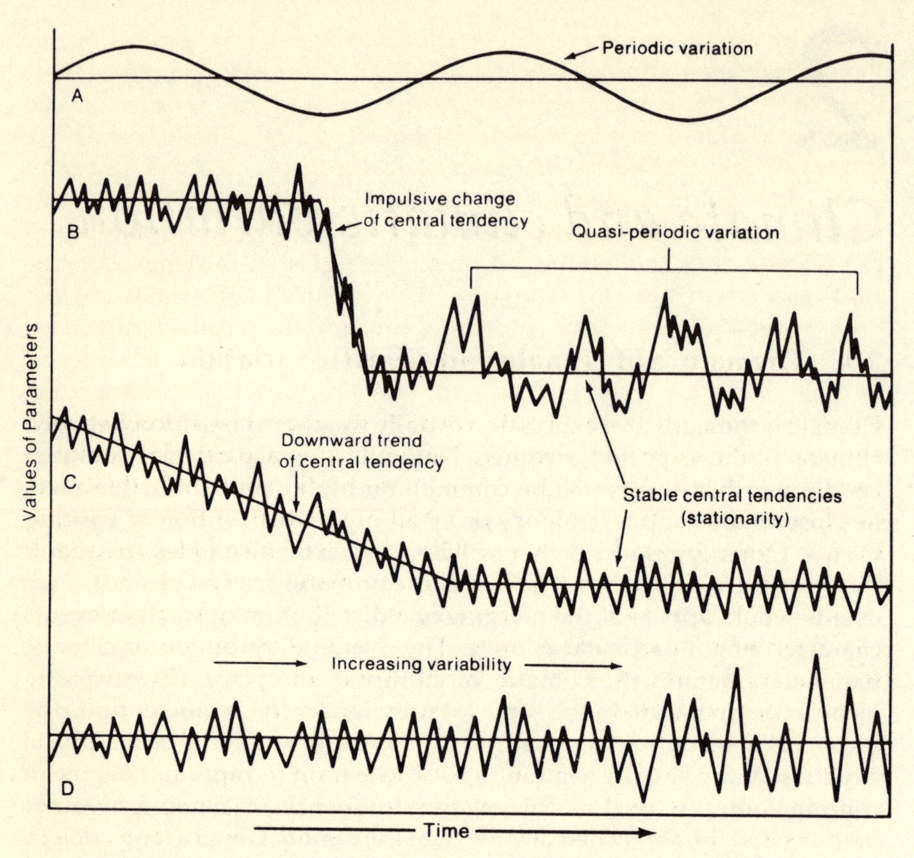

Figure 2.1 Examples of climatic variation and variability (from Hare 1979).

more or less constant or exhibit trends or impulsive changes from one mean to another (Hare 1979). Such occurrences may appear to be random in a time series but this does not necessarily mean they are not predictable. For example, a number of studies have shown that an impulsive change in the climate of some areas resulted from the eruption of Mt Agung in 1963 (Bradley & England 1978a; Yamamoto & Iwashima 1975). If a similar eruption were to occur today (similar in timing, magnitude, and chemical characteristics), it might be possible to anticipate, at least in a general sense, the (impulsive) change in climate which would ensue. More detailed studies of the relationship between major explosive eruptions and subsequent climatic conditions might enable predictions to be made, even though the eruptions themselves are non-periodic.

Finally, climatic variation may be characterized by an increase in variability without a change in central tendency, though commonly a change in variability accompanies a change in overall mean. Climatic variability is an extremely important characteristic of climate in our increasingly over-stressed world. Every year, unexpected weather events (extremes in the climate spectrum) result in hundreds of thousands of deaths and untold economic and social hardships. If climatic variability increases, the unexpected becomes more probable and the strain on social and political systems increases. Unfortunately, paleoclimatic studies generally are too imprecise to document climatic variability in the distant past, though historical, phenological, and dendroclimatic studies can shed light on this important aspect of climatic variation.

In the light of these discussions it is appropriate to consider the term "climatic change." Clearly, climates may change on different scales of time and in different ways. In paleoclimatic studies, climatic changes are characterized by significant differences in the mean condition between one time period and another. Given enough detail and chronological control, the significance of the change may be calculated from statistics describing the time periods in question. Markedly different climatic conditions between two time periods imply an intervening period of climate characterized by an upward or downward trend, or by an impulsive change in central tendency (Fig. 2.1). Many paleoclimatic records appear to provide evidence for there being distinct modes of climate, within which short-term variations are essentially stochastic (random). Brief periods of rapid, step-like, climatic change appear to separate these seemingly stable interludes (Bryson *et al.* 1970). Analysis of several thousand ^{14}C dates on stratigraphic discontinuities (primarily in pollen records from western Europe, but including data from elsewhere) lends some support to this idea (Wendland & Bryson 1974). Certain periods stand out as having been times of environmental change on a world-wide scale (Fig. 2.2). Such widespread discontinuities imply abrupt, globally synchronous climatic changes, presumably brought about by some large-scale geophysical phenomenon. In particular the period 2760–2510 years BP (the beginning of sub-Atlantic time) stands out in both palynological and archeological data as a period of major environmental and cultural change, the cause of which is not known. If such a disruption of the climate system were to recur today, the social, economic, and political consequences would be nothing short of catastrophic (Bryson & Murray 1977).

If climate is considered from a mathematical viewpoint, it is theoretically possible that a particular set of boundary conditions (solar radiation receipts, Earth surface conditions, etc.) may not give rise to a unique climatic state. In other words, two or more distinct sets of statistics ("climate") may result from a single set of controls on the atmospheric circulation (Lorenz 1968, 1970, 1976). In a practical sense, this suggests that climate (taken here to mean a particular mode of the general circulation of the atmosphere) may be essentially stable until some

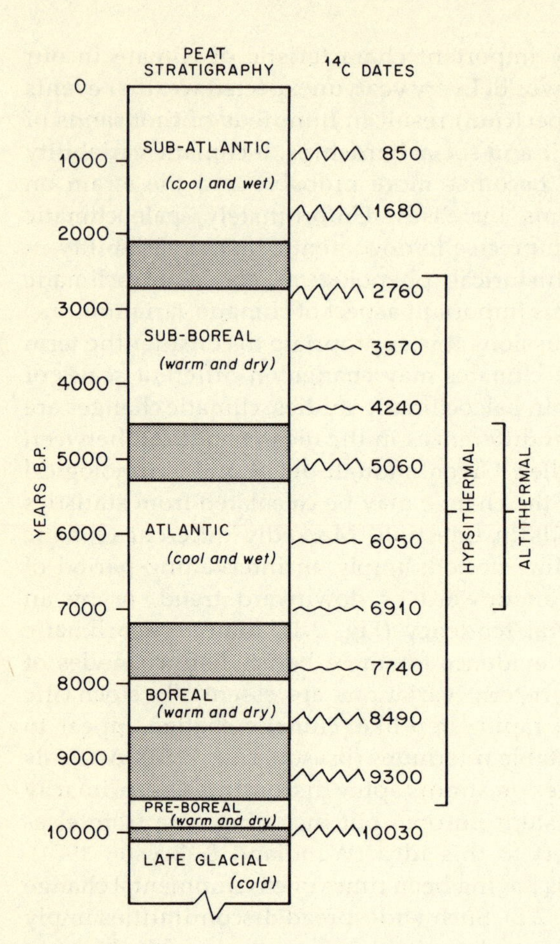

Figure 2.2 "Climatic discontinuities" revealed by analysis of over 800 ^{14}C dates on stratigraphic discontinuities in paleo-environmental (primarily botanical) records (based on data in Wendland & Bryson 1974). Major and minor discontinuities are shown by large and small jagged lines, respectively. Time limits of the Hypsithermal and the Altithermal are from Deevey and Flint (1957). The Blytt–Sernander scheme of European peat stratigraphy was developed before ^{14}C dating techniques were available. It is based on changes in peat growth which were considered to be climate-related. Radiocarbon dates now indicate that the boundaries are not precise, but vary over the ranges indicated (based on summaries by Godwin 1956 and Deevey & Flint 1957). Objective analyses of peat stratigraphy indicate that the "classical" stages of peat stratigraphy may not be of regional significance after all (except for the sub-boreal/subatlantic transition at ~2500 years BP) (Birks & Birks 1981). Nevertheless, the descriptors (atlantic, subatlantic, etc.) are still commonly used to refer to a particular time period, albeit vaguely defined, both climatically and chronologically.

external factor (e.g. a change in solar radiation output or in volcanic dust loading of the atmosphere) causes a perturbation in the system. This perturbation may be only a short-term phenomenon, after which boundary conditions return to their former state; however, the resultant climate may not be the same, even though the boundary conditions are essentially identical to those before the perturbation. One of the other "solutions" to the mathematical problem of climate may have been adopted. Such a system is said to be intransitive. If, on the other hand, there is only one unique climatic state corresponding to a given set of boundary conditions, the system is said to be transitive. If climate does operate as an intransitive system, this poses intractable problems for mathematical models of climate, and for attempts to use these for climate forecasting. There is another possibility which complicates matters further. If a minor change in boundary conditions is assumed, it is theoretically possible for there to be two or more time-dependent solutions, each with different statistics (climate) when considered over a moderate time span (i.e. the system may appear to be intransitive). However, if the time span is made sufficiently long, the different statistics converge to an essentially stable state. This is referred to as an almost-intransitive system (Lorenz 1968). In practice, this means that a single set of boundary conditions may result in different climate states over discrete time intervals. If we were observing these different states through the paleoclimatic record, they would appear to represent periods separated by a "climatic change" (implying an external causative factor) whereas they would be, in actuality, merely stages on the way to a long-term stable state.

2.2 The climate system

Although it is common to consider climate as simply a function of the atmospheric circulation over a period of time, to do so overlooks the complexity of factors which determine the climate of a particular region. Climate is the end-product of a multitude of interactions between several different subsystems – the atmosphere, oceans, biosphere, land surface, and cryosphere – which collectively make up the climate system. Each subsystem is coupled in some way to the others (Fig. 2.3) such that changes in one subsystem may give rise to changes elsewhere (Sec. 2.3).

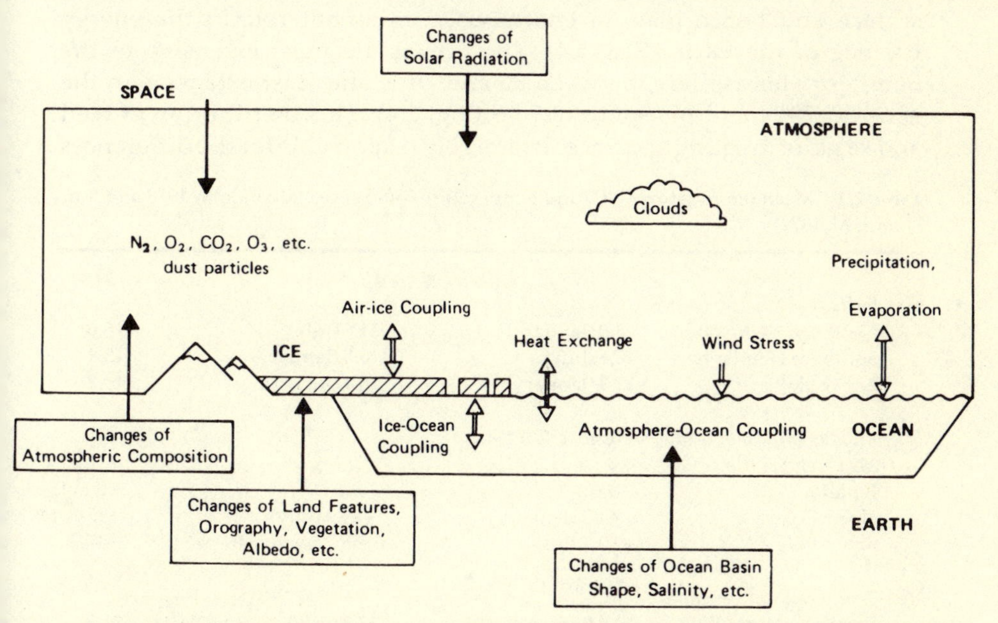

Figure 2.3 Schematic diagram of major components of the climatic system. Feedbacks between various components play an important role in climate variations (from US Committee for GARP 1975).

Of the five principal subsystems, the atmosphere is the most variable; it has a relatively low heat capacity (low specific heat) and responds most rapidly to external influences (on the order of 1 month or less). It is coupled to other components of the climate system through energy exchanges at the surface (the atmospheric boundary layer) as well as through chemical interactions which may affect atmospheric composition (Junge 1972, Jaenicke 1981, Bolin 1981). Only recently has it been possible to assess variations in atmospheric composition and turbidity through time (Hammer *et al.* 1980, Neftel *et al.* 1982). Such variations are of particular importance since they may be a fundamental cause of climatic variation in the past.

The oceans are a much more sluggish component of the climate system than the atmosphere. Surface layers of the ocean respond to external influences on a timescale of months to years, whereas changes in the deep oceans are much slower; it may take centuries for significant changes to occur at depth. Because water has a much higher heat capacity than air, the oceans store very large quantities of energy, and act as a buffer against large seasonal changes of temperature. On a large scale, this is reflected in the differences between seasonal temperature ranges of the Northern and Southern Hemispheres (Table 2.1). On a smaller scale, proximity to the ocean is a major factor affecting the climate of a region. Indeed, it is probably the single most important factor, after latitude and elevation.

At the present time, the oceans cover 71% (361×10^6 km^2) of the Earth's surface and hence play an enormously important role in the energy balance of the Earth (Sec. 2.4). The oceans are most extensive in the Southern Hemisphere, between 30 and 70°S, and least extensive in the zone 50–70°N and poleward of 70°S (Fig. 2.4). This distribution of land and sea is of great significance; it is largely responsible for the differences

Table 2.1 Mean temperatures (°C) and temperature differences (after Flohn 1978 and Van Loon *et al.* 1972).

	Extreme months		*Year*
(a) Surface			
Northern Hemisphere	8.0 (January)	21.6 (July)	15.0
Southern Hemisphere	10.6 (July)	16.5 (January)	13.4
Entire globe	12.3 (January)	16.1 (July)	14.2
(b) Middle Troposphere (300–700 mb layer)			
Mean temperatures			
Equator	−8.6	−8.6	−8.6
North Pole	−41.5 (January)	−25.9 (July)	−35.9
South Pole	−52.7 (July)	−38.3 (January)	−47.7
Temperature differences			
Equator – North Pole	32.9 (January)	17.3 (July)	27.3
Equator – South Pole	29.7 (January)	44.1 (July)	39.1

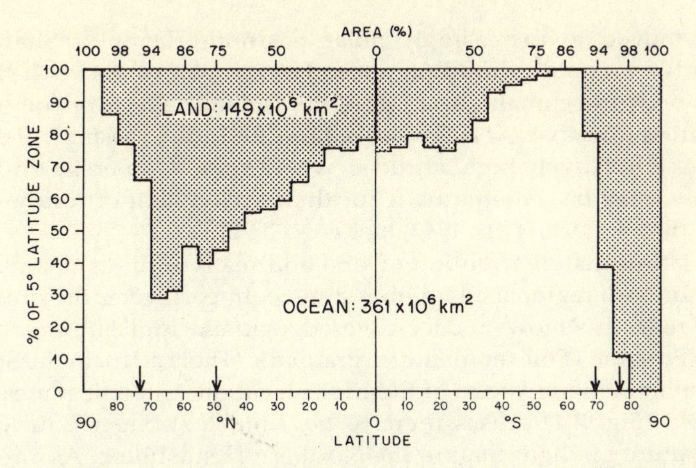

Figure 2.4 Percentage distribution of land and ocean by 5° latitude band. Land area shaded. Upper figures give percentage of hemispheric surface area equatorward of latitudes shown. Arrows indicate mean latitudinal ranges of seasonal snow cover (cf. Table 2.3).

in atmospheric circulation between the two hemispheres, and has important implications for glaciation of the Earth (Flohn 1978). On a global scale, the relative *proportions* of land and sea have changed little during the Quaternary, in spite of sea-level changes due to the growth and decay of continental ice sheets. When sea level was 100 m below present levels ocean area decreased by only 3% (though this is equivalent to a 10% increase in land-surface area). Such changes undoubtedly had regional significance; in particular, sea-level changes may have had important effects on oceanic circulation (e.g. Donn & Ewing 1968) and certainly must have influenced the degree of continentality of some areas (e.g. Barry 1982, Nix & Kalma 1972).

The oceans play a critical role in the chemical balance of the atmospheric system, particularly with respect to atmospheric carbon dioxide levels. Because the oceans contain very large quantities of CO_2 in solution, even a small change in the oceanic CO_2 balance may have profound consequences for the radiation balance of the atmosphere, and hence climate (Manabe & Wetherald 1975, Manabe *et al.* 1981). The role of the oceans in global CO_2 exchanges is of particular importance, not only for an understanding of past climatic variations but also for insight into future CO_2 trends in the atmosphere (Broecker *et al.* 1979, Bolin 1981; see Sec. 6.6).

The land surface of the Earth interacts with other components of the climate system on all timescales. Over very long periods of time, continental plate movements (in relation to the Earth's rotational axis) have had major effects on world climate (Tarling 1978, Frakes 1979). It is no coincidence that the frequency of continental glaciation increased as the plates moved to increasingly polar positions (Donn & Shaw 1975). Similarly, mountain-building episodes (orogenies) have had major effects on world climate. Apart from the dynamic effects on atmospheric circulation (Kasahara *et al.* 1979, Yoshino 1981), the presence of elevated surfaces at relatively high latitudes, where snow can persist throughout the year, may be a prerequisite for the development of continental ice sheets (Brooks 1926, Flint 1943, Ives *et al.* 1975).

The latitudinal distribution of land and sea is of fundamental significance for both regional and global climate. In particular, the presence of highly reflective snow- and ice-covered regions at high latitudes strongly affects Equator – Pole temperature gradients (Table 2.1b). In the Southern Hemisphere, the presence of the high elevation Antarctic plateau south of ~75°S (Fig. 2.4) causes there to be a much stronger Equator – Pole temperature gradient than in the Northern Hemisphere. As a result, an intense westerly circulation pattern develops above the surface layers (60% stronger, on average, than westerlies in the Northern Hemisphere; Lamb 1959). The stronger temperature gradient also results in the subtropical high pressure belt of the Southern Hemisphere being located closer to the Equator than in the Northern Hemisphere (29–35°S as compared with 33–41°N; Fig. 2.5). This difference, stemming primarily from the polar location of Antarctica and its associated low temperatures, gives rise to a basic asymmetry in the position of climatic zones in both hemispheres (Korff & Flohn 1969, Flohn 1978).

The cryosphere consists of mountain glaciers and continental ice sheets, seasonal snow and ice cover on land, and sea ice. Its importance

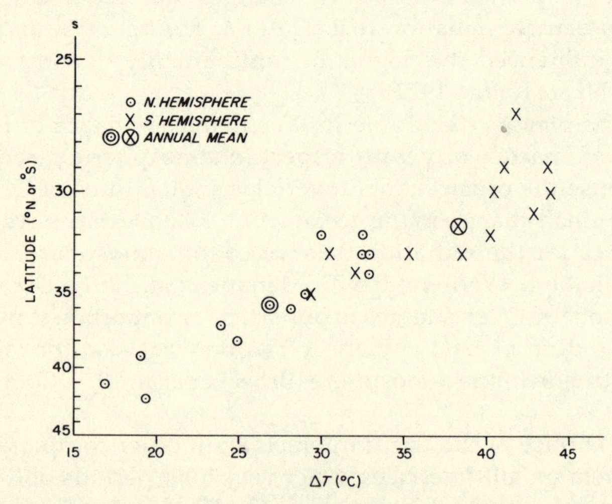

Figure 2.5 Relationship between latitude of main axis of subtropical anticyclones and hemispheric (Pole – Equator) temperature gradient in preceding month (after Korff & Flohn 1969).

in the climate system stems from the high albedo of snow- and ice-covered regions, which greatly affects global energy receipts (Kukla 1978). At present, about 8% of the Earth's surface is permanently covered by snow and ice (Table 2.2) but seasonal expansion of the cryosphere causes this figure to double (Table 2.3). The hemispheric differences are particularly profound. In the Northern Hemisphere, 4% of the total area is permanently ice-covered (mainly the Arctic Ocean (~3%) and Greenland). In winter months, sea-ice formation and snowfall on the continents results in a sixfold increase in snow and ice cover. By midwinter, 24% of the Northern Hemisphere is generally covered by snow and ice. In the Southern Hemisphere, most of the permanent ice cover is land-based, on the Antarctic continent, and seasonal changes are almost entirely due to

Table 2.2 Present extent of permanent snow and ice (glaciers and ice caps), compiled from Kukla (1978), Hughes *et al.* (1981), and Hollin and Schilling (1981).

	Area ($\times 10^6$ km^2)
Northern Hemisphere	
Greenland	1.73
Other locations	0.5
Total land-based snow and ice	2.23
Total for Northern Hemisphere	11.00
Southern Hemisphere	
Antarctica	13.0
Other locations	0.032
Total land-based snow and ice	13.032
Sea ice	4.2
Total for Southern Hemisphere	17.23
Entire globe	
Total land-based snow and ice	~15.3
Sea ice	~13.0
Total for entire globe	~28.3

Table 2.3 Seasonal changes in snow and ice cover area ($\times 10^6$ km^2); snow and ice extent based on the period 1967–74 (Kukla, 1978).

	Maximum extent			Minimum extent		
	Month	Area	Percentage (%)	Month	Area	Percentage (%)
Northern Hemisphere	February	60.1	24†	August	11.0	4†
Southern Hemisphere	October	34.0	13†	February	17.2	7†
Entire globe	December	79.1	16‡	August	42.3	8‡

† Percentage of area of hemisphere.
‡ Percentage of area of entire globe.

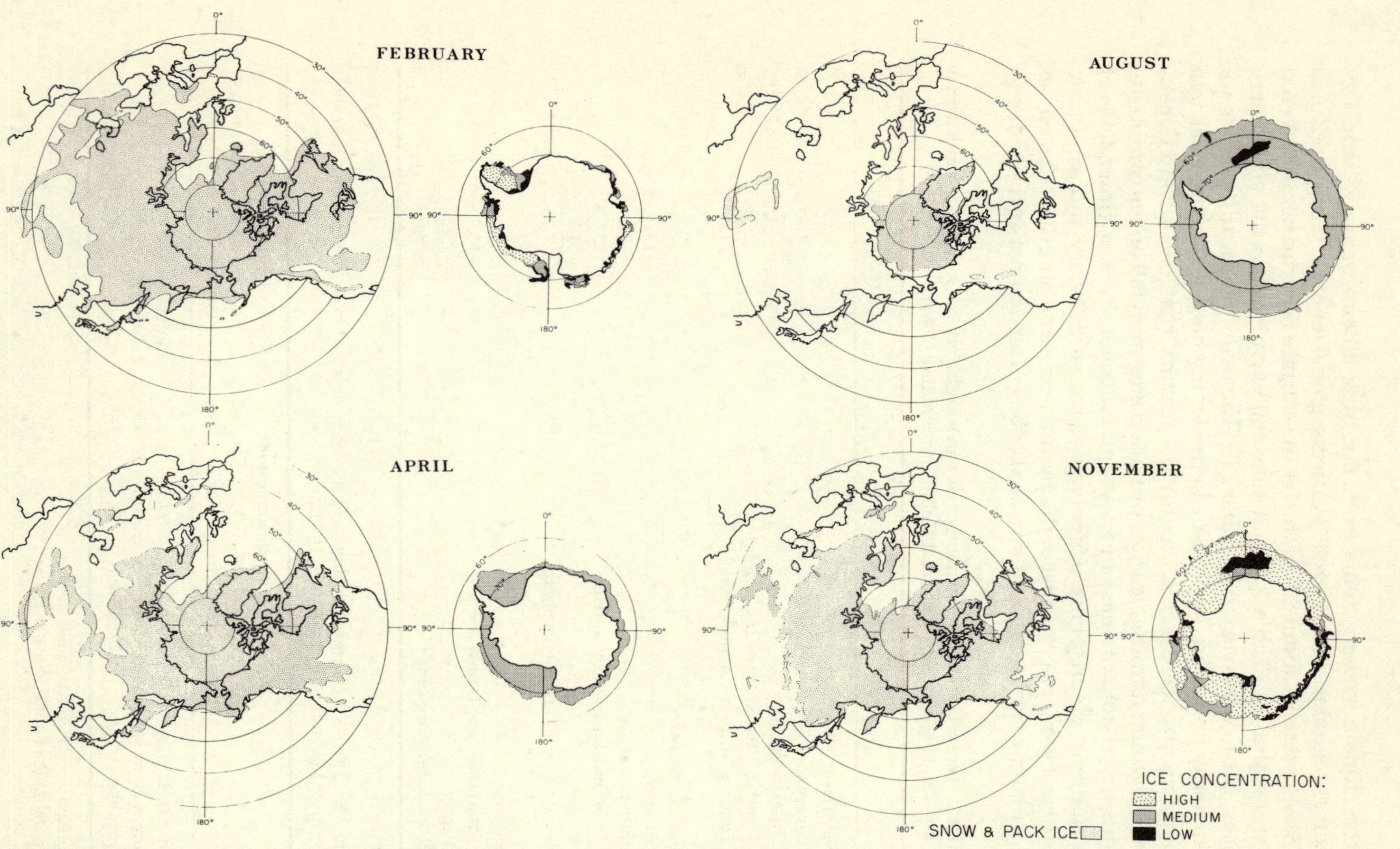

Figure 2.6 Extent of snow and ice at four intervals during the year. Note maximum *global* ice extent in November, minimum in August, cf. Table 2.3 (from Kukla 1978).

an increase in sea-ice formation (Fig. 2.6). By midwinter, 13% of the Southern Hemisphere is generally covered by snow and ice. It is of particular interest that the cryosphere, considered on a global scale, doubles in area over a relatively short period, from August to December, on average. Given the variability in seasonal timing of snow- and ice-cover changes in both hemispheres, it is quite probable that very large area increases may occur over an even shorter period, and this has important implications for theories of climatic change (Kukla 1975). Clearly, part of the cryosphere undergoes extremely large seasonal variations and hence has a very short response time. Glaciers and ice sheets, on the other hand, respond very slowly to external changes, on the timescale of decades to centuries; for large ice sheets, adjustment times may be measured in millennia.

The final component of the climate system is the biosphere, consisting of the plant and animal worlds, though vegetation cover and type are mainly of significance for climate. Vegetation not only affects the albedo, roughness, and evapotranspiration characteristics of a surface, but also influences atmospheric composition through the removal of carbon dioxide and the production of aerosols and oxygen. Absence of vegetation may result in significant increases in particulate loading of the atmosphere, at least locally, and this may of itself be a significant factor in altering climate (Bryson & Baerreis 1967, Charney *et al.* 1975). Vegetation type varies greatly from one region to another (Table 2.4). Forests and woodlands cover 28% of the continents and play a major role in the removal of atmospheric CO_2 (Woodwell *et al.* 1978). Deserts and desert scrublands occupy ~18% of the continents, and are the major sources of wind-blown dust (though cultivated lands are increasingly susceptible to wind erosion also). The response time of the biosphere varies widely, on the order of years for individual elements of the biosphere to centuries for entire vegetation communities.

Human beings are, of course, part of the biosphere and human activities play an increasingly important role in the climate system. Increases in atmospheric CO_2 concentration, changes in natural vegetation, increases in particulate loading of the lower troposphere, and possible reductions in atmospheric ozone concentrations in the stratosphere may all be attributed to man's world-wide activities (Wilson & Matthews 1971, Kellogg 1977, Munn & Machta 1979). The extent to which the climate system can adjust to these activities without drastic changes in climate or climatic variability remains uncertain. The only certainty is that mankind has become exceedingly vulnerable to any unexpected perturbations of climate (Schneider & Mesirow 1976, Schneider & Temkin 1978). Common sense argues for extreme caution in those activities which may contribute to global-scale climatic effects (Kellogg & Schneider 1974).

Table 2.4 Areas of major vegetation units of the world and estimated albedo.

			Area (× 10^6 km^2)[†]	Albedo (%)[‡]
(a) *Forest*			50.0	7–17
Comprising:	tropical rainforest	17.0		7–15
	Raingreen forest	7.5		
	Summergreen forest	7.0		
	Mediterranean sclerophyll	1.5		
	Mixed warm temperature	5.0		13–17
	Boreal	12.0		7–15
(b) *Woodland*			7.0	15–20
(c) *Dwarf and open scrub*			26.0	
Comprising:	Desert scrub	18.0		20–30
	Tundra	8.0		10–15
(d) *Grassland*			24.0	15–20
Comprising:	Tropical grassland and grassland savanna	15.0		
	Temperate grassland	9.0		
(e) *Desert*			8.5	25–44
(f) *Cultivated land*			14.0	8–20
(g) *Fresh water*			4.0	7–10
Comprising:	Swamps and marshes	2.0		
	Lakes and rivers	2.0		
Total of all vegetation units		133.5		
(h) *Permanent snow and ice*			15.3	65–90
Total continental area			148.8	

[†] From Lieth (1975).
[‡] From Kukla (1981), Kukla and Robinson (1980), and Baumgartner *et al.* (1976).

2.3 Feedback mechanisms

Interactions within the climate system often involve complex, non-linear relationships. All components of the climate system are intimately linked or coupled with all other components, such that changes in one subsystem may involve compensatory changes throughout the entire climate system. These changes may amplify the initial disturbance, or anomaly, or dampen them. Interactions which tend to amplify the disturbance are termed positive feedback mechanisms or processes; they operate in such a way that the system is increasingly destabilized. Interactions which tend to dampen the initial disturbance are termed negative feedback mechanisms or processes; they provide a stabilizing influence on the system, tending to preserve the status quo (Kellogg 1975).

Growth of continental ice sheets provides an example of positive feedback mechanisms. Whatever the initial perturbation of the climate system which led to continental ice-sheet growth in the past (see Sec. 2.6), once snow and ice persisted year round, the higher continental albedo would have resulted in lower global radiation receipts, hence lower temperature, and a more favorable environment for ice-sheet growth. Clearly, at some point other factors (such as precipitation starvation and bedrock depression) must have come into play as the ice sheet grew in size to reverse this trend toward increasing glacierization of the planet (cf. Budd & Smith 1981).

Changes in atmospheric CO_2 concentration also may induce positive feedbacks. As CO_2 levels increase, there will be an increase in the absorption of terrestrial (infra-red) radiation by CO_2; concomitantly, there will be an increase in infra-red absorption by water vapor, resulting from enhanced CO_2 infra-red emissions. Tropospheric temperatures will thus increase (the "greenhouse effect"), though the magnitude of this increase remains controversial (Idso 1980, 1982, Schneider *at al.* 1980, Crane 1981, Ramanathan 1981). As atmospheric temperatures increase, the temperature of the upper layers of the ocean may also increase, causing CO_2 in solution to be released to the atmosphere, thereby reinforcing the trend toward higher temperatures. This (rather simplistic) example of a physical – biochemical feedback is sometimes referred to as the "runaway greenhouse effect". Whether such an eventuality will occur due to anthropogenic production of excess CO_2 is doubtful. It might be argued that as temperatures increase the higher temperatures would result in more evaporation from the oceans, increased cloudiness (higher global albedo), and hence a decrease in energy to the system. In addition, higher temperatures at high latitudes, associated with increased poleward advection of moisture, might be accompanied by more snowfall, resulting in higher continental albedo (and/or a shorter snow-free period) and hence lower overall global energy receipts. Such mechanisms are examples of negative feedbacks, whereby the system tends to become stabilized after an initial perturbation.

Interactions between different parts of the climate system which are brought about by a process within the system are considered internal mechanisms of climatic variation. They involve initiation by an internal factor, such as the upwelling of cool deep-ocean water or an unusually persistent snow cover over an extensive area of the land surface, which may be amplified by other components of the climate system and eventually lead to an adjustment in the atmospheric circulation. These adjustments within the climate system may in turn alter, and perhaps eliminate, the original factor which initiated the climatic variation. Generally, such mechanisms are stochastic in nature, so that the climatic consequences are not predictable over timescales much longer than the timescale of the initiating process. By contrast, there are factors external to the climate system which may bring about ("force") adjustments in climate, but those changes have no influence on the initiating factor (Mitchell 1976). Changes in solar output and/or spectral characteristics, changes in the Earth's orbital parameters, and changes in atmospheric turbidity due to explosive volcanic eruptions are examples of external factors which may cause changes in the climate system but which are not affected by those changes (Robock 1978). Some of these mechanisms of climate variation are deterministic (predictable) since they vary in a known way. This is particularly the case with the Earth's orbital variations, which have been calculated accurately both for periods back in time and into the future (Vernekar 1972, A. Berger 1978, 1979). There is therefore an element of predictability in the consequent climatic changes, though these may, in turn, depend on the particular internal conditions of the climate system prevailing at the time of the external forcing.

2.4 Energy balance of the Earth and its atmosphere

As the Earth sweeps through space on its annual revolution around the Sun, it intercepts a minute fraction of the energy emitted by this all-important star. Because the Earth is (approximately) spherical, and rotates on an axis inclined (at present) 23.4° to the plane across which it moves around the Sun (the ecliptic), energy receipts vary greatly from one part of the globe to another. Furthermore, the pattern of energy receipts is constantly changing. These differential energy receipts are the fundamental driving force of the atmospheric circulation. If solar output is assumed to be invariant, the spatial and temporal patterns of energy receipts impinging on the outer atmosphere can be calculated (Fig. 2.7; Newell & Chiu 1981). However, for conditions near the surface of the Earth, the role of the atmosphere must be considered, since the atmosphere greatly diminishes potential solar radiation receipts. A consideration of energy exchanges in the Earth – atmosphere system also provides some insight into the potentially important factors involved in climatic variations and variability. For the sake of simplicity, it will be assumed that, for the system as a whole, energy receipts at the outer limits of the atmosphere during the course of a year are equal to 100 units (Fig. 2.8). As radiation penetrates the atmosphere, as a global average 26% is either reflected from cloud tops or scattered upward by molecules and particulate matter in the air. Because the Earth's surface is also reflective, another 4% of incoming solar radiation is returned to space, without heating the

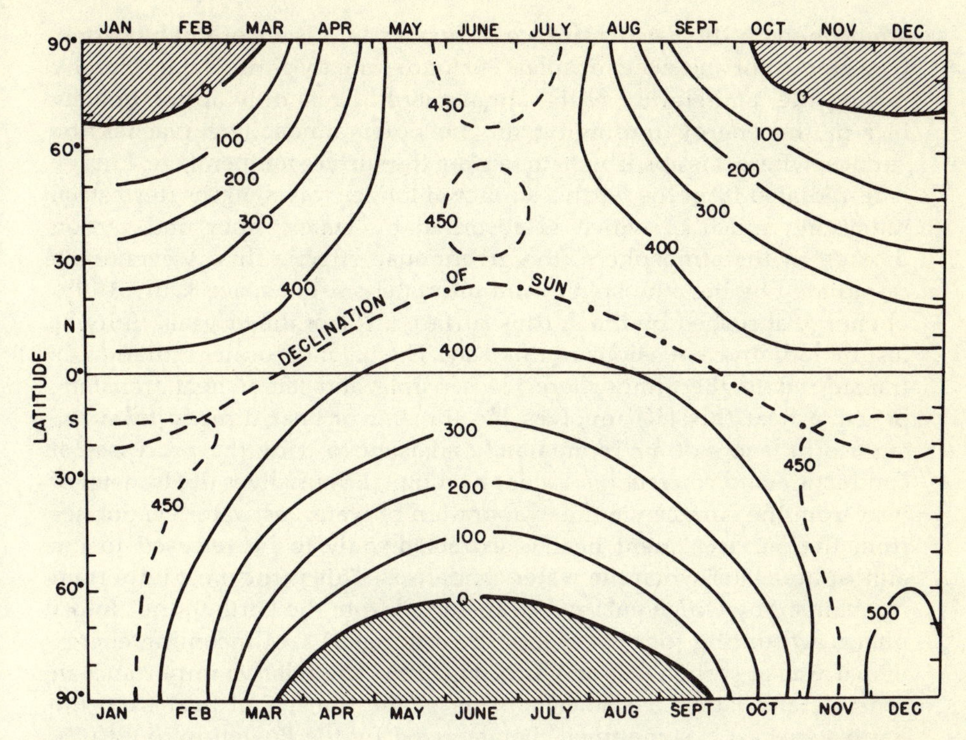

Figure 2.7 Distribution of solar radiation at top of atmosphere (in Watt hours per square meter). Apparent position of Sun overhead at noon (declination) is shown by dotted line.

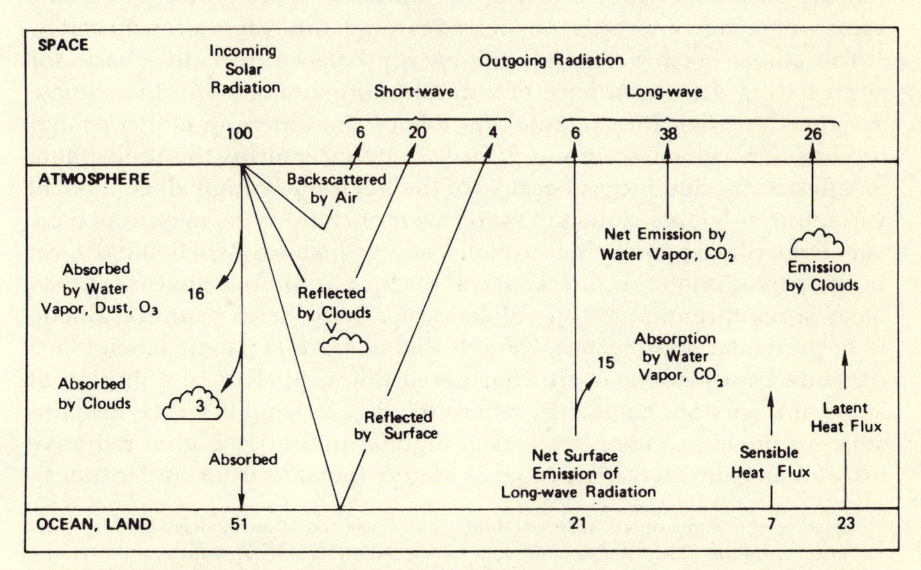

Figure 2.8 Mean annual radiation and heat balance of the atmosphere relative to 100 units of incoming solar radiation (from US Committee for GARP 1975).

atmosphere or the Earth's surface. A further 19% is absorbed by ozone, by water vapor and water droplets in clouds, and by particulates, thereby raising the temperature of the atmosphere.† Thus only approximately half of the energy impinging on the outer atmosphere reaches the surface, where it is absorbed, increasing the surface temperature. Energy is re-radiated from the Earth's surface at longer wavelengths (terrestrial radiation), much of which is absorbed by water vapor and carbon dioxide in the atmosphere (the greenhouse effect). This is eventually re-radiated by the atmosphere and ultimately lost to space. Only ~40% of energy absorbed by the Earth's surface (21% of the original units) is lost by radiative emissions in this way. The balance, or net radiation, is transferred to the atmosphere via sensible and latent heat transfers. Sensible heat flux (H) involves the transfer of heat directly from the surface to layers of air immediately adjacent to it by the processes of conduction and convection. Latent heat flux (LE) involves the transfer of heat from the surface via the evaporation of water; as water evaporates from the surface, latent heat is extracted, only to be released to the atmosphere later when the water condenses. This is the most important mechanism by which energy is transferred from the Earth to the atmosphere, accounting for over half of the original 51% of incoming energy which was absorbed by the Earth (Fig. 2.8). The relative importance of sensible and latent heat mechanisms in the transfer of heat from the Earth's surface is sometimes characterized by the Bowen ratio (H/LE); high values (≥ 10) are typical of desert areas where values of latent heat flux are very low, whereas low Bowen ratios (≤ 1) are typical of oceanic areas where most energy is transferred through the evaporation of water.

The global mean values for the energy balance provide a basis for appreciating the importance of a number of parameters in the climate system. Consider, for example, the role of cloudiness in global energy receipts. On a global scale one-fifth of all energy entering the atmosphere is reflected by cloud tops, because of their extremely high albedo. Small variations in global cloud cover, or even of cloud type, may thus have very large consequences for the global energy balance (Mitchell 1965), yet in the paleoclimatic record we have no clues as to how cloudiness may have varied through time on a global scale. At the Earth's surface, albedo is of particular significance, though this is more apparent when zonal (latitude band) averages are considered (Fig. 2.9). This distribution of snow and ice dominates this pattern (cf. Fig. 2.4) and is largely responsible for the large energy deficits at high latitudes (i.e. higher radiative losses than gains, accommodated by energy transfers from low latitudes).

†Because the atmosphere absorbs short-wave solar radiation as well as long-wave radiation from the Earth, it also emits long-wave radiation both upward and downward (counter-radiation). Overall, however, there is a net loss of long-wave radiation from the Earth to space via the atmosphere.

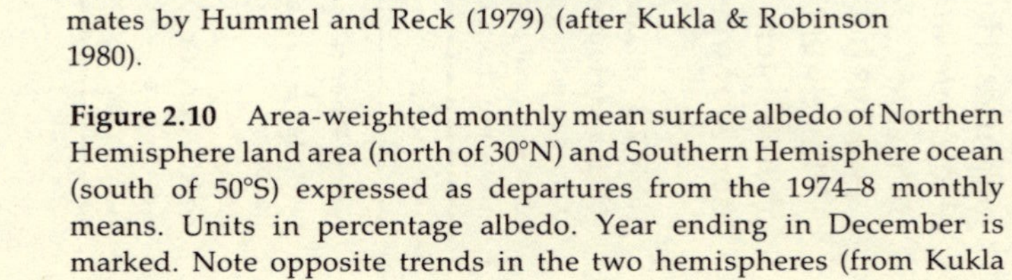

Figure 2.9 Latitudinal distribution of seasonal average surface albedo (averaged around latitude bands, i.e. zonally). ———, estimates by Kukla and Robinson (1980); o----o, estimates by Hummel and Reck (1979) (after Kukla & Robinson 1980).

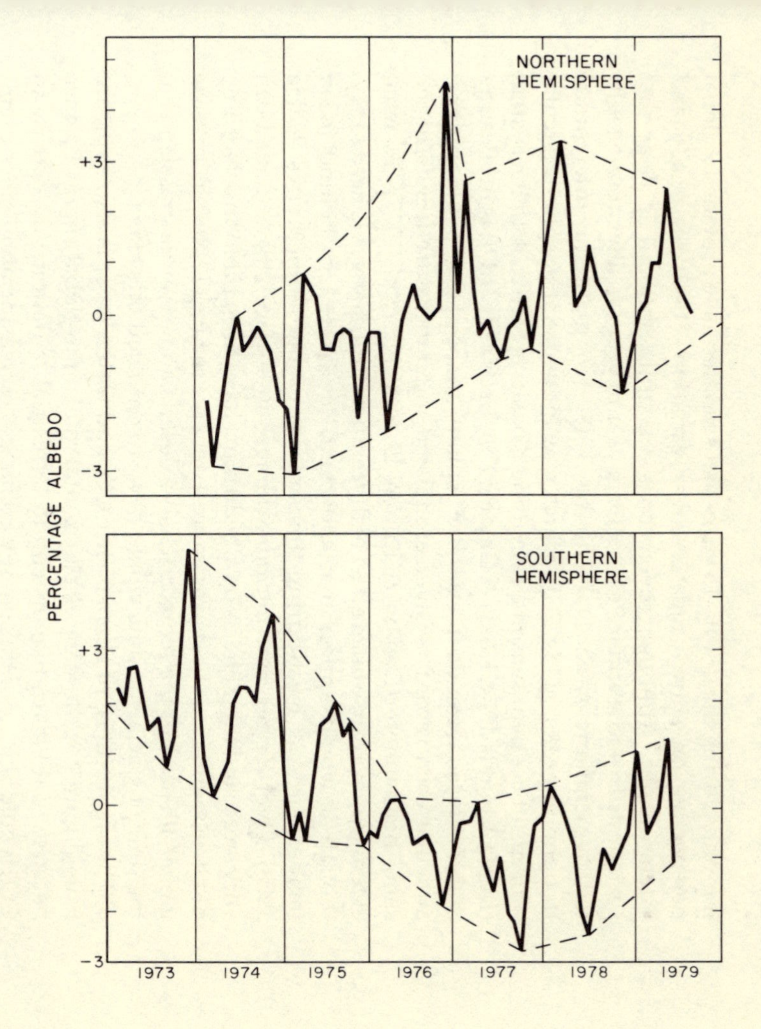

Figure 2.10 Area-weighted monthly mean surface albedo of Northern Hemisphere land area (north of 30°N) and Southern Hemisphere ocean (south of 50°S) expressed as departures from the 1974–8 monthly means. Units in percentage albedo. Year ending in December is marked. Note opposite trends in the two hemispheres (from Kukla 1979).

It is only during the last 10–15 years that satellites have provided a global perspective on snow- and ice-cover variations, both seasonally and interannually. Although the records are quite short, it is clear that variations in snow and ice extent from year to year can alter area-weighted hemispheric surface albedo by 3–4% (compare the interannual troughs, or peaks, in Fig. 2.10) which may influence atmospheric circulation in subsequent seasons, providing a positive feedback to the system (Kukla & Kukla 1974, Kukla & Gavin 1979). In this regard, it is interesting that the decadal trends in each hemisphere are opposite to one another, and that each record is inversely related to temperature trends in the same hemisphere (Damon & Kunen 1976, Kukla *et al.* 1977). To what extent this is a cause or an effect of the temperature changes is hard to say, though it would appear that snow and ice cover lags behind mean temperatures, at least during times of decreasing temperatures (Kukla 1978). Over longer time periods, changes in surface albedo have been very large, and their effects on albedo must have been profound. Not only did continental ice sheets and more extensive sea ice (Table 2.5) increase global albedo but the more extensive deserts and savanna grasslands at the time of glacial maxima would have accentuated this effect.

The significance of atmospheric CO_2 and water vapor is also apparent from Figure 2.8; these gases play a vital role in the global energy balance because of their relative opacity to terrestrial radiation. An increase in CO_2 would reinforce this energy exchange, increasing atmospheric temperatures. However, many other interactions and consequences would also ensue and it is this complexity which makes forecasts of the climatic impact of CO_2 increases so difficult (Schneider 1975, Smagorinsky 1981, Crane 1981).

This thumbnail sketch of the radiation balance of the Earth–atmosphere system is very much a simplification of reality. Most importantly, there are large regional differences in values of net radiation and of latent

Table 2.5 Maximum extent of land-based ice sheets during the Pleistocene (after Flint (1971, and Hollin & Schilling 1981). Note that not all areas experienced maximum ice cover at the same time during the Pleistocene. It is therefore not appropriate to total these values. Also seasonal snow cover and sea-ice extent are not included, so these figures represent minimum changes in the area of the overall cryosphere (cf. Tables 2.3 & 2.4).

	Area ($\times 10^6$ km^2)
North America	16.22
Greenland	2.30
Europe	7.21
Asia	3.95
South America	0.87
Australasia	0.03
Antarctica	13.81

Table 2.6 Mean latitudinal values of the heat balance components of the Earth's surface (kcal cm^{-2} yr^{-1}) (from Budyko 1978). *R* is the radiative flux of heat (radiation balance of the Earth's surface) equal to the difference of absorbed short-wave radiation and the net long-wave radiation outgoing from the Earth's surface; *LE* is the heat expenditure for evaporation (*L* is the latent heat of vaporization, *E* is the rate of evaporation); *P* is the turbulent flux of heat between the Earth's surface and the atmosphere; F_0 is the heat income resulting from heat exchange through the sides of the vertical column of a unit section going through the Earth's surface with the ambient layers. (1 kcal cm^{-2} = 69.75 W cm^{-2})

Latitude	Land			Ocean				Earth			
	R	*LE*	*P*	*R*	*LE*	*P*	F_0	*R*	*LE*	*P*	F_0
70–60°N	22	16	6	23	31	22	−30	22	20	11	−9
60–50	32	23	9	43	47	19	−23	37	33	13	−9
50–40	45	25	20	64	67	16	−19	54	45	18	−9
40–30	58	23	35	90	96	14	−20	76	65	23	−12
30–20	64	19	45	111	109	7	−5	94	75	21	−2
20–10	74	32	42	121	117	7	−3	109	95	16	−2
10–0	79	57	22	124	104	7	13	114	93	10	11
0–10°S	79	61	18	127	99	6	22	116	90	9	17
10–20	75	45	30	122	113	9	0	112	98	14	0
20–30	71	28	43	109	106	11	−8	100	88	18	−6
30–40	62	29	33	92	82	11	−1	88	76	14	−2
40–50	44	22	22	72	51	6	15	71	50	7	14
50–60	35	22	13	46	35	9	2	46	35	9	2
Earth as a whole	50	27	23	91	82	9	0	79	66	13	0

and sensible heat flux due to the geography of the earth (distribution of continents and oceans, surface relief, vegetation, and snow cover) and the basic climatic differences from one region to another (principally variations in cloud cover and type) (Budyko 1978). This is readily apparent from a consideration of annual energy balance components for the Earth's surface, shown as zonal averages in Table 2.6, and mapped in Figures 2.11–2.13. Net radiation varies from near zero at high latitudes to >140 kcal cm^{-2} a^{-1} over parts of the tropical and equatorial oceans (Fig. 2.11). On the continents, net radiation is lower than the zonal average due to higher albedo of the surface (e.g. in desert regions) or because of higher cloud amounts which reduce surface radiation receipts (Table 2.6). For the Earth as a whole (Table 2.6, bottom line) 84% of net radiation is accounted for by latent heat expenditures (66 of 79 kcal cm^{-2} a^{-1}). If we just consider the oceans, however, 90% of net radiation is utilized in evaporation compared to only 54% (27 of 50 kcal cm^{-2} a^{-1}) on the continents. In fact, in extremely arid areas, latent heat transfer may account for only 15–20% of the net radiation (cf. Figs 2.11 & 2.12). In those areas, sensible heat flux is of primary importance (Fig. 2.13). For the continents as a whole, 46% of net radiation is utilized in sensible heat transfers. Over the oceans, sensible heat flux is only important at high

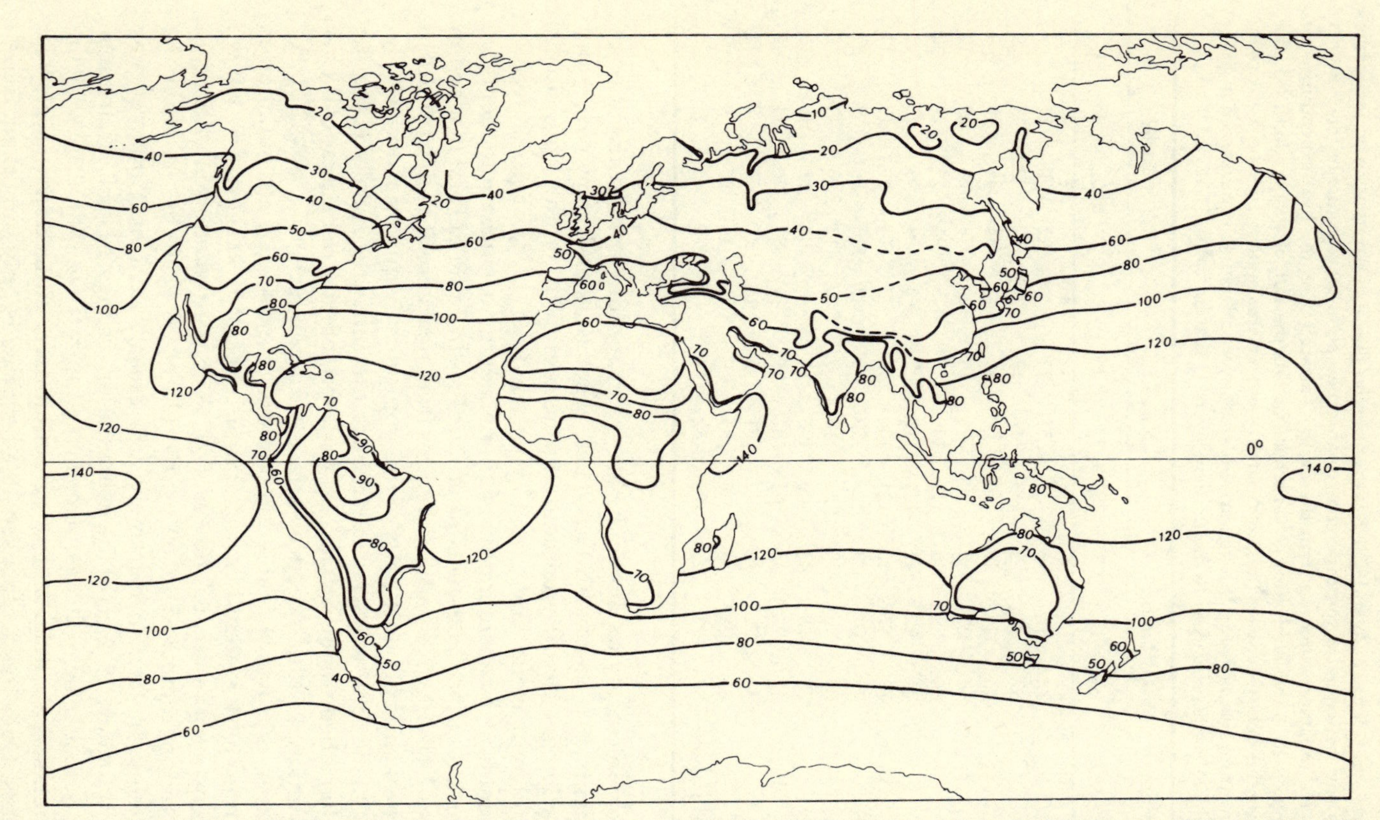

Figure 2.11 Radiative balance (net radiation, R_n) of the Earth's surface (in kilocalories per square centimeter per year). Note discontinuities at ocean/land boundaries (from Budyko 1978).

Figure 2.12 Expenditure of latent heat for evaporation (latent heat flux, L, in kilocalories per square centimeter per year) (from Budyko 1978).

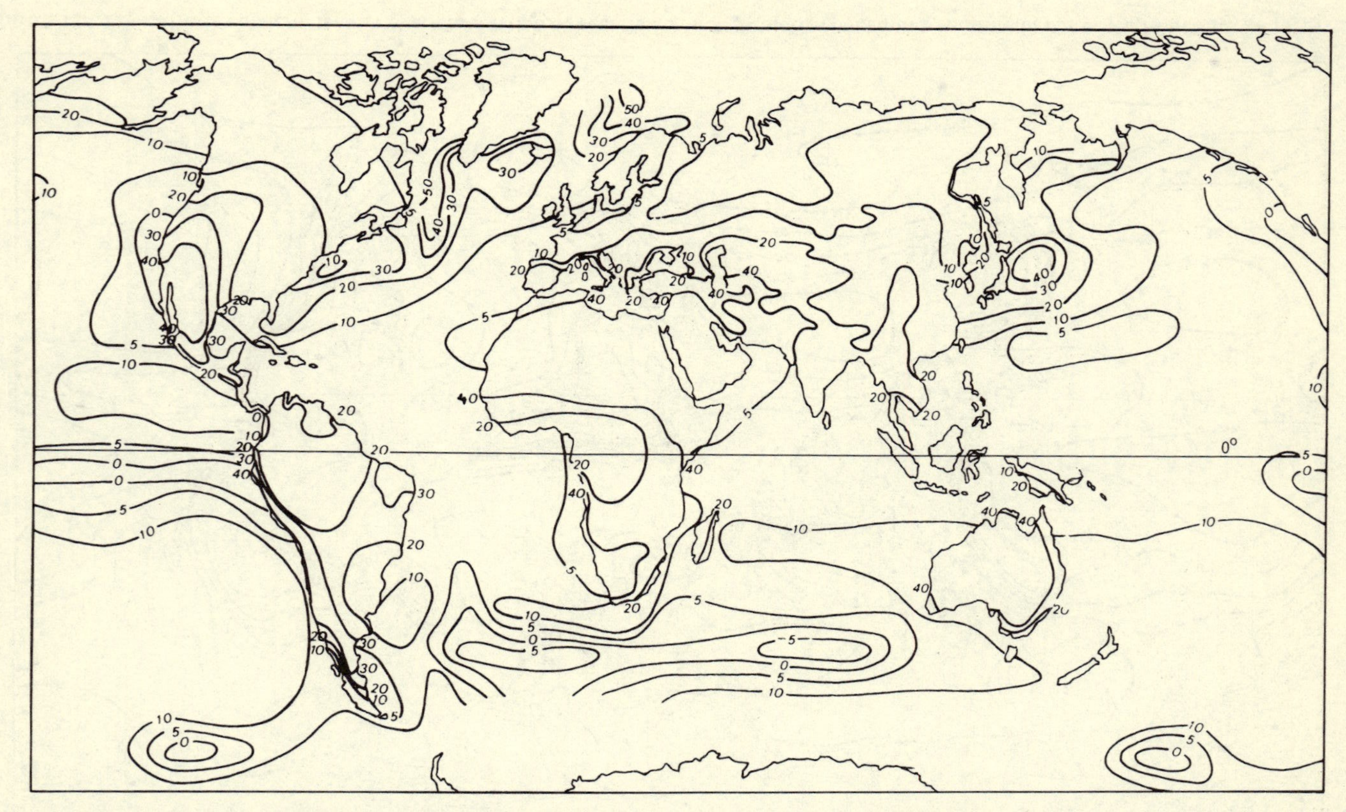

Figure 2.13 Sensible heat flux between Earth's surface and the atmosphere (in kilocalories per centimeter per year) (from Budyko 1978).

northern latitudes where northward-flowing currents bring warm water into contact with cold polar air masses (Fig. 2.13). Ocean currents themselves play a very important role in energy transport, as is clear from column 8 in Table 2.6. "Excess" heat is transferred from equatorial and tropical regions to higher latitudes where the energy thereby made available may even exceed net radiation at the surface (e.g. 60–70°N; cf. Figs. 2.11 & 2.13).

From this overview of the energy balance of different regions it is only a short step to consider how components of the energy balance of some areas may have varied in the past, and how human activities may affect the energy balance of some areas in the future. Of course, it will only be possible to do this in a crude way since the energy balance of any one site is a function of a great many variables, including parts of the climate system far from the site in question. Nevertheless, some general points can be made. Consider, for example, the vast tropical rainforests of the Amazon Basin. At the present time, net radiation in this area averages ~83 kcal cm^{-2} a^{-1} with a Bowen ratio of 0.3 (Table 2.7; Baumgartner 1979). During the last glaciation, much of the area was occupied by savanna (Brown & Ab'Sáber 1979); if modern analogies are any guide, the area would have had a much higher albedo, lower net radiation, and higher Bowen ratio. Other tropical rainforest regions also experienced similar changes in vegetation and hence in energy balance (Flenley 1979). As tropical forests today occupy more than 11% of the continental area, these changes had major consequences for the energy balance of the world as a whole. It also seems likely that modern destruction of the tropical rainforests will bring about marked changes in the energy balance of low latitudes, and perhaps of the entire Earth (Baumgartner 1979, Potter *et al.* 1975, Henderson-Sellers 1982).

Significant changes in the energy balance of today's desert regions have also occurred over the last 20 000 years. In particular, from 10 000 to 4000 years BP, much of the Sahara was occupied by savanna grassland (Sec. 8.3.2; Street 1981). Such changes would have been accompanied by a lower albedo and a much lower Bowen ratio (Table 2.7). Conversely, the destruction of desert scrub and savanna by overgrazing in marginal environments (desertification) would involve an opposite change of energy balance.

Table 2.7 Energy balance for different surfaces (in kilocalories per square centimeter per year) (after Baumgartner 1979).

	R	*L*	*H*	*a*	*H/L*
tropical rainforest	83	64	19	10	0.3
savanna	49	30	19	25	0.6
desert	53	6	47	35	8.0

The energy balance changes associated with changes in natural vegetation do not, of course, provide information on why the environmental changes occurred in the first place. However, they do provide important baseline data for computer models of the general circulation at particular time periods in the past (e.g. Williams *et al.* 1974, Gates 1976a,b; Manabe & Hahn 1977) and point to potentially important feedbacks between the atmosphere and underlying surface once the vegetation changes have occurred. The development of a particular vegetation type may, in fact, bring about changes in the energy balance which would favor persistence of the "new" vegetation type (cf. Charney *et al.* 1975).

2.5 Timescales of climatic variation

Climate varies on all timescales and space scales, from interannual climatic variability to very long-period variations related to the evolution of the atmosphere and changes in the lithosphere. Examples of known climatic fluctuations are shown in Figure 2.14. In this diagram, each row represents an expansion, by a factor of ten, of each interval on the row above it. Thus, one can envisage short-term (high-frequency) variations nested within long-term (lower-frequency) variations (Kutzbach 1976). However, in the paleoclimatic record, as we delve further and further back in time, it is increasingly difficult to resolve the higher-frequency variations. As climatic variations on the timescale of decades to centuries are of the utmost importance to modern society, increasing attention must be focused on paleoclimatic data pertinent to this problem (Ingram *et al.* 1981a).

Climatic fluctuations on different timescales may be brought about by internal or external mechanisms which operate at different frequencies (Fig. 2.15). Changes in the Earth's orbital parameters, for example, are likely candidates for climatic variations on the timescale of glacials and interglacials during the late Quaternary but can not account for climatic variations that have occurred over the last thousand years. For fluctuations on that timescale, other factors, such as volcanic dust loading of the atmosphere (Lamb 1970) or internal adjustments between different subsystems in the climate system, are more likely to be involved. Of course, it must be recognized that consideration of independent causative factors does not mean that they may not have operated together to bring about climatic fluctuations of varying magnitude at different times in the past. Climatic variations result from multivariate causes, though individual factors may account for much of the variance of climate at a particular frequency. Mitchell (1976) has pointed out that much of the variance of the climate record results from stochastic processes internal to the climate system. This includes short-period atmospheric processes

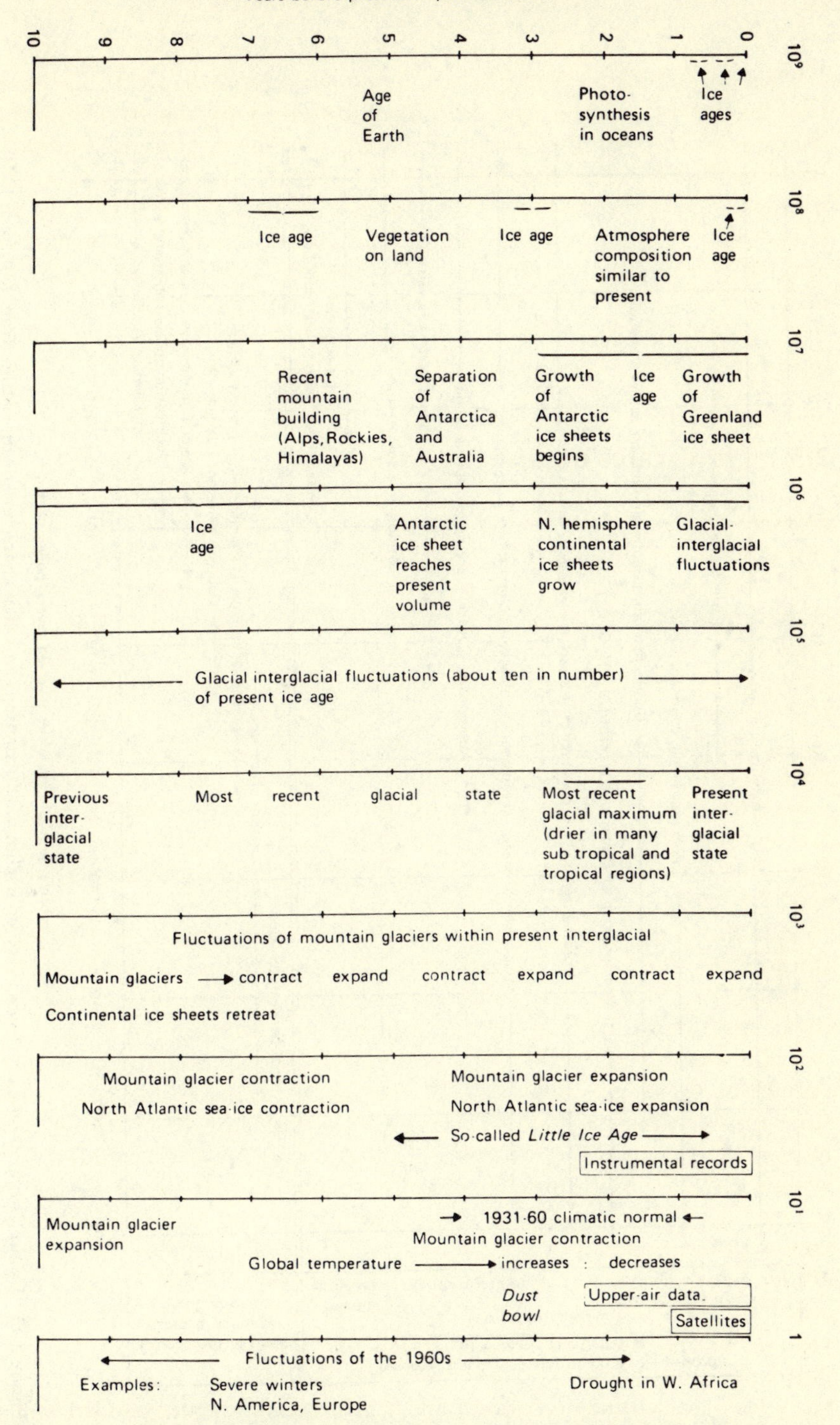

Figure 2.14 Examples of climatic fluctuations at timescales ranging from 10^9 to 1 years. Each successive column, from left to right, is an expanded version (expanded by a factor of 10) of one-tenth of the previous column. Thus, higher-frequency climatic variations are "nested" within lower-frequency changes (from Kutzbach 1974).

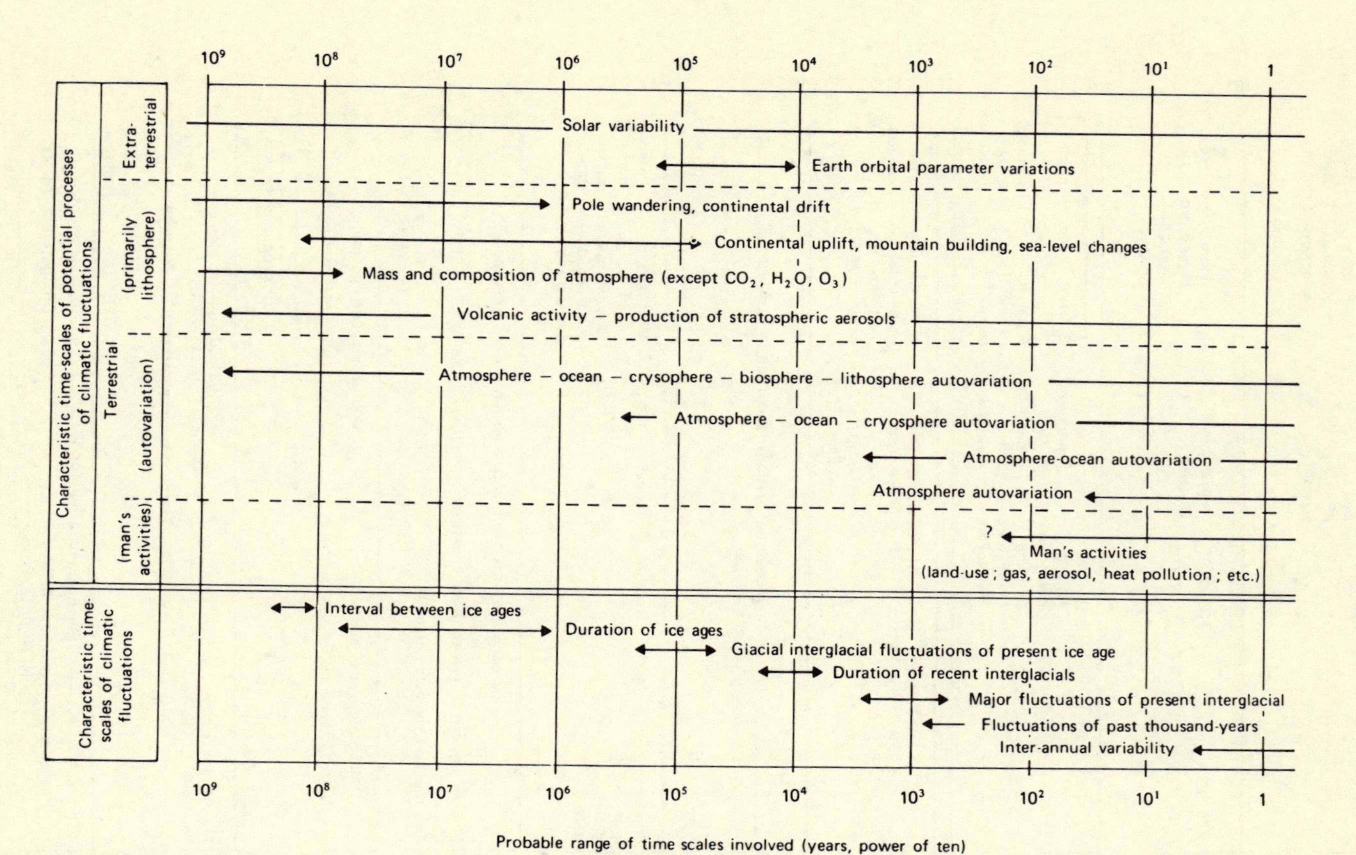

Figure 2.15 Examples of potential processes involved in climatic fluctuations and their characteristic timescales (from Kutzbach 1974).

(e.g. turbulence) with time constants on the scale of minutes or hours, to slower-acting processes or feedback mechanisms which add to climatic variance over longer timescales. However, these factors only contribute "white noise" to the climate spectrum on timescales longer than the timescale of the process in question (i.e. they contribute to the variance of climate in a random, unpredictable manner, with no effects concentrated at a particular frequency). Superimposed on this "background noise" are certain peaks in the variance spectrum of climate which correspond to external forcing mechanisms operating over a restricted time domain (i.e. they are periodic or quasi-periodic phenomena) (Fig. 2.16). These deterministic forcing mechanisms are only known to operate at a few, relatively narrow, frequencies, and although very important to

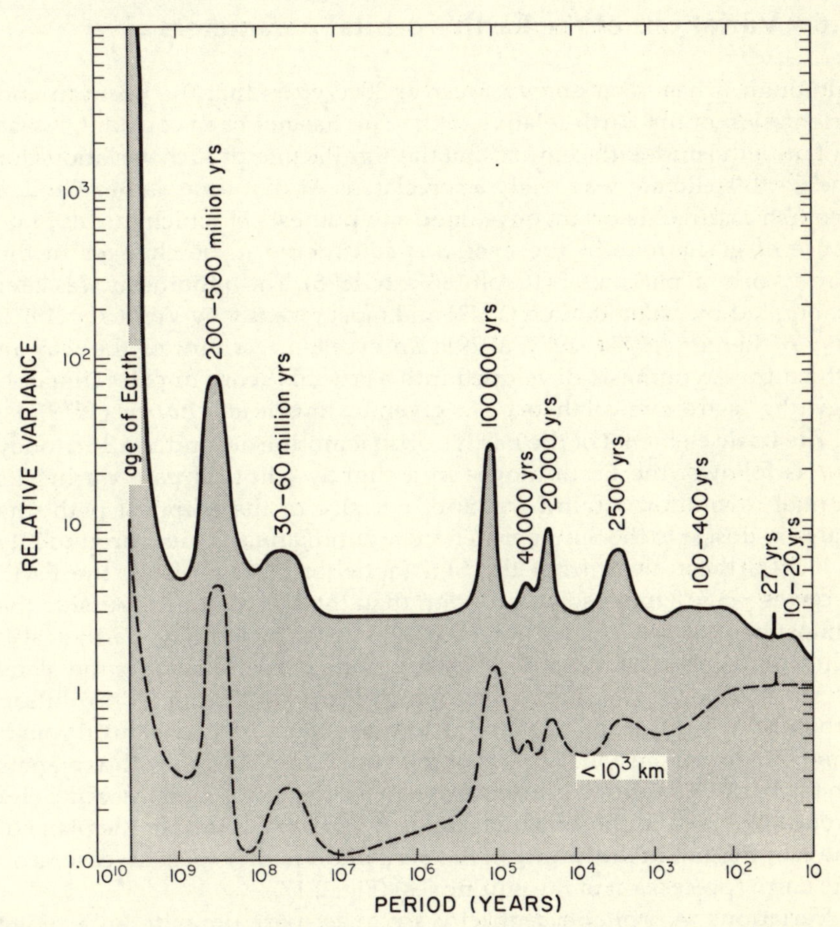

Figure 2.16 Estimate of the relative variance of climate over all periods of variation, from those comparable to the age of the Earth to 10 years. Periodic forcing factors (such as those due to orbital perturbations) appear as peaks of relative variance (after Mitchell 1976).

climatic variance at those frequencies their contribution to overall climatic variation is minor compared to the role of stochastic processes. This presents problems for both climatic predictability and the interpretation of past climatic changes (as seen in the paleoclimatic record) in terms of particular causative factors (Mitchell 1976). Nevertheless, certain external forcing mechanisms have often been called upon to account for features of the paleoclimatic record. The most important of these for climatic fluctuations in the late Quaternary are variations in the Earth's orbital parameters which many claim to be the underlying cause of glacial–interglacial cycles over at least the last million years (e.g. van Woerkem 1953). The basis for this belief is outlined below and in Section 6.8.

2.6 Variations of the Earth's orbital parameters

Although it has been known for over 2000 years that the position and orientation of the Earth relative to the Sun has not been constant, it was not until the mid-19th century that the significance of such variations for the Earth's climate was really appreciated. At that time, James Croll, a Scottish natural historian, developed a hypothesis in which the ultimate cause of glaciations in the past was considered to be changes in the Earth's orbital parameters (Croll 1867a,b, 1875). The hypothesis was later elaborated by Milankovitch (1941) and most recently by Vernekar (1972) and A. Berger (1977a, 1978, 1979). An excellent account of the way in which this hypothesis developed into a crucial theory in paleoclimatology (the "astronomical theory") is given by Imbrie and Imbrie (1979).

The basic elements of the Earth's orbital motion around the Sun today are as follows: the Earth moves in a slightly elliptical path during its annual revolution around the Sun; because of the elliptical path, the Earth is closest to the Sun (perihelion) around January 3, and around July 5 it is farthest away from the Sun (aphelion). As a result, the Earth receives ~3.5% more solar radiation than the annual mean (outside the atmosphere) at perihelion and ~3.5% less at aphelion. The Earth is also tilted on its rotational axis 23.4° from a plane perpendicular to the plane of the ecliptic (the "surface" over which it moves during its revolution around the Sun). None of these factors has remained constant through time due to gravitational effects of the Sun, Moon, and the other planets on the Earth. Variations have occurred in the degree of orbital eccentricity around the Sun, in the axial tilt (obliquity) of the Earth from the plane of the ecliptic, and in the timing of the perihelion with respect to seasons on the Earth (precession of the equinoxes) (Fig. 2.17).

Variations in orbital eccentricity are quasi-periodic with an average period length of ~95 800 years over the past 5 million years. The orbit has varied from almost circular (essentially no difference between perihelion

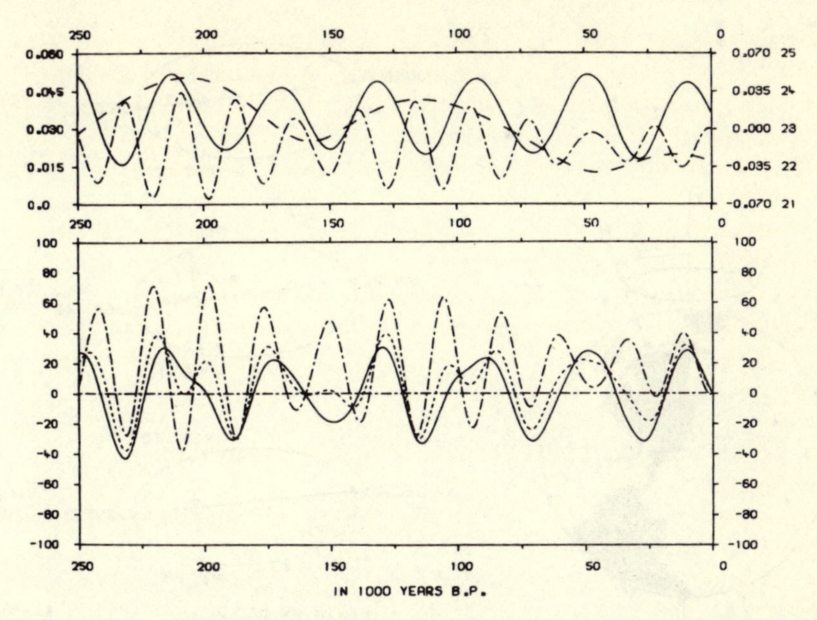

Figure 2.17 Variations of eccentricity (----), precession (–·–·–) and obliquity (——) over the last 250 000 years (upper diagram) and northern hemisphere summer solar radiation at 80 (——), 65 (........), and 10°N (–·–·–) latitude (expressed as departures from 1950 AD values) (lower diagram). Note that the radiation signal at high latitudes is dominated by the ~41 000 year obliquity cycle whereas at lower latitudes the ~23 000 precessional cycle is dominant (after A. Berger 1978).

and aphelion) to maximum eccentricity when solar radiation receipts (outside the atmosphere) varied by ~30% between aphelion and perihelion (e.g. at ~210 000 years BP; Fig. 2.17). Eccentricity variations thus affect the relative intensities of the seasons, which implies an opposite effect in each hemisphere.

Changes in axial tilt are periodic with a mean period of 41 000 years. The angle of inclination has varied from 21.8 to 24.4°, with the most recent maximum occurring about 100 000 years ago (Fig. 2.17). The angle defines the latitudes of the polar circles (Arctic and Antarctic) and the tropics, which in turn delimit the area of day-long polar night in winter, and the maximum latitudes reached by the zenith sun in midsummer in each hemisphere. Changes in obliquity have relatively little effect on radiation receipts at low latitudes but the effect increases towards the poles. As obliquity increases, summer radiation receipts at high latitudes increase, but winter radiation totals decline. This is seen in the summer radiation variations over the last 250 000 years for 65 and 80°N (Fig. 2.17), which reflect mainly the periodic changes in axial tilt. Since the tilt is the same in both hemispheres, changes in obliquity affect radiation receipts in the Southern and Northern Hemispheres equally.

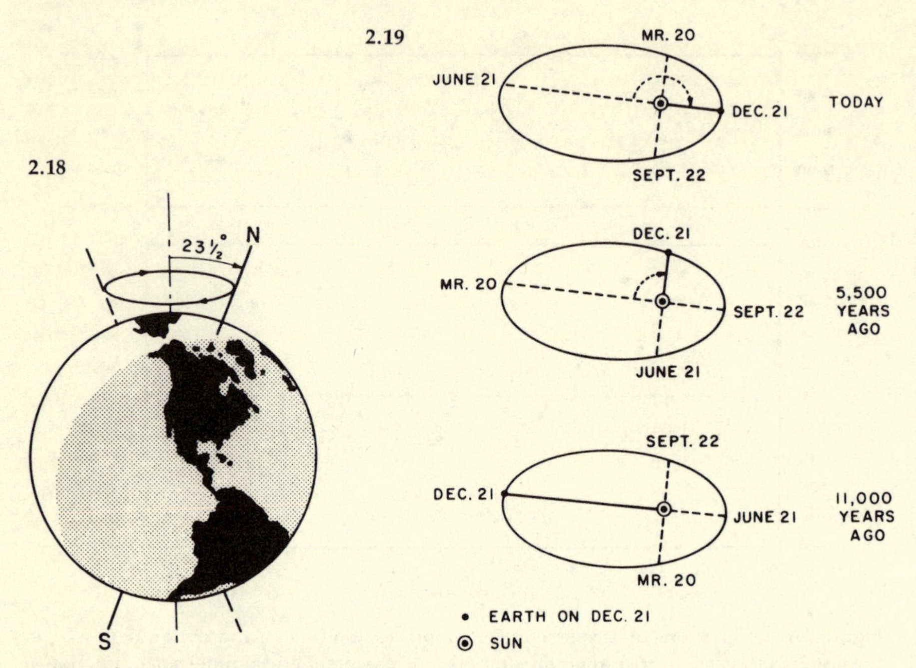

Figure 2.18 The Earth wobbles slightly on its axis (due to the gravitational pull of the Sun and Moon on the equatorial bulge of the Earth). In effect, the axis moves slowly around a circular path and completes one revolution every ~23 000 years. This results in precession of the equinoxes. (Fig. 2.19). This effect is independent of changes in the angle of tilt (obliquity) of the Earth, which changes with a period of ~41 000 years (from Imbrie & Imbrie 1979).

Figure 2.19 As a result of a wobble in the Earth's axis (Fig. 2.18), the positions of the equinox (March 20 and September 22) and solstice (June 21 and December 21) change slowly around the Earth's elliptical orbit, with a period of ~23 000 years. Thus 11 000 years ago the Earth was at perihelion at the time of the summer solstice whereas today the summer solstice coincides with aphelion (from Imbrie & Imbrie 1979).

Changes in the seasonal timing of perihelion and aphelion result from a wobble in the Earth's axis of rotation as it moves around the Sun (Fig. 2.18). The effect of the wobble (which is independent of variations in axial tilt) is to change the timing of the solstices and equinoxes relative to the extreme positions the Earth occupies on its elliptical path around the Sun (Fig. 2.19). Thus, 11 000 years ago, perihelion occurred when the Northern Hemisphere was tilted towards the Sun (mid-June) rather than in the Northern Hemisphere's midwinter, as is the case today. Obviously, precessional effects are opposite in the Northern and Southern Hemispheres. The change in precession occurs with a mean period of ~21 700 years (Fig. 2.17).

Clearly, the effects of precession of the equinoxes on radiation receipts will be modulated by the variations in eccentricity; when the orbit is near circular the seasonal timing of perihelion is inconsequential. However, at maximum eccentricity, when differences in solar radiation may amount to 30%, seasonal timing is crucial. The solar radiation receipts of low latitudes are mainly affected by variations in eccentricity and precession of the equinoxes, whereas higher latitudes are mainly affected by variations in obliquity. Since the eccentricity and precessional effects in each hemisphere are opposite, but the obliquity effects are not, there is an asymmetry between the two hemispheres, in terms of the combined orbital effects, which becomes minimal poleward of ~70°. It is also worth emphasizing that the orbital variations do not cause any overall (annual) change in solar radiation receipts; they simply result in a seasonal redistribution, such that a "low" summer radiation total is compensated for by a "high" winter total, and vice versa (A. Berger 1980).

It is important to note that the periods mentioned for each orbital parameter (41 000, 95 800, and 21 700 years for obliquity, eccentricity, and precession, respectively) are *averages* of the principal periodic terms in the equations used to calculate the long-term changes in orbital parameters. For the precessional parameter, for example, the most important terms in the series expansion of the equation correspond to periods of ~23 700 and ~22 400 years; the next three terms are close to ~19 000 years (A. Berger 1977b). When the most important terms are averaged, the mean period is 21 700 years, but some paleoclimatic records may be capable of resolving the principal ~19 000 and ~23 000 year periods separately (cf. Hays *et al.* 1976). Similarly, the mean period of changes in eccentricity is 95 800 years but it may be possible to detect separate periods of ~95 000 and ~123 000 years in long high-resolution ocean core records corresponding to important terms (or "beats" produced by interactions of important terms) in the equation (cf. Wigley 1976).

Considered together, the superimposition of variations in eccentricity, obliquity, and precession produce a complex, ever-varying pattern of solar radiation receipts at the outer edge of the Earth's atmosphere. To appreciate the magnitude of these variations and their spatial and temporal patterns, it is common to express the radiation receipts for a particular place and moment in time as a departure (or anomaly) from corresponding seasonal or monthly values in 1950. An example is shown in Figure 2.20 for the month of July at all latitudes (90°N–90°S) from 0 to 200 000 years BP (A. Berger 1979). Of particular interest are the radiation anomalies at high northern latitudes (60–70°N) which Milankovitch (1941) considered to be critical for the growth of continental ice sheets. In this zone, periods of lower summer radiation receipts would have favored the persistence of winter snow into summer months, eventually leading to the persistence of snow cover throughout the year. Such conditions may have occurred at 185 000, 115 000, and 70 000 years BP (Fig. 2.20). At these times, there was the combination of conditions which

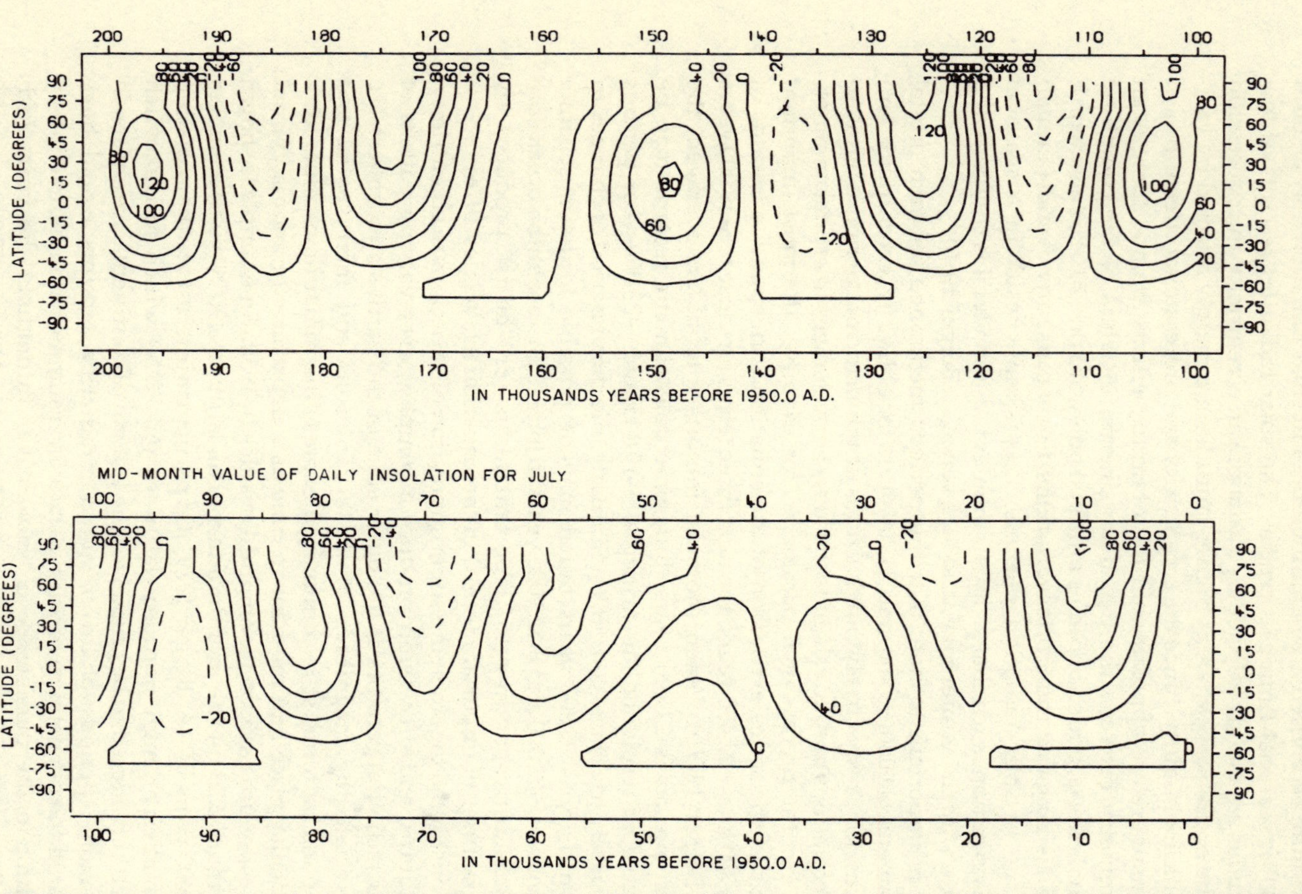

Figure 2.20 Long-term deviations of solar radiation from their AD 1950 values. Values were calculated at 10° latitude intervals from 100 000 years BP to the present day (lower diagram) and from 200 000 years BP to 100 000 years BP (upper diagram). Negative departures are shown by dotted lines. Values are given in calories per square centimeter per day (after A. Berger 1979).

Milankovitch suggested were most conducive to glaciation—minimum obliquity, relatively high eccentricity, and the Northern Hemisphere summer coinciding with aphelion (cf. Fig. 2.17). At the same time, warmer winters (i.e. Northern Hemisphere winters coinciding with perihelion) would have favored increased evaporation from the subtropical oceans, thereby providing abundant moisture for precipitation (snowfall) at higher latitudes. Stronger Equator–Pole temperature gradients in summer and winter would have resulted in an intensified general circulation and more moisture transported to high latitudes to fuel the growing ice sheets. It is of great interest, therefore, that recent ocean core analyses point to these periods as being important times of ice growth on the continents (see Sec. 6.9).

Most of Milankovitch's attention focused on the radiation anomalies in summer and winter months, but recently attention has turned to transitional months which appear to be most sensitive to changes in solar radiation receipts and to snow-cover expansion. In particular, autumn months are especially critical (Kukla 1975). To examine the monthly pattern of solar radiation change through time, A. Berger (1979) has computed month by month values of solar radiation departures from long-term means at 60°N, for the last 500 000 years (e.g. Fig. 2.21). From these calculations it is clear that not only do the monthly departures vary greatly in amplitude but the seasonal timing of the anomalies may shift very rapidly from one part of the year to another. For example, a large positive anomaly of solar radiation in June and July at ~125 000 years BP was "replaced" by a large negative anomaly in the same month by 120 000 years B.P. Such features of the record have been termed "insolation signatures" (A. Berger 1979) and are considered to be characteristic of a change from a relatively warm climate phase to a cooler one. During the last 500 000 years such signatures are observed centered at 486, 465, 410, 335, 315, 290, 243, 220, 199, 127, 105, and 84 thousand years BP, all periods which coincide remarkably well with geological evidence of deteriorating climatic conditions.

Finally, it is important to recognize that although the zone centered on 65°N may be of great importance in the actual mechanism of continental ice growth, a more fundamental control on glaciation is the atmospheric circulation which is largely a function of the Equator–Pole temperature gradients at different times of the year (stronger radiation gradients produce stronger temperature gradients). When radiation gradients are strong, a more vigorous atmospheric circulation can be expected; subtropical high-pressure systems would tend to be displaced to lower latitudes and a more intense circumpolar westerly flow would develop, leading to increased moisture flux to high latitudes. Weaker radiation gradients imply that the major axis of subtropical high-pressure cells would be displaced poleward, and a more sluggish westerly circulation

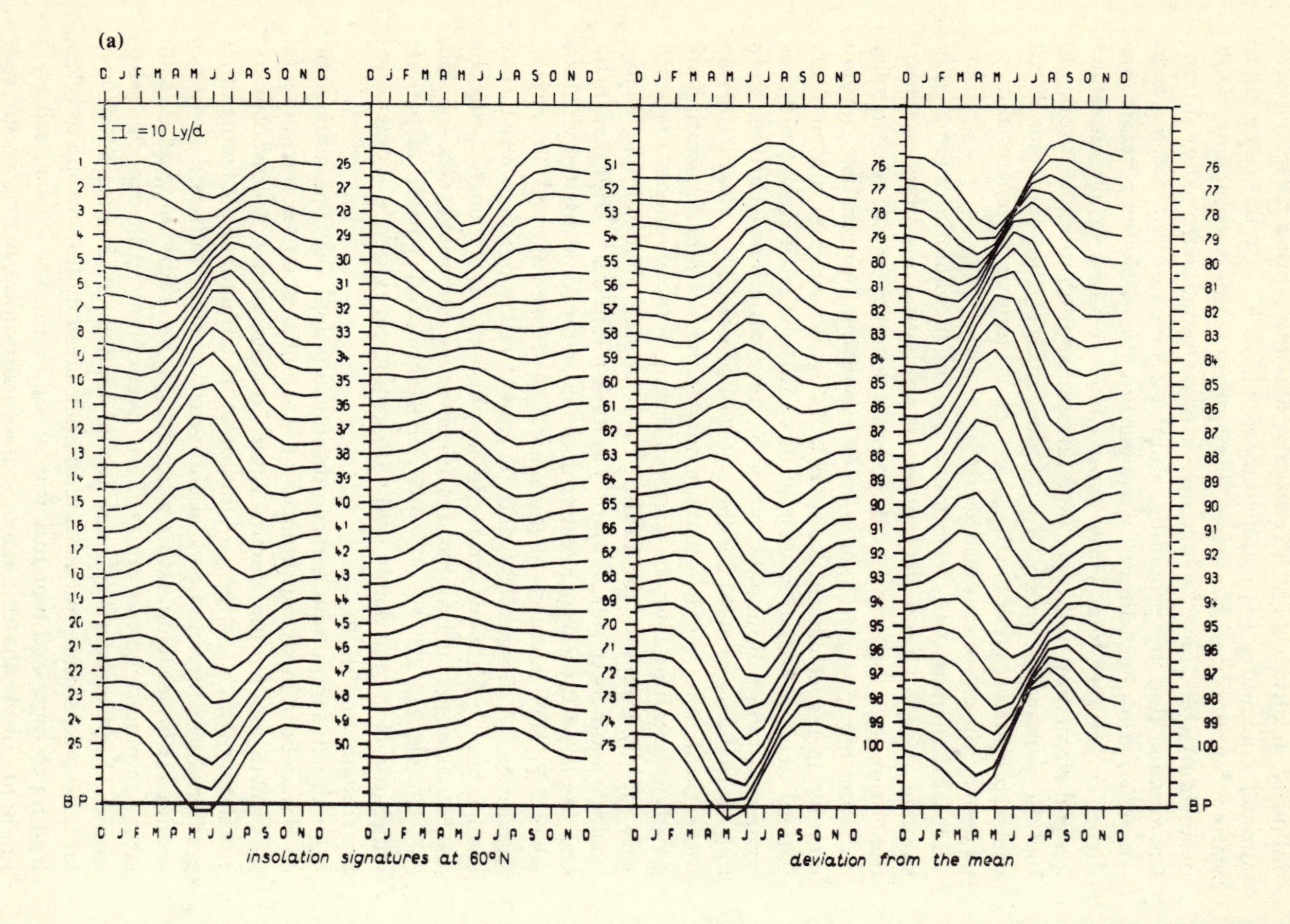
(a)
=10 Ly/d
D J F M A M J J A S O N D
BP
insolation signatures at 60° N
deviation from the mean

(b)

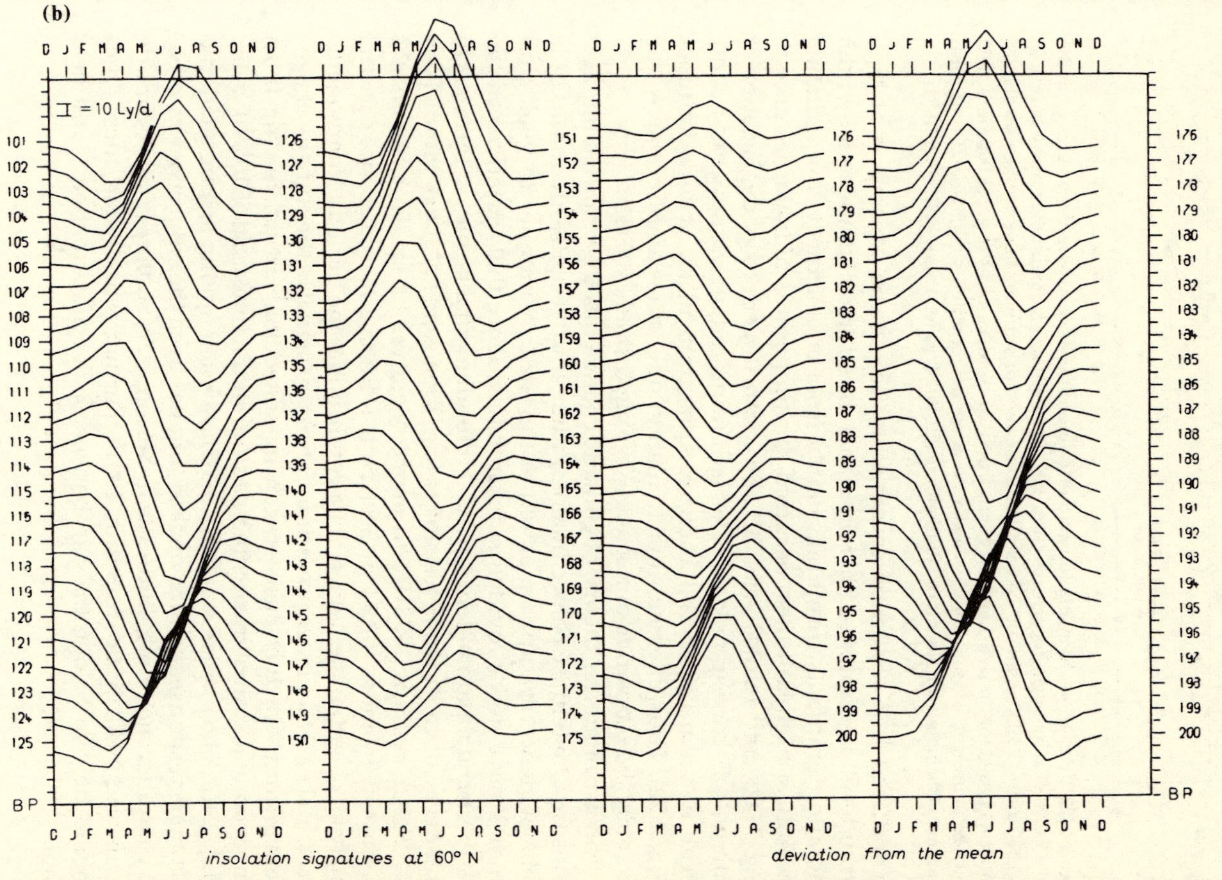

Figure 2.21 "Insolation signatures" at 60°N from 200 000 years BP to the present. Each line shows the annual cycle of deviations from mean mid-monthly insolation for the last 500 000 years. Vertical scale in upper left corner (langleys per day). Periods when large positive summer solar radiation anomalies change to large negative anomalies (insolation signatures) appear to correspond to times when the climate changed from warm to cool (from A. Berger 1979).

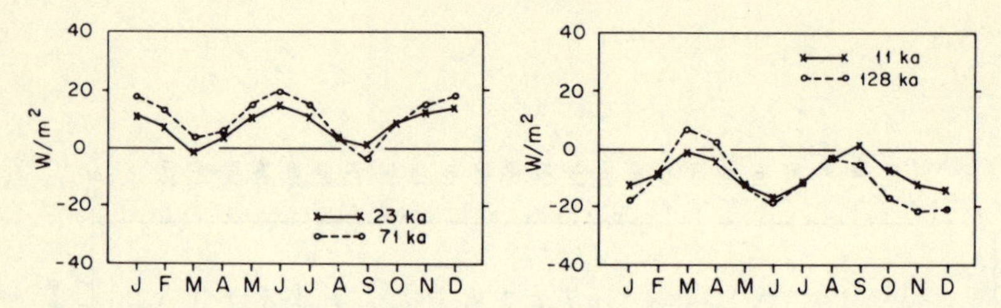

Figure 2.22 Variations in insolation gradients (monthly) expressed as departures from the last 150 000 year averages for selected time periods. Periods of maximum ice growth (e.g. at ~71 000 and ~23 000 years BP correspond to periods of stronger than average insolation gradients in all months (left-hand diagram). Times of rapid ice decay (e.g. 128 000 and 11 000 years BP) correspond to generally weaker than average gradients (right-hand diagram). Gradients calculated for Northern Hemisphere (30–90°N) (after Young & Bradley 1984).

would lead to a reduction in moisture flux to the continents at high latitudes. It is of interest, therefore, that the stronger summer and winter radiation gradients (resulting mainly from anomalously low radiation receipts at high latitudes) occurred during periods of major ice growth (e.g. 72 000 and 115 000 years BP). By contrast, periods of deglaciation or interglacials correspond to weaker latitudinal radiation gradients, mainly resulting from higher radiation receipts, particularly at high latitudes (Fig. 2.22). Thus the resulting circulation intensities amplify the overall anomaly, be it positive or negative (Young & Bradley 1984).

The astronomical theory of climatic change has tremendous implications for Quaternary paleoclimatology but, until recently, there was little field evidence to support or refute the idea. A number of geological studies have now provided substantial evidence that variations in the Earth's orbital parameters are indeed fundamental factors in the growth and decay of continental ice sheets (e.g. Broecker *et al.* 1968, Mesolella *et al.*, 1969, Hays *et al.* 1976, Ruddiman & McIntyre 1981a). This evidence is discussed in more detail in Section 6.8. Suffice it to say that spectral analysis of certain geological data confirm that the periodic components in orbital parameters do provide a significant signal to the climate-dependent record. On this basis, it can be stated that orbital forcing is an important factor in climatic fluctuations on the timescale of 10^4–10^6 years (Fig. 2.16). The precise mechanism of how such forcing is translated into a climate response remains unclear, but has been the subject of a number of computer modeling studies (e.g. Mason 1976, A. Berger 1977c, Imbrie & Imbrie 1980, Budd & Smith 1981). Further work along these lines should help to bridge the gap between paleoclimatic theory and field data.

3

Dating methods I

3.1 Introduction and overview

Accurate dating is of fundamental importance to paleoclimatic studies. Without reliable estimates on the age of events in the past it is impossible to investigate if they occurred synchronously or if certain events lead or lag others; neither is it possible to assess accurately the *rate* at which past environmental changes occurred. Strenuous efforts must therefore be made to date all proxy materials, to avoid sample contamination, and to ensure that the stratigraphic context of the sample is clearly understood. It is equally important that the assumptions and limitations of the dating procedure used are understood so that a realistic interpretation of the date obtained can be made. It is often just as important to know the margins of error associated with a date as to know the date itself. In this chapter, we discuss the main dating methods widely used for late Quaternary studies today. The review by Hedges (1979) includes some other techniques which may yet prove useful, but which at present have not progressed beyond the experimental stage.

Dating methods fall into four basic categories (Fig. 3.1): (a) *radio-isotopic methods*, which are based on the rate of atomic disintegration in a sample or its surrounding environment; (b) *paleomagnetic methods*,† which rely on past reversals of the Earth's magnetic field and their effects on a sample; (c) *organic and inorganic chemical methods*, which are based on time-dependent chemical changes in the sample, or chemical characteristics of a sample; (d) *biological methods*, which are based on the growth of an organism to date the substrate on which it is found.

Not all dating methods provide reliable *absolute* dates, but may give an indication of the *relative* age of different samples. In these cases, it may be possible to calibrate the "relative age" technique by absolute (e.g. radio-isotopic) methods, as discussed, for example, in Section 4.23.

†One could argue that paleomagnetic changes do not constitute a *method* of dating but rather a method of stratigraphic correlation. Nevertheless, the development of a reliable timescale for paleomagnetic changes (Sec. 4.1.4) has meant that paleomagnetic changes are used, *de facto*, as dated reference horizons. Since this is now common usage in Quaternary studies, it is appropriate to consider paleomagnetic change as another dating method.

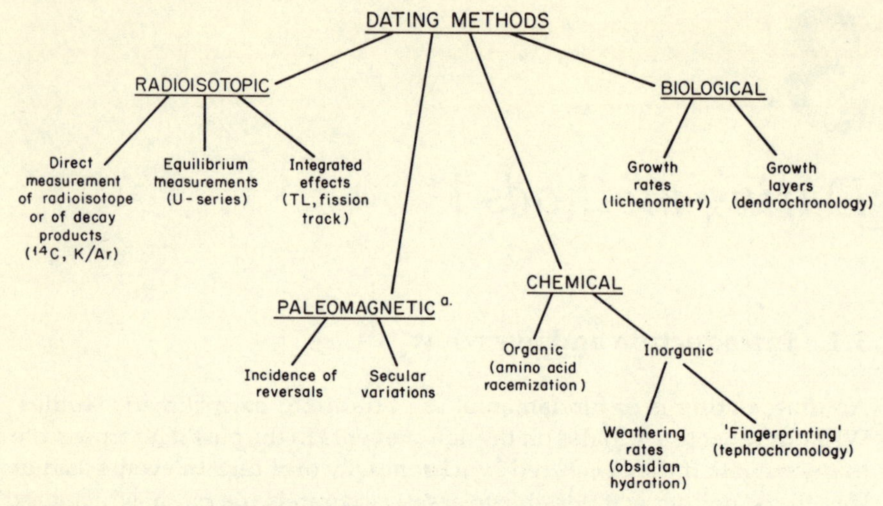

Figure 3.1 Principal dating methods used in paleoclimatic research.

In this chapter, radio-isotopic methods are dealt with; paleomagnetic, chemical, and biological methods are discussed in Chapter 4.

3.2 Radio-isotopic methods

Atoms are made up of neutrons, protons, and electrons. For any one element, the number of protons (the atomic number) is invariant, but the number of neutrons may vary, resulting in different isotopes of the same element. Carbon, for example, exists in the form of three isotopes; it always has six protons, but may have six, seven, or eight neutrons, giving atomic mass numbers (the total number of protons and neutrons) of 12, 13, and 14, designated ^{12}C, ^{13}C, and ^{14}C. Generally each element has one or more stable isotopes which account for the bulk of its occurrence on Earth. For example, in the case of carbon, ^{12}C and ^{13}C are the stable isotopes; ^{12}C is by far the most abundant form. It is estimated that the carbon exchange reservoir (atmosphere, biosphere, and the oceans) contains 42×10^{12} tons of ^{12}C, 47×10^{10} tons of ^{13}C, and only 62 tons of ^{14}C. Unstable atoms undergo spontaneous radioactive decay by the loss of nuclear particles (α or β particles‡) and, as a result, they may transmute into a new element. ^{14}C, for example, decays to nitrogen and ^{40}K decays to ^{40}Ar and ^{40}Ca. Furthermore, the decay rate is invariable so that a given quantity of the radioactive isotope will decay to its daughter product in a known interval of time; this is the basis of radio-isotopic dating methods.

‡An α particle is made up of two protons and two neutrons (i.e. a helium atom) and a β particle is an electron. Neutrons may decay to produce a β particle and a proton, thereby causing a transmutation of the element itself.

Table 3.1 Half-lives of radio-isotopes used in dating. Although the half-life of ^{14}C is calculated to be 5730 ± 40 years, by convention the "Libby half-life" of 5568 ± 30 years is used (see footnote).

^{14}C	5.73×10^3 years
^{238}U	4.51×10^9 years
^{235}U	0.71×10^9 years
^{40}K	1.31×10^9 years

Providing that the radio-isotope "clock" is started close to the stratigraphically relevant date, measurement of the isotope concentration today will indicate the amount of time which has elapsed since the sample was emplaced. The amount of time which it takes for a radioactive material to decay to half its original amount is termed its half-life. Table 3.1 lists the half-lives of some radio-isotopes which have been used in the context of dating. In the case of radiocarbon (^{14}C) the half-life is 5730 ± 30 years.§ Thus a plant which died 5730 years ago has only half its original ^{14}C content remaining in it today. After a further 5730 years from today it will have only half as much again, i.e. 25% of its original ^{14}C content, and so on (Fig. 3.2).

For a radioactive isotope to be directly useful for dating it must possess several attributes: (a) the isotope itself, or its daughter products, must occur in measurable quantities and be capable of being distinguished from other isotopes, or its rate of decay must be measurable; (b) its half-life must be of a length appropriate to the period being dated; (c) the initial concentration level of the isotope must be known; (d) there must be some connection between the event being dated and the start of the radioactive decay process (the "clock"). The relevance of these factors will be made clear in the ensuing sections.

In general terms, radio-isotopic dating methods can be considered in three groups (Fig. 3.1), those which measure (a) the quantity of a radio-isotope as a fraction of a presumed initial level (e.g. ^{14}C dating) or the reciprocal build-up of a stable daughter product (e.g. potassium–argon, and argon–argon dating); (b) the degree to which members of a chain of radioactive decay are restored to equilibrium following some initial external perturbation (uranium-series dating); (c) the integrated

§When Libby expounded the principles of radiocarbon dating (Libby 1955) he calculated a half-life for ^{14}C of 5568 years. This was the average of a number of estimates up to that time, and was adopted by all radiocarbon dating laboratories. By the early 1960s, further work had demonstrated that the original estimate was in error by 3% and the half-life was closer to 5730 years (Godwin 1962). To avoid confusion, it was decided to continue using the "Libby half-life" (rounded to 5570 years) and this practice has continued. For practical purposes it is not a significant problem since all dates are now reported in the journal *Radiocarbon* using the Libby half-life. However, when comparing "radiocarbon years" with calendar years (historical, archeological, and/or astronomical events) and dates obtained by other techniques, adjustments are necessary.

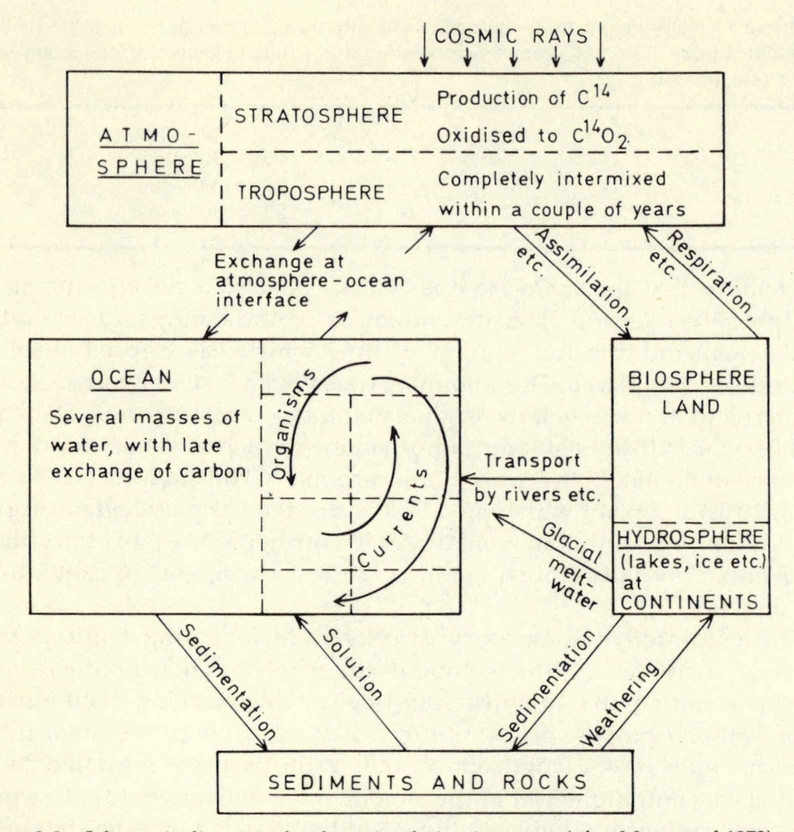

Figure 3.2 Schematic diagram of carbon circulation in nature (after Mangerud 1972).

effect of some local radioactive process on the sample materials, compared to the value of the local (environmental) flux (fission-track and thermoluminescence dating). Each of these methods will be considered separately.

3.2.1 *Radiocarbon dating*

For studies of late Quaternary climatic fluctuations, ^{14}C or radiocarbon dating has proved to be by far the most useful. Because of the ubiquitous distribution of ^{14}C, the technique can be used throughout the world and has been used to date samples of peat, wood, bone, shell, paleosols, "old" sea water, marine and lacustrine sediments, and atmospheric CO_2 trapped in glacier ice. Furthermore, the useful time-frame for radiocarbon dating spans a period of major, global environmental change which would be virtually impossible to decipher in any detail without accurate dating control. Radiocarbon dating is also ideal for dating man's development from paleolithic time to the recent historical past and it has therefore proved invaluable in archeological studies. Recently, variations in the ^{14}C content of the atmosphere have become of interest in themselves because of the implications these have for solar and/or geomagnetic variations through time and hence for climatic fluctuations.

3.2.1.1 *Principles of ^{14}C dating.* Radiocarbon ($^{14}_{6}C$) is produced in the upper atmosphere by neutron bombardment of atmospheric nitrogen atoms:

$$^{14}_{7}N + ^{1}_{0}n \rightarrow ^{14}_{6}C + ^{1}_{1}H$$

The neutrons have a maximum concentration at around 15 km and are produced by cosmic radiation entering the upper atmosphere. Although cosmic rays are influenced by the Earth's magnetic field and tend to become concentrated near the geomagnetic poles (thus causing a similar distribution of neutrons and hence ^{14}C), rapid diffusion of ^{14}C atoms in the lower atmosphere obliterates any influence of this geographical variation in production. ^{14}C atoms are rapidly oxidized to $^{14}CO_2$, which diffuses downwards and mixes with the rest of atmospheric carbon dioxide and hence enters into all pathways of the biosphere (Fig. 3.2). As Libby (1955) states, "Since plants live off the carbon dioxide, all plants will be radioactive; since the animals on earth live off the plants, all animals will be radioactive. Thus . . . all living things will be rendered radioactive by the cosmic radiation."

During the course of geological time, an equilibrium has been achieved between the rate of new ^{14}C production in the upper atmosphere and the rate of decay of ^{14}C in the global carbon reservoir. This means that the 7.5 kg of new ^{14}C estimated to be produced each year in the upper atmosphere is approximately equal to the weight of ^{14}C lost throughout the world by the radioactive decay of ^{14}C to nitrogen, with the release of a β particle (an electron):

$$^{14}_{6}C \rightarrow ^{14}_{7}N + \beta + \text{neutrino}$$

The total weight of global ^{14}C thus remains constant.† This assumption of an essentially steady concentration of radiocarbon during the period useful for dating is fundamental to the method, though it now appears that, in detail, this assumption is invalid (see Sec. 3.2.1.5).

Plants and animals assimilate a certain amount of ^{14}C into their tissues through photosynthesis and respiration; the ^{14}C content of these tissues is in equilibrium with that of the atmosphere because there is a constant

†Prior to atomic bomb explosions in the atmosphere the "equilibrium" quantity of ^{14}C was estimated to be ~62 metric tons. Since the 1950s, the amount of artificially produced ^{14}C has increased by perhaps 3–4%, though most of this has, as yet, remained in the atmosphere; consequently ^{14}C levels there have almost doubled (Aitken 1974).

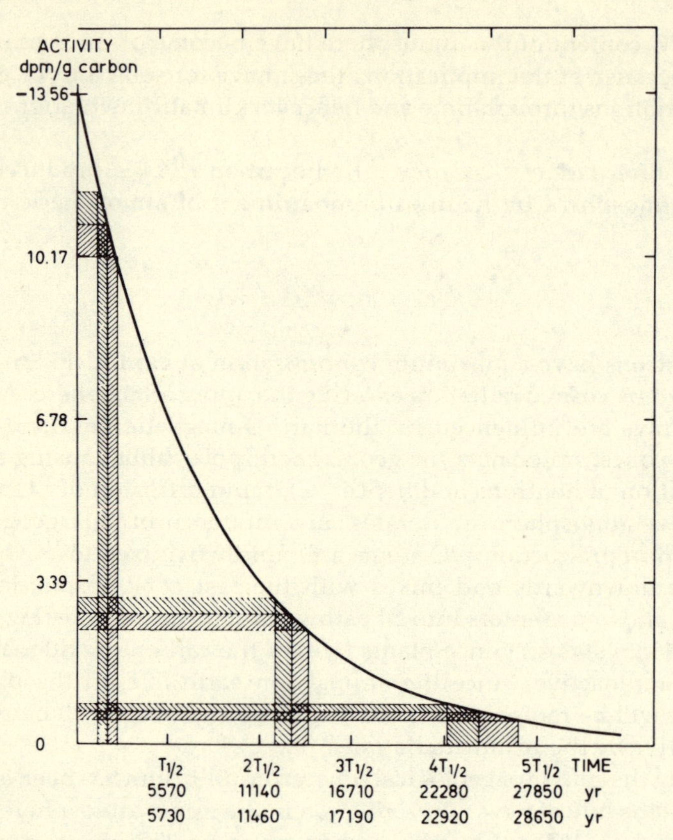

Figure 3.3 The decay of radiocarbon with time. Both the Libby and revised half-lives ($T\frac{1}{2}$) are shown on the abscissa. The ordinate scale is in disintegrations per minute per gram of carbon (after Olsson 1968).

exchange of new CO_2 as old cells die and are replaced. However, as soon as an organism dies, this exchange and replacement of ^{14}C from the atmosphere ceases. From that moment on the ^{14}C content of the organism declines as the ^{14}C decays to nitrogen, and the ^{14}C content is henceforth purely a function of time; the radioactive "clock" has been activated. Because the ^{14}C content declines at a negative exponential rate (Fig. 3.3), by the time that ten half-lives have elapsed (57 300 years) the sample contains less than 0.001% of the original ^{14}C content of the organism when it was alive (see Appendix). To put this in a more practical way, in a 1 g sample of modern carbon, decay of the radiocarbon atoms in the sample will produce about 15 β particles per minute, a rate which is relatively easy to count. By contrast, 57 300 years after an organism has died, 1 g of its carbon will produce only about 2.2 β particles *per day* (Aitken 1974). It is this ever-decreasing decay rate with increasing sample age that makes conventional radiocarbon dating so difficult; it simply becomes impossible to separate the extraneous "background" radiation from disintegrations of the sample, as the signal to noise ratio is too small.

3.2.1.2 Measurement procedures, materials and problems. Nearly all radiocarbon-dating laboratories use either proportional gas counters or liquid scintillation techniques. In the former method, carbon is converted into a gas (methane, carbon dioxide, or acetylene) which is then put into a "proportional counter" capable of detecting β particles (variations in output voltage pulses being proportional to the rate of β-particle emission). In liquid scintillation procedures, the carbon is converted into benzene or some other organic liquid and placed in an instrument which detects scintillations (flashes of light) in the liquid, produced by β-particle emissions. In both cases, stringent measures are necessary to shield the sample counters from extraneous radioactivity in the instrument components, laboratory materials, and surrounding environment, including the occasional cosmic ray penetrating the Earth's atmosphere from outer space. Indeed, the difficulty of separating the sample β-particle signal from environmental "noise" was one of the major obstacles to the development of ^{14}C dating, particularly, of older samples which have very low levels of ^{14}C anyway (Libby 1970). Lead shielding, electronic anti-coincidence counters (to alert the counter to particles entering the counting chamber from outside) and construction of laboratories beneath the ground are common strategies to help keep background radiation levels as low as possible.

One of the problems of dating very old samples by ^{14}C methods is the large sample size needed to obtain enough radiocarbon for its β activity to be counted. Technical difficulties place an upper limit on the volume of gas or liquid that can be accurately analysed; hence, for very old samples, some means of concentrating the ^{14}C is needed to reduce the volume. One solution is to concentrate a gas containing the ^{14}C (e.g. $^{14}CO_2$) by thermal diffusion, "enriching" the sample and reducing the required volume. Effectively, the gas containing the heavier isotope is encouraged to collect in the lower chamber of a thermal diffusion column; in this way, the radioactive component is concentrated, reducing the total volume of gas necessary for accurate counting. However, the procedure is time consuming; a sixfold enrichment may take up to 5 weeks! Nevertheless, this procedure has enabled samples as old as 75 000 years (13 half-lives) to be dated (Stuiver *et al.* 1978). The main limitation is that the initial sample must be large enough to yield 100 g of carbon for analysis.

More recently, it has been demonstrated that extremely low concentrations of ^{14}C can be detected by mass spectrometry. Instead of measuring the quantity of ^{14}C in a sample indirectly, by the incidence of β-particle emissions, the concentrations of individual ions are measured. Ions are accelerated in a cyclotron or tandem electrostatic accelerator to extremely high velocities; they then pass through a magnetic field which separates the different ions, enabling them to be distinguished (Stuiver 1978a). The technique is still in the experimental stages but promises to provide dates on samples with extremely low ^{14}C levels, perhaps as old as 100 000 years. Furthermore, sample sizes used in this technique are much smaller (10 mg) than in conventional ^{14}C dating, so that dates of samples as small as individual seeds, or foraminifera, may become possible. At present, several accelerators designed specifically for ^{14}C dating applications are under construction. This new technique could revolutionize dating of late Quaternary events by providing important chronological control on events as far back as the beginning of the last glaciation.

3.2.1.3 Accuracy of radiocarbon dates. It is tempting to accept a radiocarbon date as the gospel truth, particularly if it confirms a preconceived notion of what the sample age should be! Radiocarbon dates are, however, statements of probability (as are all radiometric measurements). Radioactive disintegration varies randomly about a mean value; it is not possible to predict when a *particular* ^{14}C atom will decay, but for a sample containing 10^{10}–10^{12} ^{14}C atoms a certain number of disintegrations will occur, on average, in a certain length of time. This statistical uncertainty in the sample radioactivity (together with similar uncertainty in the radioactive decay of calibration samples and "noise" due to background radiation) is inherent in all ^{14}C dates. A single "absolute" date can therefore never be assigned to a sample. Rather, dates are reported as the midpoint of a Poisson probability curve; together with its standard deviation, the date thus defines a known level of probability. A date of 5000 $\pm$ 100 years BP, for example, indicates a 68% probability that the true (radiocarbon) age is between 4900 and 5100 years BP, a 95% probability that it lies between 4800 and 5200 years BP, and a 99% probability of it being between 4700 and 5300 years BP. The range of uncertainty in a date may be reduced by the use of large samples, by extended periods of counting, and by the reduction of laboratory background noise. However, even rigorous ^{14}C analysis can not account for all sources of error and these must be considered before putting a great deal of confidence in the precise date obtained.

3.2.1.4 Sources of error in ^{14}C dating

(a) PROBLEMS OF SAMPLE SELECTION AND CONTAMINATION. It is self-evident that a contaminated sample will give an erroneous date, but it is frequently impossible to ascertain if a sample has indeed been contaminated. Some forms of contamination are relatively straightforward; modern rootlets, for example, may penetrate deep into a peat section and without careful inspection of a sample and removal of such material, gross errors may occur. More abstruse problems arise when dating materials containing carbonates (e.g. shell, coral, bone). These materials are particularly susceptible to contamination by modern carbon because they readily participate in chemical reactions with rain water and/or ground water. Most molluscs, for example, are primarily composed of calcium carbonate in the metastable crystal form aragonite. This aragonite may dissolve and be redeposited in the stable crystal form of calcite. During the process of solution and recrystallization, exchange of modern carbon takes place and the sample is thereby contaminated (Grant-Taylor 1972). This problem also exists in corals which are all aragonite. Commonly, X-ray diffraction is used to identify different carbonate mineral species, and materials with a high degree of recrystallization are discarded. However, Chappell and Polach (1972) have noted that recrystallization can occur in two different modes: one open system and therefore susceptible to modern ^{14}C contamination, and one closed system which is internal and involves no contamination. The former process tends to be concentrated around the sample margins, as one might expect, and so a common strategy is to dissolve away the surface 10–20% of the sample with hydrochloric acid and to date the remaining material. For shells thought to be very old, in which recrystallization may have affected a deep layer, the remaining inner fraction should ideally be dated in two fractions (an "outer inner" and an "inner inner" fraction) to test for consistency of results. However, even repeated leaching with hydrochloric acid may not produce a reliable result in cases where recrystallization has permeated the entire sample. It is worth noting in this connection that "infinite"-aged shells (those well beyond the range of ^{14}C dating techniques) which are contaminated by only 1% with modern carbon will have an apparent age of 37 000 years (Olsson 1974). Thus even very small amounts of modern carbon can lead to gross errors and many investigators thus consider that dates of >25 000 years on shells should be thought of as essentially "infinite" in age. While such a conservative approach is often laudable, it does pose the danger that correct dates in, say, the 20 000–30 000 year range may be overlooked. A possible remedy to this problem of dating old shells is to isolate a protein, conchiolin, present in very small quantities in shells (1–2%) and to date this, rather than the carbonate (R. Berger *et al.* 1964). Carbon in conchiolin does not undergo exchange with carbon in the surrounding environment and hence is far less likely to be contaminated. Unfortunately, to isolate enough conchiolin, very large samples (>2 kg) are needed, and these are often not available.

Similar problems are encountered in dating bone; because of exchange reactions with modern carbon, dates on total inorganic carbon or on apatite carbonate are unreliable (Olsson *et al.* 1974). As in the case of shells, a more reliable approach is to isolate and date carbon in the protein collagen. However, this slowly disappears under the influence of an enzyme, collagenase, so that in very old samples the amount of collagen available is extremely small and extra-large samples are needed to extract it. Furthermore, collagen is extremely difficult to extract without contamination and different extraction methods may give rise to different dates.

Another form of error concerns the "apparent age" or "hard-water effect" (Shotton 1972). This problem arises when the materials to be dated, such as freshwater molluscs or aquatic plants, take up carbon from water containing bicarbonate derived from old, inert sources. This is a particularly difficult problem in areas where water is contaminated by dissolved humus from old peat bogs or in areas where limestone and other calcareous rocks occur. In such regions, surface and ground water may have much lower $^{14}C/^{12}C$ ratios than that of the atmosphere due to solution of the essentially ^{14}C-free bedrock. Plants and animals existing in these environments will assimilate carbon in equilibrium with their surrounding milieu rather than the atmosphere, hence they will appear older than they are in reality, sometimes by as much as several thousand years. This problem was well illustrated by Shotton (1972) who studied a late-glacial stratigraphic section in North Jutland, Denmark. Dates on contemporaneous twigs and a fine-grained vegetable residue, thought to be primarily algal, fell neatly into two groups, with the algal material consistently 1700 years older than the terrestrial material. The difference was considered to be the result of a hard-water effect, the aquatic plants assimilating carbon in equilibrium with water containing bicarbonate from old, inert sources.

Other studies have demonstrated further complexities in that the degree of "old carbon" contamination may change over time. Thus, for example, Karrow and Anderson (1975) suggested that some lake sediments in south-western New Brunswick, Canada, studied by Mott (1975), were contaminated with old carbon shortly after deglaciation. Initial sedimentation was mainly marl derived from carbonate-rich till and carbonate bedrock, but as the area became vegetated and soil development took place the sediments became more organic and less contaminated by "old carbon". Dates on the deepest lake sediments are thus anomalously old and would appear to give a date of deglaciation inconsistent with the regional stratigraphy. This points to the more general observation that the geochemical balance of lakes may have changed through time and that the modern water chemistry may not reflect former conditions. This is particularly likely in formerly glaciated areas where the local environment immediately following deglaciation would have been quite different from that of today. One should thus interpret basal dates on lake sediments or peat bogs with caution. Equally, dates on aquatic flora and fauna, in closed lake basins which have undergone great size changes, must also be viewed in the light of possible changes in the aqueous geochemistry of the site.

It is important to recognize that not all types of contamination are equally significant; contamination by modern carbon is far more important than that by old carbon because of its much higher activity (Olsson 1974). Figures 3.4 and 3.5 show the errors associated with different percentage levels of contamination by modern and old material, respectively. It will be seen that a 5000 year old sample, 20% contaminated with 15 000 year old carbon, would give a date in error by only ~1250 years. By contrast, a 15 000 year old sample contaminated with only 3% of modern carbon would result in a dating error of about the same magnitude. Very careful sample selection is therefore needed; dating errors are most commonly the result of inadequate sampling.

(b) VARIATIONS IN ^{14}C CONTENT OF THE OCEANIC RESERVOIR. In the above section, the problem that some freshwater aquatic plants or molluscs may be contaminated by water containing very low levels of ^{14}C was discussed. In the case of marine organisms the problem is much more universal. The turnover of oceanic water bodies is much slower than that of the atmosphere. This is particularly true for water below the thermocline. At high latitudes surface waters are cooled and sink down, spreading out to form abyssal water bodies such as the North Atlantic Deep Water and the Antarctic Bottom Water. This water may remain out of contact with the atmosphere for centuries because of the overlying warmer water in mid and low latitudes. During this time, the ^{14}C content of the deep water decreases so that deep-water samples commonly give ^{14}C ages of ~1000 years and may date as much as 2500 years BP (Fig. 3.6). Indeed, the gradual decay in ^{14}C activity of oceanic water has been used to "tag" water masses and to estimate their rate of movement (Stuiver 1976; Stuiver *et al.* 1983).

In regions of upwelling, such as are found in areas of divergent surface currents, deep water depleted in ^{14}C, relative to equilibrium conditions, will be brought to the surface. This water will mix slowly with surrounding surface water resulting in an upper mixed layer with a ^{14}C activity intermediate between that of deeper water and that of surface water in equilibrium with atmospheric carbon dioxide. At present, a fairly constant ^{14}C activity, corresponding to a departure from equilibrium conditions of −4 to −5% (or an apparent water age of 320–400 years) is attained in areas well away from centers of upwelling (Krog & Tauber 1973). Thus, in the North Atlantic, modern marine shells which have grown in equilibrium with the ^{14}C content of the ocean surface waters, have an apparent age which averages around 450 years (Mangerud 1972). In

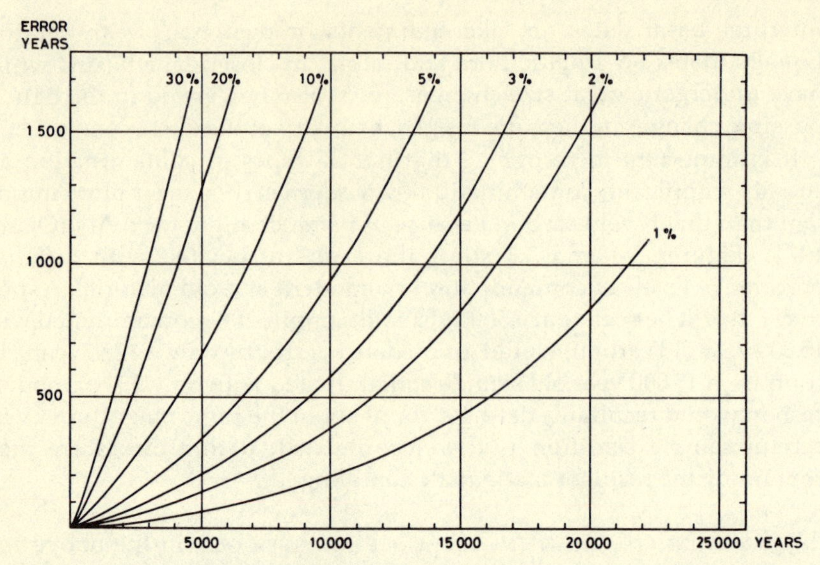

Figure 3.4 The error in a radiocarbon date if a certain fraction of the sample (indicated by each curve) is contaminated by *younger* material (having a higher ^{14}C activity). Errors expressed as age differences (abscissa) between the sample and contaminant. For example, a 16 000 year old sample contaminated with 3% modern material will yield a date in error (too young) by 1300 years (from Olsson 1974).

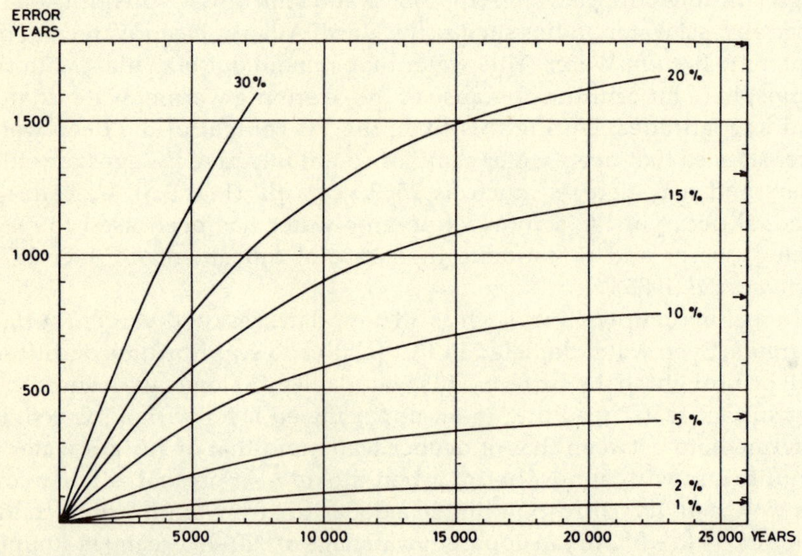

Figure 3.5 The error in a radiocarbon date if a certain fraction of the sample (indicated by each curve) is contaminated by *older* material (having a lower ^{14}C activity). Errors expressed as age differences (abscissa) between the sample and contaminant. For example a 5000 year old sample contaminated with 20% 16 000 year old material will yield a date in error (too old) by ~1300 years (from Olsson 1974).

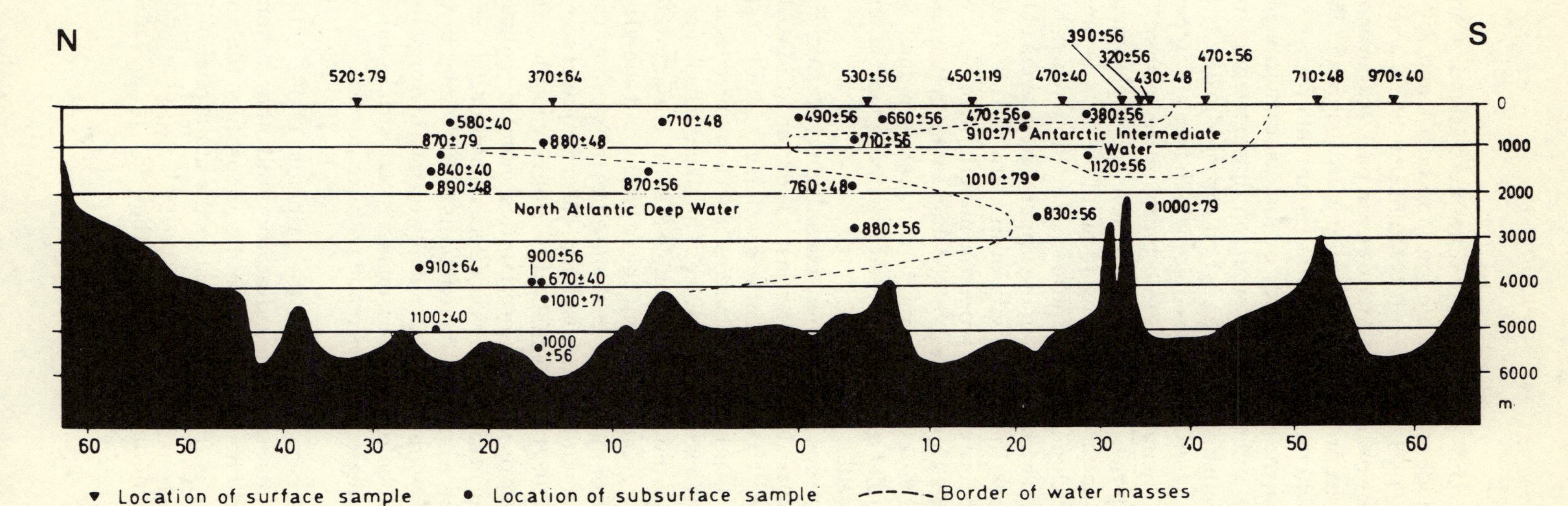

Figure 3.6 Apparent age of samples of sea water from the East Atlantic Ocean (parallel to the mid-Atlantic Ridge) (from Mangerud 1972).

eastern Greenland and south-eastern Ellesmere Island, Canada, the apparent ages of modern shells were found to be 570 and 750 years BP respectively (Hjort 1973, Mangerud & Gulliksen 1975). These higher values may reflect the influence of Arctic Ocean pack-ice cover on surface water–CO_2 exchange, or the greater mixing with deep water that occurs at high latitudes where the thermocline is absent (Bien *et al.* 1963). Whatever the reason, the resulting effect on mollusc or marine plant dates is quite significant and may necessitate an appropriate correction in dating marine organisms. However, when carbon dioxide is absorbed into the oceans a fractionation takes place which leads to an enrichment of 4–6% in the ^{14}C activity of oceanic bicarbonate relative to that of terrestrial plant material. Thus, in many regions the depletion in ^{14}C due to oceanic circulation factors, leading to an older date, is more or less balanced by the enrichment of ^{14}C due to fractionation, which would lead to a younger date (cf. Blake 1979). In the upper mixed layers of the North Atlantic, for example, the ^{14}C activity of modern marine shells is very similar to that of modern organic terrestrial material. In other areas where surface water has an apparent age older than that compensated for by fractionation (~400 years), direct comparison between dates on terrestrial and marine samples may not be possible without some adjustment in sample age being made.

Further complications arise when considering oceanic circulation changes which occurred during the last glaciation (CLIMAP Project Members 1976). The ^{14}C content of surface waters no doubt was different from that of today as a result of changes in ocean currents, centers of upwelling, and more extensive pack-ice. The likely effect such changes had is hard to assess, though it is probable that the surface waters of the North Atlantic had a greater apparent age in glacial and late-glacial times than today. The role of the ice sheets themselves in "storing" carbon and limiting equilibrium with the atmosphere is also a point of some debate, though Mangerud (1972) argues that this is unlikely to affect shell dates significantly. Because of all these uncertainties, many investigators make no correction for the oceanic reservoir effect in late-glacial marine samples, arguing that there is an insufficient basis of knowledge to do so. One must therefore be aware of whether dates on marine organisms have been adjusted or not before comparisons with terrestrial samples and/or samples from other regions can be made with confidence.

(c) FRACTIONATION EFFECTS. Basic to the principle of ^{14}C dating is the assumption that plants assimilate radiocarbon and other carbon isotopes in the same proportion as they exist in the atmosphere (i.e. the $^{14}C/^{12}C$ ratio of plant tissue is the same as that in the atmosphere). However, during photosynthesis, when CO_2 is converted to carbohydrates in plant cells, an isotopic fractionation occurs such that ^{12}C is more readily "fixed" than ^{14}C, resulting in a lower ^{14}C content than that of the atmosphere (Olsson 1974). ^{14}C "depletion" may be as much as 5% below atmospheric levels but this is not consistent amongst all organisms. The magnitude of the fractionation effect varies from one plant species to another by a factor of two to three and depends on the particular biochemical pathways evolved by the plant for photosynthesis (Lerman 1972). This is discussed in further detail in the Appendix. Fortunately, some assessment of the ^{14}C fractionation effect can be made relatively easily by measuring the ^{13}C content of a sample. ^{14}C fractionation is very close to twice that of ^{13}C (Craig 1953), a stable isotope which occurs in far greater quantities than ^{14}C and can hence be measured by a mass spectrometer. ^{13}C content is generally expressed as a departure from a Cretaceous limestone standard (Peedee belemnite, see Appendix):

$$\delta^{14}C = 2\delta^{13}C = \frac{(^{13}C/^{12}C)_{sample} - (^{13}C/^{12}C)_{PDB}}{(^{13}C/^{12}C)_{sample}} \times 10^{3}‰.$$

A change in $\delta^{14}C$ of only 1‰ (i.e. a change in $\delta^{13}C$ of 0.5‰) corresponds to an age difference of ~8 years. Consequently, if two contemporaneous samples differed in $\delta^{13}C$ by 25‰, they would appear to have an age difference of ~400 years (Olsson & Osadebe 1974). To avoid such confusion, it has been recommended that the $\delta^{13}C$ value of all samples be normalized to −25‰, the average value for wood. By adopting this reference value, comparability of dates is possible. However, there are a number of problems with this approach. First, when considering dates on samples reported in the literature,† $\delta^{13}C$ values are rarely given, often because they were not measured. Even when measured, it is rarely mentioned whether corrections have been made to $\delta^{13}C = -25‰$. This is particularly important in the case of marine shell samples which characteristically have $\delta^{13}C$ values in the range +3 to −2‰. Standardization to $\delta^{13}C = -25‰$ thus involves an age adjustment of as much as 450 years (to be *added* to the uncorrected date).‡ Often investigators are unaware that such corrections have been made and this can obviously lead to misinterpretation. This is further complicated because of the low ^{14}C content of the oceans which gives modern sea water an "apparent age" of 400–2500 years (see Sec. 3.2.1.4(b)) and results in a correction in the opposite direction from the correction for fractionation effects (a value to be *subtracted* from the uncorrected age). For the North Atlantic region, the apparent age of sea water is 450 ± 40 years (Mangerud 1972), so the fractionation and oceanic effect more or less cancel out. In other areas,

†*Radiocarbon* is the definitive source of ^{14}C dates; reference to this publication is recommended rather than the use of dates as reported in the literature.

‡Some dating laboratories correct marine shells to a ^{13}C reference level of 0.0‰ and terrestrial plant material to −25‰ (e.g. the Geological Survey of Canada).

particularly at high latitudes, the oceanic adjustment is >450 years, so the adjusted date will be lower (younger) than the original estimate, before correction for fractionation effects. Of course, for comparison of dates on similar materials within one area, these adjustments are irrelevant. However, if one wishes to compare, for example, dates on terrestrial peat with dates on marine shells, or to compare a shell date from high latitudes with one from another area, care must be taken to ascertain what corrections, if any, have been applied.

3.2.1.5 Long-term changes in atmospheric ^{14}C content. Fundamental to the principles of radiocarbon dating is the assumption that atmospheric ^{14}C levels have remained constant during the period useful for ^{14}C dating. It is now abundantly clear that this assumption is invalid, though fortunately it is not a critical problem and its magnitude can be assessed, at least for most of the Holocene. ^{14}C concentrations may result from a wide variety of factors, as indicated in Table 3.2, and it is worth noting that many of these factors may themselves be important influences on

Table 3.2 Possible causes of radiocarbon fluctuations (from Damon *et al.* 1978).

I. *Variations in the rate of radiocarbon production in the atmosphere*
- (1) Variations in the cosmic-ray flux throughout the solar system
  - (a) Cosmic-ray bursts from supernovae and other stellar phenomena
  - (b) Interstellar modulation of the cosmic-ray flux
- (2) Modulation of the cosmic-ray flux by solar activity
- (3) Modulation of the cosmic-ray flux by changes in the geomagnetic field
- (4) Production by antimatter meteorite collisions with the Earth
- (5) Production by nuclear weapons testing and nuclear technology

II. *Variations in the rate of exchange of radiocarbon between various geochemical reservoirs and changes in the relative carbon dioxide content of the reservoirs*
- (1) Control of CO_2 solubility and dissolution as well as residence times by temperature variations
- (2) Effect of sea-level variations on ocean circulation and capacity
- (3) Assimilation of CO_2 by the terrestrial biosphere in proportion to biomass and CO_2 concentration, and dependence of CO_2 on temperature, humidity, and human activity
- (4) Dependence of CO_2 assimilation by the marine biosphere upon ocean temperature and salinity, availability of nutrients, upwelling of CO_2-rich deep water, and turbidity of the mixed layer of the ocean

III. *Variations in the total amount of carbon dioxide in the atmosphere, biosphere, and hydrosphere.*
- (1) Changes in the rate of introduction of CO_2 into the atmosphere by volcanism and other processes that result in CO_2 degassing of the lithosphere
- (2) The various sedimentary reservoirs serving as a sink of CO_2 and ^{14}C. Tendency for changes in the rate of sedimentation to cause changes in the total CO_2 content of the atmosphere
- (3) Combustion of fossil fuels by human industrial and domestic activity

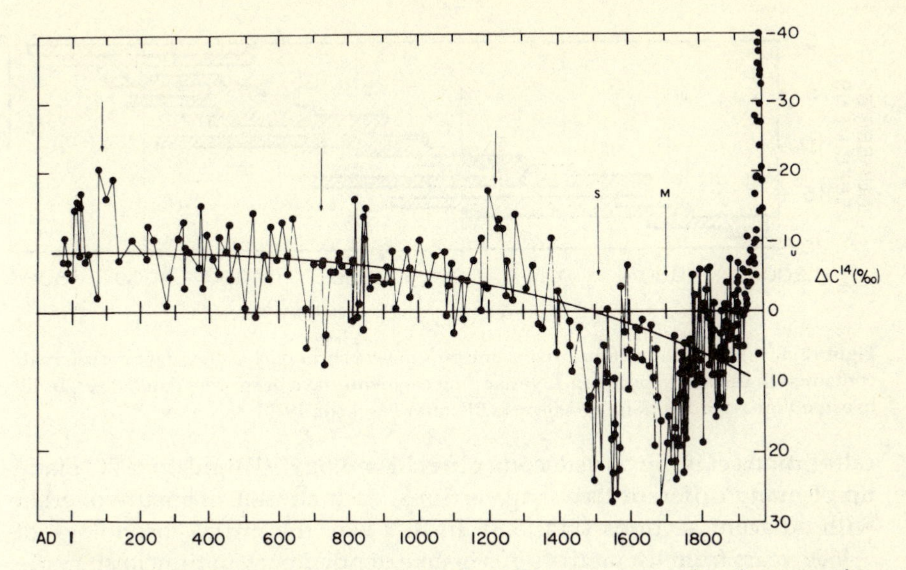

Figure 3.7 Radiocarbon variations over the last 2000 years expressed as departures from the long-term average. Trend due to geomagnetic field variation shown by curved line. Positive ^{14}C anomalies around 1500 and 1700 AD correspond to periods of reduced solar activity (the Spörer and Maunder minima, S and M respectively). At these times the increased cosmic ray flux produces more ^{14}C in the upper atmosphere. Combustion of fossil fuel has contaminated the atmosphere with ^{14}C-free CO_2, hence the large negative departures shown since ~1850 (from Eddy 1977).

climate. It is a sobering thought that fluctuations in the concentration of radiocarbon may help to explain the very paleoclimatic events to which radiocarbon dating has been applied for so many years; there could be no better illustration of the essential unity of science (Damon 1970a).

Early work on carefully dated tree rings indicated that ^{14}C estimates showed systematic, time-dependent variations (de Vries 1958). Both European and North American tree-ring samples spanning the last 400 years showed departures from the average of up to 2%, with ^{14}C maxima around AD 1500 and AD 1700 (Fig. 3.7). These secular ^{14}C variations, now known collectively as the "de Vries effect" or "Suess wiggles", appear to be closely related to variations in solar activity, as discussed further in Section 3.2.1.6, below (Suess 1980). Also seen in Figure 3.7 is the marked decline in ^{14}C activity during the last 100 years. This resulted primarily from the combustion of fossil fuel during the period (Suess 1965) causing a rapid increase in the abundance of "old" (essentially ^{14}C-free) carbon in the atmosphere (the so-called "Suess effect").

These studies generated much interest in testing the assumptions of radiocarbon dating and led to hundreds of checks being made between ^{14}C dates and corresponding wood samples, each carefully dated according to dendrochronological principles (Ch. 10). The longest tree-ring

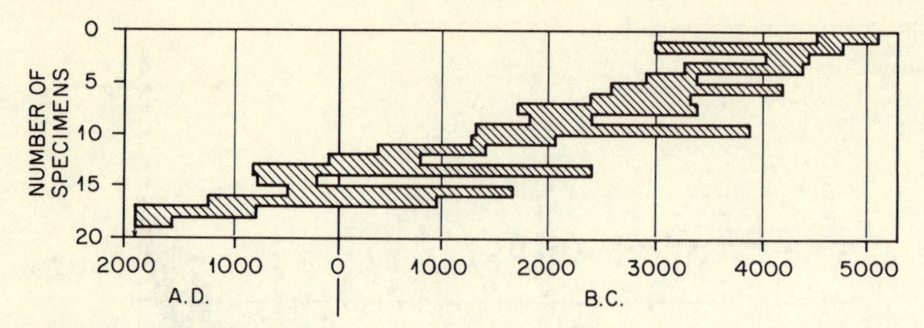

Figure 3.8 Construction of the bristlecone pine master chronology, showing time intervals contained in each tree specimen. Overlapping segments have been cross dated (see Ch. 10) to extend the record back to >7000 years BP (after Ferguson 1970).

calibration set is the "bristlecone pine chronology" (Ferguson 1970) made up of many different tree-ring sections, each chosen to ensure overlap with adjacent sections (Fig. 3.8). In this way the initial chronology of ~4600 years from living trees has been extended back in time to span the last 8681 years. Other "floating chronologies" are available, derived from radiocarbon-dated subfossil wood, the precise (dendrochronological) age of which is unknown (e.g. Pilcher *et al.* 1977, Becker 1980). Several European floating chronologies have now been accurately fixed in time by matching de Vries-type ^{14}C variations in the chronology with those observed in the well dated bristlecone pine record (e.g. Kruse *et al.* 1980). By being able to precisely match such "wiggles", recorded in wood from places thousands of kilometers apart, it is clear that the high-frequency variations have a real geophysical significance and are not simply the result of noise in the radiocarbon chronology (de Jong *et al.* 1980).

The results of bristlecone pine ^{14}C calibrations have been reported and updated many times (e.g. Damon *et al.* 1966, 1978, Suess 1970, Klein *et al.* 1982) but the different studies have now converged on a very consistent picture, as illustrated in Figure 3.9. For the last ~2500 years the ^{14}C age is very close to the dendrochronological age of wood samples (±100 years), but before 2500 years ago a systematic difference is apparent, increasing to ~1000 years (^{14}C underestimating true age) by ~6000 years BP. Calibration tables, for the conversion of ages in "radiocarbon years" to calendar years, have been prepared by Klein *et al.* (1982), enabling historical, archeological, and astronomical events in the last 8000 years to be directly compared with suitable calibrated radiocarbon dates. Attempts to extend the ^{14}C calibration to earlier periods, using varves, have so far proved to be equivocal (Tauber 1970, Stuiver 1970), perhaps mainly because of errors associated with the varve chronologies themselves (Fromm 1970). The problems of extending ^{14}C calibration to pre-Holocene times are not insurmountable, however. Stuiver (1978b) attempted a crude calibration back to 32 000 years BP, using magnetic,

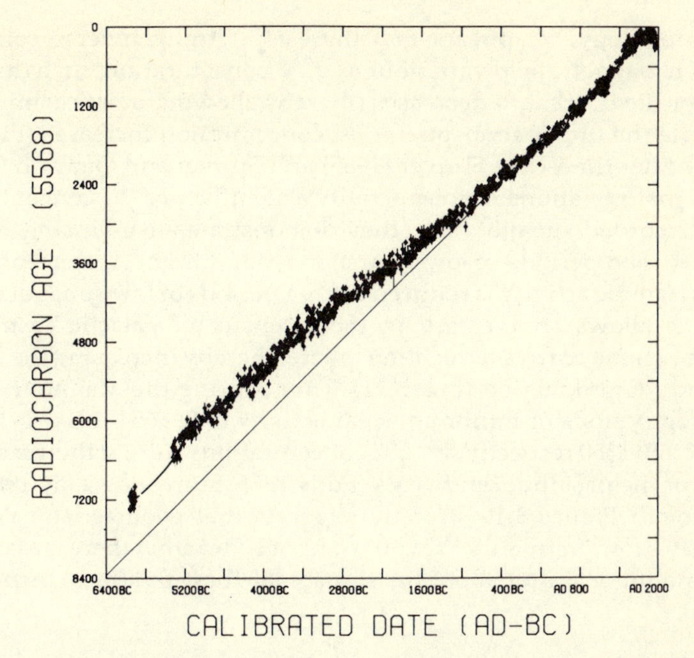

Figure 3.9 Radiocarbon dates (years BP) in relation to tree-ring dates (calendar years). Absolute correspondence is represented by the straight line. Note the systematic drift in the relationship (with ^{14}C age underestimating calendar year age) particularly before ~2500 years BP (from Klein *et al.* 1982).

thermoluminescence, and $^{230}Th/^{234}U$ dating methods to check ^{14}C dates on lake sediments. He concluded that the ^{14}C timescale has probably not varied by more than 2000 years over the interval 9000–32 000 years BP.

3.2.1.6 Causes of temporal radiocarbon variations. Table 3.2 lists some of the possible causes of ^{14}C fluctuations, and these are discussed in some detail by Damon *et al.* (1978). In general terms they can be considered in two groups: factors internal to the earth – ocean – atmosphere system (II and III in Table 3.2) and extraterrestrial factors (group I in Table 3.2). Although it is probable that all these different factors have played some part in influencing ^{14}C concentrations through time, it would appear that most of the variance in the record, as it is presently known, can be accounted for by changes in the intensity of the Earth's magnetic field (dipole moment) (Bucha 1969, 1970, Barton *et al.* 1979, Barbetti 1980) and by changes in solar activity (Stuiver & Quay 1980). The former factor is primarily related to low-frequency (long-term ^{14}C fluctuations, and the latter factor to higher-frequency (de Vries-type) fluctuations in ^{14}C. Evidence for changes in magnetic field intensity has come mainly from archeological sites through studies of thermoremanent magnetism in the minerals of baked clay (Aitken 1974). Although there are uncertainties in this chronology, it appears that there is a strong inverse correlation between magnetic field variations and ^{14}C concentration, such that as the magnetic field strength decreases (thereby allowing more cosmic rays to penetrate the upper atmosphere) ^{14}C concentration increases (Fig. 3.10). On a shorter timescale, Stuiver (1965) and Stuiver and Quay (1980) have shown that variations in solar activity also influence ^{14}C concentrations. Using recorded sunspot data, they demonstrate a convincing relationship between periods of low solar activity and high ^{14}C concentrations. Solar magnetic activity is reduced during periods of low sunspot number and this allows an increase in the intensity of galactic cosmic rays incident on the Earth's outer atmosphere, thereby increasing the neutron flux and ^{14}C production (Fig. 3.11). Thus, during the Maunder, Spörer, and Wolf periods of minimum solar activity (AD 1654–1714, 1416–1534, and ~1280–1350 respectively), ^{14}C concentrations were at their maximum levels for the past thousand years (Eddy 1976, Stuiver & Quay 1980).

Although Figure 3.10 strongly suggests that geomagnetic variation modulates low-frequency ^{14}C variations, it is clear that there is a considerable amount of scatter about the sinusoidal "best-fit" line. Furthermore,

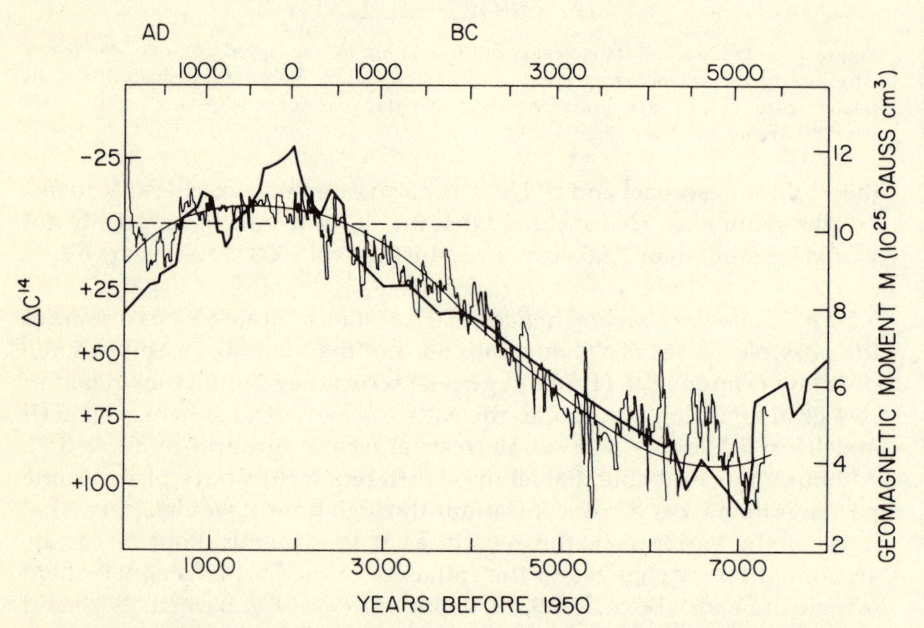

Figure 3.10 Record of deviations of relative ^{14}C concentration in the atmosphere (thin, abruptly changing line) calculated from ^{14}C dates on tree rings of known age. Note that increased relative abundances are plotted downwards from the 1890 norm which is shown as a dashed line. The thin curving line is a sinusoidal function fitted to the data. The heavy line is observed changes in the Earth's magnetic field strength according to Bucha (1970). As the magnetic field strengthened (from ~7000 to ~2000 years BP) the cosmic ray flux to the Earth was reduced, causing a decrease in ^{14}C production (modified from Eddy 1977).

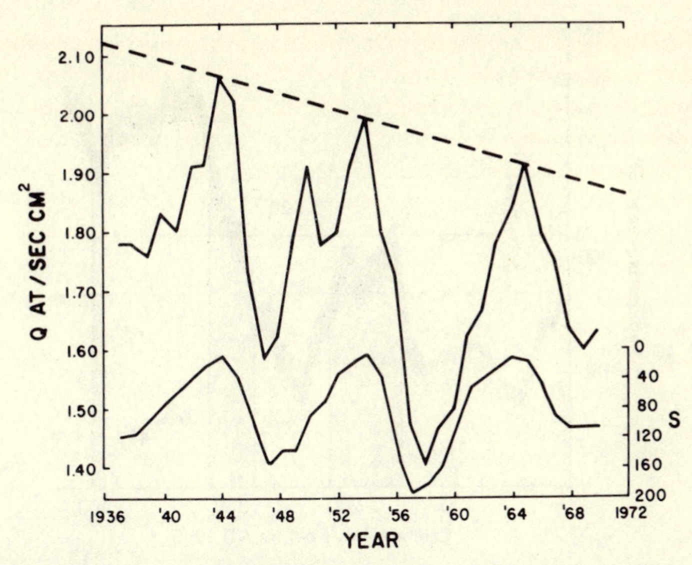

Figure 3.11 Global ^{14}C production rates, *Q*, derived from measured neutron fluxes (1937–70) in relation to sunspot numbers, *S* (plotted inversely). During periods of higher solar activity cosmic-ray bombardment of the upper atmosphere is reduced, causing ^{14}C production to decrease. The broken line represents the long-term change in ^{14}C production during solar minima (after Stuiver & Quay 1980).

the departures increase in amplitude back in time. One might attribute this simply to increasing uncertainty (noise) in the early part of the record, but Damon (1970b) argues that this is not the cause. Because of the weaker geomagnetic shielding effect at this time (only half as strong as at ~2000 years BP), the relative effect of solar modulation in ^{14}C production rates is increased. Thus, large departures from the "expected" mean values early in the record (e.g. maxima at ~5300 and 6200 calendar years BP, and minima at ~6100 calendar years BP) may be due to solar (heliomagnetic) modulation of ^{14}C production in the upper atmosphere. It is also likely that recent heliomagnetic effects, such as the ^{14}C increase during the Maunder minimum, were reduced in amplitude because of the stronger geomagnetic field intensity at that time (Eddy 1977). These views are not universally accepted, however, since some independent tree-ring chronologies indicate much smaller amplitude short-term departures from the overall long-term trend line (e.g. Pearson *et al.* 1977), in which case the apparent variability in the bristlecone pine–^{14}C calibration may indeed be due primarily to analytical noise. This is an important question and can only be resolved by more accurate ^{14}C measurements on wood samples representing time intervals as short as possible. Figure 3.12 indicates the significance of these high-frequency variations for radiocarbon dating. Very precise measurements of ^{14}C on dated wood samples have been made to calibrate ^{14}C variations back to AD 1 (single

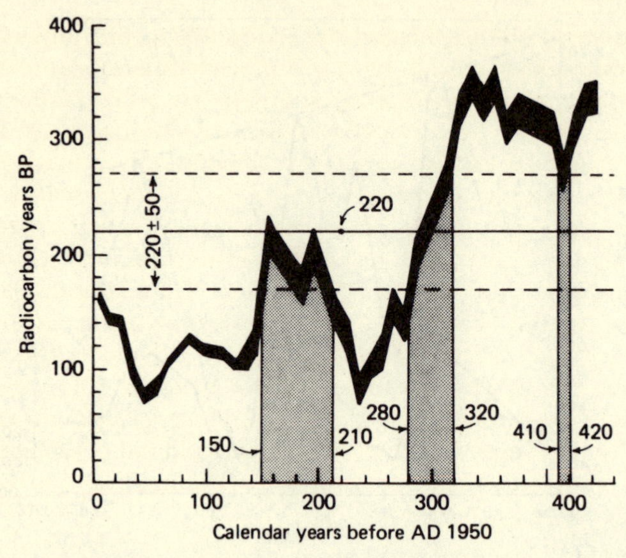

Figure 3.12 Relationship between conventional ^{14}C years and calendar years. Shaded curve is twice the counting error in the measurements. For a radiocarbon date of 220 ± 50 years BP the actual calendar year represented by the date and its counting error could be anywhere within the range of 150–210, 280–320, and 410–420 calendar years before 1950 (BP) (from Porter 1981a, after Stuiver 1978b).

year measurements for post-1820 wood, decade samples prior to that; Stuiver 1980). As illustrated, a sample which has been radiocarbon dated at 220 ± 50 years BP can not be assigned a single age range, within the probability margin of one standard deviation. Because of ^{14}C fluctuations the actual calendar age of the sample could be from 150 to 210 years BP, from 280 to 320 years BP, or even from 410 to 420 years BP (Porter 1981a). In fact, only samples from a few decades around 300 years BP are likely to yield a unique radiocarbon date. In all other cases during the last 450 years, multiple calendar dates, or a much broader spectrum of calendar ages, are derived from a single radiocarbon date (Stuiver 1978b). This raises significant problems for studies attempting to resolve short-term environmental changes (such as glacier fluctuations; Porter 1981a) and has important implications for the interpretation of all radiocarbon dates. Detailed studies of ^{14}C variations over larger intervals will be needed to document the extent of the problem in earlier periods.

3.2.1.7 Radiocarbon variations and climate. A number of authors have observed that periods of low solar activity, such as the Maunder minimum, correspond to cooler periods in the past, when glaciers advanced in many parts of the world (e.g. Denton & Karlén 1973a). As variations in radiocarbon production seem to be related to solar activity, it has also been argued that ^{14}C variations are inversely related to world-wide temperature fluctuations (Eddy 1977). This implies that solar activity, radiocarbon variations, and surface temperature are all related, perhaps through fundamental variations in the solar constant (i.e. low solar activity = high ^{14}C production rate = low temperature). If so, then the ^{14}C record itself, as a proxy of solar activity, may provide important information on the causes of climatic change. However, this is a controversial topic; several authors have shown that the correlations between radiocarbon variations and paleotemperature records are very poor when the records are examined in detail (e.g. Williams *et al.* 1981, Stuiver 1980). Furthermore, despite numerous studies utilizing the much more abundant instrumental data of the last few centuries, very few consistent and statistically significant relationships between solar activity and climate have been noted (Pittock 1978, Lansberg 1980). At this stage, therefore, the evidence relating solar activity and radiocarbon variations to surface temperatures is equivocal, an intriguing but unproven possibility.

3.2.2 *Potassium – argon dating ($^{40}K/^{40}Ar$)*

Compared to radiocarbon dating, potassium – argon dating is used far less in Quaternery paleoclimatic studies. However, potassium – argon and (more recently) argon – argon dating have indirectly made major contributions to Quaternary studies. The techniques have proved to be invaluable in dating sea-floor basalts and enabling the geomagnetic polarity timescale to be accurately dated and correlated on a world-wide basis (Dalrymple 1972; see also Sec. 4.1.4). Potassium – argon dating has also been used to date lava flows which, in some areas of the world, may be juxtaposed with glacial deposits. In this way, limiting dates on the age of the glacial event may be assigned (e.g. Löffler 1976, Porter 1979).

Potassium – argon dating is based on the decay of the radio-isotope ^{40}K to a daughter isotope ^{40}Ar. Potassium is a common component of minerals and occurs in the form of three isotopes, ^{39}K and ^{41}K, both stable, and ^{40}K, which is unstable. ^{40}K occurs in small amounts (0.012% of all potassium atoms) and decays to either ^{40}Ca or ^{40}Ar, with a half-life of 1.31 $\times 10^9$ years. Although the decay to ^{40}Ca is more common, the relative abundance of ^{40}Ca in rocks precludes the use of this isotope for dating purposes. Instead, the abundance of argon is measured and sample age is a function of the $^{40}K/^{40}Ar$ ratio. Argon is a gas which can be driven out of a sample by heating. Thus the method is used for dating volcanic rocks which contain no argon after the molten lava has cooled, thereby setting the isotopic "clock" to zero. With the passage of time, ^{40}Ar is produced and retained within the mineral crystals, until driven off by heating in the laboratory during the dating process (Dalrymple & Lanphere 1969). Unlike ^{40}C dating, $^{40}K/^{40}Ar$ dating relies on measurements of the decay product, ^{40}Ar; the parent isotope content (^{40}K) is measured in the sample.

As the abundance ratios of the isotopes of potassium are known, the ^{40}K content can be derived from a measurement of total potassium content, or by measurement of another isotope, ^{39}K. Because of the relatively long half-life of ^{40}K, the production of argon is extremely slow. Hence, it is very difficult to apply the technique to samples younger than ~100 000 years and its primary use has been in dating volcanic rocks formed over the last 30 million years (though, theoretically, rocks as old as 10^9 years could be dated by this method). Dating is usually carried out on minerals such as sanidine, plagioclase, biotite, hornblende, and olivine in volcanic lavas and tuffs. It may also be useful in dating authigenic minerals (i.e. those formed at the time of deposition) such as glauconite, feldspar, and sylvite in sedimentary rocks (Dalrymple & Lanphere 1969).

3.2.2.1 Problems of $^{40}K/^{40}Ar$ dating. The fundamental assumptions in potassium–argon dating are that (a) no argon was left in the volcanic material after formation, and (b) the system has remained closed since the material was produced, so that no argon has either entered or left the sample since formation. The former assumption may be invalidated in the case of some deep-sea basalts which retain previously formed argon during formation under high hydrostatic pressure. Similarly, certain rocks may have incorporated older "argon-rich" material during formation. Such factors result in the sample age being overestimated (Fitch 1972). Similar errors result from modern argon being absorbed on to the surface and interior of the sample, thereby invalidating the second assumption. Fortunately, atmospheric argon contamination can be assessed by measurement of the different isotopes of argon present. Atmospheric argon occurs as three isotopes, ^{36}Ar, ^{38}Ar, and ^{40}Ar. As the ratio of $^{40}Ar/^{36}Ar$ in the atmosphere is known, the specific concentrations of ^{36}Ar and ^{40}Ar in a sample can be used as a measure of the degree of atmospheric contamination, and the apparent sample age appropriately adjusted (Miller 1972).

A more common problem in $^{40}K/^{40}Ar$ dating is the (unknown) degree to which argon has been lost from the system since the time of the geological event to be dated. This may result from a number of factors, including diffusion, recrystallization, solution, and chemical reactions as the rock weathers (Fitch 1972).

Obviously, any argon loss will give a minimum age estimate only. Fortunately, some assessment of these problems and their effect on dating may be possible.

3.2.2.2 $^{40}Ar/^{39}Ar$ dating. One important disadvantage of the conventional $^{40}K/^{40}Ar$ dating technique is that potassium and argon measurements have to be made on different parts of the same sample; if the sample is not completely homogeneous, an erroneous age may be assigned. This problem can be circumvented by $^{40}Ar/^{39}Ar$ dating, in which measurements are made simultaneously, not only on the same sample, but on the same precise location within the crystal lattice where the ^{40}Ar is trapped. Instead of measuring ^{40}K directly, it is measured indirectly by irradiating the sample with neutrons in a nuclear reactor. This causes the stable isotope ^{39}K to transmute into ^{39}Ar; by collecting both the ^{40}Ar and ^{39}Ar, and knowing the ratio of ^{40}K to ^{39}K (which is a constant), the sample age can be calculated. Further details are given by Curtis (1975), who provides a very good discussion of the technique and its applications.

Actually, $^{40}Ar/^{39}Ar$ dating has no advantages over conventional $^{40}K/^{40}Ar$ dating for samples which have not been weathered, which have not been subjected to heating or metamorphism of any kind since formation, or which are free of inherited or extraneous argon. In such cases, dates from $^{40}K/^{40}Ar$ methods would be identical to those from $^{40}Ar/^{39}Ar$ methods. In practice, however, there is no way of knowing the extent to which a sample has been modified or contaminated; hence the $^{40}Ar/^{39}Ar$ method has significant advantages over $^{40}K/^{40}Ar$ because it is often possible to identify the degree to which a sample has been altered or contaminated, and thus may increase confidence in the date assigned. Furthermore, several dates can be obtained from one sample and the results treated statistically to yield a date of high precision (Curtis 1975). The advantages stem from the fact that the ^{40}K, which yields the ^{40}Ar by decay, occupies the same position in the crystal lattice of the mineral as the much more abundant ^{39}K which produces the ^{39}Ar on irradiation. Heating of the sample thus drives off the argon isotopes simultaneously. Any atmospheric argon contaminating the sample occurs close to the surface of the mineral grains, so it is liberated at low temperatures. Similarly, loss of radiogenic argon by weathering would be mainly confined to the outer surface of a mineral. In such cases the $^{40}Ar/^{39}Ar$ ratios on the initial gas samples would indicate an age which is too young (Fig. 3.13a). At higher temperatures, the deeper-seated argon from the

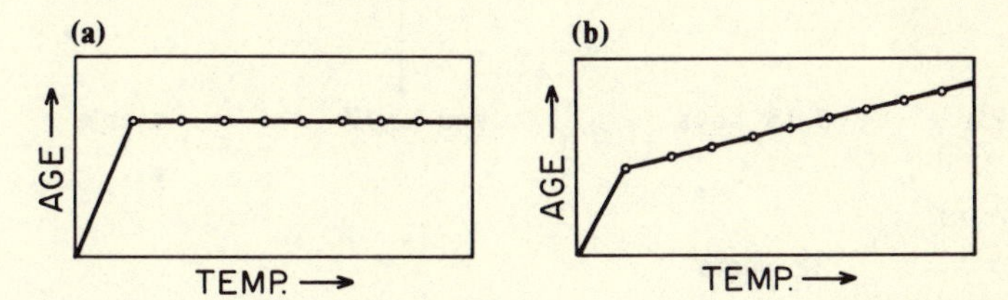

Figure 3.13 Schematic plots of $^{40}Ar/^{39}Ar$ data. Each point in (a) and (b) indicates the ages obtained for that increment of argon released as the temperature is increased in steps from 0 °C to the fusion point (~1000 °C). In (a) the data show uniform ages for all increments, the plateau indicating a precise age determination. In (b) the ages appear to be progressively older as the temperature rises, indicating loss of argon after original crystallization of the sample so that a precise age can not be determined; even the oldest age obtained is probably too young (after Curtis 1975).

unweathered, uncontaminated interiors of the crystals will be driven off and can be measured repeatedly as the temperature rises to fusion levels. If such gas increments indicate a stable and consistent age, considerable confidence can be placed in the result. By contrast, conventional $^{40}K/^{40}Ar$ dating on a sample such as that shown in Figure 3.13a would yield a meaningless age, resulting from a mixture of the gases from different levels.

Plots of "apparent age" calculated from the ratios of ^{40}Ar to ^{39}Ar at different temperatures can indicate much information about the past history of the sample, including whether the sample has lost argon since formation (Fig. 3.13b), or whether the rock was contaminated by excess radiogenic argon at the time of formation. Such interpretations are discussed further by Curtis (1975) and by Miller (1972). $^{40}Ar/^{39}Ar$ dating thus possesses considerable advantages over conventional $^{40}K/^{40}Ar$ dating methods, by providing more confidence in the resulting dates.

3.2.3 *Uranium-series dating*

Uranium-series dating is a term which encompasses a range of dating methods, all based on various decay products of ^{238}U or ^{235}U. Figure 3.14

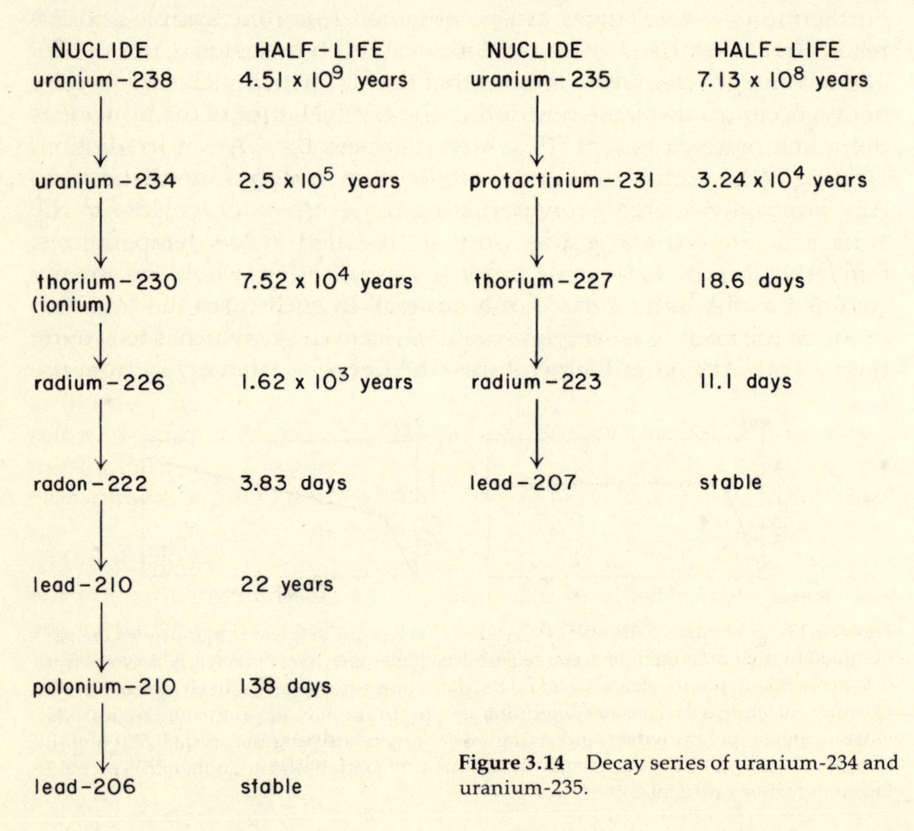

Figure 3.14 Decay series of uranium-234 and uranium-235.

illustrates the principle decay series nuclides and their respective half-lives; some intermediate products with very short half-lives (in the order of seconds or minutes) have been omitted. The main isotopes of significance for dating are ^{238}U and ^{235}U, ^{230}Th (also known as ionium), and ^{231}Pa. The ultimate product of the uranium decay series is stable lead (^{206}Pb or ^{207}Pb).

In a system containing uranium, which is undisturbed for a long period of time ($\sim 10^6$ years), a dynamic equilibrium will prevail in which each daughter product will be present in such an amount that it is decaying at the same rate as it is formed by its parent isotope (Broecker & Bender 1972). The ratio of one isotope to another will be essentially constant. However, if the system is disturbed, this balance of production and loss will no longer prevail and the relative proportions of different isotopes will change. By measuring the degree to which a disturbed system of decay products has returned to a new equilibrium, an assessment of the amount of time elapsed since disturbance can be made.

In natural systems, disturbance of the decay series is common, because of the different physical properties of the intermediate decay series products. Most important of these is the fact that ^{230}Th and ^{231}Pa are virtually insoluble in water. In natural waters these isotopes are precipitated from solution as the uranium decays, and collect in sedimentary deposits. As the isotope is buried beneath subsequent sedimentary accumulations, it decays at a known rate, "unsupported" by further decay of the parent isotopes (^{234}U and ^{235}U respectively) from which it has been separated. Thus, in sediment which has been deposited at a uniform rate, the ^{230}Th and ^{231}Pa concentrations decrease exponentially with depth. Providing that this initial concentration of the isotopes is known, the extent to which they have decayed in sediment beneath the surface can be related to the amount of time elapsed since the sediment was first deposited. Isotopic decay is expressed in terms of the activity ratios† of $^{230}Th/^{234}U$ and $^{231}Pa/^{235}U$; for the former, the useful dating range is 10 000–350 000 BP, and for the latter 5000–150 000 BP (Fig. 3.15). It is assumed that the oceanic uranium isotope composition in the past has been invariant, which is a reasonable assumption for the deep oceans but very unlikely for closed inland basins, making the technique inapplicable in such areas.

This procedure is an example of those uranium-series methods which are based on the decay of unsupported intermediate members of the series (Ku 1976). Another method with the same approach uses the $^{234}U/^{238}U$ activity ratio in coral or mollusc carbonates which were deposited in equilibrium with the uranium isotope composition of ocean

†Concentrations of elements are reported in units of "decays per minute per gram of sample". In U-series dating, these rates are considered relative to each other and are referred to as "activity ratios".

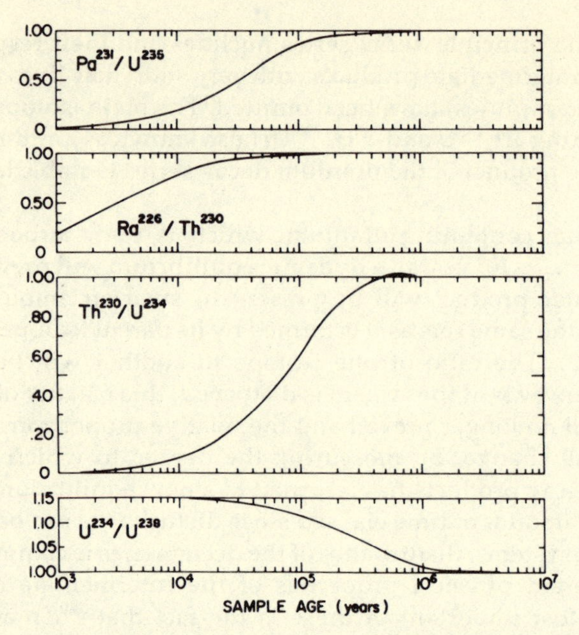

Figure 3.15 Temporal changes in activity ratios of $^{234}U/^{235}U$, $^{230}Th/^{234}U$, $^{226}Ra/^{230}Th$, and $^{231}Pa/^{235}U$ (after Broecker & Bender 1972).

water. In the oceans, the $^{234}U/^{238}U$ ratio is virtually constant at 1.14. Once the shells or coral have been isolated from the ocean (for example, by eustatic or tectonic effects) the $^{234}U/^{238}U$ ratio in them will change (^{234}U has a much shorter half-life than ^{238}U), so that eventually a new equilibrium will be reached. Variations in this ratio provide useful dating control from ~40 000 to ~1 000 000 years BP (Fig. 3.15).

On a much shorter timescale, unsupported ^{210}Pb may also be used as a chronological aid. ^{210}Pb is derived from the decay of ^{222}Rn following the decay of ^{226}Ra from ^{230}Th (Fig. 3.14). Both ^{226}Ra and ^{222}Rn escape from the Earth's surface and enter the atmosphere, where the ^{210}Pb is eventually produced. ^{210}Pb is then washed out of the atmosphere by precipitation, or settles out as dry fall-out, where it accumulates in sedimentary deposits and decays (with a half-life of 22 years) to stable ^{206}Pb. Assuming that the atmospheric flux of ^{210}Pb is constant, the decay rate of ^{210}Pb to ^{206}Pb with depth can be used to date sediment accumulation rates (e.g. Koide *et al.* 1973). It is of value only in dating sediments over the last 200 years, but this may be of particular value is isolating core-top floral and faunal elements for calibration with instrumental climatic data, to derive accurate transfer functions for paleoclimatic reconstructions. It has also proved useful in dating the upper sections of ice cores and hence allowing estimates of long-term accumulation rates to be made (Fig. 3.16; Crozaz & Langway 1966, Crozaz & Picciotto 1967).

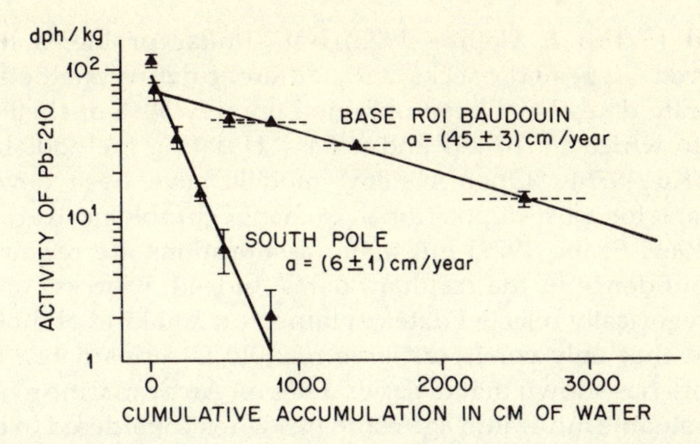

Figure 3.16 The use of lead-210 in studies of firn accumulation in Antarctica. The activity of ^{210}Pb is plotted as a function of equivalent water depth. Snow accumulation at the coastal Base Roi Baudoin is eight to nine times higher than at the South Pole (after Crozaz & Picciotto 1967).

3.2.3.1 *Methods based on the accumulation of decay products of uranium.* All the above methods relied on the physical separation of the parent and daughter isotopes, with age calculated as a function of the decay rate of the unsupported daughter isotope. A second group of methods utilizes the same principle but measures the accumulation of a daughter product rather than its decay through time (Ku 1976). These methods are commonly applied to carbonate materials (corals, molluscs, speleothems) and are based on the fact that uranium is co-precipitated with calcite or aragonite from natural waters that are essentially free of thorium and protactinium. Initial values of ^{230}Th and ^{231}Pa in the carbonates are thus negligible. Providing that the carbonate remains a closed system, the amounts of ^{230}Th and ^{231}Pa produced as the ^{234}U and ^{235}U decay will be a function of time, and of the initial uranium content of the sample. Although the two methods are complementary, ^{230}Th/^{234}U dating is more commonly used because of the low abundance and analytical difficulties of ^{231}Pa (Ku 1976). The method has been widely used to date raised coral terraces and hence to provide a chronologically accurate assessment of glacio-eustatic changes of sea level, with broad implications for paleoclimatology (e.g. Broecker & Bender 1972, Bloom *et al.* 1974).

Attempts have also been made to date molluscs in the same way as coral but the results are generally inconsistent (see, however, Szabo 1979a). The main problem is that molluscs appear to freely exchange uranium post-depositionally (i.e. they do not constitute a closed system), with the result that fossil molluscs commonly have higher uranium concentrations than their modern counterparts. Hence, the resulting thorium and protactinium concentrations are not simply a function of age (Kaufman *et al.* 1971). ^{230}Th/^{234}U dates on bone have also been attempted (Szabo & Collins 1975) but similar problems have been encountered. Repeated checks with different dating methods suggest that accurate dates have been obtained on only 50% of shell and bone samples to which $^{230}Th/^{234}U$ and $^{231}Pa/^{235}U$ dating methods have been applied (Ku 1976). "Open system" models have been developed to compensate for post-depositional exchange problems (e.g. Szabo & Rosholt 1969, Szabo 1979) but many assumptions are required which reduce confidence in the resultant dates. Indeed, Broecker and Bender (1972) categorically rejected dates obtained on any kind of molluscs and concluded that only corals can give reliable U-series dates. However, recent work has shown that U-series dates on Arctic marine molluscs can provide valuable minimum age estimates when considered in relation to amino-acid data on the same samples (Szabo *et al.* 1981).

Much more confidence can be placed in uranium-series dates obtained on carbonate samples from speleothems (stalactites and stalagmites). Such deposits are compact and not subject to post-depositional leaching. Because ^{230}Th is so insoluble, the water from which the speleothem carbonate is precipitated can be considered to be essentially thorium-free. Hence, providing that the initial uranium concentration is sufficient, measurement of the $^{230}Th/^{234}U$ ratio will indicate the build-up of ^{230}Th with the passage of time (Harmon *et al.* 1975). The main problem is to determine reliably the initial ^{234}U content, but this is not always possible (Sec. 7.6.2).

3.2.3.2 Problems of U-series dating. The major problems in U-series dating have already been alluded to briefly. First, an assumption must be made as to the initial $^{230}Th/^{234}U$, $^{234}U/^{238}U$, and/or $^{231}Pa/^{235}U$ ratios in the sample. In the deep oceans this may not be a significant problem, as modern oceanic ratios are likely to have been relatively constant over long periods of time, but in terrestrial environments this assumption is far less robust. The second problem concerns the extent to which the sample to be dated has remained a closed system through time. Recrystallization of aragonitic carbonate to calcite may provide some guidance, but as discussed in Section 3.2.1.4 this is not always reliable. At present only carbonate dates on coral seem to be consistently reliable. Reliability can be checked by obtaining activity ratios for different isotopes from the same sample. If the sample has remained "closed" the dates should all cross check and be internally consistent (Fig. 3.15). Such checks on Barbados coral samples demonstrate reproducible results (±2000–4000 years) on samples in the age range 80 000–130 000 years BP (Broecker & Bender 1972).

A third problem relates not so much to the dating methods themselves but to the interpretations commonly placed upon them. In $^{230}Th/^{234}U$ and $^{231}Pa/^{235}U$ dating methods, a plot of the log of activity ratios versus depth

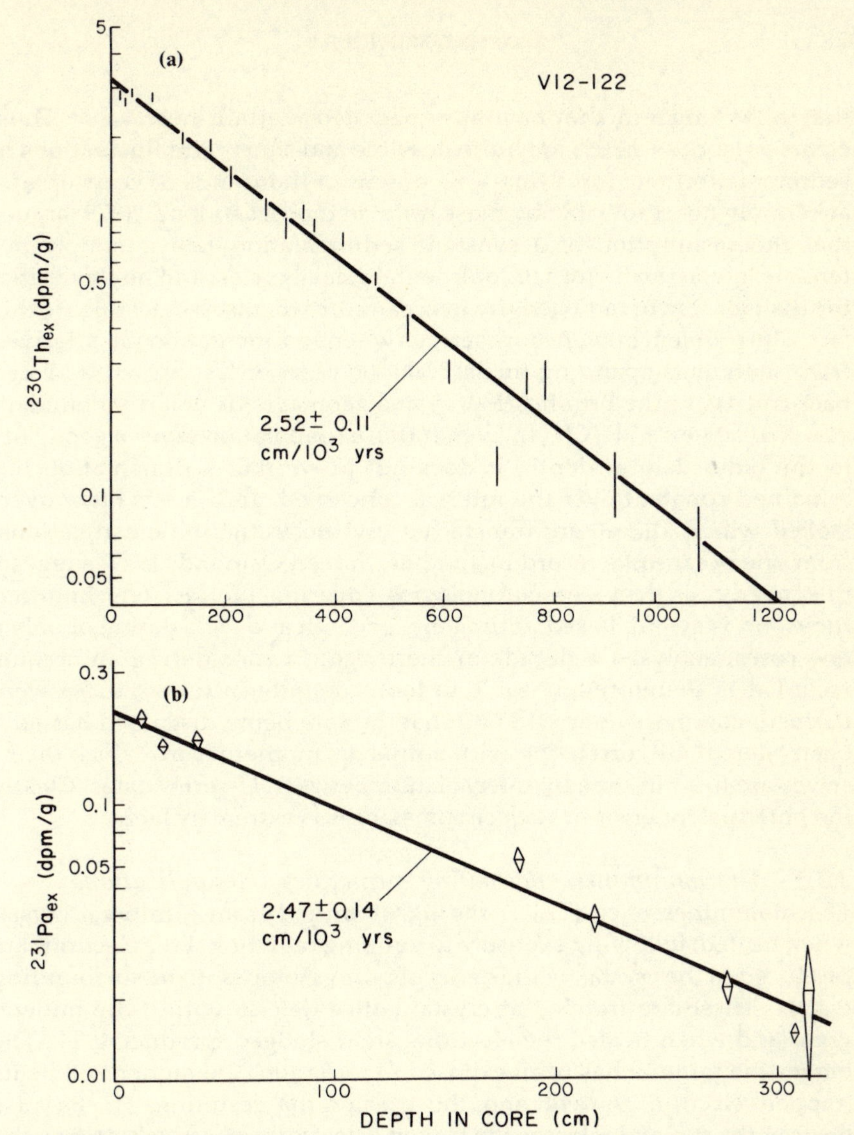

Figure 3.17 (a) Excess ^{230}Th concentrations and (b) excess ^{231}Pa concentrations versus depth in Caribbean core V12-122. As the original amounts of ^{230}Th and ^{231}Pa in freshly deposited sediment can be estimated, the extent to which they have been reduced with depth gives a measure of time since the sediment was deposited. Sedimentation rates are obtained from slopes of the best-fitting regression lines and a knowledge of the decay rate of each isotope (after Ku 1976).

is made. If the data approximate a straight line, this is taken to indicate a constant sedimentation rate (and constancy of the initial ^{230}Th and ^{231}Pa concentrations in freshly deposited sediment) (Fig. 3.17). The *mean* sedimentation rate, so derived, is then used to interpolate the ages of intermediate events between the levels sampled. Clearly, this practice in no way proves that sedimentation rates have *actually* been constant, only that in the long term they have approximated a certain mean value. Thus, errors as large as ±20% are quite possible and short-term fluctuations in sedimentation rate (on a timescale of tens of thousands of years or less) are simply not resolvable by these ratio methods. Osmond (1979) argues that the assumption of a constant sedimentation rate is simply not tenable, given the history of continental glacial cycles, and he claims that the distribution of radioactivity in ocean cores themselves testifies to this fact. The problem is even more serious when sedimentation rates derived from sediments spanning the last 300 000 years or less are extrapolated back in time to the Brunhes/Matuyama geomagnetic polarity boundary at ~700 000 years BP (Ch. 4). Even if this extrapolation gives a good "fit" to the older date at depth, it does not prove that sedimentation has remained constant over the interval concerned, and this is often overlooked when "dates" are transferred, by biostratigraphic correlations, from one ocean-core record to another. Indeed Osmond (1979) suggests that nearly all deep-sea chronologies covering the last few hundred thousand years are based, ultimately, on ^{230}Th and ^{231}Pa dating of only a few cores, analysed a decade or more ago. In considering an oceanic record it is therefore advisable to look carefully into how dates were derived; commonly one will find that the core being discussed has only been "dated" by correlation with points on another record which themselves are fixed in time by interpolation between U-series dates. Clearly the potential for error in such circumstances is extremely large.

3.2.4 Thermoluminescence dating: principles and applications

Thermoluminescence (TL) is the light emitted from a mineral crystal when heated, following exposure to ionizing radiation. Free electrons are produced in the crystal by the decay of radio-isotopes in the surrounding matrix. These are trapped at crystal lattice defects within the mineral grain and when heated the electrons are dislodged, producing TL. The longer the mineral has been exposed to radiation, the higher will be its trapped electron content and the greater the resulting TL. Because heating the mineral releases the trapped electrons, effectively setting the TL clock to zero, measurements of TL can indicate the amount of time which has elapsed since the sample was last heated (Fig. 3.18). Thus, the dating has been widely used in archeology to date pottery or baked clay samples as well as baked flints from fire hearths attributable to early man (Wintle & Aitken 1977). It has also been used to date sediment "baked" by contact with molten lava (Huxtable *et al.* 1978) and inclusions within lava (Gillot *et al.* 1979) and may prove useful in dating calcite in speleothems (Ch. 7). However, there are many problems yet to be overcome before TL dating of calcite can be considered reliable (Wintle 1978). The technique also has the potential for dating biological materials (shell or bone), but the fine-grain nature of the minerals and oxidation of

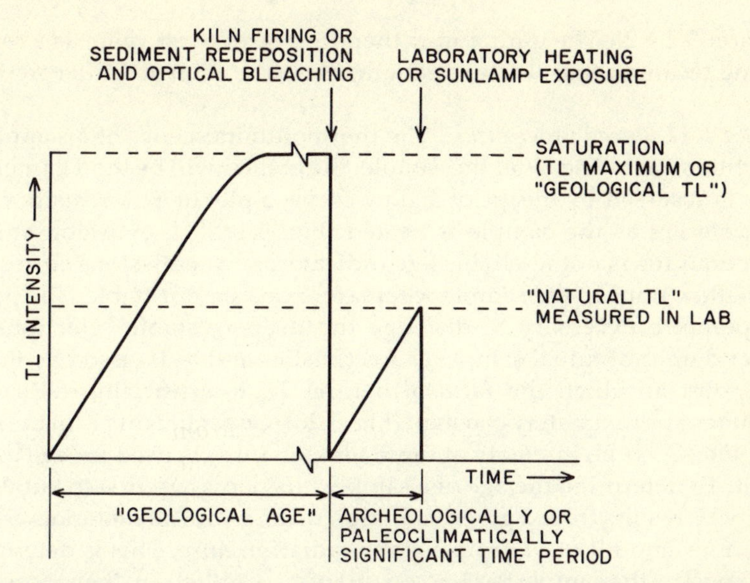

Figure 3.18 Schematic illustration of thermoluminescence (TL) intensity changes and their use in dating. TL acquired over long periods of geological time eventually leads to saturation of the sample so that no further increase in TL occurs with increasing sample age. This "geological TL" may be eliminated by heating or optical bleaching during sediment redeposition. The TL acquired subsequently is what is measured to obtain an age estimate.

organic material makes the technique very difficult (Christodoulides & Fremlin 1971).

TL dating may prove to be of most value for paleoclimatic research in its application to dating certain types of marine and terrestrial sediments. Exposure to sunlight during weathering and erosion "empties" the original (geological) TL and resets the "TL clock" to zero, prior to the accumulation of sediment (Fig. 3.18). This mechanism enables dates to be obtained on ocean sediments (Wintle & Huntley 1979, 1980), on thick loess deposits (Wintle 1981), or on paleosols. However, at present, TL dating of such sediments is in its infancy and dates are subject to large process-dependent errors (Wintle & Huntley 1982). In eastern Europe and the USSR, however, TL dates are widely used to date Quaternary geological sections, though the dating procedures and error margins are rarely given (Dreimanis *et al.* 1978). Nevertheless, it seems likely that more detailed studies of the physics of the processes involved, and improvements in laboratory techniques, will lead to a much wider use of TL dating for paleoclimatic research in the future (Aitken 1978).

The useful timescale for TL dating is probably no greater than 1 million years; accuracy may approach ±10% of the sample age, though comparison with K/Ar dates suggest that TL dates on older samples are commonly 5–15% too young, perhaps due to saturation of the electron traps (see

Section 3.2.4.2). The dating may therefore be of most value as a relative dating technique, useful for stratigraphic correlation and interpretation.

3.2.4.1 TL dating procedure. The thermoluminescence of a sample is a function of age. The older the sample, the greater will be the TL intensity. This is assessed by means of a glow curve, a plot of TL intensity versus temperature as the sample is heated (Fig. 3.19). TL emission at lower temperatures is not a reliable age indicator; such emissions correspond to shallow traps in the sample where electrons are not stable. The precise temperature necessary to dislodge the deeper "stable" electrons will depend on individual sample characteristics and is assessed by finding the point at which the ratio of natural TL to artificially induced TL becomes approximately constant (Fig. 3.20). Generally this is in the range 300–450 °C, so TL intensity at these temperatures is used for age assessment. To determine the age of a sample it is necessary first to know how much TL results from a given radiation dose, as not all materials acquire the same amount of TL from a given radiation dose. This is determined by exposing the sample to a known quantity of radiation, then measuring the corresponding increase in TL (Fig. 3.20b). The resulting increase (artificial TL) reflects the "sensitivity" of the sample to acquiring TL, which is a property specific to each individual sample. Sensitivity is expressed as TL per unit of radiation exposure or rad (radiation absorbed dose, equal to 100 ergs of energy absorbed per gram of sample). Calculations are made for a specific temperature on the glow curve (generally >300 °C). This quantity is then divided into the measured (natural) TL, at the same temperature, to obtain the radiation dose corresponding to the natural TL. The resulting value is called the equivalent dose (Q):

$$Q = \frac{\text{natural TL}}{\text{artificial TL per rad}}.$$

In theory, sample age can then be calculated simply by dividing the equivalent dose by the *measured dose rate* at the sample site:

$$\text{age (years)} = \frac{Q}{\text{dose rate (rads per year)}}.$$

The dose rate is assessed by measuring the quantity of radioactive uranium, thorium, and potassium in the sample itself, and in the surrounding matrix. Alternatively, radiation-sensitive phosphors (such as calcium fluoride) may be buried at the sample site for a year or more to measure directly the environmental radiation dosage. This problem of estimating the long-term dose rate is one of the most difficult in TL dating and a number of specific laboratory procedures have been devised to

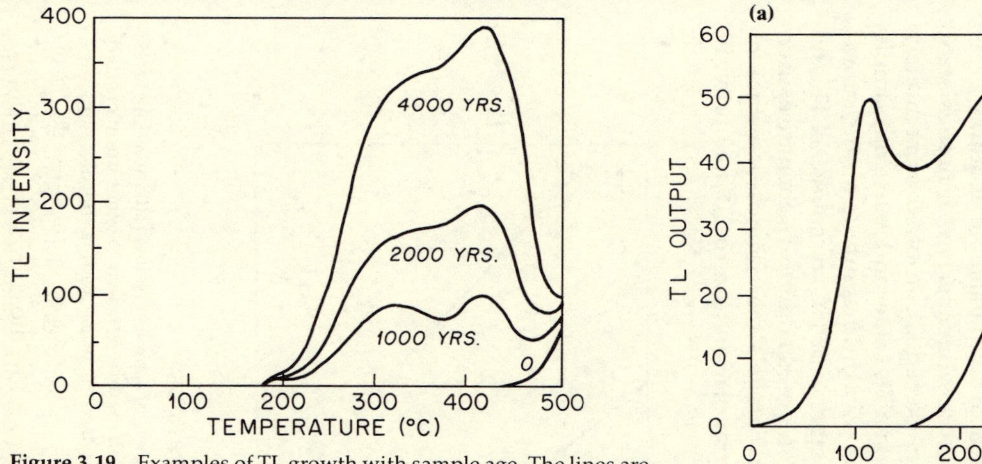

Figure 3.19 Examples of TL growth with sample age. The lines are glow curves that would have been observed with a sample taken from a pottery fragment at various lengths of time, after firing by ancient man from raw clay. Any existing TL in the raw clay would have been removed by the firing, so that at time zero only red-hot glow would have been observed (after Aitken 1974).

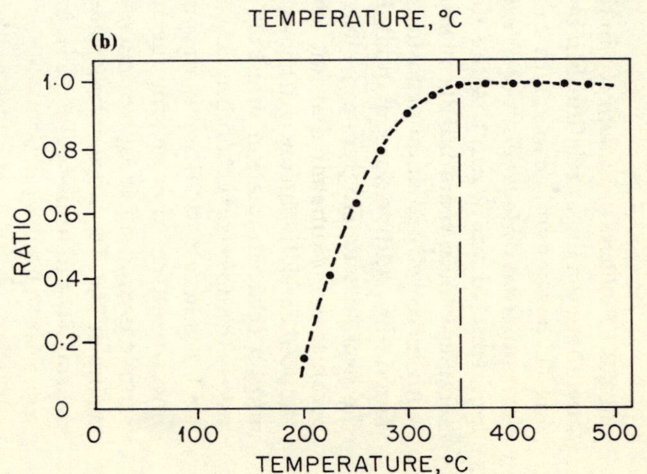

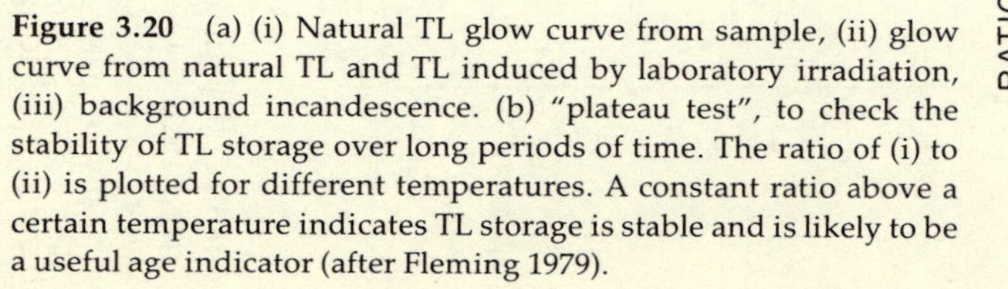

Figure 3.20 (a) (i) Natural TL glow curve from sample, (ii) glow curve from natural TL and TL induced by laboratory irradiation, (iii) background incandescence. (b) "plateau test", to check the stability of TL storage over long periods of time. The ratio of (i) to (ii) is plotted for different temperatures. A constant ratio above a certain temperature indicates TL storage is stable and is likely to be a useful age indicator (after Fleming 1979).

reduce the inherent uncertainty. Details are given in Aitken (1974) and Fleming (1979).

3.2.4.2 Problems of TL dating. In the age equation, above, it is assumed that there is a linear relationship between radiation dose and the resulting TL. It is known, however, that this is not always the case at extremely low radiation dose levels, nor at extremely high dose levels. The former problem (supralinearity) is the most significant for relatively young samples (<5000 years old). It can be assessed by plotting the change in TL with increased radiation exposure over and above the natural TL measured (the additive dose method). Extrapolation of this linear change in TL back to zero TL gives a positive intercept (I_0) on the dose axis (Fig. 3.21). It would appear that the rate at which a sample acquires TL is reduced at relatively low radiation dose levels (or perhaps is non-existent until a certain radiation threshold is exceeded). In order to accurately assess the dose corresponding to a given TL value (and hence the sample age) it is necessary to assess in some way the magnitude of this supralinearity effect. After heating the sample to 500 °C to remove all TL, the sample is exposed to known radiation doses and the TL then measured again. If the TL–dose relationship parallels that found for values of natural TL and above (Fig. 3.21), extrapolation to zero TL may provide

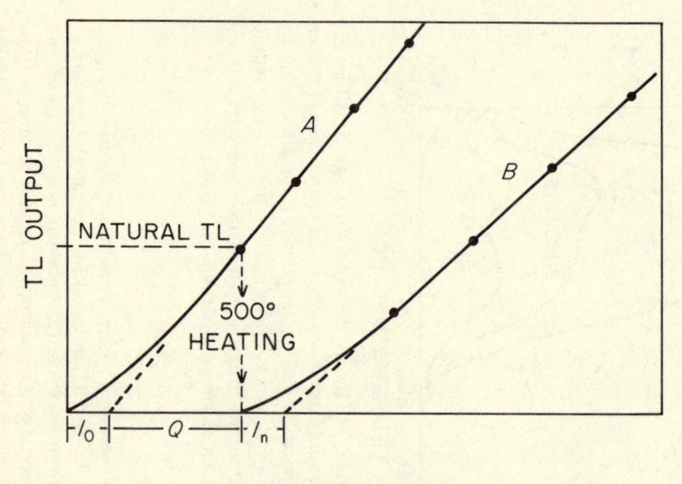

Figure 3.21 Supralinearity in radiation response of samples. At low radiation dose levels the rate at which a sample acquires TL may be reduced or non-existent until a certain radiation threshold is exceeded. This could lead to under-estimation of sample age, particularly in relatively young samples. The magnitude of the non-linearity in the TL/radiation dose relationship at low dose levels (I_0) may be assessed by heating the sample, re-exposing it to known radiation doses and extrapolating back to zero TL to obtain an estimate (I_N) of the natural supralinearity effect (I_0) (after Fleming 1979).

some measure of the supralinearity effect (I_n) (Fleming 1979). However, once a sample has been heated to 500 °C (to zero the natural TL), its sensitivity, the very characteristic one is trying to assess, may be altered (Aitken 1978). Obviously, the younger the sample age the greater will be the relative significance of this problem. At the other end of the scale, very long exposures to radiation (high doses) may result in *saturation* of the available electron traps so that further exposure will not appreciably increase the sample TL. This is quite likely to occur with samples of around a million years or more in age; very old ages indicated by TL dating are thus likely to be minimum estimates only.

A further difficulty in assessing the relationship between TL and radiation dose occurs when irradiated samples "lose" TL after very short periods of time, perhaps only a few weeks. This phenomena is called anomalous fading (Wintle 1973) and is common amongst certain minerals, particularly feldspars of volcanic origin. Unless corrected for, anomalous fading will result in underestimation of a sample age. Fortunately, it can be assessed relatively easily by storing irradiated samples in the dark and remeasuring TL periodically over a period of several months.

Perhaps the most significant problems in TL dating stem from variations in the environmental dose rate. Of particular importance is the mean water content of the sample and surrounding matrix during sample emplacement. Water greatly attenuates radiation, so a saturated sample will receive considerably less radiation in a given time period than a similar sample in a dry site; as a result the TL intensity will be much lower, giving an incorrect age indication. If the water content of the site can be assessed, this problem can be taken into account in the age calculation. Better still, if a radiation-sensitive phosphor can be placed in the environmental setting of the sample for a period of time, the effect of ground water on the radiation dose may be assessed directly. However, there is always the problem of knowing how groundwater content has varied in the past and this uncertainty is a major barrier to more accurate TL dating. Ground water may also leach away radioactive decay products, so long-term changes in groundwater content may further complicate the TL–dose relationship. Finally, it should be noted that one of the decay products of uranium is an inert gas (radon-222) which has a half-life of 3.8 days, long enough for it to escape from the sample site and effectively terminate the decay series (96% of the U-series γ-radiation energy is post-radon; Fleming 1976). Fortunately, laboratory studies have shown that relatively few soils exhibit significant radon loss either in the laboratory or *in situ*.

3.2.5 *Fission-track dating*

As already discussed in Section 3.2.3, uranium isotopes decay slowly through a complex decay series, ultimately resulting in stable atoms of lead. In addition to this slow decay, via the emission of α and β particles, uranium atoms also undergo spontaneous fission, in which the nucleus splits into two fragments. The amount of energy released in this process is large, causing the two nuclear fragments to be ejected into the surrounding material. The resulting damage paths are called fission tracks, generally 10–20μm in length. The number of fission tracks is simply a function of the uranium content of the sample and time (Faul & Wagner 1971). Rates of spontaneous fission are very slow (for ^{238}U, $10^{-16}\ a^{-1}$), but if there is enough uranium in a rock sample a statistically significant number of tracks may occur over periods useful for paleoclimatic research (Fleischer 1975).

The value of fission-track counting as a dating technique stems from the fact that certain crystalline or glassy materials may lose their fission-track records when heated, through the process of annealing. Thus, igneous rocks and adjacent metamorphosed sediments contain fission tracks produced since the rock last cooled down. Similarly, archeological sites may yield rocks which were heated in a fire hearth, thereby annealing the samples and re-setting the "geological" record of fission tracks to zero. In this respect, the environmental requirements of the sample are similar to those necessary for $^{40}K/^{40}Ar$ dating. Because different minerals anneal at different temperatures, careful selection is necessary; minerals with a low annealing temperature threshold, such as apatite, will be the most sensitive indicator of past thermal effects (Faul & Wagner 1971).

Fission tracks can be counted under an optical microscope after polishing the sample and etching the surface with a suitable solvent; the damaged areas are preferentially attacked by solvents, revealing the fission tracks quite clearly (Fleischer & Hart 1972). After these have been counted, the sample is heated to remove the "fossil" fission tracks and then irradiated by a slow neutron beam, which produces a new set of fission tracks as a result of the fission of ^{235}U. The number of induced fission tracks is proportional to the uranium content and this enables the ^{238}U content of the sample to be calculated. Sample age is then obtained from a knowledge of the spontaneous fission rate of ^{238}U. For a much more detailed discussion of the technique, and the problems of calibrating fission-track dates, see Hurford and Green (1982).

Fission-track dating may be undertaken on a wide variety of minerals in different rock types, though it has been most commonly carried out on apatite, micas, sphene and zircons in volcanic ashes, basalts, granites, tuffs, and carbonatites. It has also been widely used in dating amorphous (glassy) materials such as obsidian and is therefore useful in tephrochronological studies (see Sec. 4.4; Westgate & Briggs 1980). Its useful age range is very large, from 10^2 to 10^8 years, but error margins are very difficult to assess and are rarely given. Micro-variations in crystal uranium content may lead to large variations in the fission track count on different sections of the same sample (Fleming 1976). This potential source of error may be reduced by repeated measurements, but for samples which are old and/or contain little uranium, the labor involved in counting precludes such checks being made. Tracks may also "fade" under the influence of mechanical deformation or in particular chemical environments and thus lead to underestimation of age (Fleischer 1975). Fission-track dating has rarely been used in paleoclimatology but its use in archeology is relatively common. In most cases where fission-track dating could be used, $^{40}Ar/^{39}Ar$ dating is probably preferable, but fission-track dates may provide additional confidence in the age assigned. It is also worth noting that fission-track dating can provide useful age control in the 30 000–100 000 years' time interval, which is too young for routine $^{40}K/^{40}Ar$ and $^{40}Ar/^{39}Ar$ dating, and too old for conventional ^{14}C dating.

4

Dating methods II

4.1 Paleomagnetism

Variations in the Earth's magnetic field, as recorded by magnetic particles in rocks and sediments, may be used as a means of stratigraphic correlation. Major reversals of the Earth's magnetic field are now well known and have been independently dated in many localities throughout the world. Consequently, the record of these reversals in sediments can be used as time markers or chronostratigraphic horizons. In effect, the reversal is used to date the material by correlation with reversals dated independently elsewhere. However, as the reversal signal is an a.c. type (i.e. either normal or reversed) it is necessary to know approximately the age of the material under study to avoid miscorrelations.

In addition to aperiodic global-scale geomagnetic reversals, smaller amplitude, quasi-periodic variations of the Earth's magnetic field have also occurred. These secular variations were regional in scale (over distances of 1000–3000 km) and can be used to correlate well dated "master chronologies" with other undated records exhibiting similar paleomagnetic variations.

4.1.1 The Earth's magnetic field

The magnetic field of the Earth is generated by electric currents within the Earth's molten core. The exact mechanism of its formation is not agreed upon, but for our purposes it is sufficient to consider the field as if it were produced by a bar magnet at the center of the Earth, inclined at $\sim 11°$ to the axis of rotation (Fig. 4.1a). At the Earth's surface, we are familiar with this global field through magnetic compass variations. If a magnetized needle is allowed to swing freely, it will not only rotate laterally to point towards the magnetic pole, but also become inclined vertically, from the horizontal plane. The angle the needle makes with the horizontal is called the inclination (Fig. 4.1b). The inclination varies greatly, from near 0° at the Equator to 90° at the magnetic poles. If the needle is weighted, to maintain it in a horizontal plane, it will remain pointing towards magnetic north, and the angle it makes with true (geographical) north is called the declination (Fig. 4.1b).

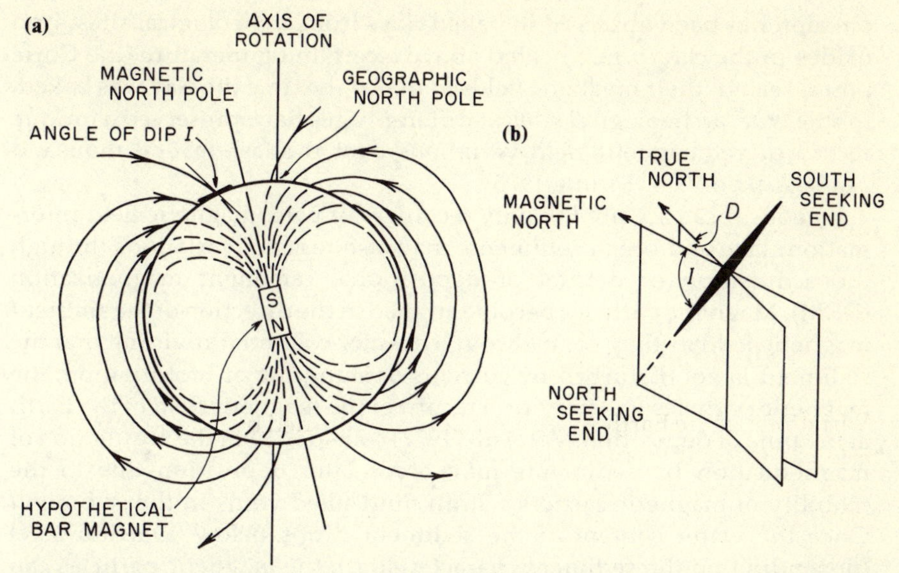

Figure 4.1 (a) The Earth's magnetic field. The main part of the Earth's magnetic field (the dipole field) can be thought of hypothetically as a bar magnet centered at the Earth's core. The lines of force represent, at any point, the direction in which a small magnetized needle tries to point. The concentration of these lines is a measure of the magnetic field strength. (b) Declination and inclination. Declination is a measure of the horizontal departure of the field from true north; inclination is a measure of dip from the horizontal. The resultant force is a vector representing declination, inclination, and field strength.

The Earth's magnetic field is considered to be made up of two components – a primary and fairly stable component (the dipole field), which is represented by the bar magnet model, and a much smaller residual or secondary component (the non-dipole field), which is less stable and geographically more variable. Major changes in the Earth's magnetic field are the result of changes in the dipole field, but minor variations may be due to non-dipole factors (see Sec. 4.1.5).

Because of the nature of the (dipole) field and the way in which it is generated, any change in its characteristics will affect all parts of the world. Records of significant magnetic field variations in a stratigraphic column (magnetostratigraphy) can thus be used directly to correlate sedimentary sequences in widely dispersed locations, regardless of whether they have common fossils or even similar facies.

4.1.2 *Magnetization of rocks and sediments*

So far we have referred to paleomagnetic variations and their usefulness without considering how such variations are recorded. It has been known for over 50 years that a molten lava will acquire a magnetization parallel to the Earth's magnetic field at the time of its cooling. This is known as thermoremanent magnetization (TRM). The same phenomenon has been observed in baked clays from archeological sites; iron oxides in the clay, when heated above a certain temperature (the Curie point) realign their magnetic fields to that at the time the clay was baked. In this way, archeological sites of different ages have preserved a unique record of geomagnetic field variations over the last several thousand years (Aitken 1974, Tarling 1975).

Igneous rocks are not the only recorders of paleomagnetic field information; lake and ocean sediments may also register variations through the acquisition of detrital or depositional remanent magnetization (DRM). Magnetic particles become aligned in the direction of the ambient magnetic field as they settle through a water column. Providing that the sediment is not disturbed by currents or slumping or bioturbation, the magnetic particles will provide a record of the magnetic field of the Earth at the time of deposition. Verosub (1977) considers that the acquisition of magnetization by sediments may occur after deposition due to the mobility of magnetic carriers within fluid-filled voids in the sediment. Once the water content of the sediment drops below a critical level (depending on the sediment characteristics), the magnetic particles can no longer rotate and magnetization becomes "locked in" to the sediment. Verosub considers that this post-depositional DRM provides a more accurate record of the ambient magnetic field than simple depositional DRM.

Unlike thermoremanent magnetization, detrital remanent magnetization is not an "instantaneous" event. Once the molten lava has cooled, perhaps in a matter of minutes, the ambient field becomes a permanent fixed record. In sediments, the record is subject to disturbance by burrowing animals that may raise the water content of the sediment enough for magnetic particles to rotate again, changing the magnetization until the sediment is sufficiently de-watered to fix the record once more. Thus, the sedimentary record of the Earth's magnetic field, although continuous, should be considered as a smoothed or average record, unlikely to record short-term variations except in unusual circumstances where sedimentation rates are sufficiently high. Furthermore, it has been demonstrated by Verosub (1975) that sediment disturbance may result in apparent reversals which would be hard, if not impossible, to detect in cores of fine-grained sediment (Fig. 4.2). This has probably been a primary cause of error in many reports of short-term variations of the Earth's magnetic field (excursions; see Sec. 4.1.5).

Finally, it is now also recognized that iron minerals in some sediments undergo post-depositional chemical changes which result in a magnetization characteristic of the Earth's magnetic field long after initial deposition. This is known as chemical remanent magnetization (CRM); identification of the minerals typically affected in a sample can provide a warning that errors may be expected.

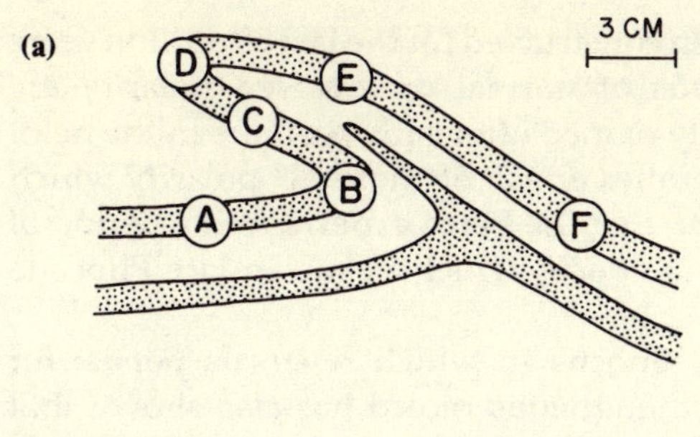

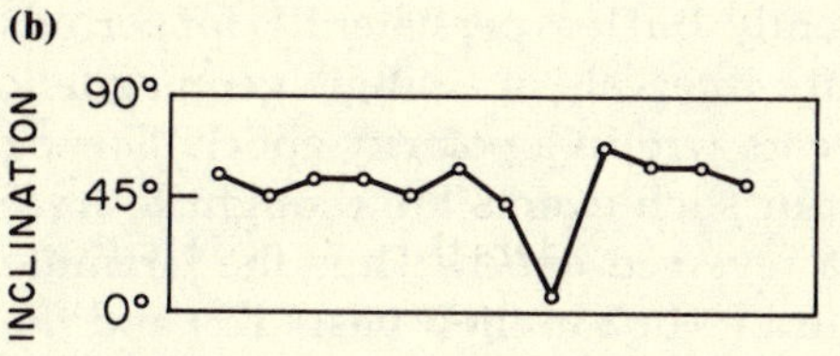

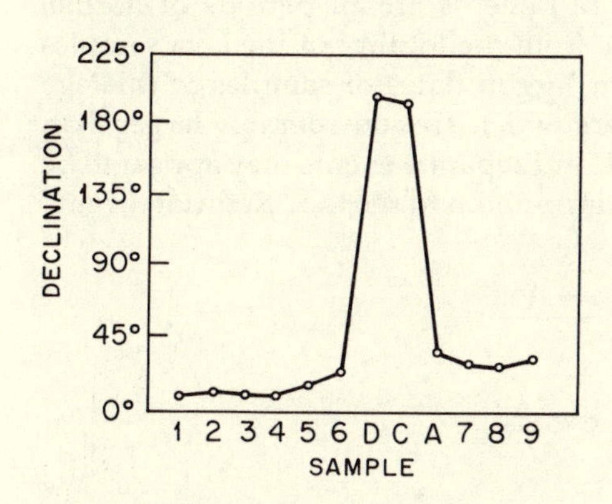

Figure 4.2 Problems of paleomagnetic stratigraphy illustrated by a varved sedimentary record. (a) A folded varved sediment sequence; shaded layers represent winter (clay) sediments; and the unshaded layer the summer (silt) deposits. (b) Paleomagnetic record obtained on a hypothetical core through the fold shown in (a), intersecting points A, C, and D. Because of sediment deformation an apparent paleomagnetic excursion is recorded. In uniform, fine-grained sediments such deformation would probably not be visible, so that the presence of an excursion might be erroneously reported (after Verosub 1975).

4.1.3 *The paleomagnetic timescale*

Most of the early work on establishing a chronology or timescale of paleomagnetic events was carried out on lava flows. It was demonstrated that at times in the past the Earth's magnetic field has been the reverse of today's and that these periods of reversal lasted hundreds of thousands of years. Potassium–argon dating methods enabled dates to be assigned to periods of "reversed" and "normal" fields so that eventually a complete chronology spanning several million years was constructed (Cox 1969). Indeed the development of this chronology went hand in hand with the theory of plate tectonics since new lavas, produced at the centers of spreading (e.g. the Mid-Atlantic Ridge) were found to record identical paleomagnetic sequences on either side of the ridge (Watkins 1972). Careful study of lava flows and sea-floor paleomagnetic anomaly patterns has so far enabled a fairly accurate chronology of reversals to be constructed back to 13 million years BP (Harrison *et al.* 1979), and more uncertain chronologies have been constructed for the last 80 million years (McDougall 1979). Major periods of normal or reversed polarity are termed polarity epochs, generally named after early workers in the field. Thus we are currently in the Brunhes epoch of "normal" polarity which began ~720 000 years BP. Prior to that the Earth experienced a period of reversed polarity, the Matuyama epoch, which began in late Pliocene times (Fig. 4.3).

In addition to major polarity epochs in which reversals persist for periods of $\sim 10^6$ years or more, the igneous record has also shown that reversals have occurred more frequently, but less persistently, for periods known as polarity events. These are intervals of a single geomagnetic polarity generally lasting 10^4–10^5 years within a polarity epoch. During the last 2 million years, three or four such events are thought to have occurred, all within the Matuyama reversed epoch. Thus the Jaramillo (0.94–0.89 million years BP) the Gilsá (~1.62 million years BP) and the Olduvai (1.91–1.76 million years BP) events are all periods of normal polarity, each one taking its name from the locality of the lava samples studied. Unfortunately, potassium–argon dates of samples of this age commonly have standard error bars which are considerably larger than the duration of events being dated, and separate events may appear to be the same. Consequently, it is not uncommon to see the "Réunion event"

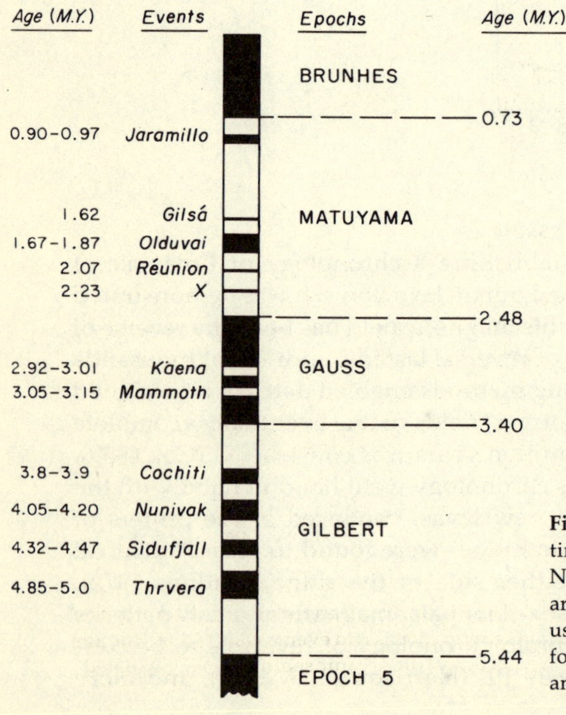

Figure 4.3 Paleomagnetic polarity timescale for the last 5 million years. Normal polarity periods in black. Dates are based on K/Ar dates on lava flows using recent revisions of time constants for potassium-40 (after McDougall 1979 and Mankinen & Dalrymple 1979).

(Fig. 4.3) considered as part of the Olduvai event, or the Gilsá event considered to extend into part of the Olduvai, as defined here. In short, the dating of these relatively brief events is uncertain and subject to change as new analyses are carried out. Figure 4.3 gives the current status of the polarity timescale as interpreted by McDougall (1979).

All of the preceding discussion has referred to studies of polarity changes observed in lavas, but the widest application of paleomagnetism to paleoclimatic studies has been in the identification of reversals in sedimentary deposits, notably in ocean cores. Studies of detrital remanent magnetization in ocean sediments may, in favorable circumstances, give a paleomagnetic record comparable even in detail with the non-marine record (Opdyke 1972; Fig. 4.4). It is common in studies of undated ocean cores to plot the polarity sequence changes with depth and to assign an age of 0.72 million years to the first major reversal (the Brunhes/Matuyama boundary). Younger ages are then derived by interpolation, assuming a zero age for the uppermost sediments and a constant sedimentation rate. Checks on certain levels, using U-series, or even ^{14}C, dates near the surface, may also be used to verify sedimentation rates based on these rather blithe assumptions. It should be noted that commonly no such checks are made and confirmation of the reality of a constant sedimentation rate may be sought by seeing if the interpolated

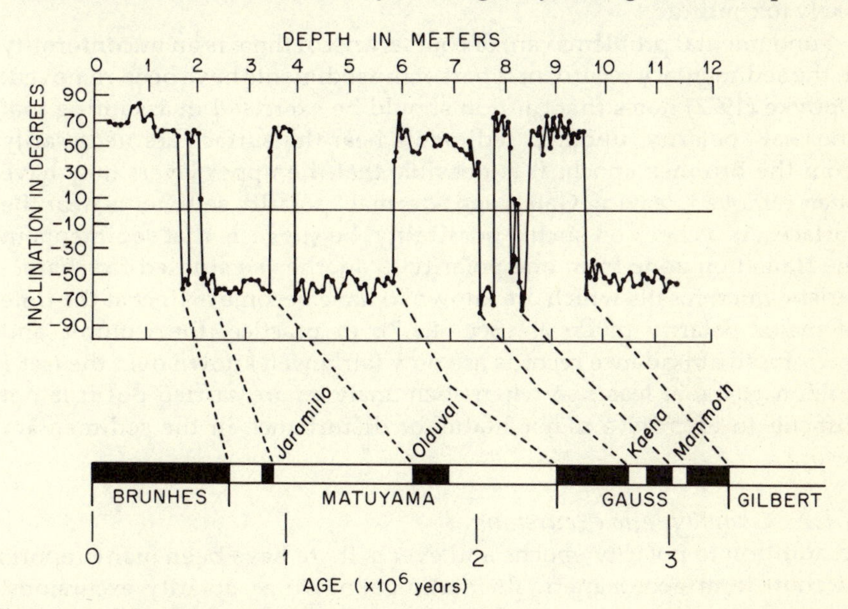

Figure 4.4 Actual paleomagnetic record (of inclination) in a North Pacific ocean core (V20-105; 39°N. 178°W). Sharp reversals in inclination are clearly seen in the record. The suggested stratigraphic sequence of paleomagnetic epochs and events is shown. As the age of reversals is known (from K/Ar dating of lava flows) the oceanic sediments can be "dated" by stratigraphic correlations (after Opdyke 1972).

age of a "known" stratigraphic feature (e.g. isotopic substage, 5e, the last interglacial peak; see Sec. 6.3.2) is determined in this way, "correctly". If so, the assumption of a constant sedimentation rate is considered to be reasonable. However, it is clear that calculating an *average* sedimentation rate for a period of 700 000 years can disguise significant variations within that period, and without good dating control on older sections of the record there is no guarantee that the rate has not varied from above average in some sections to below average in others. In the vast majority of ocean cores studied, no independent dating checks are made, other than biostratigraphic correlations at certain isotopic substages, the ages of which may themselves be questionable. Indeed, in many cores in which there is no paleomagnetic record, or in which the core does not reach the Brunhes/Matuyama boundary, dating is achieved by biostratigraphic or $\delta^{18}O$ correlations alone, by comparison with other nearby cores in which a magnetic polarity reversal was observed. The "dating" procedure for a core "B" thus becomes: paleomagnetic reversal on nearby core "A"; age of biostratigraphic features in core A by interpolation, assuming mean sedimentation rate; age of biostratigraphic features in core B by correlation between cores. Needless to say, chronologies constructed in this way are truly built on sand! Such maneuvers may lead to gross errors, but in the absence of better dating control the practice is likely to continue.

Fundamental problems can, of course, arise if there is an unconformity in the sedimentary record, or if part of the sediments have been removed. Opdyke (1972) notes that caution should be exercised in assuming that "normal" polarity, undated sediments near the surface are necessarily from the Brunhes epoch. It is possible that the upper layers may have been removed, leaving Gauss age "normal" polarity sediments near the surface. As a check on such a possibility, he suggests that sediments in the transition zone from one polarity to another be studied for characteristic microfossils which are known to have become extinct at the time of major polarity reversals (Sec. 4.1.7). In practice, the isotopic and microfossil abundance records are now fairly well known over the last 1 million years, at least; so when such analyses are carried out it is not difficult to identify a major hiatus or disturbance in the sedimentary record.

4.1.4 Geomagnetic excursions

In addition to polarity epochs and events there have been many reports of short-term geomagnetic fluctuations known as polarity excursions. Excursions are considered to be $<10^4$ years in duration and differ from events in that a fully reversed field is generally not observed, perhaps because the partial reversal is due to variations in the non-dipole component of the field. It is not yet clear whether excursions are "abortive reversals" or whether they represent normal geomagnetic behavior during a polarity epoch. In this regard it is interesting that Cox (1969), in a study of the spectrum of known reversal frequency, formulated a statistical model of geomagnetic field behavior from which he inferred the existence of short events or excursions during the last 10 million years, which were at that time undiscovered. Nevertheless, it must be stated that a great deal of controversy surrounds geomagnetic excursions and, even if they do exist, the evidence may not yet have been presented which demonstrates their existence unequivocally (Verosub & Banerjee 1977, Lund & Banerjee 1979). Consider, for example, the excursions which have been reported over the last 10–15 years. One of the earliest was the Laschamp excursion, noted in lava flows from the Puys des Laschamp in central France. Initial work showed a marked reversal in the sign of inclinations on samples from this site, but the excursion could only be dated as being between 8730 and <20 000 years BP (Bonhommet & Babkine 1969). Such an excursion, if accurately dated, could be of immeasurable value in stratigraphic correlations and so there was a tremendous impetus for other workers to search for a similar excursion within the age range 8000–20 000 years BP. Before long other excursions were reported, at 13 000–12 000, ~18 000, ~ 24 200, and 31 000–28 000 years BP, but in nearly every case subsequent studies at nearby sites could not confirm their reality. Furthermore, recent studies on the Laschamp lavas have suggested revised ages of 33 500 ± 4000 and 45 200 ± 2500 years BP (Hall & York 1978, Gillot *et al.* 1979)! In short, the evidence for any excursion since 25 000 years BP seems extremely suspect at present, and a review of the evidence for earlier excursions (e.g. the Blake excursion, ~117 000–104 000 years BP) does little to instil confidence in their reality either (Verosub & Banerjee 1977). Even if these individual events reported are correct, their absence from other sites in the same region mean that they are of little value in constructing regional chronologies or in magnetostratigraphic correlation. More rigorous criteria in the identification of excursions may reduce the number of spurious reports. In particular, for an excursion to warrant recognition it should certainly be based on observations of synchronous changes in both declination and inclination in several geographically separate cores, and the analysis should be restricted to fine-grained, homogeneous sediment. Few of the studies discussed meet these criteria; for the time being then, the reality of geomagnetic excursions must remain in doubt.

4.1.5 Secular variations of the Earth's magnetic field

A number of studies of well-dated lake sediments from different parts of the world indicate that quasi-periodic changes in declination (and to a lesser extent inclination) have occurred during the Holocene (Mackereth 1971, Creer *et al.* 1976, Thompson & Wain-Hobson 1979). These changes are of a smaller magnitude than those described as excursions and appear to be regional in extent (over distances of 1000–3000 km), presumably because they result from changes in the non-dipole component of the Earth's magnetic field.

If the changes observed can be accurately dated in one or more cores, it should be possible to construct a "master chronology" which would enable peaks and troughs in the declination record to be used as chronostratigraphic markers. These marker horizons would be valuable in estimating the age of highly inorganic sediments which do not yield enough carbon for conventional ^{14}C dating, but which do contain a clear record of declination variations. However, it is first necessary to establish a reliable, well dated magnetostratigraphic record for a "type-section" and to determine over what area this record can be considered to provide a master chronology. So far, this has only been done in western Europe (R. Thompson 1977) but work is in progress on a similar chronology for the north central United States (King 1982). It must also be clear that the undated sediments have approximately the same sedimentation rate as that of the master chronology and have not experienced any significant disturbances. If these conditions are met, the record of secular variations in declination can be used as a proxy dating method, and indeed may even be used to reinterpret ^{14}C dates on sediment (e.g. Thompson & Wain-Hobson 1979).

4.1.6 Magnetic reversals and faunal extinctions

As more and more ocean cores were studied for paleomagnetic variations, it became apparent that a significant number of microfaunal assemblages (mainly Radiolaria) either became extinct or evolved around the time of polarity reversals. In particular, a close correlation has been demonstrated between the Brunhes/Matuyama polarity boundary and faunal extinctions (Opdyke *et al.* 1966, Watkins & Goodell 1967). Extinctions are not simply local events but occur more or less synchronously throughout the geographical range of a species and can thus be widely recognized in cores spanning a wide range of latitudes (Fig. 4.5). The reasons for this correlation are unclear, but there are two schools of thought on the matter. The first hypothesis is that during a polarity reversal there is a period of reduced magnetic field intensity, during which time the dipole field may drop to zero intensity for several hundred years or more. During this period there will be an increase in cosmic radiation at the Earth's surface, which is no longer shielded by its magnetic field. The enhanced radiation will produce a higher mutation rate, causing some species to become extinct and others to develop along new evolutionary lines. Calculations of how great the increase in cosmic radiation would have been generally demonstrate that this particular hypothesis is unlikely to be important (Harrison 1968). Perhaps of more

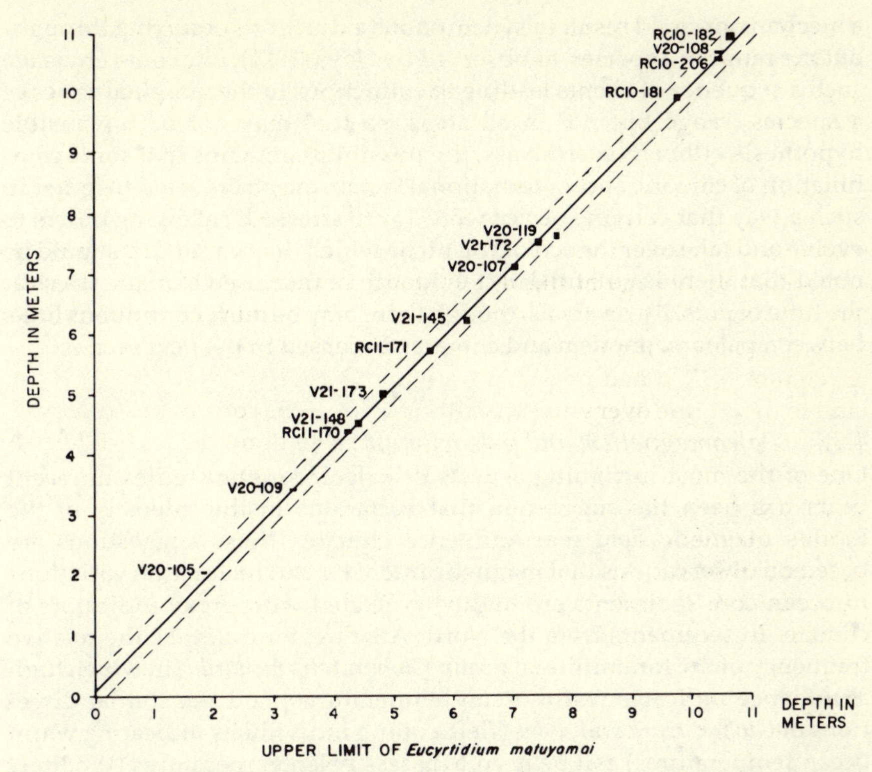

Figure 4.5 Least squares repression line (———) of the depth at which the Radiolarian *Eucyrtidium matuyamai* became extinct, against depth of the Jaramillo/Matuyama paleomagnetic reversal in the same cores. Numbers refer to cores (e.g. V20-105) which were recovered from the North Pacific Ocean; dashed lines indicate ± two standard errors of estimate. The strong correlation suggests that extinctions may occur as a result of reversals of the Earth's magnetic field. At least five other Radiolarian species became extinct around the time of paleomagnetic reversals (after Hays 1971).

significance is the idea that during a geomagnetic reversal the number of solar protons reaching the upper atmosphere would have increased, causing a reduction in the stratospheric ozone concentration and a consequent increased flux of ultraviolet radiation at the Earth's surface (Reid *et al.* 1976). This may in turn cause mutations and increase the rate of evolutionary change. However, these hypotheses are of only marginal interest to paleoclimatologists. Of more significance to our theme is the hypothesis that climatic changes associated with (resulting from?) polarity reversals may lead to faunal extinctions due to increased environmental stress on the species in question. Thus, Kennett and Watkins (1970) suggest that the increase in volcanic dust which they observe in Pacific–Antarctic cores around polarity reversals may point to increased atmospheric turbidity and colder temperatures at these times, leading to extinctions of certain species. It would seem unlikely, however, that such a mechanism could result in synchronous extinctions occurring throughout the range of a species, as observed by Hays (1971); one could envisage such a sequence of events leading to extinctions in the marginal zones of a species' range but not in all areas, so that may not be a plausible hypothesis either. Nevertheless, the possibility remains that some combination of climatic and/or radiational factors may have acted together in such a way that certain species were "over-stressed", allowing others to evolve and take over the ecological niche which they vacated. It should be noted that there is still no direct evidence of increased climatic stress at the time of polarity reversals, though there may be more continuous links between paleomagnetism and climate, discussed in the next section.

4.1.7 Paleomagnetism and paleoclimate

One of the most intriguing aspects of paleomagnetic studies in recent years has been the suggestion that variations in the intensity of the Earth's magnetic field may influence climate. These suggestions are based on observations that magnetic intensity and inclination variations in ocean core sediments are highly correlated with proxy indicators of climate. In sediments from the North Atlantic, for example, the relative frequency of the foraminiferal group *Globoratalia menardii* (in which high abundance indicates warm ocean temperatures) and the coiling directions of *Globor. truncatulinoides* (left-coiling individuals indicating warm ocean temperatures) can be used to assess paleotemperatures (Wollin *et al.* 1971). Magnetic intensity and inclination studies on parallel cores from the same area showed strong inverse correlations with the foraminiferal data, colder water temperatures corresponding to higher magnetic intensities. Subsequent work has extended the analysis to other cores, spanning the last 2 million years, and similar relationships are noted (Fig. 4.6; Wollin *et al.* 1977, 1978). Furthermore, comparison of the ocean core data with calculated variations in the Earth's orbital eccentricity shows intriguing correlations between periods of high eccentricity, low magnetic field intensity, and warm climate (Fig. 4.6). If these relationships are real, then some geophysical explanation is needed to demonstrate cause and effect. Wollin *et al.* (1978) speculate that the answer may lie far beneath the Earth's surface, at the interface between the inner liquid core of the Earth and the subsurface of the solid mantle. This interface is less ellipsoid than that of the outer surface of the Earth, so the torque due to the solar and lunar gravitational field acting on the Earth's core is smaller than that acting on the mantle. At times of maximum eccentricity, the difference in torques increases, causing perturbations in convective flow within the core to develop, weakening the dipole field. This in turn reduces the geomagnetic shield against corpuscular radiation, leading to increased radiation receipts and higher tem-

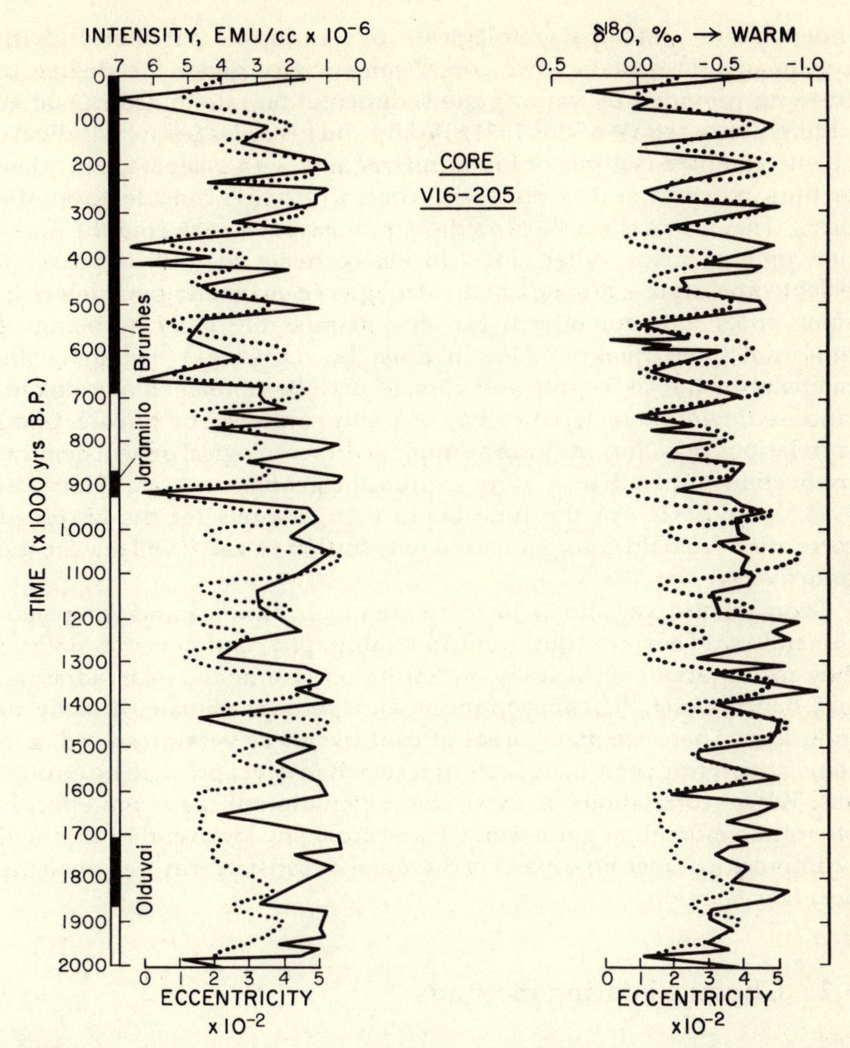

Figure 4.6 Comparison of variations in geomagnetic intensity, isotopic changes of ocean water and eccentricity of the Earth's orbit. The sedimentary record is from North Atlantic core V16-205 (15°N, 50°W). Geomagnetic intensity is inversely related to eccentricity but positively related to ice volume (and/or inversely related to sea-surface temperatures) (after Wollin *et al.* 1978).

peratures. In short, times of maximum eccentricity correspond to a weak dipole field and warmer climates.

Whether such an explanation is feasible or not remains open to debate, but there are many who would argue that the evidence for a correlation between climate and magnetic variations is spurious to start with. There are now such a large number of ocean cores which have been analyzed in various ways that it is probably possible to correlate virtually any independent micropaleontological or isotopic parameter with paleomagnetic data in *some* cores, somewhere, given the degree of freedom provided by varying the sedimentation rate to accommodate different data sets (Watkins 1971). Wollin and his colleages are not always able to compare isotopic or foraminiferal data with paleomagnetic data on the same cores, and so select other cores which they consider to be "the best." This is a reasonable procedure, but leaves unanswered the question of how many other cores do *not* correlate so well. If peaks in paleotemperature estimates match troughs in magnetic parameters in some cores, but not others, can one dismiss the latter as being of uniformly poor quality? Also, it must be recognized that magnetic intensity of the sediments and climate may be fundamentally linked, since sediment characteristics may be partly controlled by climate. Good correlations may thus indicate a simple sedimentological or geochemical connection rather than a more significant geophysical one (Harrison 1974, Kent 1982). For the time being then, reasons for the observed correlations remain ambiguous and only further research will resolve the controversy.

Geomagnetic variations have tremendous ramifications in paleoclimatology. They are significant in stratigraphy, and hence in dating; they are important in shielding the Earth from cosmic and solar radiation, and hence affect ^{14}C concentrations and perhaps climate, directly or indirectly. There are many areas of controversy as yet unresolved, and some apparently promising lines of research may yet prove to be erroneous. Where correlations do exist, the explanation of cause and effect is often little more than guesswork. There are many loose ends and it will be interesting to see how many of these can be satisfactorily tied up in the next few years.

4.2 Chemical dating methods

Two general categories of dating methods are based on chemical changes within the samples being studied. The first involves amino-acid analysis of organic samples, generally used to assess the age of associated inorganic deposits. The method may also be used to estimate paleotemperatures from organic samples of known age. The second category encompasses a number of methods which assess the amount of weathering that an inorganic sample has experienced. They are primarily used to assess the relative age of episodic deposits such as moraines or till sheets. There are a number of possible approaches (see Birkeland 1974) but one of the most widespread and well tested methods is obsidian hydration dating, an example of the general group of methods which involve measurements of weathering rinds (e.g. Colman & Pierce 1981, Chinn 1981). A third method involves the chemical and physical "fingerprinting" of volcanic ashes which often blanket wide areas after a major eruption. Chemical analyses of tephra deposits have proved to be successful in identifying unique geochemical signatures in ashes of different ages. Where the age of the tephra layers has been independently determined, the ash may be used as a chronostratigraphic horizon to date the associated deposits.

4.2.1 *Amino-acid dating*

As all living organisms contain amino acids, a dating method based on amino acids offers a tremendous range of possible applications. Since the first use of amino acids in dating fossil mollusc shells (Hare & Mitterer 1968) significant advances in the use of amino acids as a geochronological tool have been made. Both relative and absolute ages have been determined, though it is likely that almost all absolute age estimates attempted so far are in error by $>\pm 15\%$ (Williams & Smith 1977). The method has the potential of dating material ranging in age from a few thousand to several million years old. It could therefore be of great value in dating organic material well beyond the range of radiocarbon dating. Efforts have also been made to use amino acids in estimating the average temperature of a sample since deposition, and this approach may have greater potential than the use of amino-acid analysis in dating.

Amino-acid analyses require very small samples (10 mg in the case of molluscs and foraminifera, less in the case of bone). The application of amino-acid dating is thus of particular significance in dating fragmentary hominoid remains, where, in many cases, a conventional radiocarbon analysis would require destruction of the entire fossil to obtain a date (Bada & Helfman 1975). Analyses have also been carried out on samples of wood, coral, foraminifera, and marine, freshwater, and terrestrial molluscs (Schroeder & Bada 1976). Few workers are prepared to be definitive in assigning an age to the samples analyzed. More often the analysis enables relative chronologies to be established and stratigraphic sequences to be checked (Miller *et al.* 1977, 1979). It is perhaps in this application (aminostratigraphy) that amino-acid analyses offer the greatest potential. However, this is a rapidly developing field and great improvements in absolute dating may be possible (see, for example, recent studies in the volume edited by Hare *et al.* 1980). Amino-acid studies may yet become one of the most important chronological tools in Quaternary research.

4.2.1.1 Principles of amino-acid dating. Amino acids are so called because they contain in their molecular structure at least one amino group ($-NH_2$) and a carboxylic acid group ($-COOH$). These are attached to a central carbon atom, which is also linked to a hydrogen atom ($-H$)

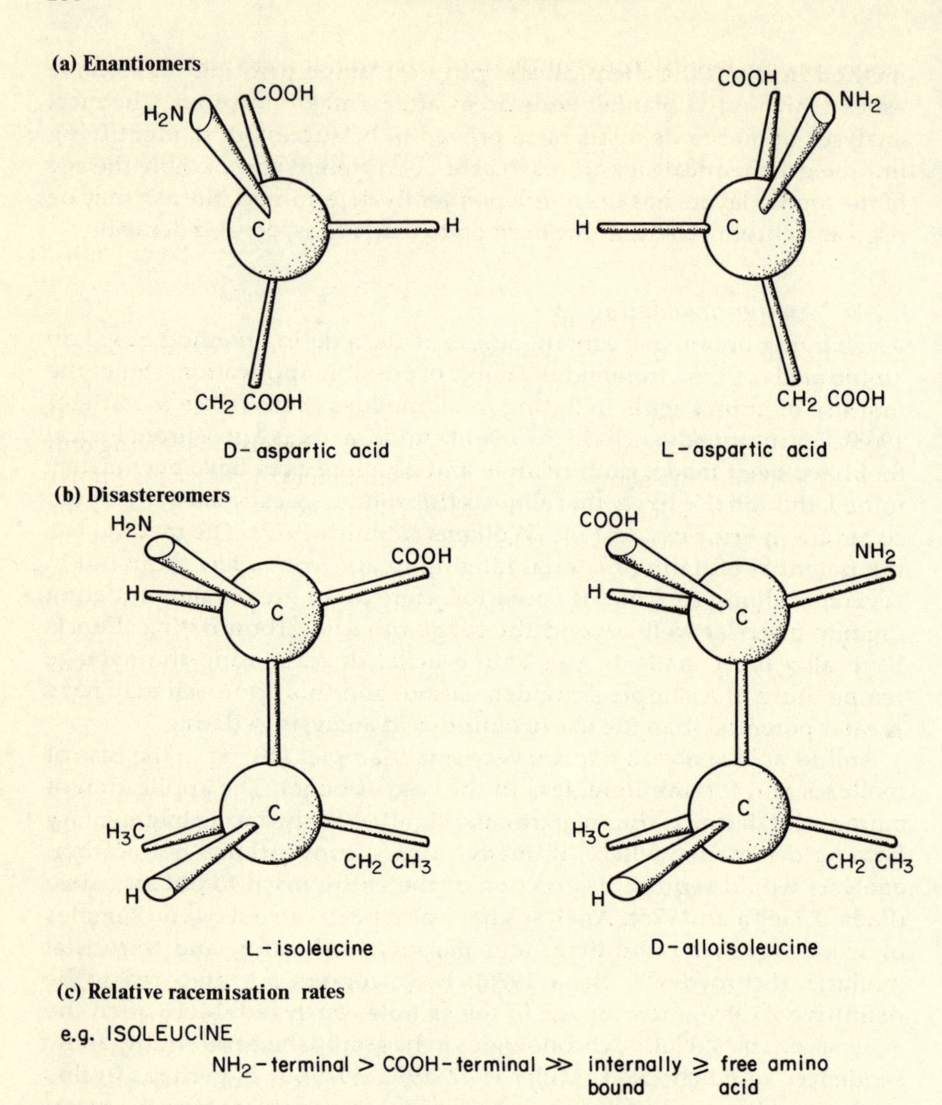

Figure 4.7 (a) An example of enantiomers (D-aspartic acid and L-aspartic acid). (b) An example of diastereomers (L-isoleucine and D-alloisoleucine). (c) Relative rates of racemization depending on whether the amino acid is internally bound, terminally bound or free.

and a hydrocarbon group (—R) (Fig. 4.7a). If all atoms or groups of atoms attached to the central carbon atom are different, the molecule is said to be chiral or asymmetric. The significance of this is that chiral molecules can exist in two optically different forms (stereoisomers), each being the mirror image of the other (Fig. 4.7b). These optical isomers or enantiomers have the same physical properties and differ only in the way in which they rotate plane-polarized light. The relative configuration of enantiomers is designated, by convention, D or L (*dextro* or *levo*) and virtually all amino acids in living organisms occur in the L configuration. Interconversion to the D configuration takes place by a process known as racemization. The extent of the racemization (expressed by the enantiomeric ratio, D : L) increases with time after the death of the organism. This can be measured by gas or liquid chromatographic methods.

Not all amino acids have only one chiral carbon atom. Several amino acids (e.g. isoleucine) contain two chiral carbon atoms, which means that they can exist as four stereoisomers – a set of mirror image isomers (enantiomers) and a set of non-mirror image isomers (diastereomers) (Fig. 4.7b). Interconversion of L-isoleucine could thus theoretically produce all four stereoisomers. However, in diagenetic processes only one of the two chiral atoms undergoes interconversion, thereby producing only one other isomer (D-alloisoleucine, a diastereomer) by a process known as epimerization† (Schroeder & Bada 1976, Rutter *et al.* 1979). Diastereomers have physical properties which are sufficiently different that they can be separated by ion-exchange chromatography. Several different amino acids have been used to assess the age of a sample, particularly aspartic acid, leucine, and isoleucine. Epimerization of isoleucine is an order of magnitude slower than aspartic acid racemization so it is potentially of more value in dating older samples or those from warmer climates where epimerization and/or racemization rates are faster.

Unfortunately, unlike radionuclide decay rates, racemization and epimerimation rates are sensitive to a number of environmental factors, particularly temperature and pH (although in most fossils only the temperature variable is significant). In addition, racemization and epimerization rates vary depending on the type of matrix in which the amino acids are found (shell, wood, bone, etc.). In carbonate fossils rates vary from one genus to another, so it is important to compare amino-acid ratios derived from analyses on similar genera (Miller & Hare 1975, King & Neville 1977). Racemization rates also depend on how amino acids are bound to each other, or if they are free (unbound). When an amino acid is bound together with others, its position in the molecule may be internal or terminal (Fig. 4.7c). If it is terminally bound it may be attached to other amino acids by a carbon atom or a nitrogen atom. Racemization rates are fastest when the amino acid is terminally bound and slowest when the individual amino acid is free, having been separated from the rest of the molecule by hydrolysis. Racemization rates of internally bound amino acids are intermediate between rates of terminally bound and free amino acids (Fig. 4.7c). What this means is that, as the peptide is hydrolyzed, at some point each amino acid will become terminally bound before being eventually split off (free). In the terminally bound

†For our purposes racemization and epimerization of amino acids can be considered as essentially equivalent processes.

position, racemization rates are greatest, so the probability is high that the free amino acid, when released, will already be racemized. When samples are analyzed, the free and bound amino acids are separated and D : L ratios are calculated for each fraction. Consequently the D : L ratios in the free fraction are higher than in the bound fraction. It is thus important to note whether analyses reported in the literature are based on bound, free, or total (free and bound) amino acid content, since the resulting ratios can vary by an order of magnitude (Table 4.1).

Table 4.1 Temperature sensitivity of amino-acid reactions in dated early Postglacial mollusc samples (Miller & Hare 1980).

Location	^{14}C age (radiocarbon years)	MAT (°C)[†]	Species[‡]	Allo : Iso[§] Total	Free
Washington	13 010	+10	*H.a.*	0.078	0.27
Denmark	13 000	+7.7	*H.a.*	0.053	0.21
Maine	12 230	+7	*H.a.*	0.050	0.21
New Brunswick	12 500	+5	*H.a.*	0.043	0.18
South-eastern Alaska	10 640		*H.a.*	0.040	0.15
Anchorage	14 160	+2.1	*M.t.*	0.034	0.16
Southern Greenland	13 380	−1	*H.a.*	0.027	≤0.09
Southern Baffin Island	10 740	−7	*M.t.*	0.024	≤0.1
Spitzbergen	11 000	−8	*M.t.*	0.022	≤0.1
Northern Baffin Island	10 095	−12	*H.a.*	0.020	ND
Somerset Island	9 000	−16	*H.a.*	0.018	ND
Modern	0	—	*H.a.*	0.018	ND

[†] Mean annual temperature of the past one to five decades based on records of the nearest representative weather station.
[‡] *H.a.* = *Hiatella arctica*; *M.t.* = *Mya truncata*. Hydrolysis rates in *Mya* are not directly comparable with those in *Hiatella*. For most localities, three or more separate values were analyzed; ratios given here are mean values.
[§] Ratio of D-alloisoleucine to L-isoleucine; ND = no detectable concentration of alloisoleucine.

By far the most significant factor affecting the rate of racemization is temperature, specifically the integrated temperature of the sample since deposition (Table 4.1). Racemization rates more or less double for a 4–5 °C increase in temperature, so the thermal history of a sample becomes of critical importance to the apparent age. An uncertainty of only ±2 °C is equivalent to an age uncertainty of ±50%, so this is clearly a major source of error in assessing the absolute age of a sample (Bada 1972). Thermal histories are rarely known to within ±2 °C, even in isolated environments, such as in caves or in the deep oceans. However, the temperature dependence of racemization rates may be put to advantage if the sample age is known independently (e.g. by ^{14}C dating). In such cases, the relative amount of racemization can indicate mean temperature of the sample since deposition (Bada *et al.* 1973) or the extent of a step-change in temperature (Schroeder & Bada 1973; see Sec. 4.2.1.4).

There are basically three approaches used in amino-acid geochronological work. Two of these aim at producing an estimate of absolute age and the third uses enantiomeric ratios as simply a stratigraphic tool. At present the latter, less ambitious, approach is the most successful.

4.2.1.2 Absolute age estimates based on amino-acid ratios. Absolute sample ages are estimated by either calibrated or uncalibrated methods (Williams & Smith 1977). The *uncalibrated method* is based on high-temperature laboratory experiments which attempt to simulate in a short period of time the lower-temperature and much slower processes which occur in samples in nature. The general amino-acid racemization reaction is as follows:

$$\text{L-amino acid} \underset{k_2}{\overset{k_1}{\rightleftharpoons}} \text{D-amino acid}$$

where k_1 and k_2 are rate constants for the forward and reverse reactions. In the high-temperature studies, racemization rates are determined by sealing a modern sample (of the same species as the fossil under investigation) in a tube and heating it for known lengths of time in a constant temperature bath. In this way genus-specific rate constants for the amino acid in question can be determined at different (elevated) temperatures. These are then plotted on an Arrhenius plot in which the log of the rate constant forms the ordinate and the reciprocal of the absolute temperature the abscissa (Fig. 4.8). If the calculated rate constants fall on a straight line, extrapolation is made (beyond the experimental results) to obtain the rate constants applicable at lower temperatures (e.g. Miller & Hare 1980). Providing that the mean temperature of the sample since its deposition is known, racemization rate constants can be obtained for that temperature from the Arrhenius plot; sample age can then be calculated from the measured D : L ratio (for the appropriate equations, see Williams & Smith 1977 p. 102). One might question whether such high-temperature, short-term laboratory kinetic studies accurately reflect the low-temperature, long-term diagenetic changes which occur in fossils. However, there is an increasing body of evidence that this is not a problem and that the high-temperature results can be extrapolated to a real life situation (e.g. Fig. 4.8; Miller & Hare 1980). The real difficulty concerns the problem of knowing accurately the thermal history of the sample, since slight errors in this parameter lead to large errors in absolute age estimates (McCoy 1981). As a result, the ages derived by means of this uncalibrated

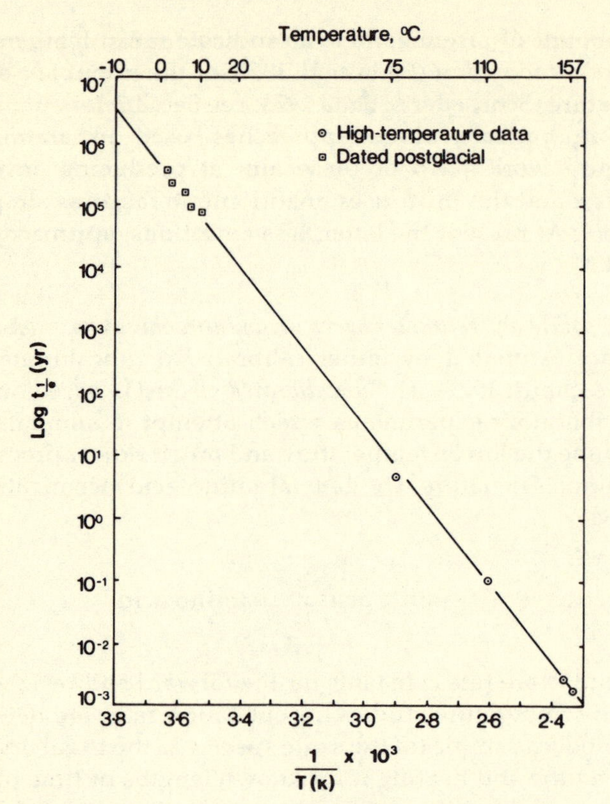

Figure 4.8 Arrhenius plot of isoleucine epimerization in *Hiatella arctica* derived from pyrolysis experiments at 75, 110, 152, and 157 °C and dated early postglacial samples (after Miller & Hare 1980).

method are considered to be the least reliable, and although age estimates based on this method have been published (e.g. Bada *et al.* 1970), there is much evidence that the results are in error (Wehmiller & Hare 1971, Williams & Smith 1977).

A more fruitful approach, though not entirely free of the problems discussed above, is to derive rate constants empirically by the measurement of D : L ratios *in situ*, in fossil samples of known age (the *calibrated method*). Other samples at the same site can then be dated, if it is assumed that they have experienced essentially the same mean temperature as the fossil used for calibration (Bada & Schroeder 1975). A Postglacial fossil calibration sample is thus not suitable for assessing the age of older "glacial age" samples since their thermal histories will be quite different. Bada and Schroeder (1975) claim that a calibration sample which has been radiocarbon dated at 20 000 years BP is best for calibrating bones too old for radiocarbon dating, the assumption being that such a sample has experienced a "representative" mean temperature typical of older samples, though this is clearly not a universally realistic assumption. The selection of a calibration sample thus presents difficulties, but they are not insuperable. Bada *et al.* (1974a) have, for example, obtained sample ages from aspartic acid racemization in bones which agree with ^{14}C dates on the same samples within 15% over an age range of 40 000–8000 years BP (average "error", ~7%). However, other aspartic acid dates on undated paleo-Indian bones (derived via the calibrated approach) have created much controversy, and the technique has been subjected to disparaging scrutiny (Bada *et al.* 1974b, Bender 1974, Williams & Smith 1977). Further independent checks using other dating methods should

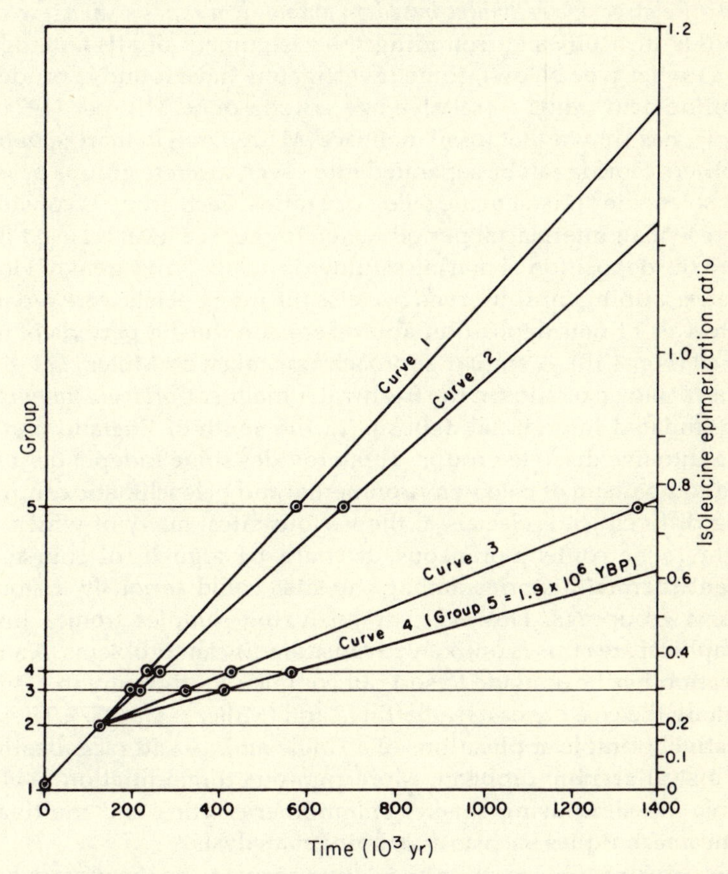

Figure 4.9 Estimates of the age of British interglacial deposits based on total allo : iso amino-acid ratios in the freshwater mollusc *Corbicula fluminalis.* The deposits can be grouped into five discrete units with average ratios ranging from 0.02 to 0.76. Unit 1 is modern; unit 2 is last interglacial (125 000 years BP). By making different assumptions about the age of unit 3 the ages of older units (4 and 5) can be postulated. For example, curve 2 assumes the age of unit 3 is 210 000 years BP. Curve 3 assumes it has an age of 310 000 years BP and curve 4 is based on the hypothesis that unit 3 is 410 000 years old. Curve 1 assumes that reversible first-order kinetics are followed exactly (after Miller *et al.* 1979).

resolve some of these criticisms. The calibrated approach has also been used on carbonate materials to give minimum age estimates for Pleistocene molluscs. By assuming (from stratigraphic considerations) that two sets of interglacial samples were deposited during marine isotope stages 5 and 7 (~124 000 and ~210 000 years BP, respectively; see Sec. 6.3.2), the ages of older molluscs could be estimated from their D : L ratios (Miller *et al.* 1979). By altering the assumptions made for the ages of the two youngest interglacial samples, various other possibilities could be presented for the ages of the older samples (Fig. 4.9).

4.2.1.3 Relative age estimates based on amino-acid ratios. In view of the numerous difficulties surrounding the assignment of absolute ages to fossil samples (see below), some investigators have found it prudent to use amino-acid ratios as relative age criteria only. Mitterer (1974), for example, has shown that fossil molluscs (*Mercenaria*) in marine deposits of southern Florida can be separated into seven discrete groups based on D-alloisoleucine : L-isoleucine (allo : iso) ratios. Each group is considered to represent an interglacial period, when higher sea levels caused flooding and the deposition of marine sediments in low-lying areas of Florida. Discrete groupings are observed because the interglacials were separated by intervals of non-deposition and/or erosion during periods of lower sea level (Fig. 4.10). A similar approach was taken by Miller *et al.* (1979), who used allo : iso ratios in the freshwater mollusc *Corbicula fluminalis* to group undated interglacial deposits in the south of England and East Anglia into five discrete groups. This provides some independent basis for the comparison of paleo-environmental and paleoclimatic conditions during different interglacials at the various sites, many of which were thought to be contemporaneous. It could be argued, of course, that different thermal histories among the sites could seriously distort the apparent groupings. However, by analyzing samples from a limited geographical area this is unlikely to cause significant problems. A similar application has been made to sort out complex stratigraphy in a limited area along the coast of eastern Baffin Island (Miller *et al.* 1977). These are all relatively simple applications of a single amino-acid racemization to solve a stratigraphic problem. More rigorous differentiation is almost certainly possible using several enantiomeric ratios and multivariate statistical techniques such as discriminant analysis.

At the present time, relative age dating seems to be the most practical application of amino-acid racemization and/or epimerization processes. It is still subject to problems of contamination, leaching, and possibly thermal differences amongst sites, but these are relatively minor problems compared to those associated with absolute age determinations. As more work is carried out, this situation may change. In particular, the calibration of amino-acid ratios seems to have great potential.

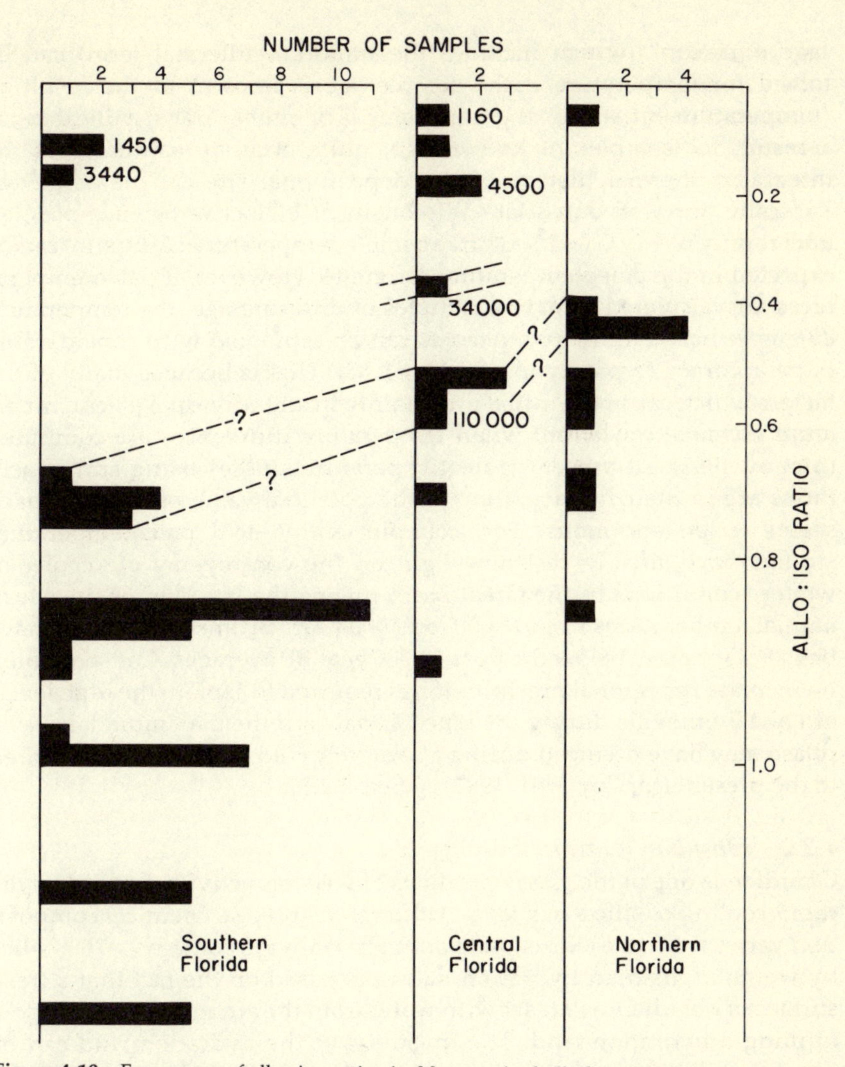

Figure 4.10 Frequency of allo : iso ratios in *Mercenaria* shells from upper Cenozoic strata of Florida. Ages are shown where a sample has been dated independently. Seven discrete groupings are noted; differences in diagenetic temperatures (of ~3 °C) between northern and southern Florida produce different ratios in stratigraphically equivalent groups (dashed lines). The seven groups represent seven interglacial sea-level rises (after Mitterer 1974).

4.2.1.4 Paleotemperature estimates from amino-acid racemization and epimerization. Although amino-acid analyses are being increasingly used in stratigraphic studies, perhaps the most significant application of amino-acid ratios is in paleotemperature reconstruction. As noted above, a major barrier to accurate age estimates from amino-acid ratios is a knowledge of the integrated thermal history of the sample. However, the "age equation" (which includes the important thermal term) can be solved for temperature if the sample age is known. In the resulting "temperature equation," the time value is of only minor significance; as a result, for samples of known age, quite accurate estimates of the integrated thermal history of the depositional site can be achieved. Typically, for well dated late Wisconsin or Holocene age samples, an uncertainty of ~3 °C (~1% of the absolute temperature of the site) can be expected in the paleotemperature estimates. However, if paleotemperatures are calculated from two samples of differing age, the temperature *difference* between the two periods can be estimated with considerably more accuracy (typically to within ±1 °C). This is because many of the factors which cause the initial uncertainty in an individual paleotemperature estimate cancel out when temperature differences are computed (McCoy 1981). At present, paleotemperature studies using amino-acid ratios are in their infancy, though the potential value of this approach seems to be enormous. For example, amino-acid paleotemperature studies have already cast new light on the controversy of "cooler or wetter" conditions in the Great Basin during the late Wisconsin. Mean annual temperatures from 16 000 to 11 000 years BP are estimated to have been 9 °C or more below the post-11 000 year BP averages. Consequently, no increase in regional precipitation is required to explain the high levels of Lake Bonneville during the Late Glacial, and the maximum lake level phase may have occurred during a relatively cold, dry climate compared to the present (McCoy 1981, 1982; cf. Sec. 8.2.3).

4.2.2 Obsidian hydration dating

Obsidian is one of the glassy products of volcanic activity, formed by the rapid cooling of silica-rich lava. Although its precise chemical composition varies from one extrusion to another, it always contains >70% silica by weight. Obsidian hydration dating is based on the fact that a fresh surface of obsidian will react with water from the air or surrounding soil, forming a hydration rind. The thickness of the hydration rind can be identified in thin sections cut normal to the surface; a distinct diffusion front can be recognized by an abrupt change in refractive index at the inner edge of the hydration rind. Hydration begins after any event which exposes a fresh surface (e.g. cracking of the laval flow on cooling, manufacture of an obsidian artifact, or glacial abrasion of an obsidian pebble); thus, providing one can identify the type of surface or crack in the rock, it is possible to date the event in question.

As one might expect, hydration rind thickness is a (non-linear) function of time; hydration rate is primarily a function of temperature, though chemical composition of the sample is also an important factor. For this reason, it is necessary to calibrate the samples within a limited geographical area against a sample of known age and similar chemical composition. These are difficult criteria to meet in a paleoclimatic context but are somewhat easier in archeological studies, where obsidian hydration dating has been most widely applied (Michels & Bebrich 1971). Obsidian was widely traded in prehistoric time and often the precise source of the material can be identified and its diffusion throughout a geographical area can be traced. If samples can be found in a ^{14}C-dated stratigraphic sequence, hydration rinds can be calibrated, providing an empirically derived hydration scale for the site. This can then be used to clarify stratigraphy elsewhere, where radiocarbon-dated samples are unavailable.

Obsidian hydration may also be used to date glacial events if obsidian has been fortuitously incorporated into the glacial deposits. Glacial abrasion of obsidian fragments creates radial pressure cracks normal to the surface and shear cracks subparallel to the surface. The formation of such "fresh" cracks allows new hydration surfaces to develop, and these effectively "date" the time of glacial activity. Hydration rinds resulting from glacial abrasion can then be compared with rinds which have developed on microfractures produced when the lava cooled initially. This event can be dated by potassium – argon isotopic methods (Ch. 3), providing independent calibration for the primary hydration rind thicknesses. Pierce *et al.* (1976), for example, analyzed obsidian pebbles in two major moraine systems in the mountains of western Montana. Dates on two nearby lava flows indicated ages of 114 500 ± 7300 and 179 000 ± 3000 years BP. Hydration rinds on cracks produced during the initial cooling of these flows averaged 12 and 16 μm respectively. These points enabled a graph of hydration thickness versus age to be plotted (Fig. 4.11). It was then possible to estimate the age of hydration rinds produced on glacially abraded cracks in the moraine samples. Two distinct clusters of hydration rind thicknesses enabled glacial events to be distinguished, at 35 000–20 000 and 155 000–130 000 years BP (Fig. 4.11). Although the dates are by no means precise, they do at least indicate the important fact that the earlier glacial event predated the

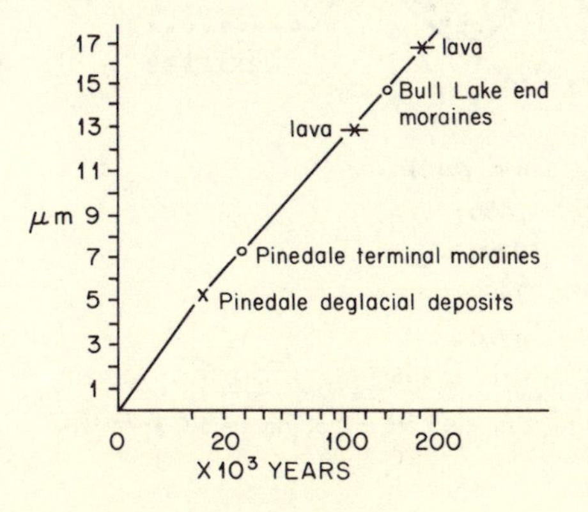

Figure 4.11 Rate of obsidian hydration in the West Yellowstone Basin, Montana. Crosses indicate hydration rinds on cooling cracks in potassium–argon-dated rhyolite lava flows and on cracks (produced by glacial abrasion) in Pinedale deglacial deposits (dated by radiocarbon). Circles indicate hydration rinds on samples from undated moraine systems. Age calibration line adjusted slightly to take into account temperature differences between the sites and variations in hydration rate due to climatic change (after Pierce *et al.* 1976).

Sangamon interglacial (~125 000 years BP), a point of some controversy in the glacial history of the western United States.

Obsidian hydration dating methods are limited by the problems of independent (radio-isotopic) calibration and of variations in sample composition, and temperature over time. Temperature effects are particularly difficult to evaluate. It is really necessary to produce a calibration curve for each area being studied, and this is not always possible. Nevertheless, where the right combination of conditions is found, obsidian hydration methods can provide a useful time-frame for events which might otherwise be impossible to date.

4.2.3 *Tephrochronology*

Tephra is a general term for airborne pyroclastic material ejected during the course of a volcanic eruption (Thorarinsson 1981). Extremely explosive eruptions may produce a blanket of tephra covering vast areas, in a period which can be considered as instantaneous on a geological timescale. Tephra layers thus form regional isochronous stratigraphic markers. Tephras themselves may be dated directly, by potassium–argon or fission-track methods, or indirectly by closely bracketing radiocarbon dates on organic material above and below the tephra layer (e.g. Naeser *et al.* 1981). In favorable circumstances, organic material incorporated within the tephra may provide quite precise time control on the eruption event (e.g. Lerbemko *et al.* 1975, Blinman *et al.* 1979). Providing that the dated tephra layer can be uniquely identified in different areas, it can be used as a chronostratigraphic marker horizon to provide limiting dates on the sediments with which it is associated. For example, a tephra layer of known age provides a *minimum* date on the material over which it lies and a *maximum* date on material superimposed on the tephra. If a deposit

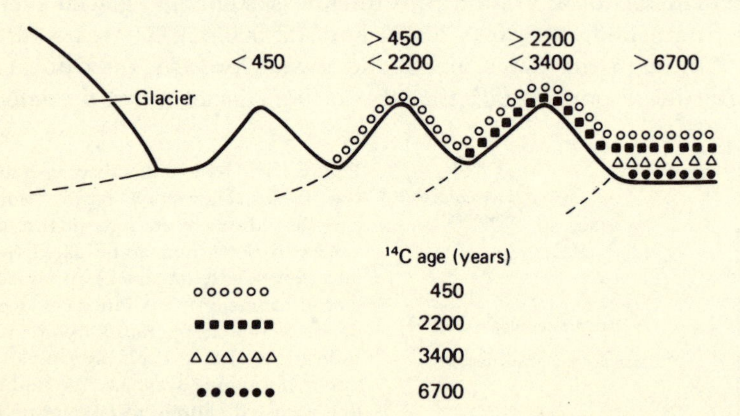

Figure 4.12 The use of tephra to date glacial deposits. If tephra age is known and tephra can be uniquely identified, ages can be used to "bracket" timing of glacial advance (after Porter 1981a).

is sandwiched between two identifiable tephra layers of known age, they provide bracketing dates for the intervening deposit (Fig. 4.12). A prerequisite for such tephrochronological applications is that each tephra layer be precisely identified. This has been the subject of much study both in the field and in the laboratory. In the field, stratigraphic position, thickness, color, degree of weathering, and grain size are important distinguishing characteristics. In the laboratory, a combination of petrographic studies and chemical analyses are generally used to identify a unique tephra signature (Kittleman 1979, Westgate & Gorton 1981). Multivariate analysis is commonly employed on the various parameters measured to provide optimum discrimination between the tephras being studied (e.g. Borchardt *et al.* 1970).

In many volcanic regions of the world, tephrochronology is a very important tool in paleoclimatic studies. In north-western North America, explosive eruptions have produced dozens of widely distributed tephra layers (Table 4.2). Some, such as the Pearlette "O" ash, covered almost the entire western United States and probably had a significant impact on hemispheric albedo (Bray 1979). Others were more local in extent; around Mt Rainier, for example, at least ten tephra layers have been identified spanning the interval from 8000 to 2000 years BP (Mullineaux 1974). Because of the eruption frequency and widespread distribution of tephra in this area, tephrochronological studies have proved to be invaluable in understanding its glacial history (e.g. Porter 1979). Tephrochronology has provided valuable time control in paleoclimatic studies of many areas, including studies of marine sediments containing ash layers (e.g. Kennett & Huddleston 1972). The two volumes edited by Sheets and

Table 4.2 Some important tephra layers in North America (after Porter 1981a and Westgate & Gorton 1981). An asterisk in column 3 indicates that an age is given in radiocarbon years.

Tephra layer	Source	Approximate age
Katmai	Mt Katmai, Alaska	AD 1912
Mt St Helens, Set T	Mt St Helens, Washington	AD 1800
Mt St Helens, Set W	Mt St Helens, Washington	450*
White River East	Mt Bona, South-eastern Alaska	1250*
White River North	Mt Bona, South-eastern Alaska	1890*
Bridge River	Plinth-Meager Mt, British Columbia	2350*
Mt St Helens, Set Y	Mt St Helens, Washington	3400*
Mazama	Crater Lake, Oregon	6720*
Glacier Peak B	Glacier Peak, Washington	11 250*
Glacier Peak G	Glacier Peak, Washington	12 750–12 000*
Pearlette O	Yellowstone National Park	600 000 ± 100 000
Bishop	Long Valley, California	700 000 ± 100 000
Pearlette S	Yellowstone National Park	1 200 000 ± 40 000
Pearlette B	Yellowstone National Park	1 900 000 ± 100 000

Grayson (1979) and by Self and Sparks (1981) provide numerous examples of the importance of tephrochronological studies in paleoclimatic research. It is also of particular importance in understanding the causes of climatic variation to have as complete a record as possible of major explosive eruptions. There is a great deal of evidence to implicate such eruptions in causing climatic deterioration and periods of glacial advance (Bray 1974, 1976, 1977, Porter 1981b).

4.3 Biological dating methods

Biological dating methods generally use the size of an individual species of plant as an index of the age of the substrate on which it is growing. They may be used to provide minimum age estimates only, since there is inevitably a delay between the time a substrate is exposed and the time it is colonized by plants, particularly if the surface is unstable (as, for example, in an ice-cored moraine). Fortunately this delay may be short and not significant, particularly if the objective is simply relative age dating.

4.3.1 Lichenometry

Lichens are made up of algal and fungal communities living together symbiotically. The algae provide carbohydrates via photosynthesis and the fungi provide a protective environment in which the algal cells can function. Morphologically, lichens range from those with small bush-like thalli (foliose lichens) to flat disc-like forms which grow so close to a rock surface as to be inseparable from it. These crustose lichens commonly increase in size radially as they grow and this is the basis of lichenometry, the use of lichen size as an indicator of substrate age (Locke *et al.* 1979). Lichenometry has been most widely used in dating glacial deposits in tundra environments where lichens often form the major vegetation cover and other types of dating methods are inapplicable (Beschel 1950, 1961, Benedict 1967). The technique may also be used to date lake-level (and perhaps even sea-level) changes, glacial outwash, and trim-lines, rockfalls, talus stabilization, and the former extent of permanent or very persistent snow cover.

4.3.1.1 Principles of lichenometry. Lichenometry is based on the assumption that the largest lichen growing on a rock substrate is the oldest individual. If the growth rate of the particular species is known, the maximum lichen size will give a minimum age for the substrate, since all other thalli must be either late colonizers or slower growing individuals (i.e. those growing in less than optimum conditions). Lichen size dates the time at which the freshly deposited rocks become stable, since an unstable substrate will prevent uninterrupted lichen growth. Growth rates can be obtained by measuring maximum lichen sizes on substrates of known age, such as gravestones, historic or prehistoric rock buildings, or moraines of known age (perhaps dated independently by historical records or radiocarbon). It is also possible to measure growth directly by photographing or tracing lichens of varying sizes every few years on identifiable rock surfaces (Miller & Andrews 1973, Ten Brink 1973). Generally, the maximum diameter of the lichen thallus is measured on individuals which have shown fairly uniform radial growth.

Growth rates vary from one region to another so it is necessary to calibrate the technique for each study site, but the general form of the growth curve is now fairly well established. After initial colonization of the rock surface, growth is quite rapid (known as the great period); growth then slows to a more or less constant rate (Fig. 4.13; Beschel 1950). Different lichens grow at different rates and indeed some species may approach senescence whilst other species are still in their great period of growth. The black foliose lichen *Alectoria minuscula*, for example, rarely exceeds 160 mm in diameter on rock surfaces in Baffin Island; lichens of this size represent a substrate age of ~500–600 years BP. By contrast, *Rhizocarpon geographicum* has only just entered its period of linear growth by this time (at ~30 mm diameter) and will continue to grow at a nearly constant rate for thousands of years after that. In fact, it has been estimated that a 280 mm thallus of *Rh. geographicum* on eastern Baffin Island dates its substrate at ~9500 ± 1500 years BP (Miller & Andrews 1973). Similarly, a 480 mm *Rh. alpicola* thallus in the Sarek mountains of Swedish Lapland is thought to have begun its growth following deglaciation of the region ~9000 years BP (Denton & Karlén 1973b). Different lichens may thus be selected to provide optimum dating resolution over different timescales. However, in view of its ubiquity, ease of recognition, and useful size variation over the last several thousand years, the lichen *Rh. geographicum* has been most commonly used in lichenometrical studies (Fig. 4.14; Locke *et al.* 1979).

Once a growth curve for the species in question has been established, measurements of maximum lichen sizes on moraines and other geomorphological features can be used to estimate substrate age (Fig. 4.13).

4.3.1.2 Problems of lichenometry. There are innumerable problems in lichenometry; indeed they are so prevalent that two leading proponents of the method were forced to admit that "the biological and ecological foundations of lichenometry are so tenuous as to almost preclude its use" (Webber & Andrews 1973)! Nevertheless, the technique has such great attractions for dating events which would otherwise be undateable that the potential difficulties are not considered to be so great as to invalidate

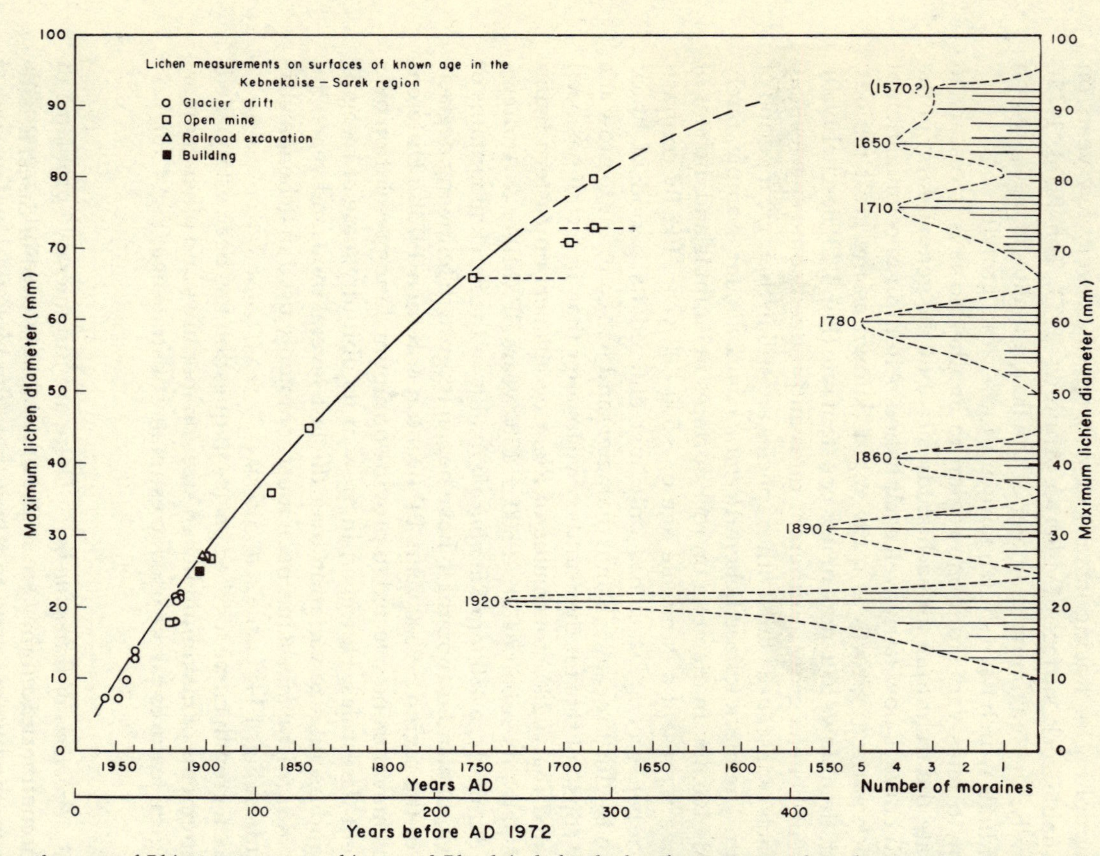

Figure 4.13 Growth curve of *Rhizocarpon geographicum* and *Rh. alpicola* for the last four centuries based on measurements in the Kebnekaise and Sarek mountains, Swedish Lappland. Bar graph at right shows frequency of moraines as characterized by the appropriate maximum thallus diameters. Several discrete moraine groups are evident (after Denton & Karlén 1973b).

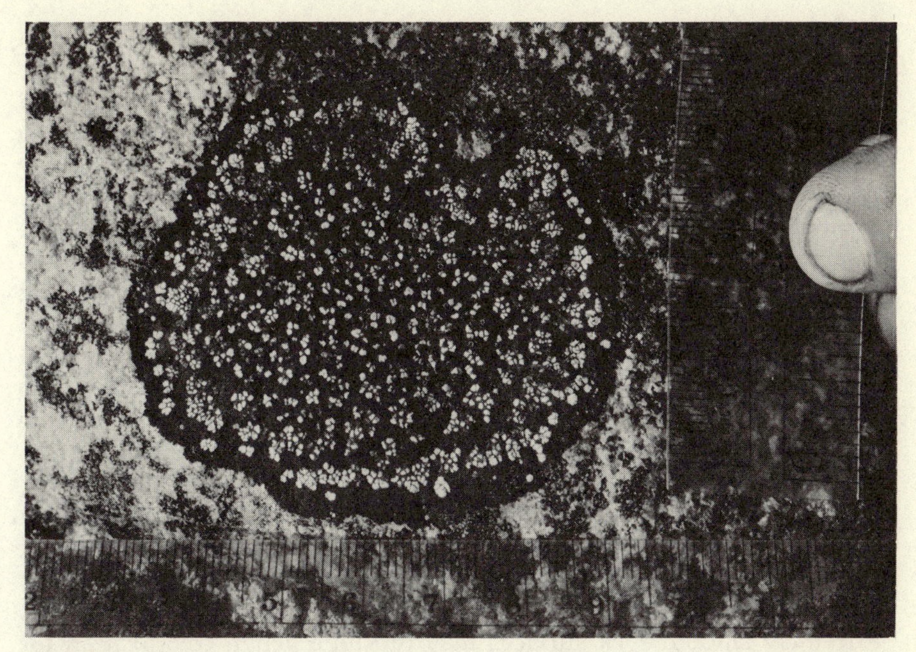

Figure 4.14 *Rhizocarpon geographicum*, a lichen which grows radially, enabling substrates to be dated if growth rate of lichen is known (photograph kindly provided by G. H. Miller).

the method. There are three general areas of uncertainty, relating to biological, environmental, and sampling factors (Jochimsen 1973).

BIOLOGICAL FACTORS. Lichens are exceedingly difficult to identify to species level in the field and most users of lichenometry have no training in lichen taxonomy. Indeed, lichen taxonomy is itself a contentious subject, which compounds the users' difficulties. *Rhizocarpon geographicum* is exceedingly similar to *Rh. superficiale* and *Rh. alpicola* (King & Lehmann 1973), and doubtless many investigations have been based on a mixture of observations (Denton & Karlén 1973b). This presents no problem, of course, providing that the different species grow at similar rates, but generally such factors are not well known. What evidence there is suggests that growth rates may vary between species (e.g. Calkin & Ellis 1980, Innes 1982). Lichen dispersal and propagation rates are also inadequately understood. Many lichens propagate their algal and fungal cells independently so that it may take some time for two individuals to find each other and form a new symbiotic union. In other cases, lichens are propagated when part of the parent breaks off the rock substrate and is blown or washed away to a new site. In either case, there may be a significant delay between the exposure of a fresh rock surface and colonization by lichens. Furthermore, even when lichen cells become established, decades may elapse before the thallus becomes visible to the naked eye. As time passes, rock surfaces may become virtually covered in lichens and inevitably this results in competition between individuals; indeed some lichens appear to secrete a chemical which inhibits growth in their immediate vicinity (Ten Brink 1973). Such factors seem likely to reduce growth rates as rocks become heavily lichen-covered and this may give the erroneous impression of a relatively young age for the substrate.

Finally, as lichens become very old, growth rates may decline. Little information is available on senescence in lichens and unfortunately this corresponds to the part of the growth curve where there is the least dating control. Often growth rates beyond a certain age (i.e. the final dated control point) are assumed to continue at a constant rate, whereas in all probability the rate declines with increasing lichen age. This will lead to (possibly large) underestimates of substrate age; such errors can be avoided if extrapolation of growth rates is not attempted.

ENVIRONMENTAL FACTORS. Lichen growth is dependent on substrate type (particularly surface texture) and chemical composition (e.g. Porter 1981c). Rocks which weather easily, or are friable, may not remain stable long enough for a slow-growing lichen to reach maturity. Conversely, extremely smooth rock surfaces may preclude lichen colonization for centuries and possibly many never support lichens. Extremely calcareous rocks may also inhibit growth of certain lichens. Measurements should thus be restricted to lichens growing on similar lithologies whenever possible.

Climate is a major factor affecting lichen growth rates; comparison of growth rates from different areas suggests that slower growth rates are found in areas of low temperature, short growing seasons, and low precipitation (Fig. 4.15). However, both macro- and microclimatic factors are of significance. In particular, lichens require moisture for growth and the frequency of small precipitation amounts, even from fog and dew, may be of more significance than annual precipitation totals. Radiation receipts are also important because they largely determine rock temperatures. Generally it is impossible to equate such factors on those rocks used to calibrate the lichen growth curve with rocks which are eventually to be dated. Commonly, calibration will be carried out on buildings or gravestones in a valley bottom, whereas the features to be dated are hundreds of meters higher than the calibration site. Similar problems may be encountered along extensive fjord systems where conditions at the fjord mouth are less continental than at the fjord head (Fig. 4.16). Lichen growth is far slower in the more continental locations, even over distances as short as 50 km, probably due to the lower frequency of coastal fogs and generally drier climates inland. Increasing elevation also appears to be significant in reducing growth rates, even though moisture availability might be expected to increase (Miller 1973, Porter 1981d).

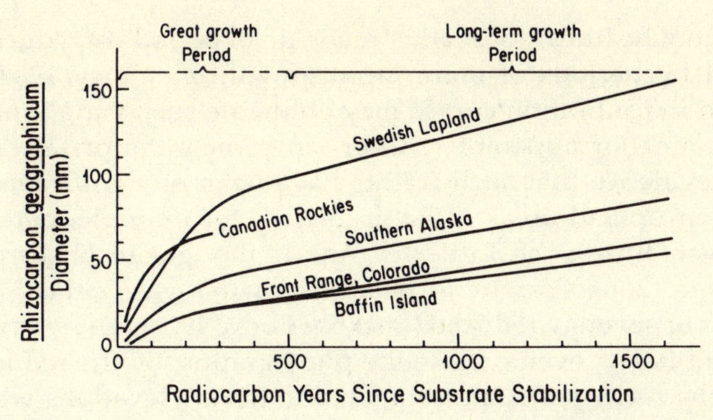

Figure 4.15 Growth rates of *Rhizocarpon geographicum* in different areas of the world (after Calkin & Ellis 1980).

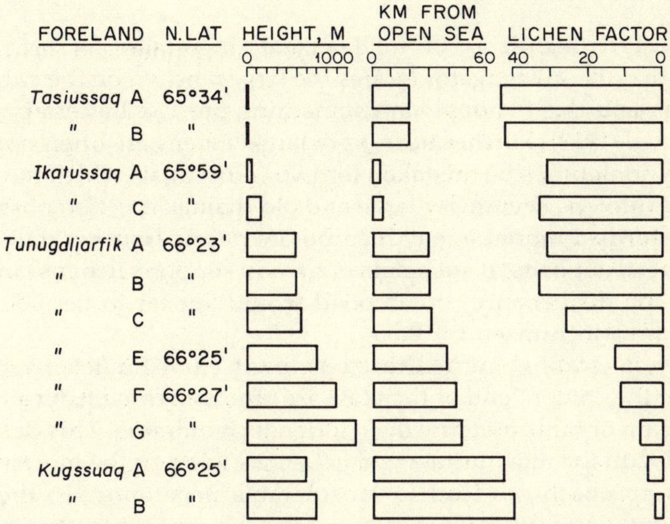

FORELAND		N.LAT	HEIGHT, M (0–1000)	KM FROM OPEN SEA (0–60)	LICHEN FACTOR (40–20–0)
Tasiussaq	A	65°34'			
"	B	"			
Ikatussaq	A	65°59'			
"	C	"			
Tunugdliarfik	A	66°23'			
"	B	"			
"	C	"			
"	E	66°25'			
"	F	66°27'			
"	G	"			
Kugssuaq	A	66°25'			
"	B	"			

Figure 4.16 Maximum size of century-old thalli of *Rhizocarpon tinei* in millimeters ("lichen factor") in relation to continentality in glacier forelands of West Greenland. Lichen growth is reduced with elevation and distance from open sea. Latitudinal differences are insignificant (after Beschel 1961).

Presumably, this is offset by longer-lying snow and reduced growing season due to lower temperatures (Flock 1978). All these factors may complicate the construction of a simple growth curve for a limited geographical area. Further problems arise due to the possible influence of long-term climatic fluctuations. Apart from the general effect of lower temperatures in the past, it is quite probable that in the high elevation and/or high latitude sites where lichenometry is most widely used, periods of cooler climate resulted in the persistence of snow banks which would have reduced lichen growth rates (Koerner 1980). Growth curves may thus not be linear, but rather made up of periods of reduced growth separated by periods of more rapid growth (cf. Curry 1969). Lack of resolution in calibration curves may obliterate such variations, but this could account for apparent "scatter" in some attempts at calibration. There is evidence that such factors have been of significance in some regions; on upland areas of Baffin Island, for example, persistence of snow cover during the Little Ice Age is thought to have resulted in "lichen-free zones", where lichen growth was either prevented altogether or severely reduced (Locke & Locke 1977). These zones can be recognized today, even on satellite photographs, by the reduced lichen cover of the rocky substrate compared to lower elevations where snow cover was only seasonal (Andrews *et al.* 1976). Similarly, attempts to date moraines which have periodically been covered by snow for long intervals would give erroneously young ages for the deposits (Karlén 1979).

SAMPLING FACTORS. It is of fundamental importance in lichenometric studies that the investigator locates the largest lichen on the substrate in question, and this is not always something one can be certain of doing (Locke *et al.* 1979). Furthermore, very large lichens are often not circular and may sometimes be mistaken for two individuals which have grown together into one seemingly large and old thallus. It is also possible that a newly formed moraine may incorporate debris from rockfalls or from older glacial deposits; if such debris already supports lichens, and if they survive the disturbance, the deposit would appear to be older than it actually was (Jochimsen 1973).

Finally, in establishing a calibrated growth curve for lichens, reference points at the "older" end of the scale are often obtained from a radiocarbon date on organic material overridden by a moraine. This date is then equated with the maximum-sized lichen growing on the moraine today. Such an approach can lead to considerable uncertainty in the growth curve. First, dates on organic material in soils overridden by ice may be very difficult to interpret (e.g. Matthews 1980). Secondly, there may be a gap of several hundred years between the time organic material is overridden by a glacial advance and the time the morainic debris becomes sufficiently stable for lichen growth to take place. This would lead to overestimation of lichen age in a calibration curve.

A consideration of all these factors indicates that caution is needed in using lichenometry as a dating method, even for relative age dating. Nevertheless, if consideration is given to the possible pitfalls, it can provide useful age estimates. Most problems would result in only minimum-age estimates on substrate stability, but, in some cases, overestimation of age could result. Although it is worth being aware of the potential difficulties of the method, it is unreasonable to expect that all of the problems, discussed above, would subvert the basic assumptions of the method all of the time, and often the potential errors can be eliminated in various ways. Lichenometry is thus likely to continue to play a role in dating rocky deposits in Arctic and alpine areas, and hence make an important contribution to paleoclimatic studies.

4.3.2 *Dendrochronology*

Although not a widely used dating technique in paleoclimatology, the use of tree rings for dating environmental changes has proved useful in some cases (e.g. Sigafoos & Hendricks 1961). The concepts and methods used in dendrochronological studies are discussed in more detail in Chapter 5 and need not be repeated here. Basically, dendrochronological studies are used in two ways: (a) to provide a minimum date for the substrate on which the tree is growing (e.g. an avalanche track or deglaciated surface) and (b) to date an event which disrupted tree growth but did not terminate it. The former application is straightforward but to obtain a *close* minimum date assumes that the "new" surface is colonized very rapidly. This is highly probable in the case of avalanche zones (indeed, young saplings may survive the event) but in deglaciated areas surface instability due to subsurface ice melting and inadequate soil structure may delay colonization for several decades. It is also worth noting that, unlike lichens, which may live for thousands of years, trees found on moraines are rarely more than a few hundred years old. Even when very old trees are located and dated, it can not be assumed that they represent first generation growth. For example, Burbank (1981) found that a moraine on which the oldest tree was dated as being ~750 years old, by dendrochronological methods, could be dated tephrochronologically as older than 2500 years BP.

A more widespread application of dendrochronology involves the study of growth disturbance in trees. When trees are tilted during their development they respond by producing compression or reaction wood on the lower side of the tree in order to restore its natural stance. This causes rings to form eccentrically after the event which tilted the tree; the event can be accurately dated by identifying the year when growth changes from concentric to eccentric (Lawrence 1950, Burrows & Burrows 1976). Such techniques have been used in dating the former occurrence of avalanches (Potter 1969, Carrara 1979) and hurricanes (Pillow 1931) and the timing of glacier recession (Lawrence 1950). They have also been successfully applied in more strictly geomorphological applications, in studying stream erosion rates and soil movements on permafrost (Shroder 1980).

5

Ice cores

5.1 Introduction

The accumulation of past snowfall in the polar ice caps and ice sheets of the world provides an extraordinarily valuable record of paleoclimatic and paleo-environmental conditions. These conditions are studied by detailed physical and chemical analyses of ice and firn (snow which has survived the summer ablation season†) in cores recovered from very high elevations on the ice surface. In such locations (known as the dry snow zone; Benson 1961) snow melt and sublimation are essentially zero, so that snow accumulation has been continuous for a very long period of time, perhaps as much as several hundred thousand years (Dansgaard *et al.* 1973). The snowfall provides a unique record, not only of precipitation amounts *per se*, but also of air temperature, atmospheric composition (including gaseous composition and particulate content), the occurrence of explosive volcanic eruptions, and even of past variations in solar activity (Table 5.1). At present, over 80 cores of more than 100 m in length have been recovered from ice sheets, ice shelves, and glaciers in both hemispheres; these are listed in Table 5.2 and are shown in Figures 5.1 and 5.2. In a number of cases, cores have been recovered which penetrated right through the ice body being drilled and a few contain debris from the ice/sub-ice interface (e.g. S. Herron & Langway 1979, Gow *et al.* 1979, Koerner & Fisher 1979).

Paleoclimatic information has been obtained from ice cores by three

†The metamorphism of snow crystals to firn, and eventually to ice, occurs as the weight of overlying material causes crystals to settle, deform, and recrystallize, leading to an overall increase in unit density. When firn is buried beneath subsequent snow accumulations, density increases as air spaces between the crystals are reduced by mechanical packing and plastic deformation until, at a unit density of about 0.83 kg m^{-3}, interconnected air passages between grains are sealed off into individual air bubbles (M. M. Herron & Langway 1980). At this point, the resulting material is considered to be ice. The depth of this transition varies considerably from one ice body to another, depending on surface temperature and accumulation rate; for example, it does not occur until about 68 m depth at Camp Century, Greenland, and ~100 m at Vostok, Antarctica; thus "ice cores" *sensu stricto* are actually firn cores near the surface (cf. Table 2.2 in Paterson 1981). For our purposes, this distinction is not very important and the term ice core will henceforth be used to refer to both ice and firn core sections.

Table 5.1 Potential paleoclimatic information from ice cores.

	Analysis
Paleotemperatures	
Summer	Melt layers
Annual?	$\delta^{18}O$, crystal size
Interhemispheric difference, leads and lags	$\delta^{18}O$
Paleoaccumulation (net)	Seasonal signals
Ice sheet stability/flowlines	Gas content
Volcanic activity (intensity and frequency)	Acidity; trace elements
Tropospheric turbidity	Microparticle content, trace elements
Atmospheric composition: long-term and man-made changes	CO_2 content, trace elements
Solar activity/geomagnetic field strength	NO_3^- content

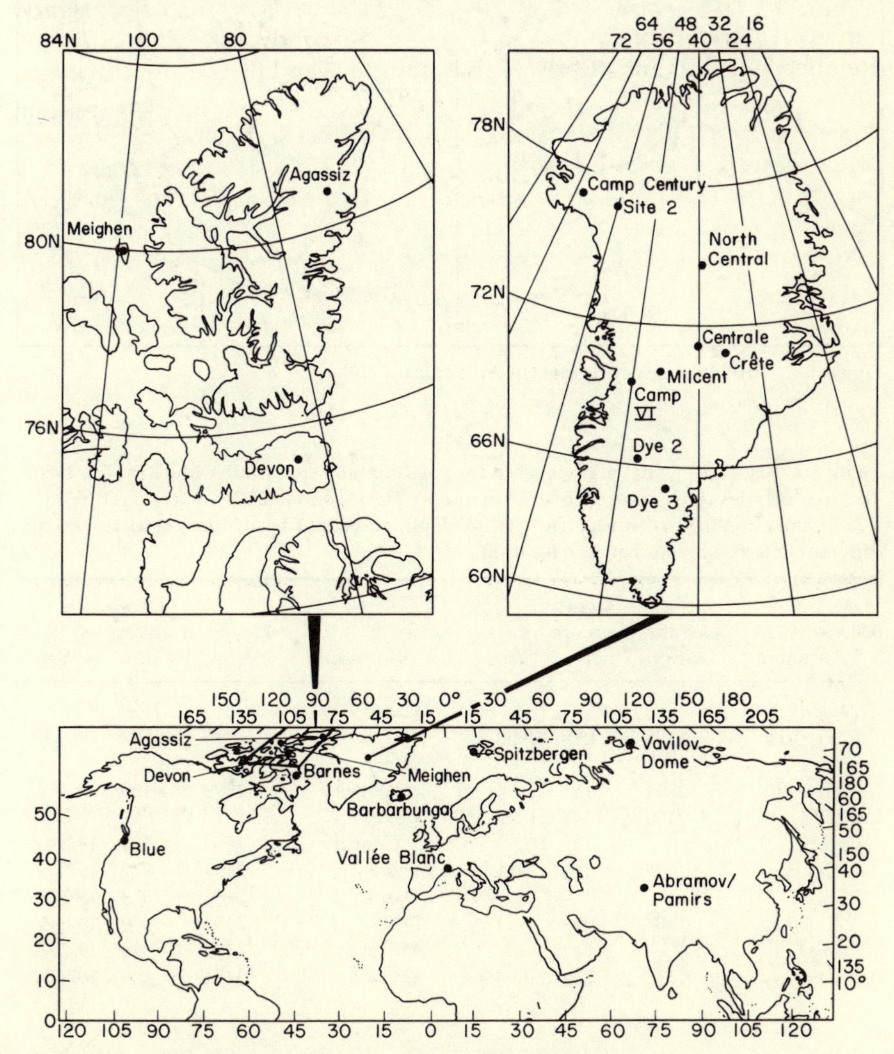

Figure 5.1 Principal ice coring sites of the Northern Hemisphere.

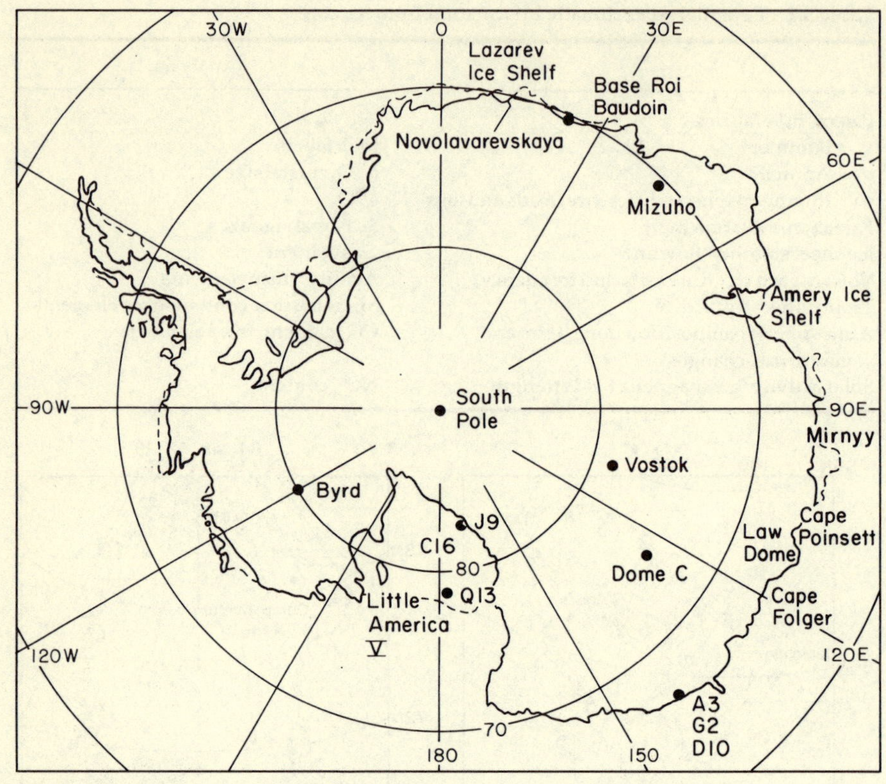

Figure 5.2 Principal ice coring sites of Antarctica.

Table 5.2 Ice cores (>100 m) recovered for paleoclimatic study (after MacKinnon, 1980). This list includes those cores recovered in (more or less) complete sections and returned to a laboratory for further study. The 100 m depth threshold is arbitrary and represents different lengths of record at each location.

0000/99 Site name	Elevation (m)	Depth of core (m)	Year	Site name	Elevation (m)	Depth of core (m)	Year
Greenland				Camp Century	1885	1387	1963/6
Camp VI	1598	125	1950	Camp Century	1890	100	1977
Station Centrale	2994	151	1950	Camp Century	1890	101	1977
Site 2	2100	305	1957	Dye 3	2480	372	1971
Camp Century	1885	186	1961	Dye 3	2480	378	1975
Camp Century	1885	238	1962	Dye 3	2480	220	1978/9
Camp Century	1885	264	1963	Milcent	2450	398	1973
				Crete	3174	405	1974

Table 5.2 continued

Site name	Elevation (m)	Depth of core (m)	Year
Dye 2	2200	101	1974
North Central 1	2941	100	1977
North Central 2	2941	109	1977
North Central	2941	102	1977
Canada			
Meighen Ice Cap	260	121	1965
Devon 71	1800	230	1971
Devon 72	1800	299	1972
Devon 73	1800	299	1973
Barnes Ice Cap T061	745	223	1976
Barnes Ice Cap T081	643	160	1977
Barnes Ice Cap	862	300	1978
Mer de Glace Agassiz	1820	337	1977
Mer de Glace Agassiz	1850	149	1979
USA			
Blue Glacier	1325	137	1962
Mount Logan	5300	103	1980
Iceland			
Bardarbunga	2000	104	1969
Bardarbunga	2000	415	1972
Spitzbergen			
Ice Divide, Grenfiord and Fritof Glaciers	450	211	1975
West Spitzbergen		211	1976
Lomonosov Glacier		210	1977
Alps			
Vallée Blanc	3500	187	
Monte Rosa	4450	124	1982
USSR			
Abramov Glacier, Pamirs	~4400	110	1972
Abramov Glacier, Pamirs	~4400	106	1974
Pamirs	~3000	137	1974
Vavilov Dome, Severnaya Zemlya	~780	450	1978
Severnaya Zemlya	~780	459	1978
Antarctica			
Amery Ice Shelf G-1	50	310	1968
Law Dome (Cape Folger) SGA	375	320	1969
Law Dome, Summit, SGD	1390	385	1969
Law Dome, Cape Poinsett, SGJ	400	112	1972
Law Dome	600	113	1972
Law Dome	375	348	1974
Law Dome, BHQ	940	419	1977
Law Dome Summit, BHQ	1390	475	1977
Base Roi Baudoin	39	116	1961
A3	220	106	1964/6
G2	112	106	1964/6
D-10	270	304	1974
Dome C	3240	905	1977/8
Dome C	3240	180	1978/9
Mizuho	2230	148	1972
Mirnyy	200	371	1956/8
Mirnyy	1000	250	1969
Vostok	3500	952	1970–3
Vostok	3500	905	1974
Vostok 1	2940	105	1974/5
Vostok	2940	430	1977–9
Vostok	3500	104	1979/80
Vostok	3500	100	1979/80
Vostok	3500	2083	1983

Table 5.2 continued

Site name	Elevation (m)	Depth of core (m)	Year	Site name	Elevation (m)	Depth of core (m)	Year
Novolaz-arevskaya	500	330	1974/5	J-9	60	416	1978
				J-9	60	100	1974
Novalaz-arevskaya	500	810	1976/7	J-9	60	152	1976
				J-9	60	330	1976
Lazarev Ice Shelf	40	357	1975	J-9	60	171	1977
				Q-13	60	100	1977
Lazarev Ice Shelf	40	447	1975	C-16	60	100	1977
				South Pole	2912	100	1974
Little America V	42	256	1958	South Pole	2912	112	1978
				South Pole	2912	237	1980/2
Byrd	1524	308	1958	South Pole	2912	202	1981/2
Byrd	1524	227	1965/7				
Byrd	1524	335	1967/8	*Peru*			
Byrd	1524	2164	1968	Quelccaya Ice Cap	5500	162	1983
Byrd	1524	354	1969/70				
Byrd	1524	366	1971/2				

main approaches. These involve the analysis of (a) stable isotopes of water; (b) dissolved and particulate matter in firn and ice; (c) the physical characteristics of the firn and ice, and of air bubbles contained in the ice. Each approach has also provided a means of estimating the age of ice at depth in ice cores (Sec. 5.3).

5.2 Stable isotope analysis

The study of stable isotopes (particularly deuterium and ^{18}O) has become a major focus of paleoclimatic research in recent years. Currently, most work has been on stable isotope variations in ice and firn, and in the tests of marine fauna recovered from ocean cores. However, increasing attention is being placed on other natural isotope recorders, such as speleothems (stalactites and stalagmites), tree rings, and peat (Gray 1981). In this section a brief introduction to the theory behind stable isotope work is provided and applications to ice-core analysis are discussed. The importance of stable isotopes in other branches of paleoclimatic research are dealt with in Chapters 6, 7, and 10.

Water is the most abundant compound on Earth. It is the primary compound in all forms of life, is perhaps the most important agent in weathering, erosion, and geological recycling of materials, and, of course, plays a crucial role in the global energy balance. The study of "fossil water," either directly, in the form of firn and ice, or indirectly through materials deposited from solution in "fossil water" (e.g. speleothems),

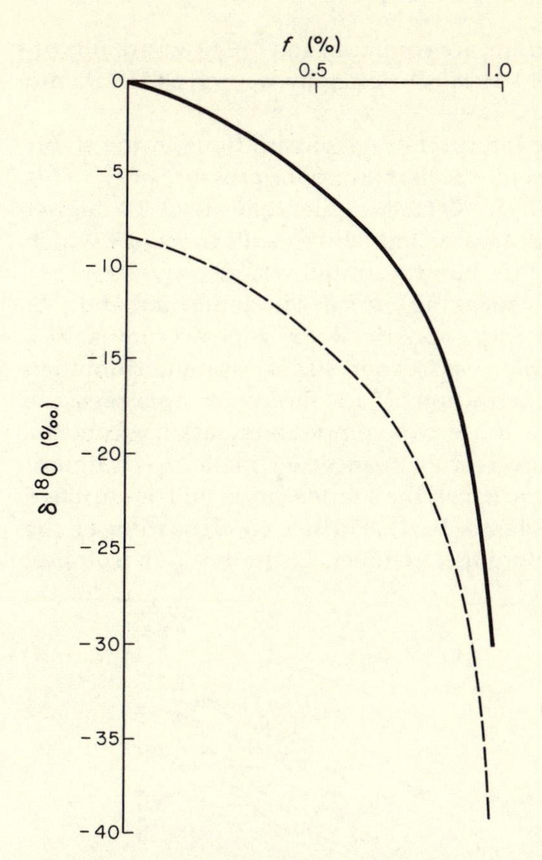

Figure 5.3 Diagrammatic plot of the $^{18}O/^{16}O$ ratio ($\delta^{18}O$) in a water (liquid–vapor) system in which the liquid is removed as it is formed by progressive condensation. Isotopic equilibrium between vapor and liquid is assumed (i.e. an equilibrium Rayleigh condensation process). ----, δ_{vapor}, ———, δ_{liquid}. f = percentage of the original water vapor condensed (from Epstein & Sharp, *J. Geol.* **67**, © 1959 by the University of Chicago.

thus has important implications in many aspects of paleo-environmental reconstruction.

In common with most other naturally occurring elements, the constituents of water, oxygen and hydrogen, may exist in the form of different isotopes. Isotopes result from variations in mass of the atom in each element. Every atomic nucleus is made up of protons and neutrons. The number of protons in the nucleus of an element (the atomic number) is always the same, but the number of neutrons may vary, resulting in different isotopes of the same element. Thus, oxygen atoms (which always have eight protons), may have eight, nine, or ten neutrons, resulting in three isotopes with atomic mass numbers of 16, 17, and 18 respectively (^{16}O, ^{17}O, and ^{18}O). In nature these three stable isotopes occur in relative proportions of 99.76% (^{16}O), 0.04% (^{17}O), and 0.2% (^{18}O). Hydrogen has two stable isotopes, ^{1}H and ^{2}H (deuterium), with relative proportions of 99.984% and 0.016% respectively. Consequently, water molecules may exist as any one of nine possible isotopic combinations with mass numbers ranging from 18 ($^{1}H_2\,^{16}O$) to 22 ($^{2}H_2\,^{18}O$). However, as water with more than one "heavy" isotope is very rare, generally only four major isotopic combinations are common, and only two are important in paleoclimatic research ($^1H^2H^{16}O$, generally written as HDO, and $^1H_2{}^{18}O$).

The basis for paleoclimatic interpretations of variations in the stable isotope content of water molecules is that the vapor pressure of $H_2{}^{16}O$ is higher than that of $HD^{16}O$ and $H_2{}^{18}O$ (10% higher than HDO, 1% higher than $H_2{}^{18}O$). Evaporation from a water body thus results in a vapor which is poorer in deuterium and ^{18}O than the initial water; conversely, the remaining water is (relatively speaking) *enriched* in deuterium and ^{18}O. At equilibrium, for example, atmospheric water vapor contains 10‰ (parts per thousand or per mille) less ^{18}O and 100‰ less deuterium than mean ocean water. When condensation occurs, the lower vapor pressure of HDO and $H_2{}^{18}O$ results in these two compounds passing from the vapor to the liquid state more readily than water made up of lighter isotopes. Hence, *compared to the vapor*, the condensation will be enriched in the heavy isotopes (Dansgaard 1961). Further condensation of the vapor will continue this preferential removal of the heavier isotopes,

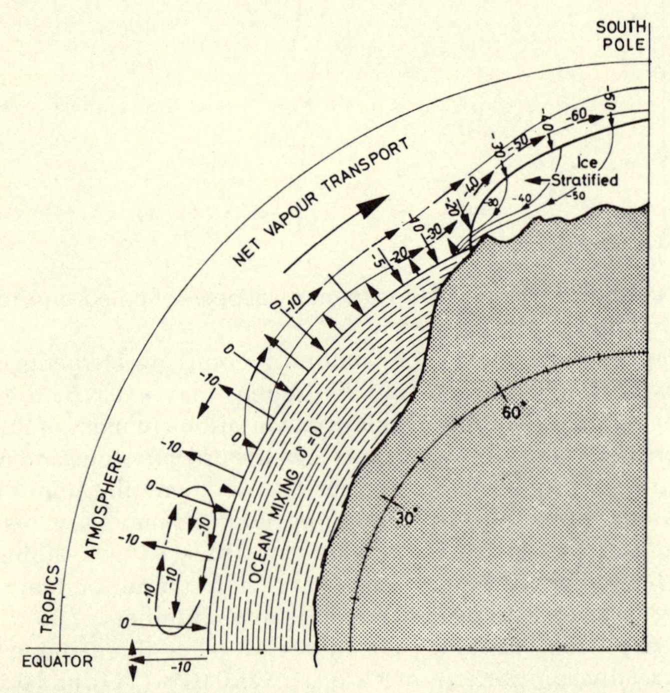

Figure 5.4 Schematic diagram to illustrate isotopic depletion of water vapor en route to Greenland ice sheet. As air mass cools, precipitation produced is preferentially enriched in ^{18}O, leaving the remaining vapor relatively depleted. Consequently, with further condensation, the precipitation contains less and less ^{18}O (i.e. lower $\delta^{18}O$ values). This isobaric effect is accentuated by uplift (adiabatic) effects over the ice sheet itself, so that lowest $\delta^{18}O$ values are found in the ice-sheet interior (after Dansgaard *et al.* 1971, Robin 1977).

leaving the vapor more and more depleted in HDO and $H_2{}^{18}O$ (Fig. 5.3). As a result, continued cooling will give rise to condensate with increasingly lower HDO and $H_2{}^{18}O$ concentrations than when the condensation process first began. The greater the fall in temperature, the more condensation will occur and the lower will be the heavy isotope concentration, relative to the original water source (Fig. 5.4). Isotopic concentration in the condensate can thus be considered as a function of the temperature at which condensation occurs (subject to certain reservations noted below in Sec. 5.2.2).

5.2.1 *Stable isotopes in water: measurement and standardization*

In the majority of paleoclimatic studies using stable isotopes, oxygen is generally the element of primary interest. Isotopes of hydrogen are less abundant in nature and are more difficult to analyze in the laboratory. In oxygen isotope work, the water sample is isotopically exchanged with carbon dioxide of known isotopic composition:

$$^{1}H_2{}^{18}O + {}^{12}C^{16}O_2 \rightleftharpoons {}^{1}H_2{}^{16}O + {}^{12}C^{16}O^{18}O$$

The relative proportions of ^{16}O and ^{18}O in carbon dioxide from the sample are then compared with the isotopic composition of a water standard (Standard Mean Ocean Water or SMOW†) and the results expressed as a departure ($\delta^{18}O$) from this standard, thus

$$\delta^{18}O = \frac{(^{18}O/^{16}O)_{sample} - (^{18}O/^{16}O)_{SMOW}}{(^{18}O/^{16}O)_{SMOW}} \times 10^3‰.$$

All measurements are made using a mass spectrometer and reproducibility of results within $\pm 0.1‰$ is generally possible.

A $\delta^{18}O$ value of -10 therefore indicates a sample with an $^{18}O/^{16}O$ ratio 1% or 10‰ less than SMOW. Under our present climate, the lowest $\delta^{18}O$ value ever recorded in natural waters is $-57‰$, in snow from the highest and most remote parts of Antarctica (Morgan, 1982).

5.2.2 *Oxygen-18 concentration in atmospheric precipitation*

In Section 5.2, and Figure 5.3, the isotopic composition of water in equilibrium with water vapor was considered. In reality, we cannot

†In order that isotopic analyses in different laboratories be comparable, a universally accepted standard is used, known as SMOW (Standard Mean Ocean Water; Craig 1961b). This is not an actual oceanic water sample, but is based on a US National Bureau of Standards distilled water sample (NBS-1). However, the zero point on the SMOW scale has been adjusted so that it is more or less equivalent ($-0.1‰$) to the isotopic composition of real ocean water (measured in samples from depths of 500–200 m in the Atlantic, Pacific, and Indian Oceans; Epstein & Mayeda 1953). Isotopic studies based on carbonate fossils use as a standard a Cretaceous belemnite from the Peedee Formation of North Carolina (PDB-1). Carbon dioxide released from PDB-1 $\simeq +0.2‰$ relative to CO_2 equilibrated with SMOW (Craig 1961b).

consider the process to be always at equilibrium between vapor and condensate, nor can the process be considered to occur in isolation. Exchanges between atmospheric water vapor, water droplets in the air, and water at the surface (which may be isotopically "light") do occur continuously, so this complicates any simple temperature – isotope effect which we might expect to find (e.g. Koerner & Russell 1979). Specifically, the ^{18}O content of precipitation depends on (a) the ^{18}O content of the water vapor at the start of condensation (this could be very low if evaporation occurred over an inland lake or ice body where ^{18}O concentrations are less than mean ocean water); (b) the amount of moisture in the air compared to its initial moisture content; (c) the degree to which water droplets undergo evaporation en route to the ground and whether any of this re-evaporated vapor re-enters the precipitating air mass (Ambach *et al.* 1968); (d) the temperature at which the evaporation and condensation processes take place. In spite of these complications, empirical studies have demonstrated that geographical and temporal variations in isotopes do occur, reflecting temperature effects due to changing latitude, altitude, distance from moisture source, season, and long-term climatic fluctuations (Dansgaard *et al.* 1973, Koerner & Russell 1979). As any interpretation of ice-core isotopic records is rooted in an evaluation of these factors, it is important to consider them each in turn.

5.2.3 Geographical factors affecting stable isotope concentrations

As already noted, one of the principle factors affecting $\delta^{18}O$ values is the temperature at which condensation takes place. Unfortunately, this is a difficult parameter to measure as the elevation at which condensation occurs in a cloud is not known and conditions within clouds are constantly changing due to turbulence and latent heat release. Early work by Dansgaard (1953) and Epstein (1956) demonstrated isotopic changes in precipitation falling during the passage of warm and cold fronts but the first attempt to relate upper air temperatures to the isotopic composition of precipitation directly was that of Picciotto *et al.* (1960) (Fig. 5.5). They found a relationship between the mean temperature within precipitating clouds at King Baudoin Station, Antarctica (Fig. 5.2), and the isotopic composition of snowfall at the surface, thus

$$\delta^{18}O = 0.9T + 6.4,$$

where T is in the range $+5\,°C$ to $-30\,°C$.

Their results appeared to be independent of cloud height or precipitation type, but correlations were poor with surface temperature because of strong temperature inversions over the ice surface. Subsequently, Aldaz and Deutsch (1967) conducted a similar study at the South Pole, in which isotopes in snow samples collected during the course of a year were compared with temperatures at the surface and up to 500 mb. They found

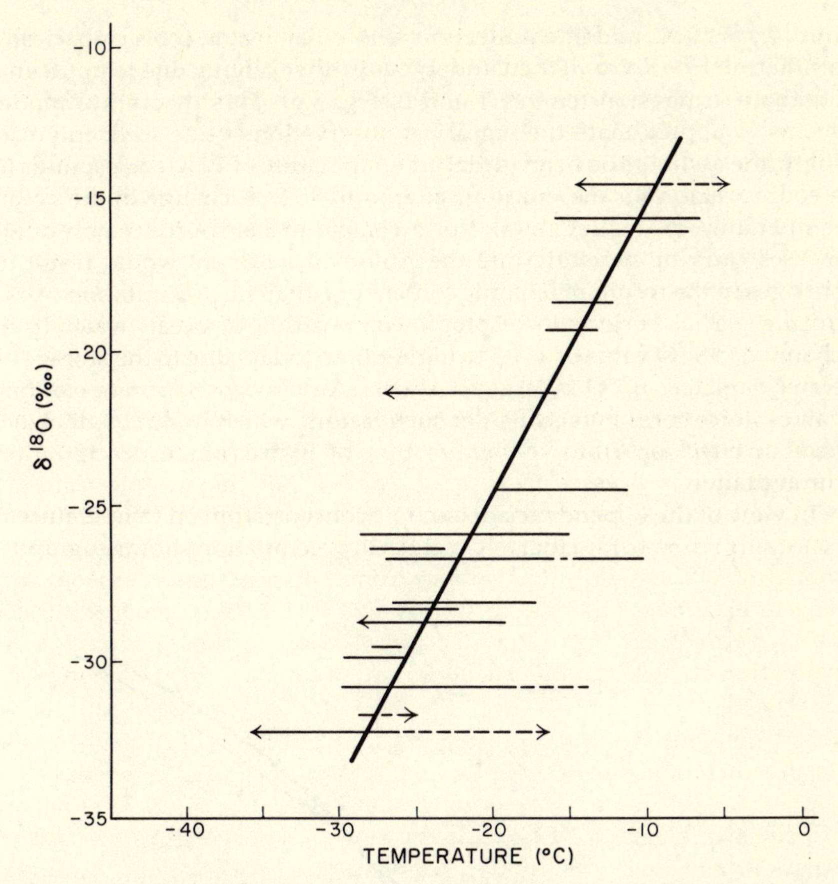

Figure 5.5 Isotopic composition ($\delta^{18}O$) in snowfall compared to the corresponding temperature in the precipitating cloud (after Picciotto *et al.* 1960).

a relationship between $\delta^{18}O$ values and "the locus of all the effective condensation level temperatures" (t) such that

$$\delta^{18}O = 1.4t + 4.0,$$

where t is in the range $-25\,°C$ to $-50\,°C$.

Both this regression and that of Picciotto *et al.* (1960) compare favorably with theoretical $\delta^{18}O$–temperature relationships (Fig. 5.6) calculated assuming a common moisture source (of relatively constant temperature throughout the year) and undergoing equilibrium Rayleigh condensation† en route to the South Pole (Aldaz & Deutsch 1967). With these assumptions, the best fit to the empirical data occurs when moist air,

†Slow condensation with immediate removal of the condensate from the vapor after formation.

initially at 10 °C, and in equilibrium with ocean water, cools isobarically at 1000 mb to −5 °C and then undergoes further cooling due to uplift and adiabatic expansion (curves 1 and 2, Fig. 5.6). This theoretical model seems to approximate the empirical observations quite well; interestingly, the assumption of an initial air temperature of 10 °C corresponds to a source region for the moist air at around 45°S. A change in the ocean temperature of this region and/or a change in the source region of air masses carrying moisture into the Antarctic continent would result in changes in the resultant isotopic content of Antarctic precipitation. Also, during glacial periods the isotopic composition of ocean water itself changed (a $\delta^{18}O$ value of 1–1.5‰ higher than today) due to the storage of water depleted in ^{18}O in large ice sheets. Any interpretation of isotopic values in ice cores must consider such factors, which undoubtedly have had an effect on isotopic composition of high-latitude precipitation through time.

In view of the dependency of isotopic concentration on temperature it is not surprising to find that $\delta^{18}O$ values in precipitation show geographi-

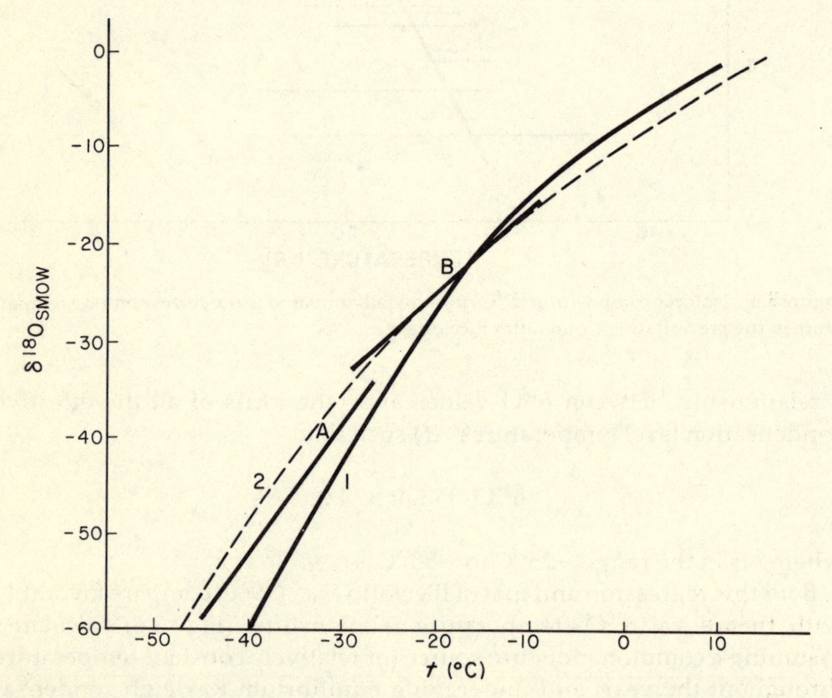

Figure 5.6 Relationship between oxygen isotope ratio ($\delta^{18}O$) and temperature of condensation level in samples of Antarctic precipitation. Curves A and B are based on empirical observations at King Baudoin base and Amundsen–Scott Station. Curves 1 and 2 are theoretical using different assumptions about the fractionation of ^{18}O at very low temperatures (after Aldaz & Deutsch 1967).

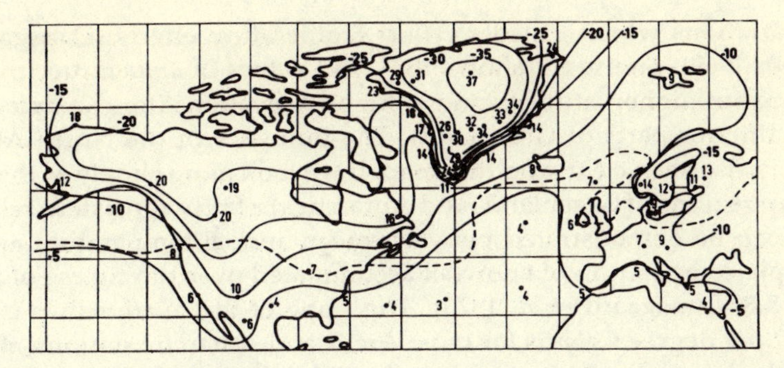

Figure 5.7 Geographical pattern of $\delta^{18}O$ values of annual precipitation, based on sample periods of 2–10 years. There is a general latitudinal decline in $\delta^{18}O$ values which is accentuated by higher elevations over the inland regions of Greenland. The influence of the Gulf Stream is clearly seen. The dashed line represents the maximum extent of the Wisconsin continental ice and of permanent pack ice (after Dansgaard & Tauber 1969).

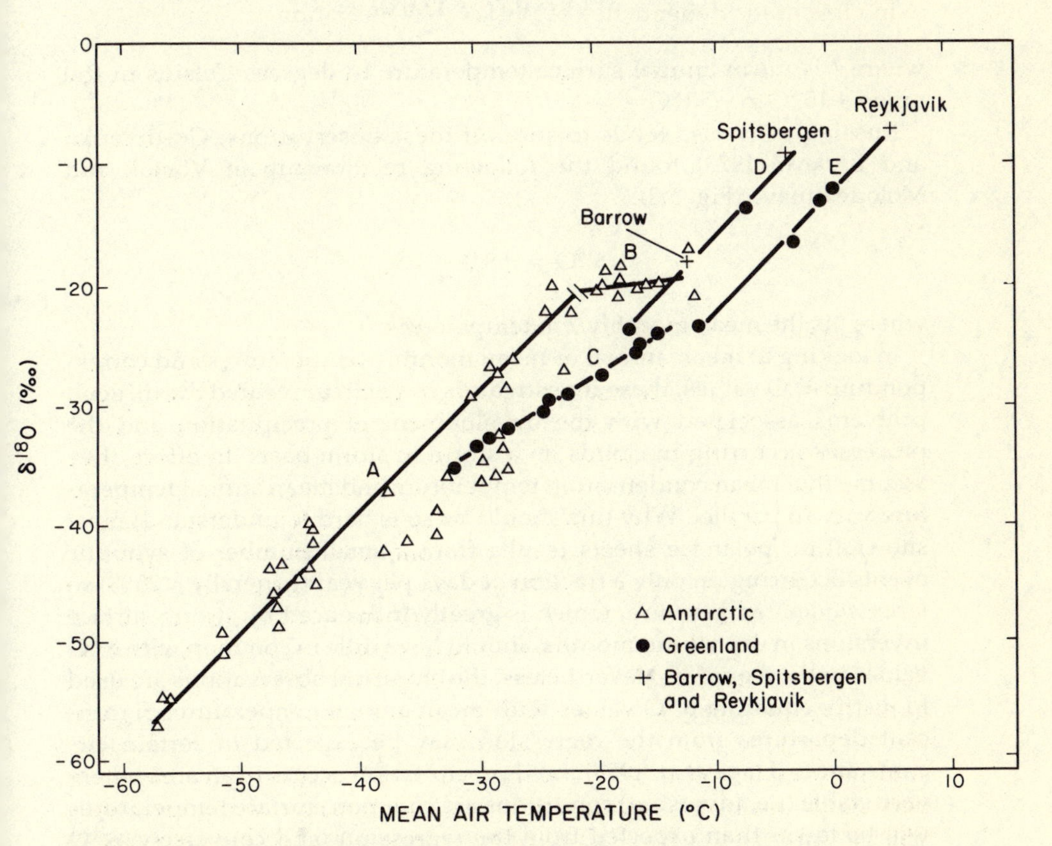

Figure 5.8 Relationship between $\delta^{18}O$ and mean annual temperatures, based on temperatures 10 or 12 m below the surface in firn. Antarctic data shown as triangles; Greenland and North Atlantic data shown as circles (after Dansgaard *et al.* 1973).

cal variations which broadly reflect temperature effects (Dansgaard & Tauber 1969). Figure 5.7 shows that $\delta^{18}O$ values of *annual* precipitation decrease at higher latitudes and at higher elevations (for example in the high interior parts of Greenland). The influence of the North Atlantic Drift is also seen clearly on this map. If we look more closely at the polar ice cap regions of Greenland (and Antarctica) a fairly consistent relationship can be demonstrated between mean annual temperature and the isotopic composition of snowfall accumulated over the course of a year (Fig. 5.8; Dansgaard *et al.* 1973). The slope of the relationship is ~1‰ ($\delta^{18}O$) per degree Celsius for most Arctic and Antarctic stations, though at high elevations of interior Greenland the slope is closer to ~0.6‰ per degree Celsius. These figures bracket Dansgaard's earlier estimates on the $\delta^{18}O$–temperature relationship, based on a geographically more extensive data set (Dansgaard 1964),

$$\delta^{18}O = 0.7t - 13.6‰,$$

where t is mean annual surface temperature in degrees Celsius in the range +15 °C to −50 °C.

Russian work also tends to support these observations; Gordiyenko and Barkov (1973) found the following relationship at Vostok and Molodezhnaya (Fig. 5.2):

$$\delta^{18}O = 0.84t - 7.5,$$

where t is the mean monthly 2 m temperature.

In looking at mean annual or mean monthly temperatures and corresponding $\delta^{18}O$ values, these investigators have circumvented the difficult problems associated with the development of precipitation and the processes occurring in clouds on a storm to storm basis. In effect, they assume that mean condensation temperature and mean annual temperature vary in parallel. Why this should be so is hard to understand; most snowfall on polar ice sheets results from a small number of synoptic events occurring on only a fraction of days per year (generally <25%) so *mean annual* temperature, which is greatly influenced by strong surface inversions in dry winter months, should have little in common with $\delta^{18}O$ values in the ice cores. Nevertheless, the empirical observations are used to justify equating $\delta^{18}O$ values with mean annual temperature. Significant departures from the regression may be expected in certain circumstances (Hage *et al.* 1975): (a) if precipitation occurs in an area where very stable (i.e. inversion) conditions are common, surface temperatures will be lower than expected from the regression (and conversely, $\delta^{18}O$ values will appear anomalously high); (b) if local precipitation is derived from water which was re-evaporated from a source with an already low

$\delta^{18}O$ content (e.g. freshwater lake or snow cover), the mean annual $\delta^{18}O$ value may fall below the regression line, perhaps by as much as 10‰ for complete re-evaporation of precipitated water) (cf. Koerner & Russell 1979).

Although these factors may be important in some areas, in general the $\delta^{18}O$ record is considered a proxy of temperature variations through time. However, temperature is dependent on many factors which in turn affect $\delta^{18}O$, so it is not possible to convert $\delta^{18}O$ variations directly to variations in temperature. The most important of these factors are considered below, before long-term changes in $\delta^{18}O$ are examined.

5.2.3.1 Latitudinal effects on $\delta^{18}O$. As Figure 5.7 indicates, there is obviously a latitudinal influence on $\delta^{18}O$. Lower $\delta^{18}O$ values are found as a result of the loss of heavy isotopes in water condensed en route to high latitudes. This is sometimes referred to as an isobaric effect, implying a systematic change brought about by overall cooling at a particular level in the atmosphere, rather than cooling brought about by a change in elevation (adiabatic cooling). Using data from western Greenland, Dansgaard (1961) estimated a decrease in $\delta^{18}O$ of ~0.93‰ per degree of latitude as one moves northward. However, this rate is not constant throughout the world, depending very much on the climatology of the place in question. $\delta^{18}O$ values probably decreased more rapidly with latitude during glacial periods, when Equator – Pole temperature gradients were considerably stronger (Wilson & Hendy 1971) but the magnitude of this difference is hard to assess.

5.2.3.2 Effects of elevation and distance from moisture source on $\delta^{18}O$. Some of the earliest work on $\delta^{18}O$ and δD demonstrated a decrease in the heavy isotope concentration of water with increasing elevation, due to adiabatic cooling of the precipitating air mass as it is lifted aloft. On large ice sheets, this effect is superimposed on a "distance from moisture source" factor which results in lower $\delta^{18}O$ concentrations at increasing distance from oceanic moisture sources (e.g. Koerner 1979; Fig. 5.9). Hence, at high elevations in central Antarctica, thousands of kilometers from the southern oceans, atmospheric precipitation has the lowest heavy isotope concentrations of any natural water occurring today (Morgan 1982). In Greenland, Dansgaard (1961) has estimated the combined inland – elevation effect as equivalent to a decrease in $\delta^{18}O$ of 0.62‰ per 100 m increase in elevation in a direction perpendicular to the coastline. Using this figure he was able to crudely estimate the altitude of deposition of ice in icebergs calved off the western coast of Greenland and, by dating the ice, then estimate the rate of ice flow in different glacier systems (Scholander *et al.* 1962). It is quite likely, however, that the altitude – inland effect at any one location has not been constant over time. During glacial periods,

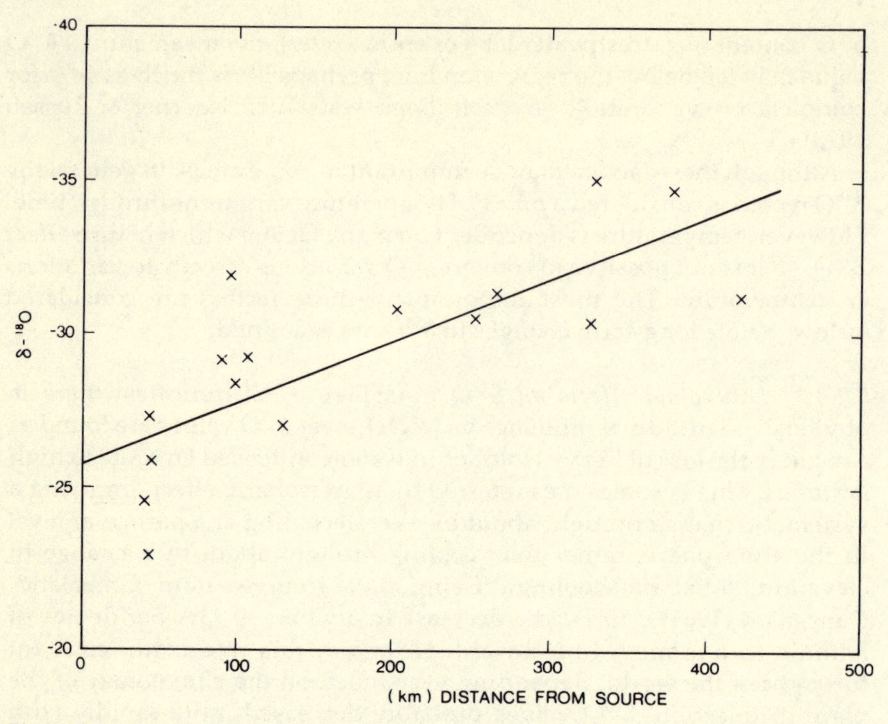

Figure 5.9 Relationship between $\delta^{18}O$ and the distance from moisture source in the eastern Queen Elizabeth Islands, North West Territories, Canada (based on snowfall from August 1973 to May – June 1974) (after Koerner 1979).

ice thicknesses gradually increased, resulting in lower $\delta^{18}O$ value at the surface because of the increase in elevation. Secondly, more extensive pack-ice during glacial periods would effectively increase distance to moisture sources, leading to even lower $\delta^{18}O$ values in isolated continental interiors (cf. Kato 1978). Finally, a fall in mean temperature would cause a relatively larger change in $\delta^{18}O$ because of the curvilinear nature of the relationship (Fig. 5.6).

5.3 Dating ice cores

One of the biggest problems in any ice-core study is determining the age – depth relationship. Many different approaches have been used (Table 5.3) and it is now clear that fairly accurate timescales can generally be developed for the last 10 000–12 000 years (see below). Prior to that, there is increasing uncertainty about ice age. For example, in a recent reassessment of the timescale originally proposed for the Camp Century, Greenland, ice core (Dansgaard *et al.* 1969, 1971) Dansgaard *et al.* (1982)

Table 5.3 Methods used in dating ice cores (after Hammer *et al.* 1978).

Method	Time range under favorable conditions (years)	Accuracy under favorable conditions (%)	Comments
(a) *Radio-isotopic*			
^{210}Pb	150	10	
^{32}Si and ^{39}Ar	1 000	10	10^3 kg ice needed
^{14}C (conventional)	25 000	5	$>10^4$ kg ice needed
^{14}C (accelerator techniques)	70 000	5	$<10^2$ kg ice needed
(b) *Seasonal changes*			
Trace substances (Na, K, Mg, Al)	15 000	1–5	Accumulation rate >0.25 m ice per year needed
$\delta^{18}O$	15 000	1–5	Accumulation rate >0.25 m ice per year needed
Microparticles	10 000	5–10	Range and accuracy depend strongly on location
(c) *Reference horizons*			
Tritium or gross β activity	20–30	~1	1 kg ice needed
Specific electrolytic conductivity	10 000	5	Correlation with eruption of known age
Volcanic ash or dust	1000+	5	Correlation with nearby eruption of known age
Reflection horizons	10 000+	?	Correlation with volcanic event of known agent
(d) *Glacier dynamics*			
Theoretical flow model	10 000+	?	Several major assumptions necessary; steady state ice sheet

propose that the level formerly considered to be ~60 000 years in age be reinterpreted as ~115 000 years old. Similarly Duval and Lorius (1980) suggest that the Antarctic core from Dome C may span ~53 000 years, whereas Lorius *et al.* (1979) had proposed a "basal date" of between 27 000 and 32 000 years BP. Such revisions are, of course, perfectly legitimate, but the large age differences which various approaches may produce illustrate how difficult it is to be definitive about the long-term ice-core chronologies. This is discussed further in Section 5.4. At present, the age uncertainties beyond 10 000–12 000 years BP are so great that extreme caution is needed in interpreting those chronologies which have been published so far.

Many of the methods which have been investigated in order to improve the dating of ice cores have themselves produced important paleoclimatic information. Some of the principal methods used and their paleoclimatic implications are now reviewed.

5.3.1 *Radio-isotopic methods*

The decay of ^{210}Pb (washed out from the atmosphere) has been used successfully in studies of snow accumulation over the last 100–200 years, providing an important perspective on the very short accumulation records available in remote parts of Antarctica and Greenland (e.g. Crozaz *et al.* 1964, Crozaz & Langway 1966). Some dates have been obtained from the radio-isotopes ^{32}Si and ^{39}Ar, but these isotopes have relatively short half-lives, making them useful only for the last 1000 years of record, and the technique is only considered accurate to $\pm$10–20% (Dansgaard *et al.* 1966, Oeschger *et al.* 1977). ^{14}C dating has been attempted on ice as old as 6000 years BP (Paterson *et al.* 1977) but very large amounts of ice (generally several tons) must be melted at depth, in fracture-free ice, to ensure a large enough sample. Although this has been accomplished, it is logistically very difficult and the samples may easily be contaminated (Oeschger *et al.* 1966, 1977). Consequently, most ^{14}C dates which have been obtained on ice samples are generally reported with fairly large standard errors. New accelerator techniques of ^{14}C dating (see Sec. 3.2.1.2) will enable samples to be obtained from far less ice, making radiocarbon dating an extremely valuable method of dating ice as old as 70 000–100 000 years. There is also the prospect that isotopes of chlorine and beryllium (^{36}Cl and ^{10}Be with half-lives of 310 000 and 1.5 $\times 10^6$ years respectively) may be used in combination to provide a crude estimate ($\pm$10 000 years?) of ice age near the base of ice sheets (Dansgaard 1981). At present, however, apart from ^{210}Pb analysis, radio-isotopic dating of ice and firn is not a routine operation and other stratigraphic techniques are generally preferred.

5.3.2 *Seasonal variations*

Certain components of ice cores show quite distinct seasonal variations which enable annual layers to be detected. These can then be counted to provide an extremely accurate timescale for the last few thousand years. Where uncertainties exist in one seasonal chronology, a comparison with other parameters enables accurate cross-checking to be accomplished, thereby reinforcing confidence in the timescale produced (Hammer *et al.* 1978).

5.3.2.1 Seasonal variations in $\delta^{18}O$. Because of the greater cooling which occurs in winter months, much lower ^{18}O concentrations are found in winter snow than in summer snow. This results in a very strong seasonal signal which can be used as a chronological tool, providing accumulation rates are reasonably high (>25 cm water equivalent per year), wind scouring of snow is not severe, and no melting and refreezing of snow and firn has occurred. In effect, the annual layer thickness can be identified by counting each couplet of high and low $\delta^{18}O$ values from the top of the core downward (Fig. 5.10). Unfortunately, at increasing depths in polar ice sheets the amplitude of the seasonal signal is reduced until it is eventually obliterated. In the upper layers, where density is <0.55 g cm^{-3}, this results from isotopic exchange between water vapor and firn. In lower, denser layers, where air channels are closed off, obliteration results from diffusion of water molecules within the ice. This process is accelerated due to thinning by plastic deformation as the annual layers approach bedrock; thinning increases isotopic gradients in the ice, making molecular diffusion more effective in obliterating the seasonal variations (Fig. 5.10). Recent work by Johnsen (1977) raises the possibility of reconstructing the "lost" seasonal variations. Johnsen models the processes which cause the record to become smoothed, then reverses the model to deconvolute the $\delta^{18}O$ profile (Fig. 5.11). Further work along these lines may provide a new way of checking the chronology of deep ice cores which at present can only be crudely "dated" by flow models (see Sec. 5.3.4 below).

In cores where seasonal isotopic differences are still preserved down to dense firn and ice layers, further smoothing due to molecular diffusion is so slow that the signal may then be preserved for thousands of years. Thus, at Camp Century in Greenland, if parts of the ice core were not missing or damaged (which they are) it would theoretically be possible to count annual layers back to 8300 years BP, without deconvolution of the isotopic record. This is not possible in most of Antarctica, though, because of low accumulation rates (generally <25 cm water equivalent per year) which result in the seasonal signal being "lost" at relatively shallow depths. In many cases, removal of seasonal (or indeed annual) accumulation by wind scouring may occur, destroying any seasonal signal entirely. On temperate glaciers and ice caps, where snow-melt and percolation of meltwater takes place, it is also impossible to detect a reliable seasonal isotopic signal. In these conditions, seasonal differences in both δD and $\delta^{18}O$ are rapidly smoothed out (within a few meters of the surface) due to isotopic exchange as the ice recrystallizes (Árnason 1969).

5.3.2.2 Seasonal variations in microparticles and trace elements. Detailed studies of microparticulate matter and trace element concentrations in ice samples from Antarctica and Greenland reveal periodic variations which

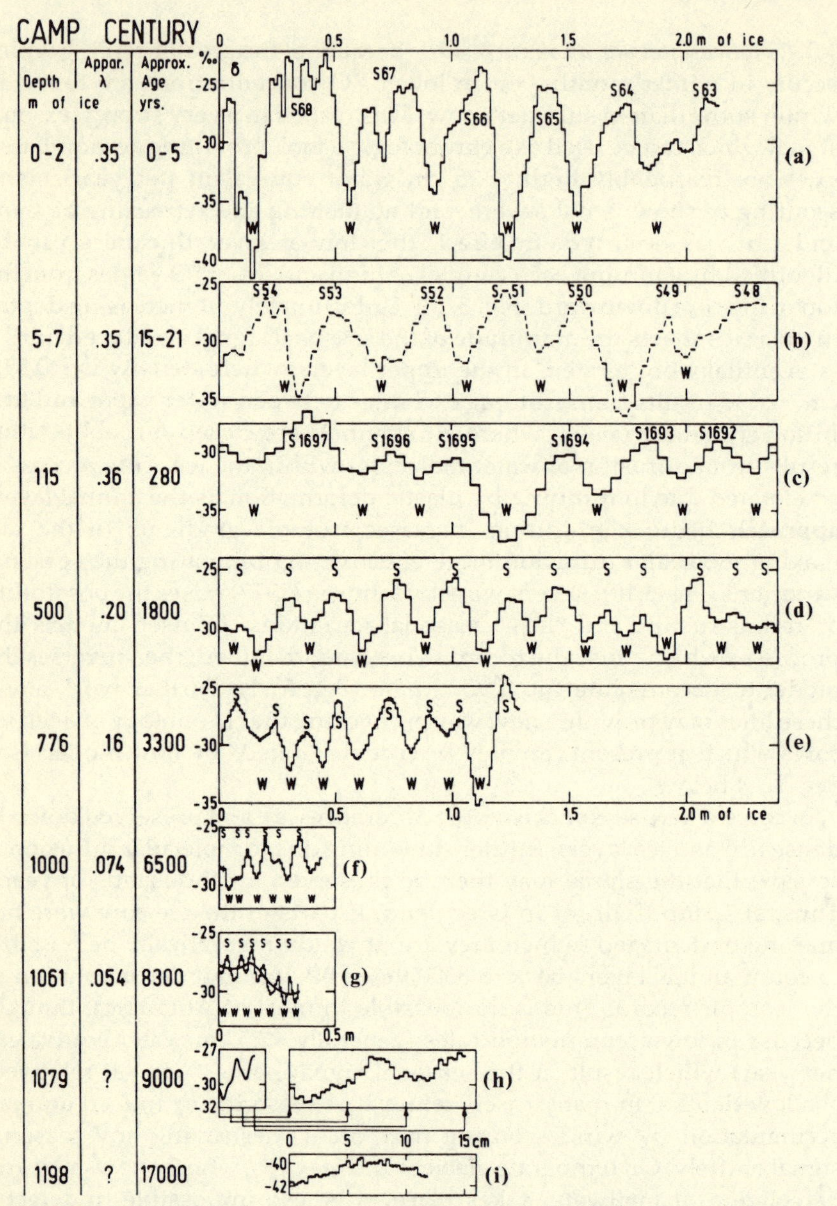

Figure 5.10 $\delta^{18}O$ variations in snow and firn at different depths in ice core from Camp Century, Greenland. S and W indicate interpretations of summer and winter layers, respectively. As the ice sinks towards the base of the ice sheet, the annual layer thickness (λ) is reduced due to plastic deformation. Within a few years, short-term $\delta^{18}O$ variations are obliterated by mass exchange in the porous firn. With increasing age, the amplitude of the seasonal δ cycle is reduced to ~2‰. As annual layers become thinner, the seasonal $\delta^{18}O$ gradients increase and molecular diffusion in the ice smoothes out the intra-annual variations. Eventually, seasonal differences are obliterated entirely (after Johnsen *et al.* 1972).

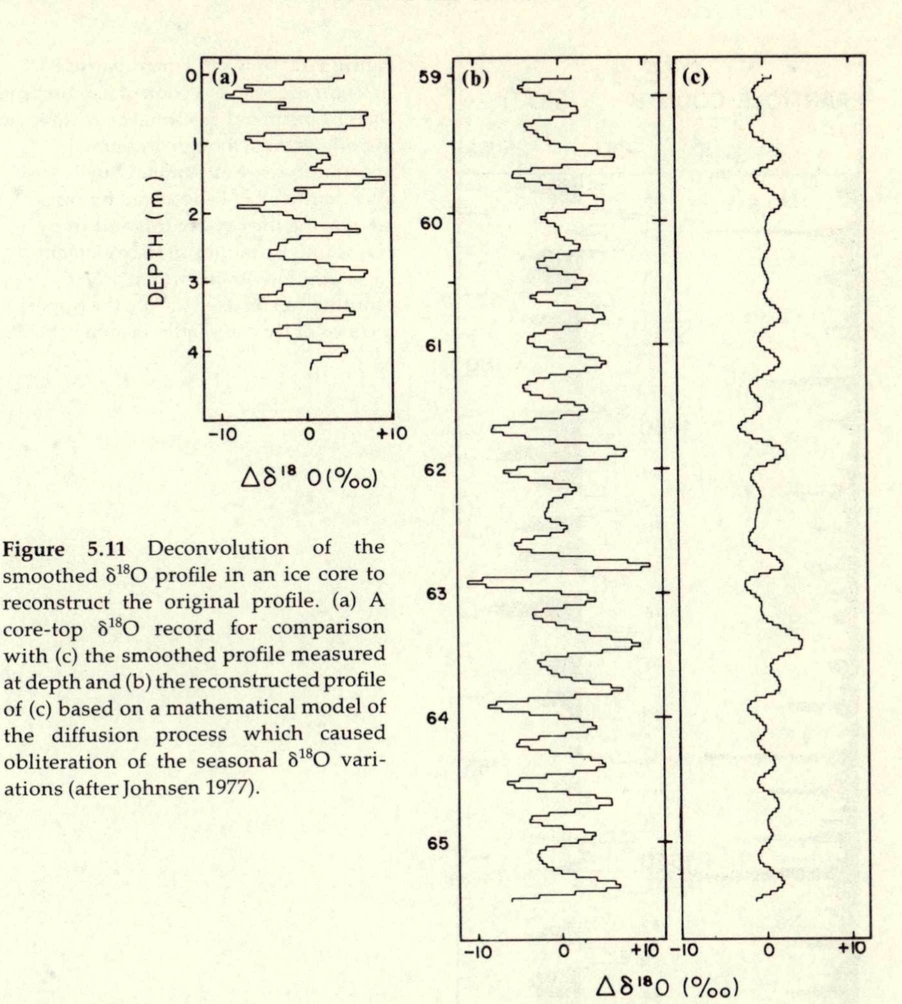

Figure 5.11 Deconvolution of the smoothed $\delta^{18}O$ profile in an ice core to reconstruct the original profile. (a) A core-top $\delta^{18}O$ record for comparison with (c) the smoothed profile measured at depth and (b) the reconstructed profile of (c) based on a mathematical model of the diffusion process which caused obliteration of the seasonal $\delta^{18}O$ variations (after Johnsen 1977).

appear to be seasonal. For example, Hamilton and Langway (1967) examined clay-sized microparticle frequency in Greenland ice and estimated that the number of microparticles increased to a maximum in late winter–early spring, presumably as a result of a more vigorous atmospheric circulation at this time of year. Conversely, microparticle frequency minima were generally observed in autumn. Their work has been confirmed by Hammer (1977a) who found this seasonal variation in microparticle concentration in Greenland cores ranging from Dye-2 at 66°N to Camp Century at 77°N (Fig. 5.12). A similar seasonal variation in sodium, calcium, magnesium, and aluminum ions has been measured by Langway *et al.* (1977); spring concentrations of these ions are five to ten times greater than at other times of the year.

Compared to the diffusion rate of water molecules, which leads to obliteration of the seasonal $\delta^{18}O$ record at depth, diffusion of micropartì-

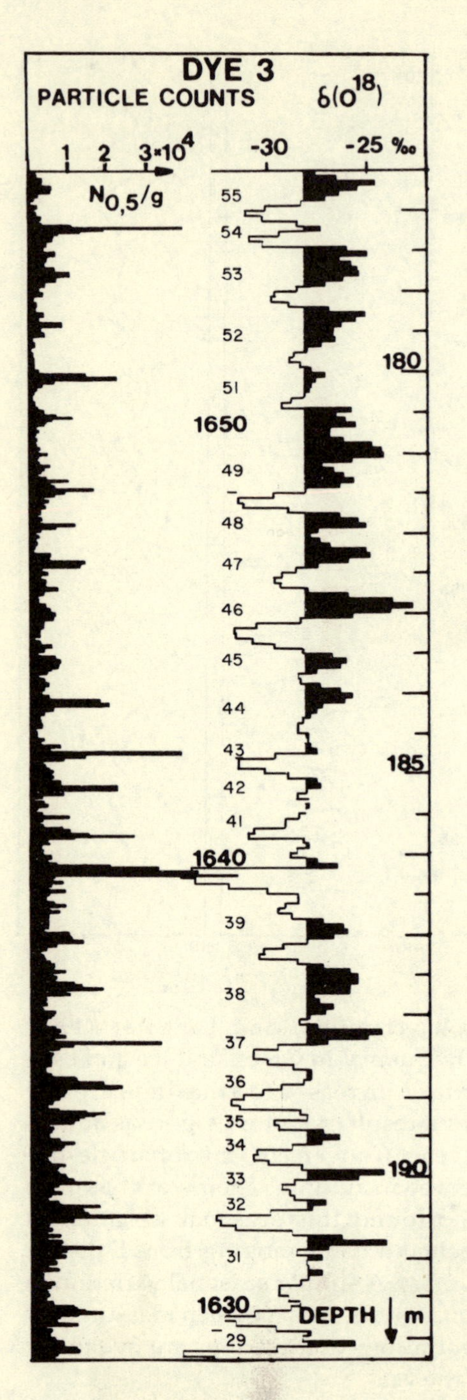

Figure 5.12 $\delta^{18}O$ and microparticle concentrations in a section of ice core from Dye-3, Greenland. Seasonal peaks are seen in both records, though occasional uncertainties are apparent. Usually such uncertainties can be resolved by cross-referencing the two records and/or by examining variations in trace element concentration. In this way seasonal counting can be used to date the upper sections of ice cores (after Hammer 1977a).

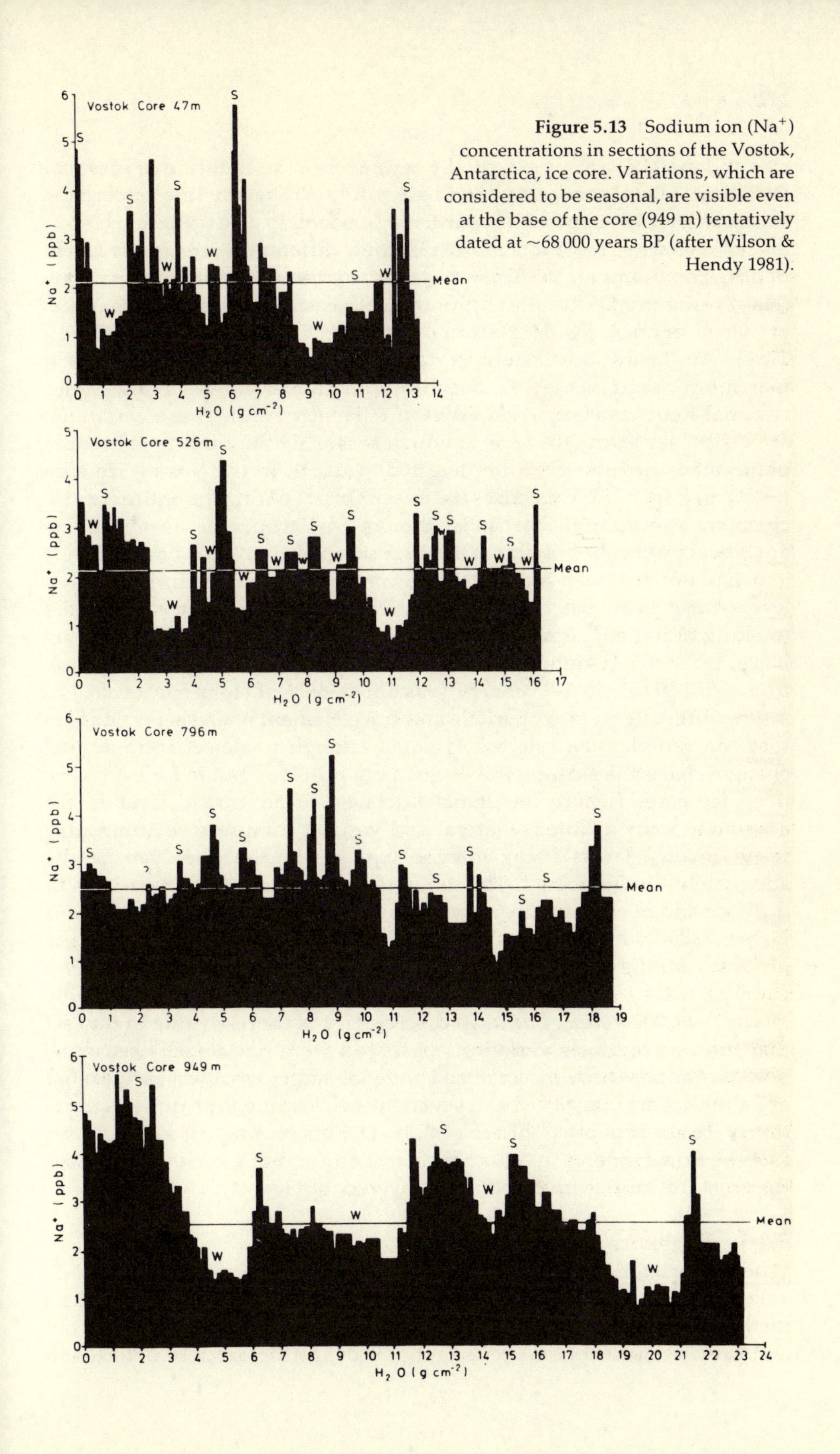

Figure 5.13 Sodium ion (Na^+) concentrations in sections of the Vostok, Antarctica, ice core. Variations, which are considered to be seasonal, are visible even at the base of the core (949 m) tentatively dated at ~68 000 years BP (after Wilson & Hendy 1981).

cles and metallic ions is essentially zero. Hence the counting of seasonal variations may allow dating of ice back to late Wisconsin time, or perhaps even earlier. This approach is particularly useful in areas where accumulation rates are so low that seasonal isotopic differences are rapidly lost at depth. For example, in Antarctica, the concentration of sodium ions (Na^+) varies markedly, due to pronounced seasonal changes in the influx of marine aerosols (M. M. Herron & Langway 1979, Warburton & Young 1981). At Vostok, in eastern Antarctica, Na^+ concentrations reach a maximum in summer layers, due to sublimation of snow, leaving higher residual ionic concentrations (Wilson & Hendy 1981). These variations are visible far below the level at which seasonal $\delta^{18}O$ variations become obliterated, and can even be detected at depth in the Vostok ice core (~950 m) (Fig. 5.13). This raises the possibility of dating the entire core by chemical stratigraphic methods, though this would be an enormous analytical project (involving at least six samples from each annual layer) and has not yet been attempted. Nevertheless, samples from selected levels have been used to calculate accumulation rates through time (making minor adjustments for thinning of the layers at depth). On this basis, Wilson and Hendy (1981) estimated that accumulation at Vostok from ~50 000 to ~13 000 years BP was only ~60% of Holocene levels.

One difficulty in microparticle and trace element analysis is to ensure that the sample size selected is small enough to detect intra-annual changes. Near the surface, this is not a big problem, but in ice from very deep ice cores (where the actual thickness of an annual layer is not accurately known) intense lateral and vertical compressive strain may result in dust layers being merged together so that they can not be adequately distinguished. This is particularly true if the strain rates of dirty ice and of clean ice are very different, as suggested by Koerner and Fisher (1979). Such problems may have led to the great differences in ages proposed for the base of the Byrd ice core by Thompson *et al.* (1975) using microparticles (~27 000 years) and by Johnsen *et al.* (1972) using a flow model (~84 000 years). The large discrepancy is probably due to the fact that the microparticle variations observed were not actually seasonal, because the core sections analyzed were too large; what was interpreted as annual layers may have been several years, leading to an underestimation of ice age at depth (Johnsen *et al.* 1976). This does not, of course, mean that the flow model is necessarily correct either, but merely emphasizes the great difficulties involved in dating very old ice.

5.3.3 *Reference horizons*

Where characteristic layers of known age can be detected, these provide valuable chronostratigraphic markers against which other dating methods can be checked. On the short timescale, radioactive fallout from atmospheric nuclear bomb tests in the 1950s and 1960s can be detected in firn by measuring the tritium content (or gross β activity). As the timing of the first occurrence of these layers is fairly well known (spring 1953 in Greenland and February 1955 over much of Antarctica, reaching maximum levels in 1963) they can be used as marker horizons for snow accumulation studies, facilitating regional surveys of net balance over the last few decades (Crozaz *et al.* 1966, Picciotto *et al.* 1971, Koerner & Taniguchi 1976).

On a much longer timescale, other reference horizons have resulted from major explosive volcanic eruptions. Violent eruptions may inject large quantities of dust and gases (most importantly hydrogen sulfide and sulfur dioxide) into the stratosphere where they are rapidly dispersed around the hemisphere. The gases are oxidized photochemically and dissolve in water droplets to form sulfuric acid, which is eventually washed out in precipitation. Hence, after major explosive volcanic eruptions, the acidity of snowfall increases to levels significantly above background values (Hammer 1977b). By identifying highly acidic layers resulting from eruptions of known age, an excellent means of checking seasonally based chronologies is available (Fig. 5.14). Variations in electrical conductivity (a measure of acidity) along a 404 m core from Crête, Central Greenland, reveal a record which closely matches eruptions of known age (Hammer *et al.* 1978, 1980). The core was originally dated by a combination of methods, primarily seasonal counting (Hammer *et al.* 1978). This enabled the acidity record to be checked against historical evidence of major eruptions during the last 1000 years (e.g. Lamb 1970), confirming that the timescale developed was extremely accurate. Once major acidity peaks have been identified they can be used as critical reference levels over the entire ice sheet. For example, the highest acidity levels in the last 1000 years resulted from the eruption of Laki, Iceland, in 1783. The only acidity peak of greater magnitude in the last 2000 years resulted from another Icelandic eruption (Eldgja) at AD 934 ± 2 (Hammer 1980). These peaks can be easily identified both in the laboratory and in the field, providing excellent chronological checks. There is also evidence that the major acid layers noted in the Crête ice core correspond to internal reflection horizons observed on radio-echo soundings through the Greenland ice sheet (the horizons resulting from changes in the dielectric constant of ice when acidic). Thus, direct chronostratigraphic correlations between cores may be inferred using observed reflection horizons (Hammer 1980, Millar 1981).

Construction of an acidity index as a proxy of volcanic activity obviously has other important paleoclimatic implications (Fig. 5.15). It is of particular interest that the Crête record shows that the most volcanically quiet period over the last 1500 years (Fig. 5.14) was from ~AD 1110 to AD1250 (840–700 years BP), a time corresponding well with the Medieval warm period (Lamb 1965; cf. Sec. 5.4.2 & Ch. 11). By contrast, the higher

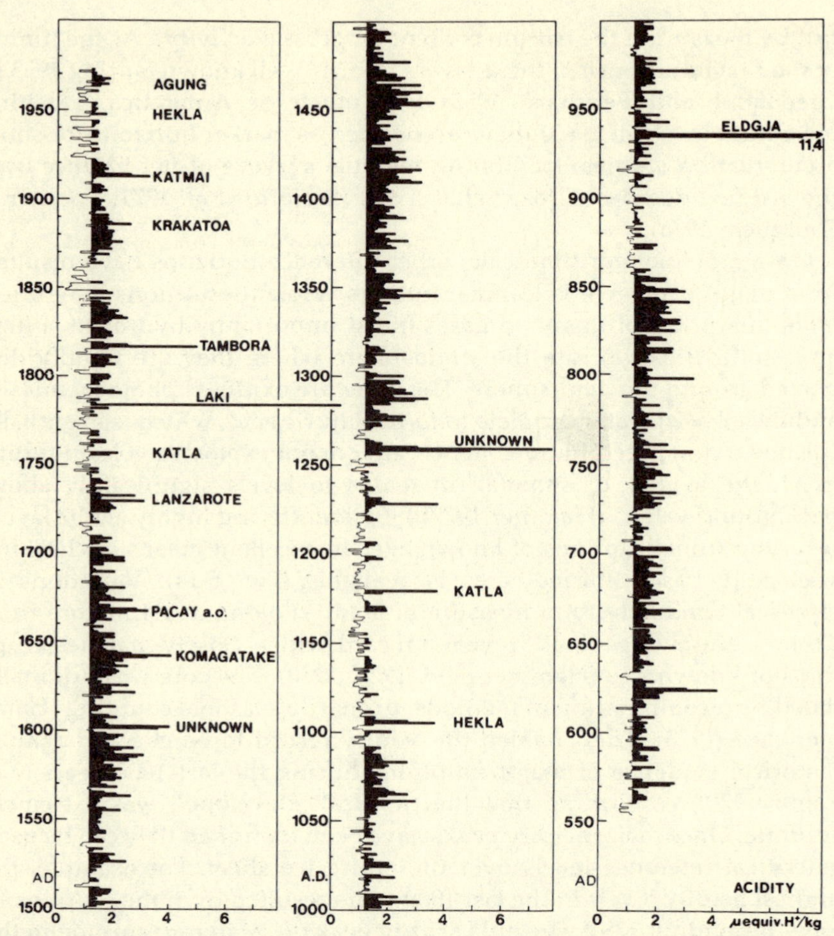

Figure 5.14 Mean acidity of annual layers from AD 553 to AD 1972 in the ice core from Crête, central Greenland. Acidities above the background (1.2 μequiv H^+ per kilogram of ice) are due to fallout of acids, mainly H_2SO_4, from volcanic eruptions north of ~20°S. The ice core is dated with an uncertainty of ±1 year in the past 900 years, increasing to ±3 years at AD 553, which makes possible the identification of several large eruptions known from historical sources e.g. Laki, Iceland, 1783; Tambora, Indonesia, 1815; Hekla, Iceland, 1104). Also seen is the signal from the Icelandic volcano Eldgja, which was known to have erupted shortly after AD 930. Note the low level of volcanic activity recorded from AD 1100 to AD 1250 and from AD 1920 to AD 1960. Considerably higher levels of volcanic activity occurred from AD 550 to AD 850 and from AD 1250 to AD 1750 (after Hammer *et al.* 1980).

levels of volcanic activity from 1300 to 1500 and from 1550 to 1700 may have played an important role in the cooler conditions of the "Little Ice Age" (Lamb 1982). Also apparent is the relative absence of major volcanic eruptions from 1920 to 1960, which was a period of exceptional warmth throughout the Northern Hemisphere. Such observations are intriguing, but perhaps simplistic, as there are many occasions when high volcanic

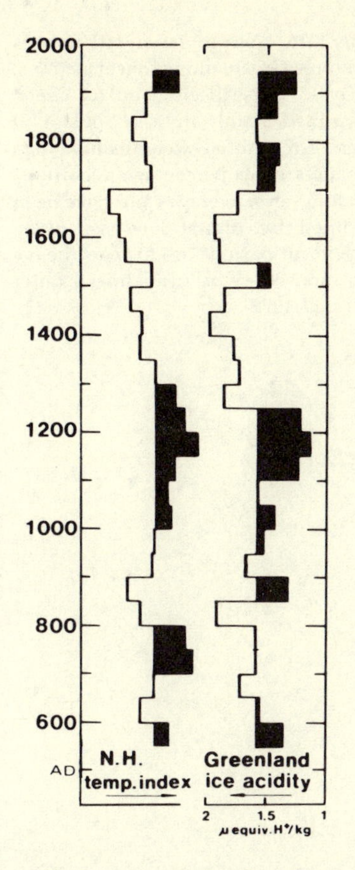

Figure 5.15 Correlation between volcanic activity index (based on ice core acidity record in Fig. 5.14) and a generalized index of Northern Hemisphere temperatures (note that scales are reversed). This is based on a composite of $\delta^{18}O$ in the Greenland ice sheet, tree-ring widths in the White Mountains of California, and central England temperatures (derived from historical and botanical sources prior to the start of continuous instrumental records in the 17th century). Because of the uncertainties in the interpretation of these records the composite only gives a crude relative temperature scale. However, there is a general correlation between periods of below average volcanic activity (black) and above average temperatures (black) and vice versa (after Hammer *et al.* 1980).

activity corresponds to exceptionally warm conditions, even in Greenland (cf. Sec. 5.5). Nevertheless, it is clearly of great paleoclimatic importance to obtain longer records of volcanic activity, spanning the entire Holocene if possible, from both Greenland and Antarctica. This will greatly improve our understanding of the role played by volcanic activity in modulating climatic variation on the timescale of decades to millennia.

Perhaps the best approach to identifying annual layers is a composite one, using $\delta^{18}O$ profiles, microparticles, and other reference horizons, such as the dates of known volcanic eruptions and associated microparticle and/or electrolytic conductivity peaks in the ice. In this way, questionable sections of, say, the $\delta^{18}O$ record may be resolved by reference to the microparticle record. This multi-parameter approach is the one advocated by Hammer *et al.* (1978) and was used to date very precisely the Crête core from central Greenland back to AD 548. They estimate their chronology is accurate to within ±2 years back to AD 1177 and to within ±3 years to AD 548. Similar studies on the cores from Milcent (to AD 1177) and Dye-3 (to AD 1245) have enabled accumulation rates to be calculated, given certain assumptions about vertical strain since deposition, density

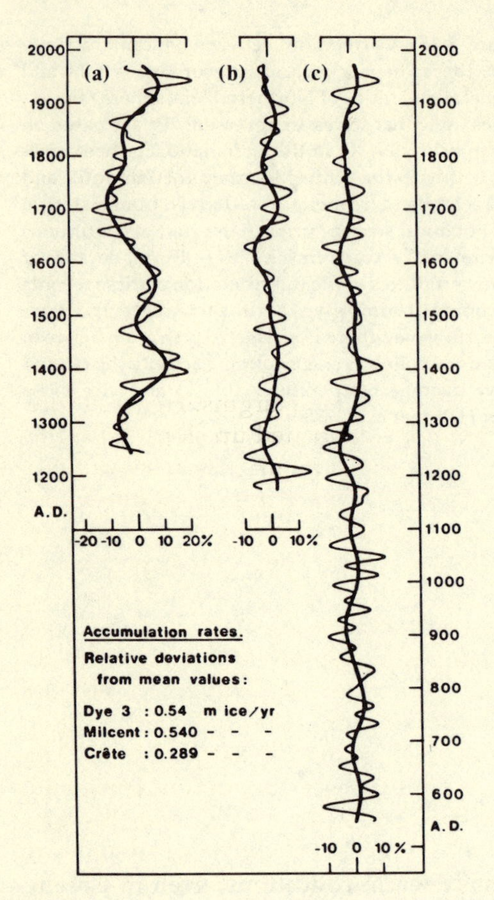

Figure 5.16 Accumulation rate records from three Greenland ice sheet stations, (a) Dye-3, (b) Milcent, and (c) Crête, determined mainly from seasonal $\delta^{18}O$ cycles. The annual accumulation data are expressed as percentage departures from long-term averages and have been smoothed by digital low-pass filters with cut-off periods of 120 years (heavy lines) and 30 years (thin lines). (after Reeh *et al.* 1978).

variations down core, and accumulation rate deviations up stream (Reeh *et al.* 1978). Figure 5.16 shows the three records expressed as relative departures from long-term means, smoothed by low-pass filters. The Dye-3 record indicates two periods of above average accumulation, from ~1380 to 1610 and from ~1870 to 1970, with particularly heavy accumulation around AD 1400 (also seen in the Crête and Milcent records). The Dye-3 record shows markedly drier intervals prior to ~1380 and from ~1610 to ~1870, with the period around 1700 being the "driest". Data from Milcent and Crête are less variable and have little in common with the Dye-3 record. A slightly wetter interval around 1300–1520 may correlate, in part, with the (somewhat later) wet interval at Dye-3, but it is probable that different moisture sources are responsible. Certainly, the Crête record gives no support for there having been any significant long-term change in accumulation at the summit of the Greenland Ice Sheet over the last 1400 years. This is similar to the conclusion reached by Koerner (1977b) after preliminary work on the Devon Island Ice Cap cores.

5.3.4 *Theoretical models*

Dating ice at great depth poses severe problems which can not be easily resolved by the methods described above. At present, the only method widely used to date pre-Holocene ice is to calculate ice age at depth by means of a theoretical ice-flow model (Dansgaard & Johnsen 1969). The model used to provide a preliminary timescale for the Camp Century, Byrd Station, and Vostok cores is shown schematically in Figure 5.17. It is a modified version of Nye's model of glacier flow (Nye 1959), incorporating a non-uniform vertical deformation (strain) rate and taking into account that the ice is frozen to its base. Ice formed at the ice-sheet summit will thin at depth by plastic deformation, and over long periods of time will be carried out towards the ice-sheet margins. Hence, a core from site C in Figure 5.17 will contain ice deposited up slope, with the oldest and deepest ice originating at the summit. Because summit temperatures are cooler, if the core is not recovered from the highest part of the ice sheet, $\delta^{18}O$ values at depth must be corrected for this altitude effect which will be present regardless of whether any long-term climatic fluctuation has occurred.

Because of the nature of ice flow and deformation, most of the time period recorded in an ice core is found in the lowest few meters (Fig. 5.18). Consider, for example, the model used by Dansgaard *et al.* (1971) in a first attempt to date the Camp Century ice core. This predicted an age of 30 000 years at 1300 m, 60 000 years at 1350 m and 157 000 years at 1373 m (Fig. 5.18). Clearly, such a quasi-exponential age – depth function could lead to relatively large errors if off by only a few meters, and this was certainly possible considering the assumptions which were made in their flow model for the sake of simplicity. These were as follows: (a) the

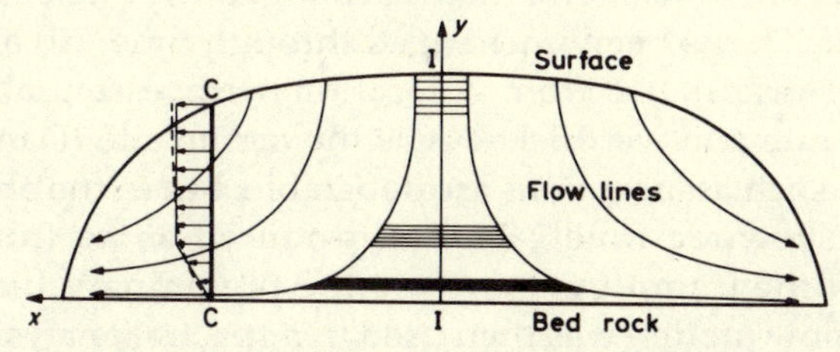

Figure 5.17 Schematic diagram to illustrate the approach of dating deep ice by means of a theoretical flow model. Line C—C represents an ice core taken away from the ice sheet summit (as, for example, Camp Century, Greenland). Ice particles deposited on the surface will follow lines that travel closer to the base, the farther inland that they were deposited. Ice formed near the divide will be plastically deformed (thinned) with depth as indicated in the profile D–D. The dashed curve along the ice core (C–C) shows the calculated horizontal velocity profile. The ice sheet is considered to be frozen to a horizontal base, for modeling purposes. $\delta^{18}O$ values are lowest at the ice sheet summit and so a core at C–C can be expected to have lower $\delta^{18}O$ values at depth regardless of any change in climate (after Dansgaard *et al.* 1973).

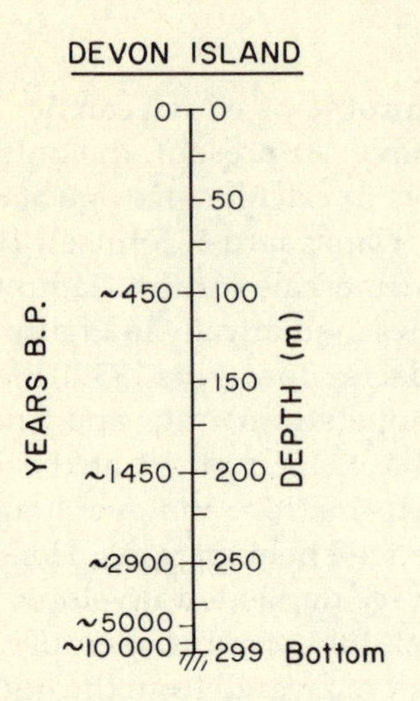

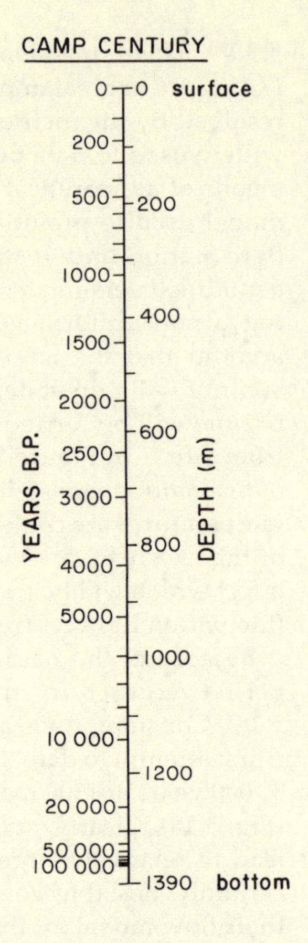

Figure 5.18 Relationship between depth and estimated age in ice cores from Devon Island Ice Cap and Camp Century Greenland. Chronologically, most of the record is obtained from the lowest few meters of each core. The time–depth relationship shown here for Camp Century is based on a theoretical flow model (Dansgaard *et al.* 1971); this provided only a rough estimate and several revisions have been made since (these are shown in Fig. 5.19). However, these revisions only significantly affect the interpretation of ice age below ~1250 m depth (after Dansgaard *et al.* 1971 and Patterson *et al.* 1977).

rate of accumulation has remained unchanged from that of the present day; (b) ice thickness has remained unchanged (cf. Fisher 1979); (c) the flow pattern of ice has been unchanged through time; (d) all sections of the core have originated from a location with essentially the same accumulation rate and ice thickness as the coring site (Dansgaard *et al.* 1969). Clearly, such assumptions are debatable, but in the absence of any other method, a crude timescale was produced as an initial working hypothesis. In the Camp Century case the preliminary timescale (predicted by the flow model) was then used in a spectral analysis of the δ^{18}O record from different sections of the core (Dansgaard *et al.* 1971). In the core section thought to represent the Holocene, a persistent periodicity, centered on 350 years, was observed. As no known cycle with this frequency had been reported (though few records have enough resolution to identify such a periodicity if it existed), it was assumed that the "real" signal being identified was a 405 year period of solar activity, noted in ^{14}C analysis of tree rings by Suess (1970). The ice-core record could be made to "fit" this periodicity by assuming an overall accumulation rate

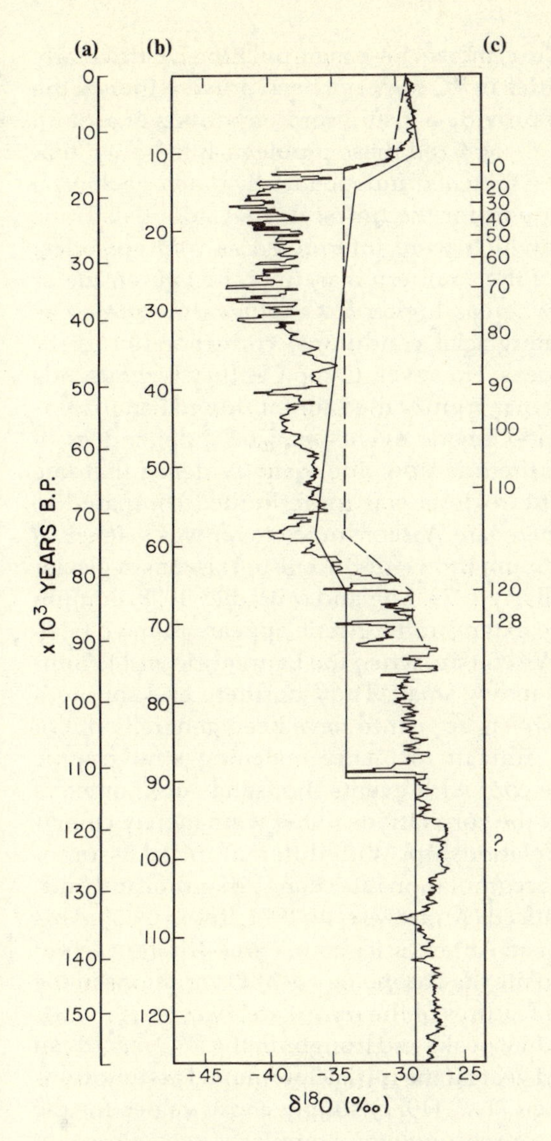

Figure 5.19 $\delta^{18}O$ record from Camp Century, Greenland showing several revisions of the timescale. (a) the original timescale developed by Dansgaard *et al.* (1969), based on a theoretical model of ice flow. This was subsequently revised (b) based on periodic signals in $\delta^{18}O$ (see text). (c) This timescale shows the most recent interpretation of ice age with depth, based on correlations with the isotopic record deep-sea sediments (Dansgaard *et al.* 1982). The isotopic data is plotted here linearly with respect to timescale (b). In Figure 5.22 it is plotted linearly with respect to timescale (c) and the lowest section is omitted. Solid and dashed lines show estimated $\delta^{18}O$ changes which might have occurred due to ice-thickness changes according to Robin (1977) and Dansgaard *et al.* (1971) respectively (after Dansgaard *et al.* 1971).

~15% lower than in recent years. Further analysis of the entire core record revealed oscillations of varying frequency centered on a period of around 2000 years back to 45 000 years BP but decreasing to around 4000 years before 100 000 years BP. These were not understood, but assumed to reflect a 2400 year period in ^{14}C concentrations, identified in tree rings (though by no means fully substantiated). Accordingly, the entire chronology was adjusted to a timescale which would give a 2400 year periodicity in the $\delta^{18}O$ values; this adjusted timescale is shown in Figure 5.19 with the older, preliminary scale for comparison.

It is, of course, not difficult to criticize these manipulations, particularly when the so-called periodicities in ^{14}C have not been proven (indeed the ^{14}C record is far too short to provide a firm record of periods of around 2400 years). Dansgaard *et al.* recognized these problems but argued that the resulting $\delta^{18}O$ chronology matched independently dated geological events, providing strong support for the timescale they adopted. Thus, extremely low $\delta^{18}O$ values (which were interpreted as cold episodes) seemed to match advances of the southern margin of the Laurentide or Fennoscandian Ice Sheets, whereas higher $\delta^{18}O$ values (interpreted as recording interstadial or interglacial conditions) corresponded to ice margin retreat in the same areas. However, Camp Century is thousands of kilometers from the southern margin of the Laurentide and Fennoscandian Ice Sheets, where glacio-climatic events may be independent of High Arctic events. In fact, there is now abundant evidence that late Wisconsin ice cover in Arctic regions was quite limited compared to more southerly latitudes, where late Wisconsin ice cover was extensive, in many areas reaching maximum limits for the entire Wisconsin glaciation (Flint 1971, Andrews & Barry 1978, England & Bradley 1978, Boulton 1979, Boulton *et al.* 1982). Ice extent in the Arctic appears to have been most extensive in the early Wisconsin, when the Laurentide and Fennoscandian Ice Sheets were relatively small. Thus, northern and southern edges of the Wisconsin ice sheets appear to have been generally out of phase. It is therefore very difficult to justify matching stratigraphic features in a High Arctic ice core with events thousands of kilometers away, and it is probable that the correlations noted were merely coincidental, implying no causal relationship. With different initial assumptions (particularly regarding accumulation rate changes) significantly different timescales can be produced (Andrews *et al.* 1974, Robin 1977). This is illustrated in Figure 5.20 for an Antarctic ice core. Three different sets of assumptions are used to calculate the chronology of $\delta^{18}O$ variations in the Byrd Station core. The broad features of the record are, of course, invariable, but the timing of individual peaks and troughs in the $\delta^{18}O$ record can be varied by several thousand years if the initial flow-model assumptions are altered (Fig. 5.20). Johnsen *et al.* (1972) simply chose values for the model which result in a Byrd core chronology as similar as possible to that at Camp Century. Obviously, the Byrd record can not be considered as an *absolute* chronology of $\delta^{18}O$ variations in Antarctica; better dating methods are needed in order to more accurately "tune" the Byrd core.

Similar problems exist in dating the lower sections of other cores and at present it is best to consider the timescales for older $\delta^{18}O$ data as very preliminary (cf. Sec. 5.4). At this stage, one can only conclude that beyond ~10 000 years BP an accurate timescale has yet to be developed, either for Camp Century or for other deep ice cores. This point is considered further in the next two sections.

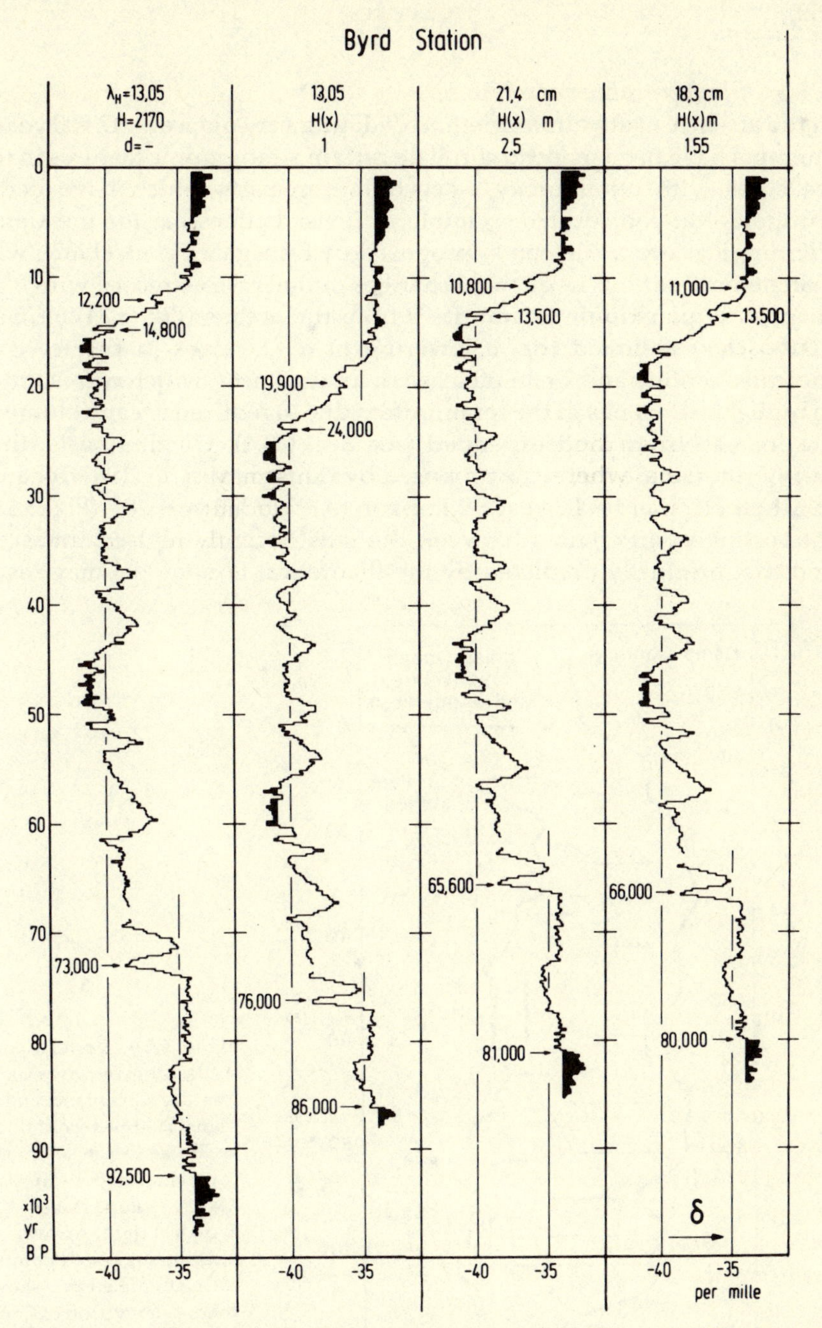

Figure 5.20 The $\delta^{18}O$ record from Byrd Station, Antarctica. A flow model like that used initially to date the Camp Century core was used, with different assumptions about annual accumulation (λ_H), thickness of the ice sheet (H), and divergence of ice flow (d) (shown at top of each diagram). Changing the initial assumptions has the effect of "stretching" or "shrinking" the isotopic record relative to the age scale at the left, as indicated by dates on characteristic features of the record (after Johnsen *et al.* 1972).

5.3.5 *Stratigraphic correlations*

In the absence of any direct method of dating very old ice (>12 000 years) attempts have been made to correlate certain stratigraphic features in the ice cores with other proxy paleoclimatic records which have better chronological control. For example, a revised timescale for the Camp Century ice core was recently proposed by Dansgaard *et al.* (1982), who matched major (low-frequency) changes in the ice-core record with $\delta^{18}O$ changes in planktonic foraminifera from the oceans (Fig. 5.21) (cf. Sec. 6.3.2). They assumed that a lowering of $\delta^{18}O$ values in the ice core indicates cooling and/or an increase in ice thickness, which corresponds to higher $\delta^{18}O$ values in the foraminifera due to reduced ocean volume as the ice sheets on land expanded (see Sec. 6.3.1). On this basis, they re-interpreted the timescale proposed by Dansgaard *et al.* (1969), changing their original 60 000 year BP horizon to ~115 000 years BP (Fig. 5.19). Interestingly, differences between the most recently revised timescale and that originally predicted by the theoretical flow model may result

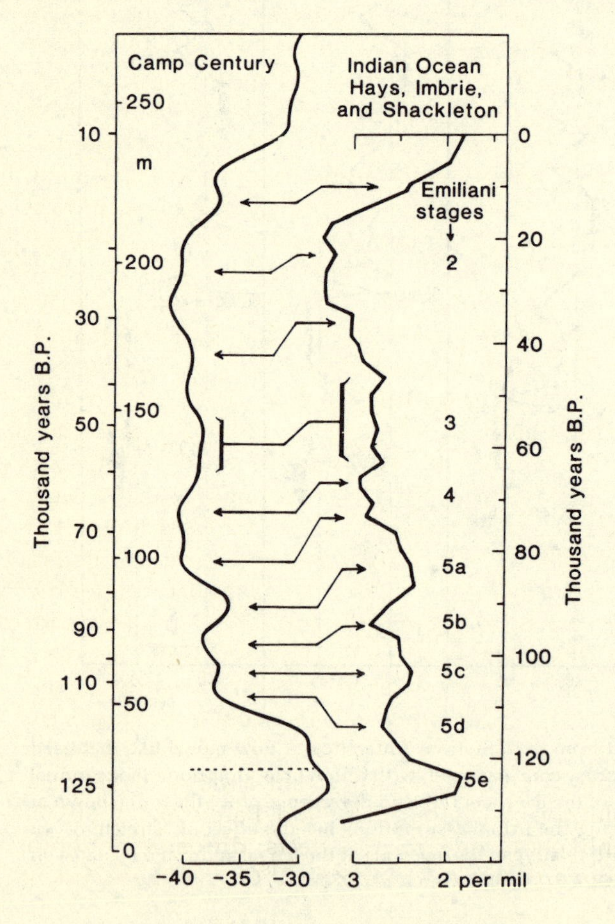

Figure 5.21 Correlation of the isotopic record of Camp Century, Greenland, and the faunal isotope record from an Indian Ocean sediment core (Hays *et al.* 1976). Marine isotope stages shown at right. Arrows indicate suggested points of correlation. Left axis shows proposed revision of Camp Century ice core timescale; right scale shows marine core timescale. Depth scale for ice core (inner left axis) is in meters above the base (after Dansgaard *et al.* 1982).

from the assumption of a constant accumulation rate which was used in the model calculations. Thus, Dansgaard *et al.* (1982) note that their new chronological model implies higher accumulation rates in the intervals 125 000–115 000, 80 000–60 000, and 40 000–30 000 years BP, all times when the oceanic $\delta^{18}O$ record indicates periods of major global ice volume increase. This is an intriguing by-product of the correlation attempted, but should only be considered as such until more definitive dating of the ice core is achieved.

5.4 Long-term $\delta^{18}O$ records of polar ice cores

The Camp Century isotopic record has been dealt with at length because it is the most widely referenced and has been discussed in the most detail. However, there are now several other ice cores which appear to extend back into pre-Holocene time, notably the cores from Byrd Station, Law Dome, and Vostok in Antarctica and from the Devon Island and Agassiz Ice Caps in the Canadian Arctic. In all of these cores, the problem of dating ice at depth is still unresolved. Figure 5.22 shows the published profiles from several sites according to timescales thought to be most appropriate by the different investigators working on each core. Alternative timescales for the Byrd, Vostok, and Camp Century cores have been proposed by Robin (1977) by taking into account the probable reduction in accumulation rate during the peak glacial times, as well as expected flow and deformation changes.

All the cores shown in Figure 5.22 appear to extend back into the last glacial period and cores from Byrd, Camp Century, and the Devon Island Ice Cap may include ice from before the last glaciation, though there is some dispute as to what this isotopically lighter basal ice actually represents. In the case of Camp Century, Robin (1977) suggests that because of significant changes in ice thickness and extent in the vicinity of present-day Camp Century, since the last interglacial, it is unlikely that ice near the base (see Fig. 5.19) represents a continuous stratigraphic sequence. He argues that the lowest ice layer in the core may contain a mixture of local ice and ice from further inland. This may account for the sharp peaks in the $\delta^{18}O$ record such as that seen around 90 000 years BP (using the timescale of Dansgaard *et al.* 1971, shown as B in Fig. 5.19). Similar "spikes" of very low $\delta^{18}O$ values have been observed in the Devon Island ice cores (Fig. 5.22). Such events were considered by Dansgaard *et al.* (1972) to represent very abrupt, major climatic changes in the past, possibly rapid cooling resulting from a surge of the East Antarctic Ice Sheet (Hollin 1980). On the other hand, Paterson *et al.* (1977) argue that these are unlikely to be climatic effects; diffusion at depth would have smoothed out any abrupt changes in $\delta^{18}O$, so if these spikes

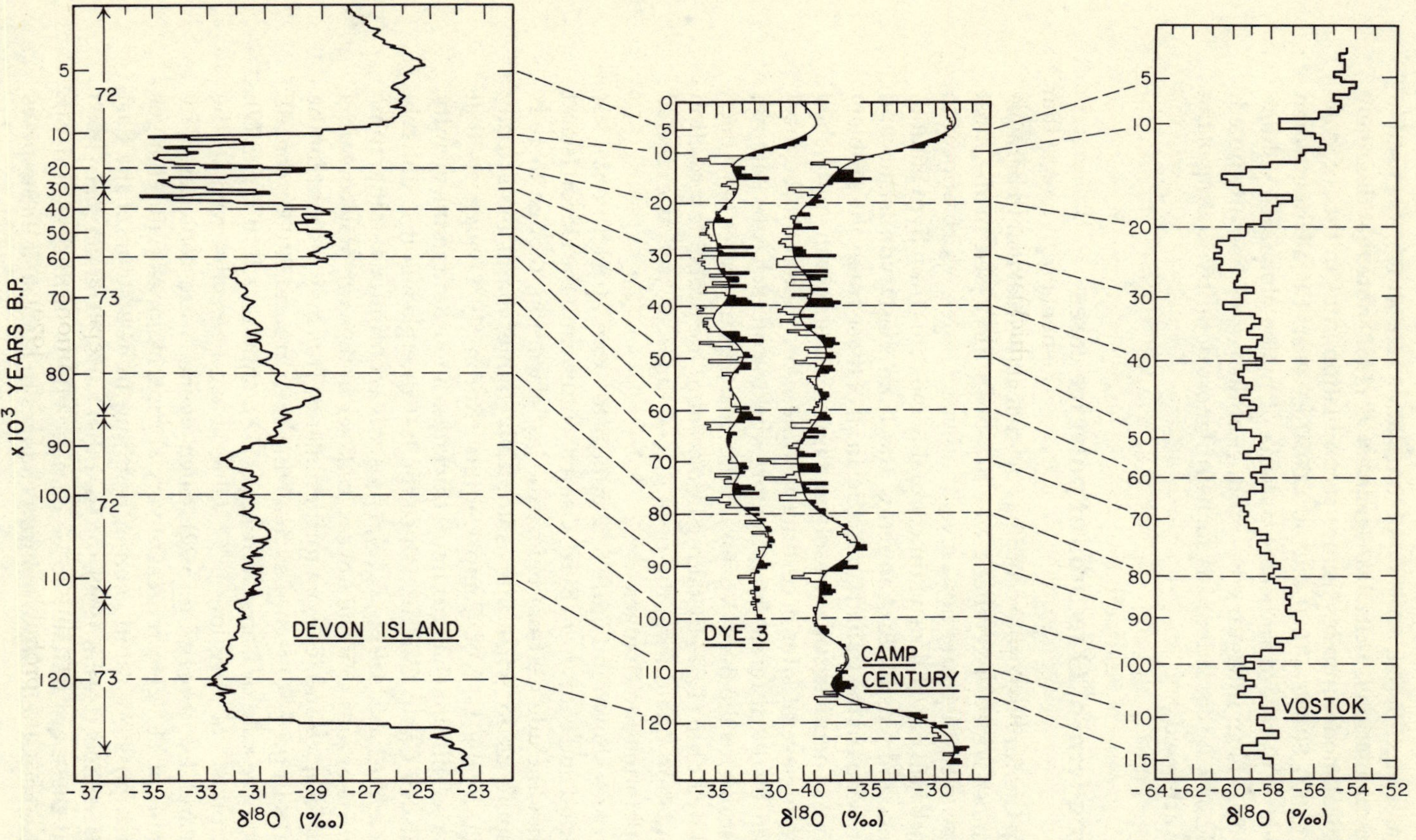

Figure 5.22 $\delta^{18}O$ records from four long ice cores compared for the last ~120 000 years of record. *Timescales may not be correct*. The Camp Century chronology is based on cross correlations with the marine isotope record. The Dye 3 and Devon Island Ice Cap chronologies are based on matching of major stratigraphic features with those of Camp Century. The Devon Island record was originally matched with the Camp Century chronology shown as (b) in Figure 5.19 (Paterson *et al.* 1977). Following the revision of this timescale by Dansgaard *et al.* (1982), the corresponding chronology for Devon Island is as shown here. However, the earlier chronology is preferred by the Devon Island research group. The Vostok chronology is based on theoretical calculations of ice flow assuming variable accumulation rates above the ~68 000 BP level and constant accumulation below that

were climatic events they would indeed have to have been catastrophic! A more likely explanation is that variations in vertical strain rate at depth due to disturbances close to the uneven bedrock cause thinning and "pinching out" of horizontal layers, perhaps leading to "gaps" in the very old ice record. In fact, Koerner and Fisher (1979) estimate that as much as 30% of the originally continuous isotopic record is missing from each of the Devon Island ice cores in the lowest 6 m (i.e. in the pre-Holocene sections). Abrupt changes in the lowest few meters of both cores do not match, though they are from sites only 27 m apart. Possibly faulting or shearing has occurred at depth, but there is no crystallographic evidence of shear planes at the appropriate levels (Paterson *et al.* 1977). Based on the Devon Island ice-core studies, Koerner and Fisher (1979) suggest that ice cores from relatively small ice sheets (<400 m thick) will have slightly disturbed paleoclimatic records beyond 5000 years BP and discontinuous records beyond 10 000 years BP.

As mentioned already, whether climatic changes have occurred or not, ice at depth is likely to have lower $\delta^{18}O$ values than that near the surface. This is because the deep ice may have originated from locations upstream of the ice-core site in areas where the temperatures are colder and the snow contains less $\delta^{18}O$ (unless the ice core was taken from the summit of an ice cap which has always been the highest part of the ice cap, e.g. the Devon Island ice core). Attempts at interpreting the $\delta^{18}O$ record in terms of long-term temperature change may therefore involve adjusting the $\delta^{18}O$ profile to take this altitude effect into account. This has been done in Figure 5.19, according to estimates by the ice-core analysts themselves or by Robin (1977). The principal result of these calculations is to reduce the apparent change in temperature from late Wisconsin to Postglacial time, because much of the difference in $\delta^{18}O$ is due to the ice sheets in late Wisconsin/Weichselian time having been thicker. How much thicker is a matter of debate. Recent work on air-bubble volumes in ice by Raynaud and Lebel (1979) suggests that ice at Camp Century was ~800 m thicker during the late Wisconsin maximum (see Sec. 5.5), an estimate similar to that determined by Fisher (1979) based on a comparison of the $\delta^{18}O$ records in Camp Century and Devon Island ice cores. Barkov *et al.* (1977) estimate that ice thickness at Vostok (Antarctica) has not varied by more than 100 m during the last glaciation, whereas at Byrd station the change may have been as much as 300 m (Robin 1977). Taking such factors into account, Robin (1977) estimates that temperatures over inland Antarctica were 6–8 °C lower during the last glacial maximum compared to the present. However, estimates of temperature change over this timescale are resisted by Dansgaard *et al.* (1969), who emphasize that there are a multitude of problems involved, viz. (a) the deeper strata may have originated inland where different climatic conditions existed; (b) ice-sheet thickness has probably changed (Fisher 1979); (c) the isotopic composition of sea water has changed because of the build-up of ice sheets with low $\delta^{18}O$ concentrations; (d) the ratio of summer to winter precipitation has possibly changed over time; (e) the main meteorological wind patterns may have changed; (f) the flow pattern in the accumulation area may have changed. Much more work is needed to resolve these problems before reliable estimates of temperature change over glacial–interglacial timescales can be derived from ice-core isotopic records.

5.4.1 The Holocene $\delta^{18}O$ record

Although the problems of dating and interpreting the isotopic records at depth are legion, nearer the surface we can have much more faith in both the timescale and in direct temperature estimates. Certainly, for recent time intervals, points (a), (b), (c) and (f) above can be considered constant. All of the cores show a marked and unmistakable increase in $\delta^{18}O$ values at the end of the last glaciation and this means that they contain detailed records covering the last 10 000–12 000 years. Within the Holocene, maximum temperatures (maximum $\delta^{18}O$ values) were recorded at Camp Century from 4000–5000 years BP and at Devon Island temperatures peaked around 4800 years BP (Fig. 5.23; Fisher & Koerner 1981). In Antarctica, the Byrd record shows a relatively warm interval from 7600 to 4500 years BP (Johnsen *et al.* 1972) but this is not as significant or as constant as in the Northern Hemisphere records (Epstein *et al.* 1970). At Vostok, highest isotopic values occur between 7500 and 5500 years BP (Barkov *et al.* 1977). Following the Hypsithermal, temperatures declined, with minimum $\delta^{18}O$ values at Vostok from 3500 to 1500 years BP. At Byrd Station an increase in $\delta^{18}O$ of ~1.5‰ from ~4000 to 2000 years BP is thought to be due to a decrease in ice thickness (perhaps by 100–150 m) following the Hypsithermal (Johnsen *et al.* 1972). In the Northern Hemisphere, $\delta^{18}O$ data (corrected where necessary for changes in ice-sheet elevation) indicate that temperatures probably declined by 1 °C from 5000 to 3000 years BP (Paterson *et al.* 1977). During this time, the Meighen Ice Cap was formed (Koerner & Paterson 1974). A slightly warmer interval from 2500 to 1000 years BP can also be seen in the Devon Island and Camp Century records (Fig. 5.23) and it is interesting that this corresponds to a long period of negative mass balance seen in the Meighen Island ice core (Koerner & Paterson 1974).

5.4.2 The $\delta^{18}O$ record of the last 1000 years

Detailed studies of $\delta^{18}O$ variations of the last millennia have been made by Johnsen *et al.* (1970), Dansgaard *et al.* (1975), Paterson *et al.* (1977), Hammer *et al.* (1978), Fisher and Koerner (1981), and Gordiyenko *et al.* (1981). Over this timescale, dating of the ice cores is very precise and many of the variables which made climatic interpretations of the long-term $\delta^{18}O$ record difficult can be considered constant for this interval.

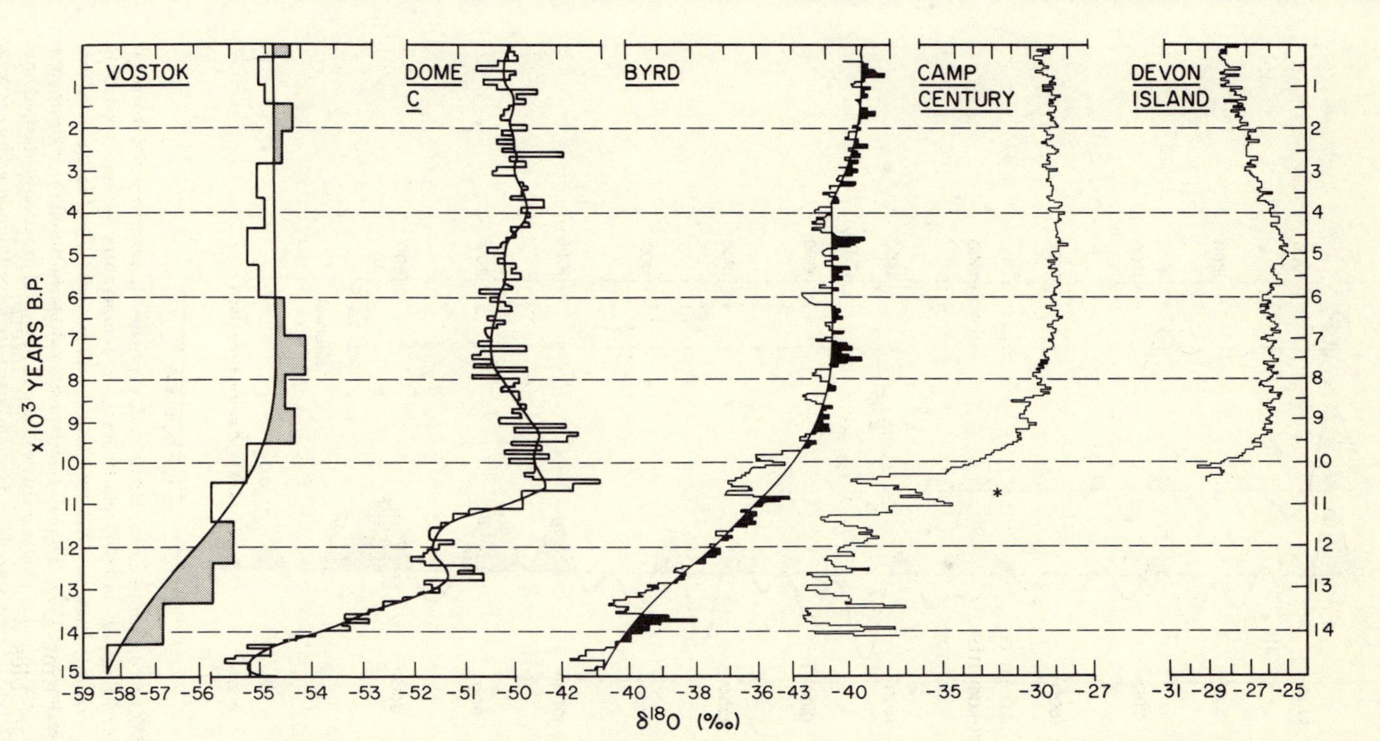

Figure 5.23 $\delta^{18}O$ records of four ice cores compared for the last 10–15 000 years of record. *Timescales may not be correct* and errors may be large towards the bottom of these core sections. Of the records shown, the Camp Century record is quite precise back to ~8500 BP; below this level the timescale is uncertain, but the section indicated by a star is thought to correspond to the Bolling/Allerod interstadial, widely observed in European paleoclimatic records (Dansgaard *et al.* 1983). The Devon Island Ice Cap record is considered to be accurate to within ±5% back to 5300 BP but is increasingly uncertain beyond that, based on cross correlation with the Camp Century record (Paterson *et al.* 1977). The Vostok, Dome C and Byrd (Antarctica) records are all based on simple flow models (Barkov *et al.* 1977, Lorius *et al.* 1979, Johnsen *et al.* 1972, respectively).

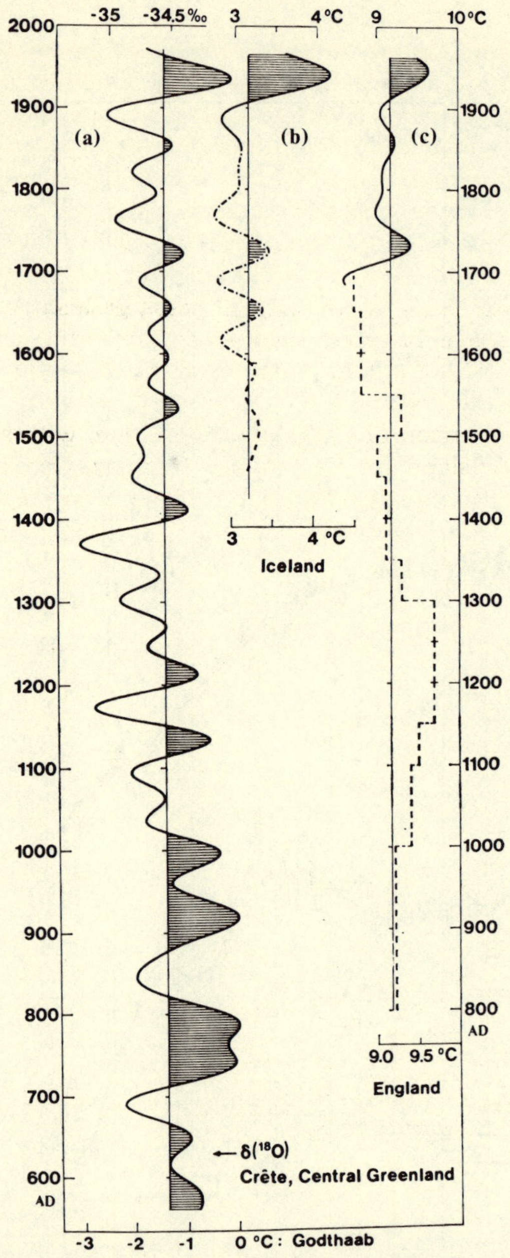

Figure 5.24 (a) $\delta^{18}O$ record from Crête, Greenland, compared to reconstructed temperatures from (c) central England and (b) Iceland. Icelandic temperature estimates based on instrumental observations back to ~AD 1850 and on calibrated sea-ice data before that (dashed line) (Bergthorsson 1969). English temperature estimates based on instrumental records back to AD 1698 (Manley 1953) and historical and botanical proxy data (dashed line) prior to that (Lamb 1965). Crête isotopic and Icelandic sea-ice record smoothed by a 60 year low-pass digital filter (after Dansgaard *et al.* 1975).

Hence direct temperature interpretations may be reasonable. As an example, Figure 5.24 shows a smoothed version of the Crête $\delta^{18}O$ record from Greenland interpreted in terms of mean annual temperatures at Godthaab (now renamed Nuuk); calibration was based on a regression of Godthaab temperature and $\delta^{18}O$ data for the past ~100 years. The reconstructed record is similar to long-term (instrumentally recorded) temperature data from central England (Manley 1953, 1974) back to the 17th century and shows good correlation with paleotemperatures estimated for Iceland (back to ~AD 1600), based on historical sea-ice data (Bergthorsson 1969). Paleotemperature estimates for central England, based on historical and botanical data (Lamb 1965), show a much poorer correlation with the Crête record.

A further comparison of $\delta^{18}O$ data from Crête, with similar records from the Devon Island Ice Cap, Camp Century, Greenland, and the Lomonosov Plateau, Spitsbergen (Fig. 5.25), shows considerable diver-

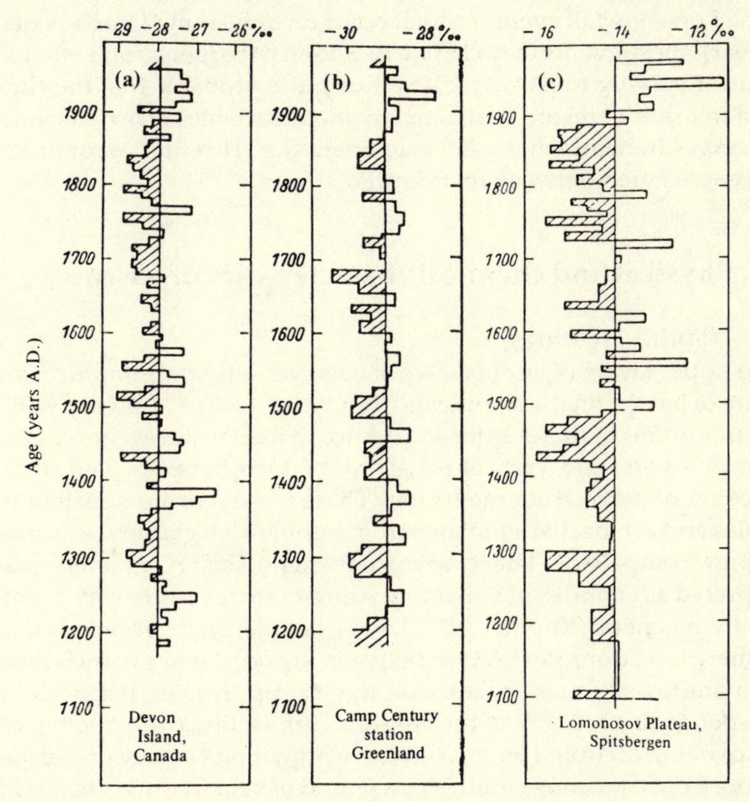

Figure 5.25 $\delta^{18}O$ records of the last 1000 years from (a) Devon Island Ice Cap, (b) Camp Century, Greenland, and (c) the Lomonosov Plateau, Spitsbergen (after Gordiyenko *et al.* 1983). Devon Island data from Paterson *et al.* (1977); Camp Century data from Johnsen *et al.* (1970). Shading indicates values lower than those at the top of the cores.

sity in the records. Most show a long period when snowfall was depleted in $\delta^{18}O$ during the 18th and 19th centuries and in some cases this is also characteristic of the 17th century record (e.g. Camp Century, Devon Island). A "warmer" interval during the 16th century is apparent in most records. Prior to that it becomes increasingly difficult to generalize, the only widespread feature being a relatively "cold" period around AD 1300 preceded by a warmer episode around AD 1250. Of course, it may be that looking for correspondence over such a large area is naïve and that there is no *a priori* reason to expect a uniform paleotemperature signal. Indeed, there is much evidence that the atmospheric circulation pattern over the North Atlantic favors inverse correlations between temperatures in Greenland and western Europe–Scandinavia (Van Loon & Rogers 1978). One could perhaps argue that each record therefore faithfully records paleotemperatures at the different sites whether they correlate or not. Until more proxy data become available this must remain a possibility (cf. Wigley 1978). However, it should not be forgotten that the $\delta^{18}O$ record is based on snowfall events (which occur on only a small fraction of days per year); these events may change in seasonal frequency and result from storms of varying trajectories. It is thus quite probable that the climatic signal in the $\delta^{18}O$ record is affected by these variables at least as much as by changes in hemispheric mean temperature. This may account for the differences noted between the records.

5.5 Physical and chemical characteristics of ice cores

5.5.1 Physical features

In the upper layers of subpolar ice sheets, various stratigraphic features that are of paleoclimatic significance can be seen (e.g. Langway 1970). The most important of these is the occurrence of melt features (horizontal ice lenses or layers, and vertical ice glands) which have resulted from the refreezing of percolating meltwater. These ice layers are deficient in air bubbles and can be distinguished from bubble-rich glacier ice formed by dry-snow compaction. The relative frequency of melt phenomena may be interpreted as an index of maximum summer temperatures or of summer warmth in general (Koerner 1977a). For example, on the Devon Island Ice Cap there is a strong correlation between the occurrence of melt features in firn and overall mass balance on the ice cap. A mass balance of zero corresponds to about 7% of the area of a core section (from the top of the ice cap) being occupied by melt layers; a higher percentage of melt layers in a core from the summit indicates a period of negative mass balance (i.e. a net reduction in the mass of the ice sheet in a particular period; Koerner 1977a). Long-term variations in melt layers may therefore be interpreted in terms of mass balance changes (Fig. 5.26). On this basis it appears that

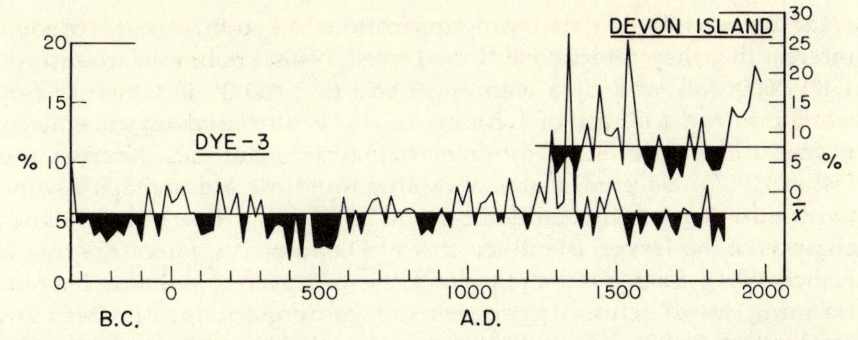

Figure 5.26 Melt features (percentage of melt features occupying a core section) in the Devon Island Ice Cap and Dye-3, Greenland, ice cores. Shading emphasizes values below the mean for ~1280–1860 AD. In the case of Devon Island, modern studies indicate values above this line signify periods of negative mass balance and vice versa (after Fisher & Koerner 1981, Herron *et al.* 1981).

the mass balance of the ice cap has been negative for approximately the last 100 years. Prior to that it had a positive balance for ~300 years, corresponding to the cooler period of the Little Ice Age. Warmer conditions prevailed from ~AD 1360 to AD 1560, with maximum mass losses being reached around AD 1530 (Koerner & Fisher 1981). It is interesting that the long cold period from 1560 to 1850 is also apparent in the Camp Century and Devon Island ice core $\delta^{18}O$ records.

Similar studies of melt features have been carried out on a 900 m core from Dye-3 in southern Greenland, enabling a longer (2200 year) record of summer warmth to be reconstructed (Fig. 5.26; Herron *et al.* 1981). Again the cooler period from ~1530 to 1860 is apparent and it is interesting that this is also recognizable in $\delta^{18}O$ records from Devon Island, Camp Century, and Dye-3 ice cores. It was clearly a widespread and major period of low temperatures. By contrast, the period from ~AD 950 to AD 1530 was exceptionally warm, reaching a maximum in the period 1313–1344. Summer temperatures were even lower than during the Little Ice Age from ~300 BC to 100 BC and from AD 350 to AD 550, though this is not apparent from the $\delta^{18}O$ records alone.

Hibler and Langway (1977) have studied the record of melt phenomena in another core from Dye-3 by means of spectral and maximum entropy analysis over the interval AD 1230–1950. They find that the record of melt features is coherent with $\delta^{18}O$ data at both high and low frequencies, even though the $\delta^{18}O$ is thought to represent mean annual temperatures (see Sec. 5.2.3) whereas the melt record is an index of summer temperatures only. Using the observed harmonics of both $\delta^{18}O$ and melt phenomena, Hibler and Langway attempted to forecast variations in temperature through the early part of the 21st century. Their results are based on apparent periodicities, of unknown cause, which account for very little variance in the original record, so it is probably wise not to place too much faith in their prognostications. Nevertheless, it is of some interest that they forecast continued (post-1940s) cooling to around AD 1980–2000, followed by a warming trend to ~AD 2040, which is very similar to the forecast of Johnsen *et al.* (1970) based on an entirely different $\delta^{18}O$ record and different statistical techniques. Koerner and Fisher (1981) also predict annual cooling to ~1990 and perhaps beyond (with summer cooling bottoming out 20–50 years later) based on an analysis of the Devon Island ice-core $\delta^{18}O$ and melt records (Koerner & Fisher 1981). Such predictions, and the prospect of enhanced global warming due to excess atmospheric CO_2 concentrations from fossil fuel combustion, led Broecker (1975) to speculate that the early part of the 21st century would be so warm as to warrant the term "super-interglacial." Whether this scenario emerges, only time will tell.

Although the record of melt phenomena in polar firn is the most important source of direct paleoclimatic information from the physical characteristics of ice cores, other types of densitometric and morphological information may be interpreted in climatic terms (Langway 1970). For example, Koerner (1968) was able to document the paleoclimatic history of the stagnant Meighen Ice Cap from a study of variations in crystal size, bubble fabric, and crystallographic *c*-axis orientation in an ice core drilled to bedrock. His conclusions were subsequently reinforced by $\delta^{18}O$ data (Koerner & Paterson 1974) which point to a post-Hypsithermal age for the ice cap, with ice growth from 4500 to ~2500 years BP followed by a prolonged period of ablation to ~AD 660. The ice cap then had a generally positive balance until the end of the 19th century, since which time it has lost considerable mass.

Stratigraphic studies may be complicated by the removal of entire seasonal horizons by katabatic (downslope) winds. In Antarctica, this is a particular problem (e.g. Watanabe *et al.* 1978) and it appears now that it may be of more significance in Greenland and the Canadian Arctic than hitherto supposed (Fisher *et al.* 1983). Nevertheless, careful interpretations of density variations, cross-checked with $\delta^{18}O$ measurement, may enable accumulation rates over time to be estimated (e.g. Gonfiantini *et al.* 1963).

5.5.2 *Gas content of ice cores*

Another important component of ice cores which is of paleoclimatic significance is the atmospheric gas content, as the air pores are closed off during the densification of firn to ice. This volume depends primarily on the atmospheric pressure, and to a lesser extent on the surface air temperature, prevailing at the formation site (Raynaud & Lorius 1973). This has important implications for ice-core studies; changes in ice-sheet thickness (surface elevation), particularly between late Wisconsin and Holocene time, result in quite large changes in $\delta^{18}O$, thereby complicat-

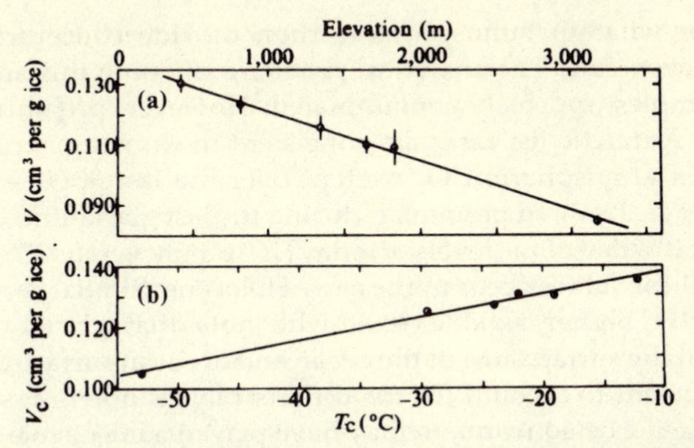

Figure 5.27 (a) Total gas content ($\bar{V}$) of Antarctic and Greenland ice versus the elevation of formation site; (b) derived pore volume at pore close-off (Vc) versus the temperature at the formation site. These relationships have been used to estimate the former elevation at which ice formed, and hence to assess ice sheet thickness at different periods in the past (after Raynaud & Lebel 1979).

ing any direct climatic interpretation of the $\delta^{18}O$ data (see Sec. 5.4). The analysis of total gas volume in ice at depth can help identify the magnitude of elevation changes during past climatic periods. Figure 5.27 shows the empircial relationship between surface elevation, surface temperature, and gas volumes in ice forming today at Antarctic and Greenland sites (Raynaud & Lebel 1979). Using these relationships and measured gas volumes in bubbles from Camp Century ice of late Wisconsin age, Raynaud and Lebel estimate that the ice sheet at Camp Century was ~800 m thicker during the late Wisconsin than during the Holocene. Such a change would account for about half the increase of 11‰ in $\delta^{18}O$ at the end of the last glaciation observed in the Camp Century ice core. The suggestion that the elevation of the Camp Century core site has changed over time is also made by Fisher (1979) in a comparison of $\delta^{18}O$ records from the Devon Island Ice Cap and Camp Century ice cores. He estimates that the Camp Century site was at least 500 m higher in pre-Holocene times. Alternatively, a change in ice-sheet geometry may have altered the flow of the ice sheet, causing ice from nearer the center of the ice sheet (currently >1500 m above Camp Century) to move towards the core site during Wisconsin time (Paterson 1977). This suggestion is supported by Raynaud and Whillans (1979), who argue that the ice-sheet geometry was different during the last glaciation as a result of lower sea level affecting the position of the ice-sheet margin in Melville Bay. Whatever the reason, it is clear that apparent changes in ice-sheet elevation are an important factor in the interpretation of ice-core $\delta^{18}O$ records, and further studies of gas content in other cores are needed to provide more information on the subject.

The gas content of ice cores has recently been studied to try and determine whether atmospheric carbon dioxide concentrations have changed over time. The analytical problems of doing this are very great as the samples are easily contaminated. However, preliminary results from two Antarctic ice cores are consistent in showing very significant changes in atmospheric CO_2 content over the last 30 000–50 000 years (Delmas *et al.* 1980). In particular, during the last glaciation, CO_2 content was only 50% that of the levels of today (160 p.p.m. versus 330 p.p.m.; Fig. 5.28). Highest values occur in the early Holocene. Similar results are also reported by Berner *et al.* (1980), who note that glacial–interglacial changes in the surface area of the ocean and of ocean surface temperature are insufficient to account for the different levels noted. It seems likely that changes in the biosphere may have played a major role in affecting global CO_2 content (cf. Shackleton 1977b). Such large changes of a gas which is so important in the global energy balance have profound implications for theories of climatic change. Future studies of this important matter will be watched with great interest.

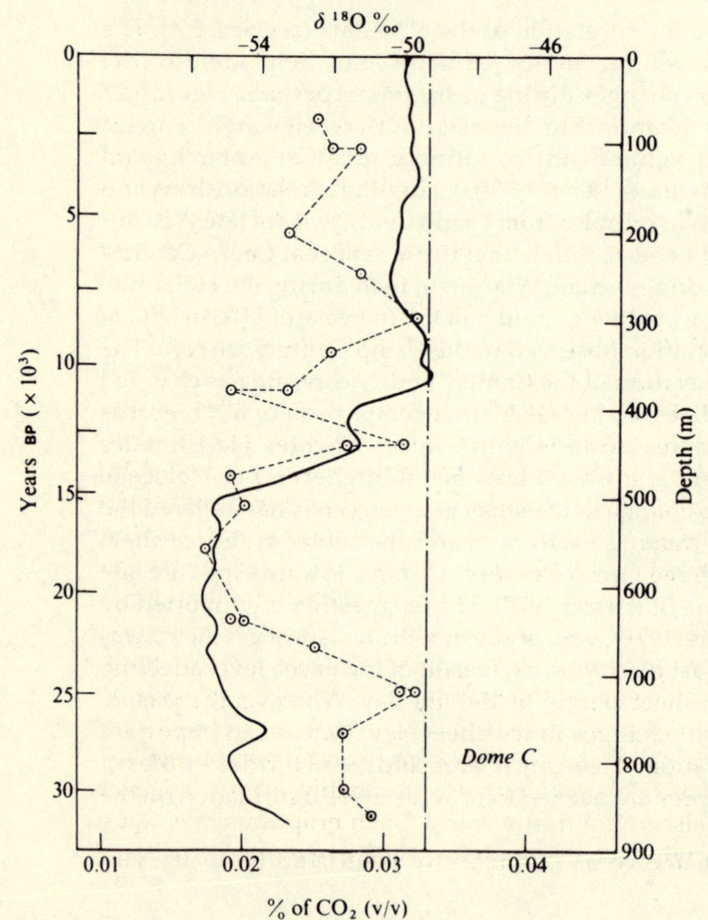

Figure 5.28 Percentage of carbon dioxide in the air of bubbles in the Dome C (Antarctic) ice core (dashed line), $\delta^{18}O$ values in the core also shown as solid line. Timescale is that proposed by Lorius *et al.* (1979). Present-day atmospheric CO_2 content (0.033%) is shown as a dotted–dashed line. Although the ice-core CO_2 data show considerable scatter, atmospheric CO_2 content during the coldest part of the last glacial period (~20 000–15 000 years BP) seems to have been only about half that of today. Similar results were obtained from studies of the Camp Century (Greenland) and Byrd (Antarctica) ice cores (Neftel *et al.* 1982).

5.5.3 *Long-term changes in particulate matter in ice cores*

Seasonal variations of particulate matter in ice cores have already been discussed with reference to dating. In addition to these high-frequency variations, very large changes in dust content are observed in pre-Holocene ice from deep ice cores (Fig. 5.29). Particle concentrations in late-glacial strata are three times greater than in Holocene ice in the Byrd Station core, six times greater in the Dome C core, 12 times greater in the Camp Century core, and 20 times greater in the Devon Ice Cap core (Thompson & Mosley-Thompson 1981, Koerner 1977a). Furthermore, trace-element analysis of the particulate matter shows that the larger particles are predominantly of wind-blown, continental origin (i.e. loess) (Petit *et al.* 1981). In Greenland ice, concentrations of calcium, aluminum, and silicon are an order of magnitude greater than in Antarctic ice and are attributed to wind erosion of areas of glacial outwash and exposed continental shelves (especially along the Siberian and Alaskan coasts) during the period of maximum eustatic sea-level lowering (Fig. 5.30). Particulate concentrations are lower in Antarctica due to the isolated position of the continent relative to other land masses. Nevertheless, late Wisconsin ice still shows higher aluminum concentrations, indicating increased transport of continental dust at that time. In addition, a greater influx of marine aerosols is indicated by an increase in both sodium and chlorine concentrations. Collectively, these results suggest a much stronger atmospheric circulation during late-glacial times (cf. Wilson & Hendy 1971), resulting in a higher global dust loading of the troposphere. This effect of high wind speeds would have been enhanced by the more extensive areas of arid and semi-arid land during late Wisconsin time (Sarnthein 1978). It is of interest to note that the main increase in particulate concentration is only seen in ice estimated to be between 24 000 and 12 000 years of age (Cragin *et al.* 1977, Koerner & Fisher 1981). $\delta^{18}O$ values were very low for tens of thousands of years prior to this period, so it seems unlikely that Wisconsin cooling was caused by particulate loading of the troposphere. It seems more plausible that the presence of major ice sheets increased the vigor of the atmospheric circulation due to stronger Equator – Pole temperature gradients and that this, plus strong katabatic winds off the ice sheets, produced the general increase in turbidity recorded in the ice cores. Whether there was also an increase in volcanic dust is debatable. Early studies of the Byrd station ice core showed that occasional volcanic dust and ash bands are particularly abundant towards the bottom of the core, where $\delta^{18}O$ values are lowest (Gow & Williamson 1971; Fig. 5.31). Although the coarser ash may have originated from nearby volcanoes in Marie Byrd Land, Thompson (1977) demonstrated that much of the finer dust probably originated from explosive eruptions elsewhere in the world. Such eruptions inject dust into the stratosphere, where it is spread poleward, eventually settling out

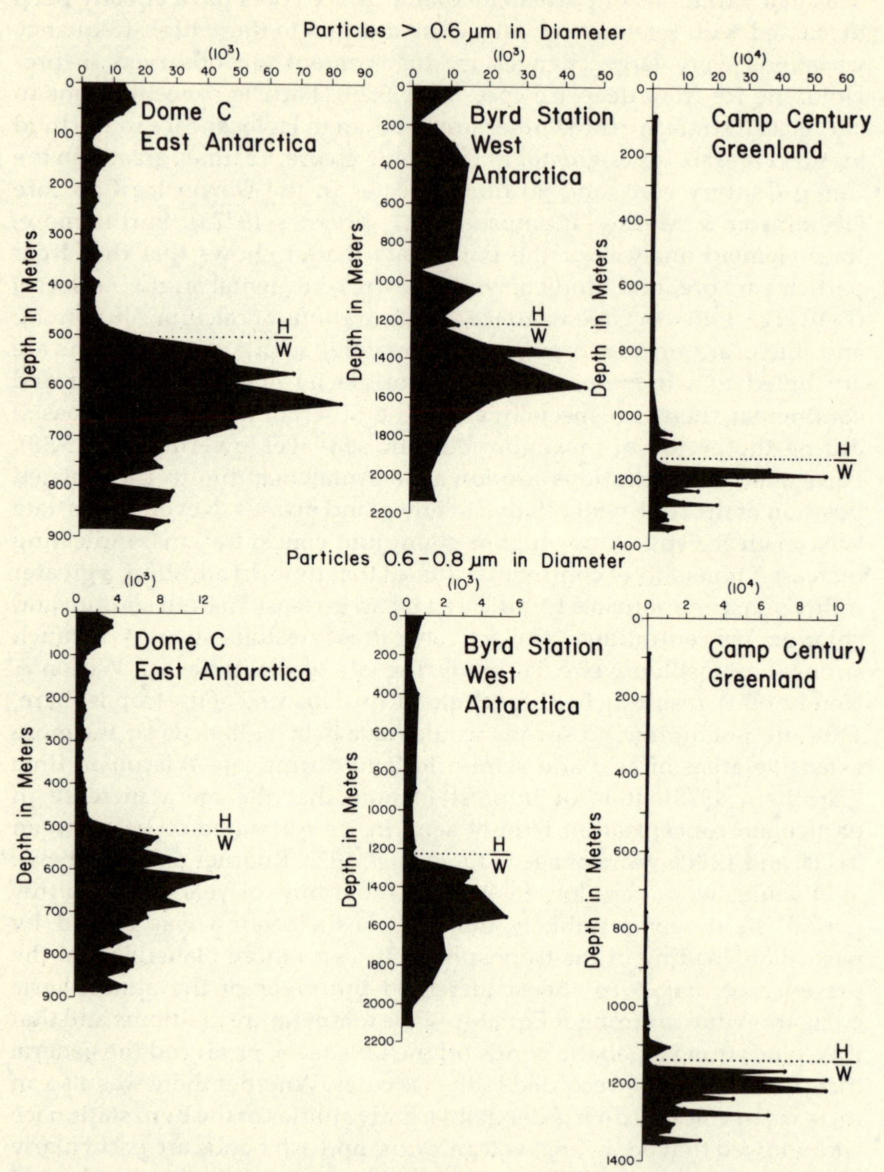

Figure 5.29 The average concentration of particles with diameters greater than or equal to 0.6 μm in samples from Antarctic and Greenland ice cores. The timing of the transition to ice of Holocene age is designated *H/W*, based on an abrupt shift to higher $\delta^{18}O$ values observed in the ice cores. Note that the abscissa scale for Camp Century is increased by one order of magnitude to accommodate the much higher particulate levels in that core. Ice dating from the last glacial period clearly contains much higher levels of particulate matter than Holocene ice (after Thompson & Mosley-Thompson 1981).

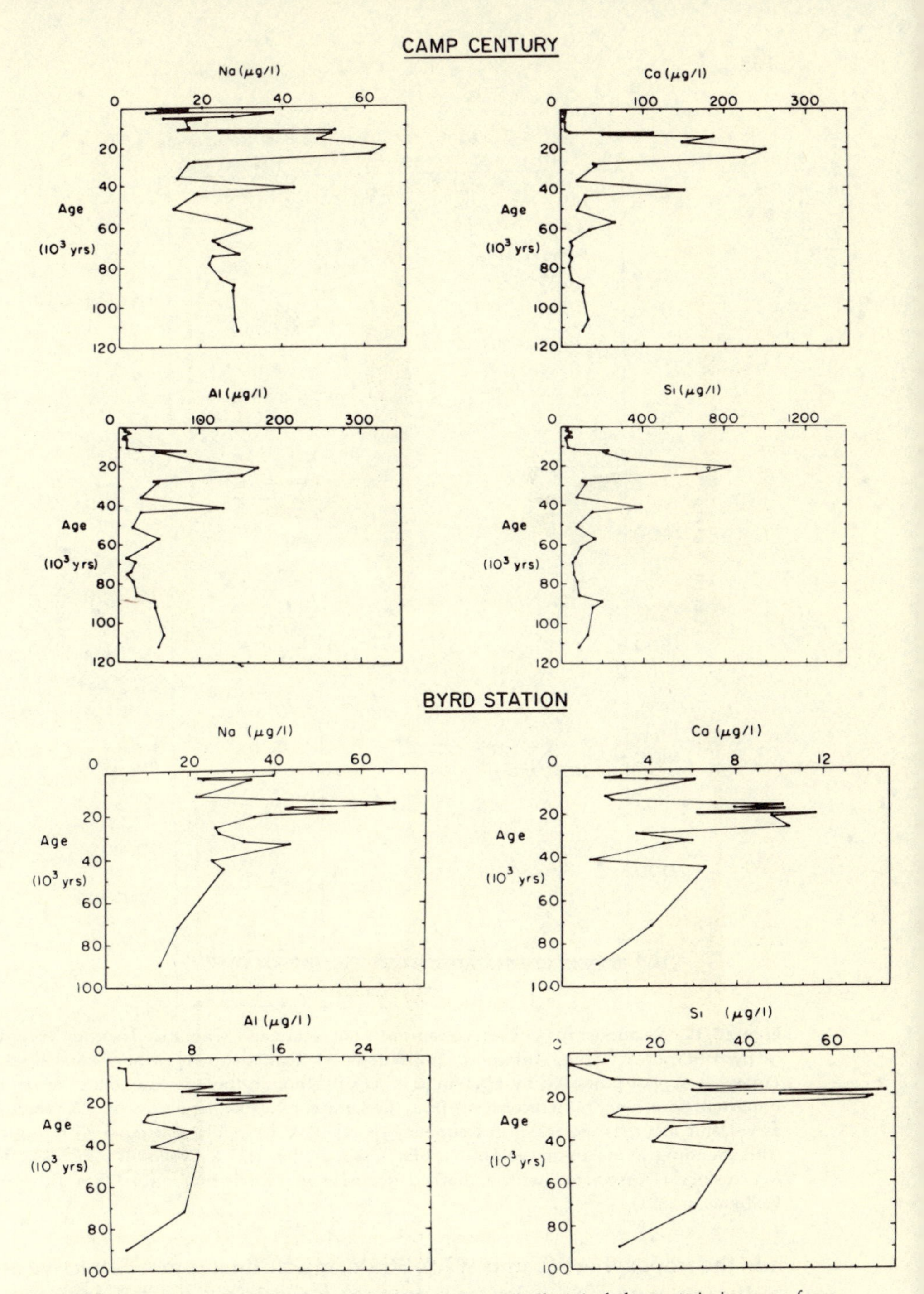

Figure 5.30 Variations in the concentration of certain chemical elements in ice cores from Camp Century, Greenland, and Byrd Station, Antarctica. Timescales are based on original estimates by Dansgaard *et al.* (1971) and Johnsen *et al.* (1972) (see text). Note differences in abscissa scales for each element. During late Wisconsin times, the records show massive increases in airborne aluminum, silicon, and calcium cations, due to stronger atmospheric circulation, more extensive arid lands, and large areas of exposed continental shelf, especially on the margins of the Arctic Ocean (after Cragin *et al.* 1977).

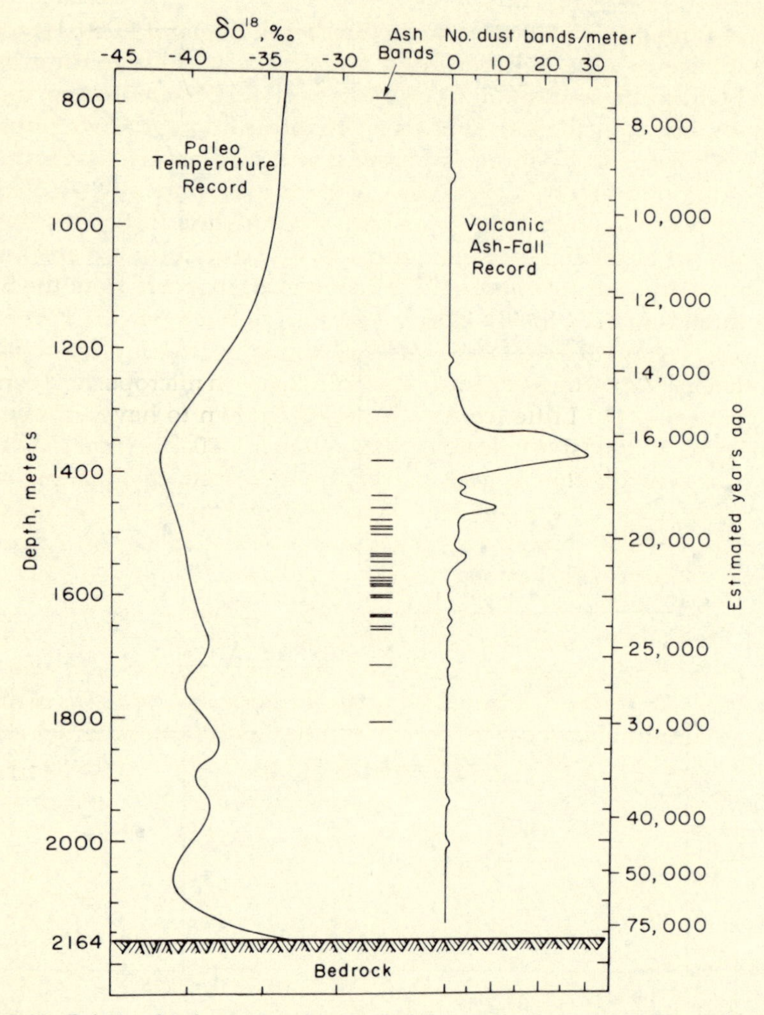

Figure 5.31 Relationship between volcanic ash-fall record and generalized isotopic record in Byrd Station ice core, Antarctica. Individual ash bands shown as horizontal lines. Timescale is that proposed by Epstein *et al.* (1970) (though the absolute values are not important here, since both records are from the same core). Note the decline in $\delta^{18}O$ values as volcanic ash falls increased in frequency (post 30 000 years BP) and the $\delta^{18}O$ minima corresponding to maximum volcanic ashfall (estimated at ~17 000 years BP). Increase in $\delta^{18}O$ values is associated with a marked decrease in ashfall frequency (after Gow & Williamson 1971).

into the troposphere (Lamb 1970). Stratospheric dust is more effective at scattering incoming solar radiation and reducing surface radiation receipts than tropospheric dust. Polar regions are quite vulnerable to the effects of volcanic dust, particularly in the summer melt season when solar radiation passes through the greatest depth of atmosphere, and the surface is illuminated continuously (Bradley & England 1978). Hence the coincidence in time between high volcanic dust concentration in the Byrd Station core and low $\delta^{18}O$ values is particularly interesting, though perhaps only coincidental. There is no evidence from Arctic ice cores that volcanic dust concentrations increased during the last glaciation, though the higher non-volcanic particulate content may mask any such signal. Certainly, in Greenland ice cores spanning the last 1000 years there is virtually no correlation between microparticle concentration and known volcanic eruptions (Hammer 1977b). However, in a core from the South Pole extending back to AD 1056 ± 90, peaks in microparticle concentration do correspond to known explosive eruptions (Mosley-Thompson & Thompson 1982). There is also a general increase in microparticle concentration during the Little Ice Age, a period known to have experienced high levels of explosive volcanic activity (Lamb 1970, Hammer *et al.* 1980). It is thus possible that at least part of the increased dust loading of the atmosphere during the last glaciation recorded in polar ice cores was volcanic in origin, though the bulk was very probably loessic. As such, it was a symptom of glaciation rather than a cause.

5.5.4 *Long-term changes in nitrate ions in ice cores*

The question of whether solar activity has remained constant over time is of great interest to paleoclimatologists. Recently, the analysis of nitrate ion concentration in polar ice cores has revealed variations which closely parallel changes in solar activity (sunspot number) in recent years (Fig.

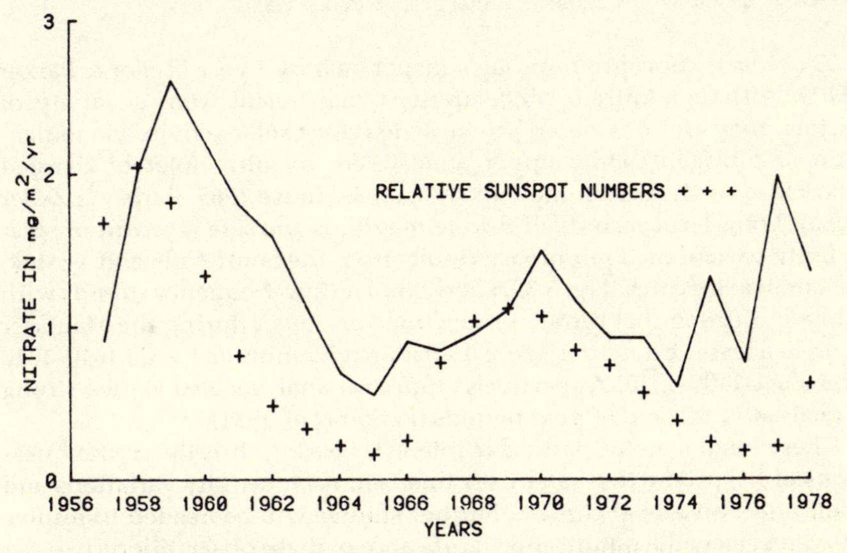

Figure 5.32 Nitrate levels in annual layers from a glaciological pit at the South Pole in relation to relative sunspot numbers. Nitrate levels peak 1 year after sunspot maxima (after Parker & Zeller 1980).

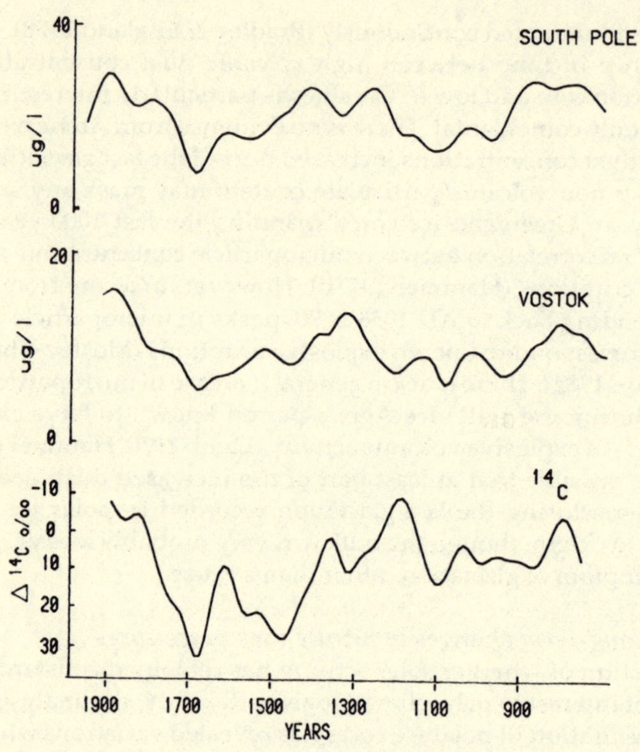

Figure 5.33 Smoothed (low-frequency) records of nitrate concentration in ice cores from the South Pole and Vostok, Antarctica (after Zeller & Parker 1981).

5.32). Nitrate concentrations lag sunspot data by 1 year (Zeller & Parker 1981). Although nitrate concentrations may result from a variety of factors, they are considered primarily to reflect solar activity via ionization of nitrogen in the upper atmosphere by ultraviolet or charged particle emissions from the Sun (Wilson & House 1965, Parker & Zeller 1980). Long-term records of nitrate may thus provide a record of solar activity variations. Preliminary results from the South Pole and Vostok, Antarctica, ice cores (Fig. 5.33) show similar (low-frequency) trends with some indication that nitrate production was lower during the Maunder (and to a lesser extent the Spörer) solar activity minima (~AD 1640–1710 and ~AD1400–1510, respectively). Spectral analysis also shows strong signals at 11, 22, and 66 year periods (Parker *et al.* 1981).

These results are of particular interest, bearing directly on the questions of ^{14}C production rates over time and solar activity variations and their effect on global climate. Further studies will be needed to resolve more precisely the nature and significance of these observations.

6

Marine sediments

6.1 Introduction

Occupying more than 70% of the Earth's surface, the oceans are a very important source of paleoclimatic information. Between 6 and 11 billion metric tons of sediment accumulate in the ocean basins annually, and this may be indicative of climatic conditions near the ocean surface or on the adjacent continents. Sediments are composed of both biogenic and terrigenous materials (Fig. 6.1). The biogenic component includes the remains of planktonic (near surface-dwelling) and benthic (deep-water) organisms which provide a record of past climate and oceanic circulation (in terms of surface water temperature and salinity, dissolved oxygen in deep water, nutrient or trace element concentrations, etc.). By contrast, the nature and abundance of terrigenous material mainly provides a record of humidity – aridity variations on the continents, or the intensity and direction of winds blowing from land areas to the oceans, and other modes of sediment transport to, and within, the oceans (fluvial erosion, ice-rafting, turbidity currents, etc.).

Over the past decade there has been an explosive growth in paleoclimatic work based on the analysis of ocean cores, fuelled primarily by the CLIMAP research group (Climate: Long-range Investigation Mapping and Prediction). The contribution of this project to paleoclimatic research has been immense; not only has the oceanic proxy record been studied and mapped over large areas (e.g. McIntyre *et al.* 1975, 1976, CLIMAP Project Members 1976), but many important insights about events on the continents have been gained (e.g. rates and timing of ice-sheet growth and decay; Ruddiman & McIntyre 1981a). An improved understanding of the chronology of continental glaciation has led to tests of hypotheses of climatic change (Hays *et al.* 1976) as well as suggested revisions of terrestrial chronologies (Kukla 1977). Many aspects of paleoclimatic research based on ocean cores remain controversial, but this is, perhaps, not entirely surprising. Revolutions, scientific or otherwise, are never without controversy.

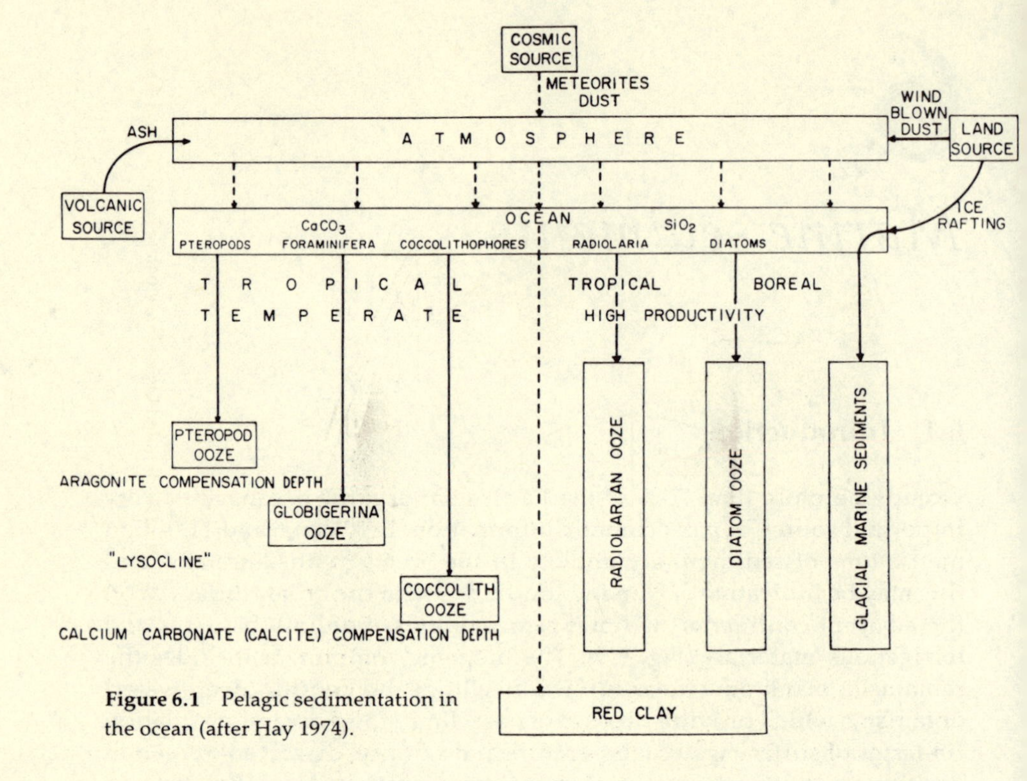

Figure 6.1 Pelagic sedimentation in the ocean (after Hay 1974).

6.2 Paleoclimatic information from biological material in ocean cores

Paleoclimatic inferences from biogenic material in ocean sediments derive from assemblages of dead organisms (thanatocoenoses) which make up the bulk of all but the deepest of deep-sea sediments (biogenic ooze). However, thanatocoenoses may not be representative of the biocoenoses (assemblages of living organisms) in the overlying water column. Selective dissolution of thin-walled specimens at depth (see Sec. 6.6), differential removal of easily transported species by scouring bottom currents, and occasional contamination by exotic species transported over long distances by large-scale ocean currents all contribute elements of uncertainty. Because of these problems, sediments over much of the ocean floor are unsuitable for paleoclimatic reconstruction. This is illustrated in Figure 6.2 for foraminiferal studies (Ruddiman 1977a), though it should be noted that in some areas unsuitable for foraminiferal preservation the remains of other organisms, such as diatoms or Radiolaria, may provide a useful record (e.g. Hays 1978, Sancetta 1979).

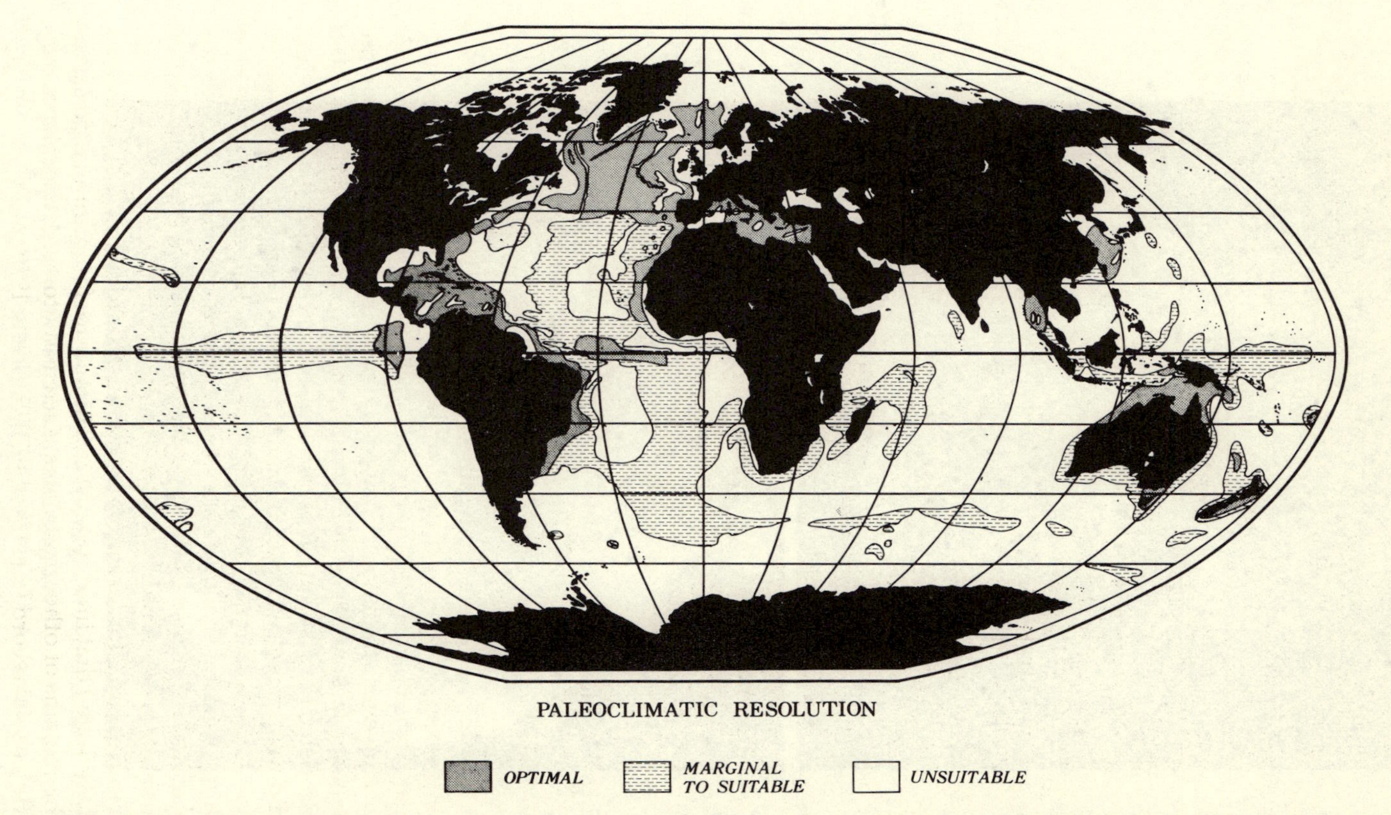

Figure 6.2 Regions considered to be optimal, marginally suitable, and unsuitable for accurate and detailed Quaternary paleoclimatological studies based on foraminifera (after Ruddiman 1977a).

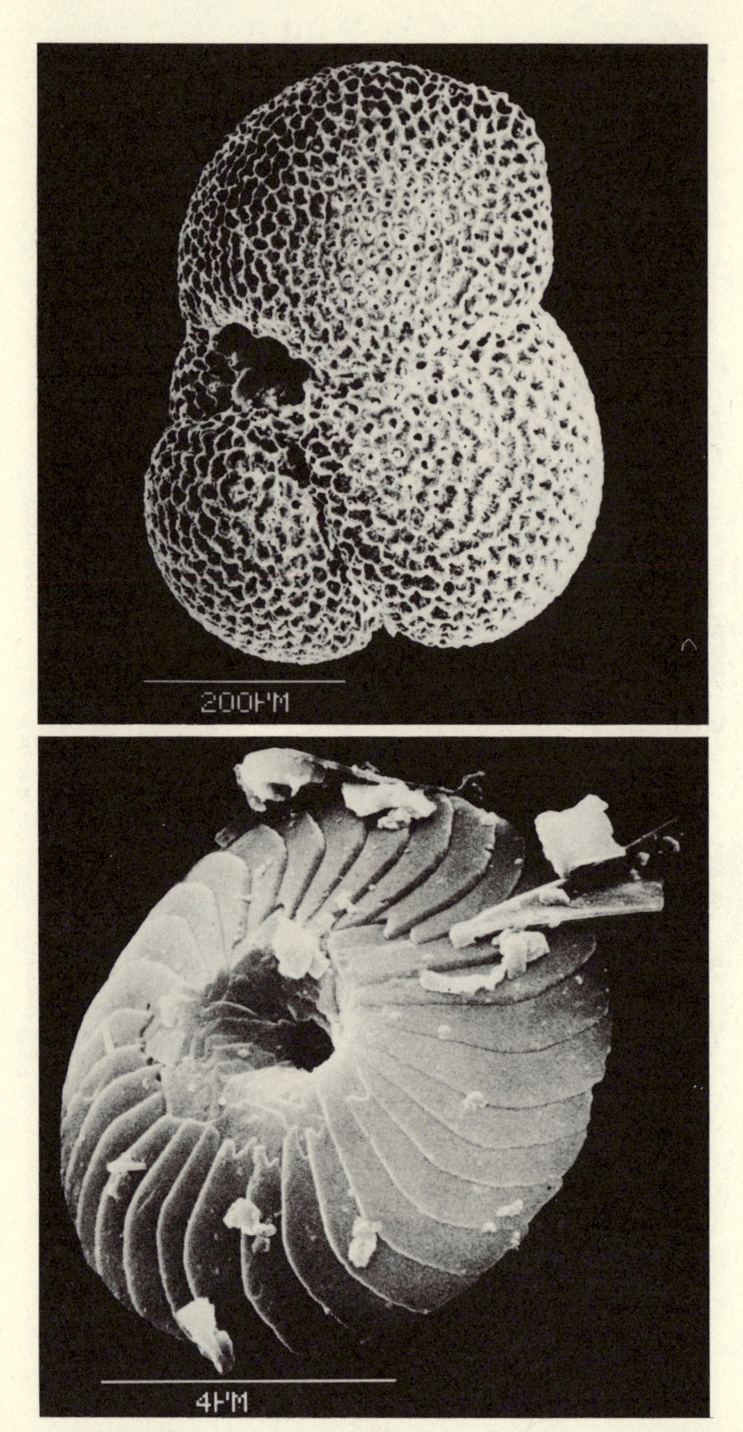

Figure 6.3 Two examples of calcareous tests used in paleo-oceanographic studies. The foraminifera *Globigerinoides sacculifer* (above) and (below) the coccolith *Cyclococcalithus leptoporus* (photograph kindly provided by W. Ruddiman, Lamont–Doherty Geological Observatory).

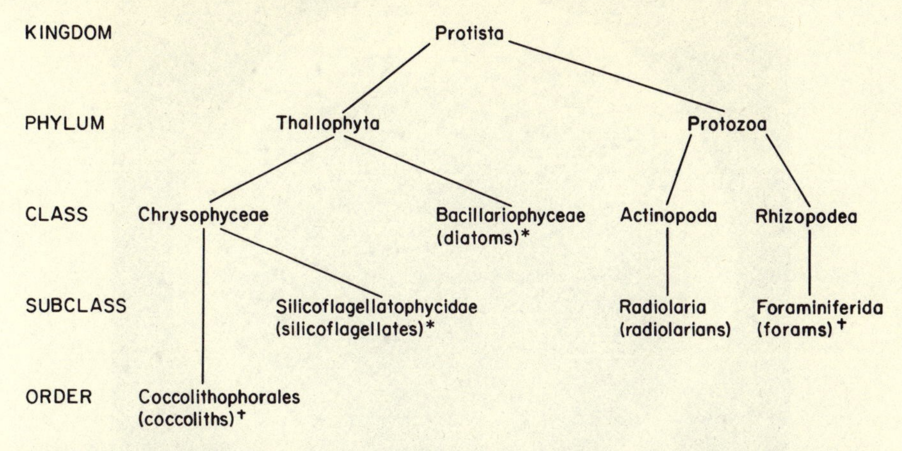

Figure 6.4 Taxonomic relationships of the main marine organisms used in paleoclimatic reconstructions. Asterisks indicate siliceous tests; the dagger indicates calcareous tests.

Biogenic oozes are primarily made up of the calcareous or siliceous skeletons or tests of marine organisms. These may have been planktonic (passively floating, near-surface, 0–200 m, organisms) or benthic (adapted to living in deep water). For paleoclimatic purposes, the most important calcareous materials are the tests of foraminifera (a form of zooplankton) and the much smaller tests, or test fragments, of coccolithophores (unicellular algae) known informally as coccoliths (Figs. 6.3 & 6.4). These are sometimes grouped with other minute forms of calcareous fossils and referred to as calcareous nannoplankton (*nanno* = dwarf) or simply nannoliths (Haq 1978). The most important siliceous materials are the remains of radiolarians and silicoflagellates (zooplankton) and diatoms (algae) (Haq & Boersma 1978; Fig. 6.5). By studying the morphology of the tests, individuals can be identified to species level and their ocean-floor distribution can then be related to environmental conditions (generally temperature and salinity) in the overlying water column (Fig. 6.6). However, it should be noted that the species assemblage in the sediment is a composite of all the species living at different depths in the water column as well as species with only a seasonal distribution in that particular area. The depth habitats of many species are not well known and it is believed that some species live at different water depths at different times in their life cycles. This is of particular significance to isotopic studies of tests since the oxygen isotope composition is a function of the water temperature (and to some extent salinity) at which the carbonate is secreted (Sec. 6.3). If test walls are secreted at varying depths through the lifespan of an individual, simple correlations with surface water temperatures and salinities may not be meaningful (Duplessy *et al.* 1981).

Paleoclimatic influences from the remains of calcareous and siliceous

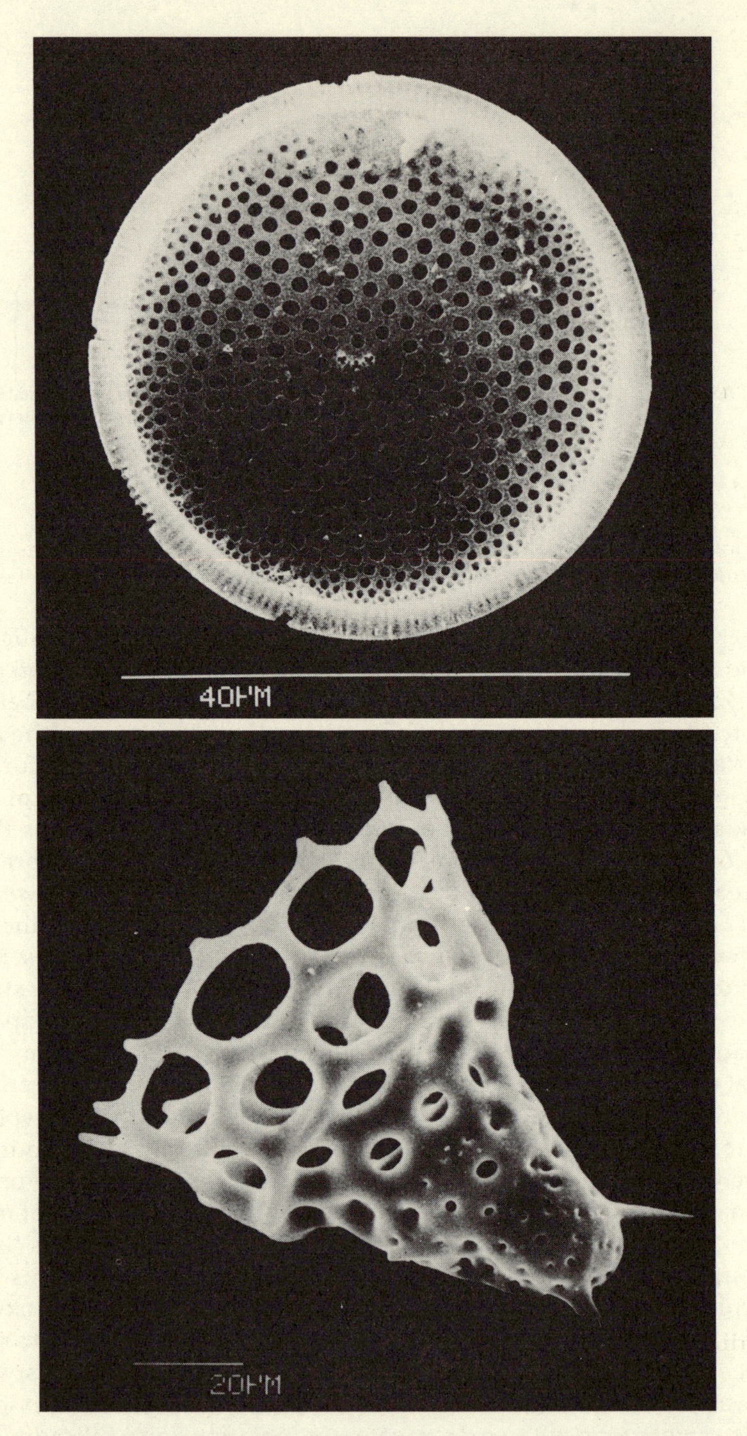

Figure 6.5 Two examples of siliceous tests used in paleo-oceanographic studies. The radiolarian *Cycladophora davisiana* (above) and (below) the diatom *Thalassiosira trifulta* (photograph kindly provided by W. Ruddiman, Lamont–Doherty Geological Observatory).

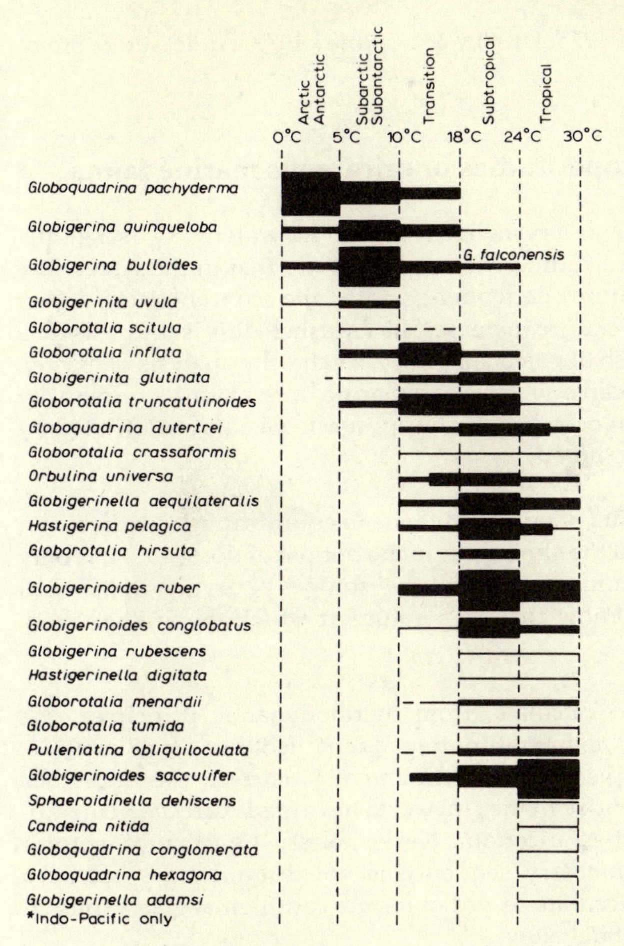

Figure 6.6 Sea-surface temperature ranges of some contemporary planktonic foraminifera, illustrating their temperature dependence. Organisms marked with an asterisk are Indo-Pacific only. Width of lines indicates relative abundance (after Boersma 1978).

organisms have resulted from basically three types of analysis: (a) the oxygen isotopic composition of calcium carbonate in the tests (Hecht 1976); (b) the relative abundance of warm- and cold-water species (e.g. Ruddiman 1971) or quantitative interpretations of species assemblages and their spatial variations through time (Imbrie & Kipp 1971); (c) morphological variations in a particular species resulting from environmental factors (ecophenotypic variations; Kennett 1976). Most work along these lines has concentrated on the Foraminiferida. In the following sections, therefore, the focus will be on foraminiferal studies. Paleoclimatic studies of coccoliths, radiolarians, diatoms, and silicoflagellates have mainly been in terms of relative abundance changes and morphological variations. Isotopic studies of diatoms have been attempted (e.g. Labeyrie 1974) but are not yet routinely carried out. Oxygen isotope variations in coccoliths provide useful paleotemperature estimates and may provide even more reliable data than that based on foraminifera alone (Margolis *et al.* 1975, Dudley & Goodney 1979; Anderson & Steinzetz 1981).

6.3 Oxygen isotope studies of calcareous marine fauna

If calcium carbonate is crystallized slowly in water, ^{18}O is slightly concentrated in the calcium carbonate relative to that in the water. The process is temperature dependent, with the concentrating effect diminishing as temperature increases. In a nutshell, this is the basis for a very important branch of paleoclimatic research – the analysis of oxygen isotopes in the calcareous tests of marine microfauna (principally foraminifera, but also coccoliths). The approach was first enunciated by Urey (1947, 1948) who noted,

> 'If an animal deposits calcium carbonate in equilibrium with the water in which it lies, and the shell sinks to the bottom of the sea . . . it is only necessary to determine the ratio of the isotopes of oxygen in the shell today in order to know the temperature at which the animal lived' (Urey 1948).

He then went on to calculate, from thermodynamic principles, the magnitude of this temperature-dependent isotopic fractionation. Although the principle of Urey's argument is correct, the numerous complications which arise in the real world have made direct paleotemperature estimates rather uncertain. Nevertheless, the value of isotopic variations in a sedimentary sequence is not diminished since such changes provide an accurate record of former continental glaciations, as discussed in more detail below.

6.3.1 *Isotopic composition of the oceans*

The oxygen isotopic composition of a sample is generally expressed as a departure of the $^{18}O/^{16}O$ ratio from an arbitrary standard:†

$$S = \frac{(^{18}O/^{16}O)_{\text{sample}} - (^{18}O/^{16}O)_{\text{standard}}}{(^{18}O/^{16}O)_{\text{standard}}} \times 10^3.$$

The resulting values are expressed in per mille (‰) units; negative values represent lower ratios in the sample (i.e. less ^{18}O than ^{16}O and therefore isotopically "lighter") and positive values represent higher ratios in the sample (more ^{18}O than ^{16}O and therefore isotopically "heavier").

† See footnote on page 127.

Empirical studies relating the isotopic composition of calcium carbonate deposited by marine organisms to the temperature at the time of deposition have demonstrated a relationship as follows:‡

$$T = 16.9 - 4.2(\delta_c - \delta_w) + 0.13(\delta_c - \delta_w)^2,$$

where T is water temperature in degrees Celsius, δ_c is the per mille difference between the sample carbonate and the SMOW standard, and δ_w is the per mille difference between the $\delta^{18}O$ of water in which the sample was precipitated and the SMOW standard (Epstein *et al.* 1953, Craig 1965).

For modern samples, δ_w can be measured directly in oceanic water samples; in fossil samples, however, the isotopic composition of the water is unknown and can not be assumed to have been as it is at the site today. In particular, during glacial periods the removal of isotopically light water from the oceans to form continental ice sheets (Ch. 5) led to an increase in the $^{18}O/^{16}O$ ratio of the oceans as a whole. Thus, the expected increase in δ_c of foraminiferal tests during glacial periods due to decreasing temperatures is complicated by the increase in δ_w of the ocean water at these times. How much of the increase in δ_c is the result of variations in δ_w is controversial. Emiliani (1955, 1966) has shown that isotopic variations in foraminiferal tests undergo quasi-periodic changes corresponding to glacial and interglacial periods (Fig. 6.7). In the Caribbean and Equatorial Atlantic, the amplitude of these variations§ is relatively constant at 1.8‰. If this was due entirely to temperature changes, it would correspond to a glacial–interglacial difference of ~8 °C. However, Emiliani estimates that only ~70% of the maximum isotopic change in δ_c was due to changes in temperature (i.e. 5–6 °C), the other 30% resulting from changes in the isotopic composition of ocean water (to +0.5‰ during maximum glaciations). Some support for his hypothesis is provided by independent (palynological and geological) evidence indicating that temperature changes on adjacent land areas were of a similar magnitude (Emiliani 1971a). However, if Emiliani is correct, the ^{18}O content of the continental ice sheets could not have been very low (he estimates an average value of −15‰ for ice in Greenland and Antarctica and only −9‰ for North American and Scandinavian ice sheets). Such values are considered to be completely untenable by Dansgaard and Tauber (1969), who argue that the present isotopic composition of

‡The precise form of the relationship depends on the particular technique used in analysis and on the temperature at which fractionation occurs (for further discussion, see Craig 1965 and Shackleton 1974).

§Glacial–interglacial $\delta^{18}O$ differences are minimum estimates due to the effects of sediment mixing (bioturbation) in cores, which tends to smooth out extremes in the record (Shackleton & Opdyke 1976).

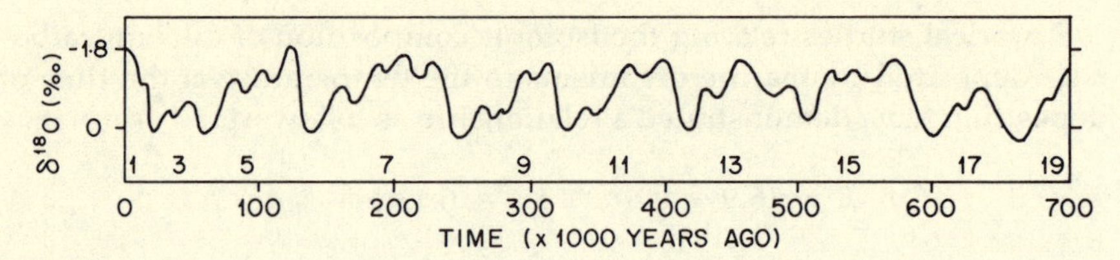

Figure 6.7 Generalized oxygen isotopic record of the last 700 000 years based on Caribbean and Atlantic deep-sea cores. This was originally considered to be a paleotemperature record of the upper layers of the ocean, but is now generally considered to be primarily a record of global ice volume. Note that values are plotted inversely on the ordinate; more negative values correspond to reduced global ice volume (from Emiliani 1978).

precipitation in the world, and the measured isotopic composition of glacial-age precipitation (in ice cores), point to much lower ^{18}O concentrations in continental and polar ice sheets (cf. Fig. 5.7). On this basis, they estimate the isotopic composition of glacial age ocean water as +1.2‰, thus accounting for ~70% (not 30%) of the observed isotopic change in foraminiferal carbonate at that time. A similar conclusion was reached earlier by Shackleton (1967) who analyzed benthic foraminifera to show that bottom water from the Caribbean, Atlantic, and Pacific registered an increase in $\delta^{18}O$ during glacial times, similar to that of surface water (between +1.4 and +1.6‰). Bottom waters today (derived from cold, dense, polar water spreading through the deep ocean basins) are relatively close to the freezing point, so the bulk of the $\delta^{18}O$ increase could not be due to lower temperatures; rather, the evidence points to at least 70% of the increase resulting from changing isotopic composition of the oceans (Duplessy 1978). Hence, it would appear that the isotopic changes recorded in foraminiferal tests are not due entirely to temperature changes but are indicative of changing terrestrial ice volumes (Shackleton 1967). Emiliani's isotopic temperature record (Fig. 6.7) can therefore be considered as a paleoglaciation record (Dansgaard & Tauber 1969). With this interpretation, it would appear that there have been several major continental glaciations during the last 425 000 years, and perhaps as many as 21 glaciations during the entire Pleistocene (van Donk 1976).

Changes in the isotopic composition of ocean water through time are not the only complications affecting a simple temperature interpretation of δ_c. Urey's initial hypothesis developed from a consideration of calcium carbonate precipitated inorganically, where the carbonate forms in isotopic equilibrium with the water. However, in the formation of carbonate tests by living organisms, metabolically produced carbon dioxide may be incorporated; in such cases, the carbonate would not be formed in isotopic equilibrium with the water, and the resulting isotopic composition would differ from the thermodynamically predicted value (Duplessy *et al.* 1970a, Vinot-Bertouille & Duplessy 1973, Shackleton *et al.* 1973). This was termed the vital effect by Urey (1947). The contribution of metabolic carbon dioxide to the test carbonate differs from one species of foraminifera to another. Modern samples of *Globigerinoides ruber*, for example, give isotopic values 0.5‰ lighter than expected from thermodynamic principles alone (based on analysis of water from their modern habitat). This is equivalent to a temperature error of ~2.5 °C (Shackleton *et al.* 1973). On the other hand, not all forams exhibit this unfortunate characteristic. For example, samples of *Pulleniatina obliquiloculata* and *Uvigerina* spp. (a benthic foram) appear to be in isotopic equilibrium with surrounding water (Shackleton 1974). In other species, where isotopic equilibrium is not achieved, there is evidence that the vital effect remains constant over time (Duplessy *et al.* 1970a). It is thus possible to circumvent this particular problem by careful selection of the species being studied, or by assessing its specific vital effect, and adjusting the measured isotopic values accordingly.

Another complication in calculating water temperatures from the isotopic composition of carbonate tests is the problem of variations in depth habitat of foraminifera. Even if the ice effect and vital effects are known, there is still some uncertainty as to whether foraminifera lived at the same depth from glacial to interglacial times. Water temperatures in the upper few hundred meters of the ocean change rapidly with depth, particularly outside the Tropics (Table 6.1), so small variations in depth habitat can be equivalent to a change in temperature of several degrees Celsius (i.e. a change perhaps as large as the glacial to interglacial change at the surface of the ocean). It is thus critical to know what factors control depth habitat of foraminifera, and in particular the depths at which tests are secreted (Emiliani 1971b). Several studies have concluded that water density (a function of temperature and salinity; Fig. 6.8) is of prime importance to individual species, since the same species may be found in different areas living at different depths, but in water of the same temperature and salinity (Emiliani 1954, 1969, Hecht & Savin 1972). During glacial periods, when the oceans were more saline (due to the removal of water to the continental ice sheets), foraminifera may have migrated upwards in the water column, to a zone of warmer water, in order to maintain a constant density environment. Conversely, they may have migrated downwards (to cooler water) in interglacials (Fig. 6.8). Clearly, such vertical migrations would result in isotopic paleotemperature estimates of glacial to interglacial temperature differences, considerably *less* than the changes actually occurring in the water column (Savin & Stehli 1974). Hence, if this model is correct, any residual paleotemperature signal obtained (after correcting for ice and vital effects) would have to be considered a minimum estimate only. This problem may, however, be a relatively minor one compared to the important effect of variations in depth habitat of foraminifera during their life cycle. There is now convincing evidence that although the tests of living forams contain

Table 6.1 Mean vertical temperature distribution (°C) and temperature gradients in the three oceans between 40°N and 40°S (from Defant, 1961).

	Atlantic Ocean		Indian Ocean		Pacific Ocean		Mean	
Depth (m)	Temperature (°C)	Gradient (°C/100 m)	Temperature (°C)	Gradient (°C/100 m)	Temperature (°C)	Gradient (°C/100 m)	Temperature (°C)	Gradient (°C/100 m)
0	20.0	2.2	22.2	3.3	21.8	3.1	21.3	2.8
100	17.8	4.4†	18.9	4.7†	18.7	4.4†	18.5	4.5†
200	13.4	1.8	14.3	1.6	14.3	2.6	14.0	2.0
400	9.9	1.5	11.0	1.2	9.0	1.2	10.0	1.3
600	7.0	0.7	8.7	0.9	6.4	0.65	7.4	0.75
800	5.6	0.35	6.9	0.7	5.1	0.4	5.9	0.5
1000	4.9	0.20	5.5	0.4	4.3	0.4	4.9	0.35
1200	4.5	0.15	4.7	0.3	3.5	0.2	4.2	0.22
1600	3.9	0.12	3.4	0.15	2.6	0.1	3.3	0.12
2000	3.4	0.08	2.8	0.09	2.15	0.05	2.8	0.07
3000	2.6	0.08	1.9	0.03	1.7	0.03	2.1	0.05
4000	1.8		1.6		1.45		1.6	

† Maximum gradient.

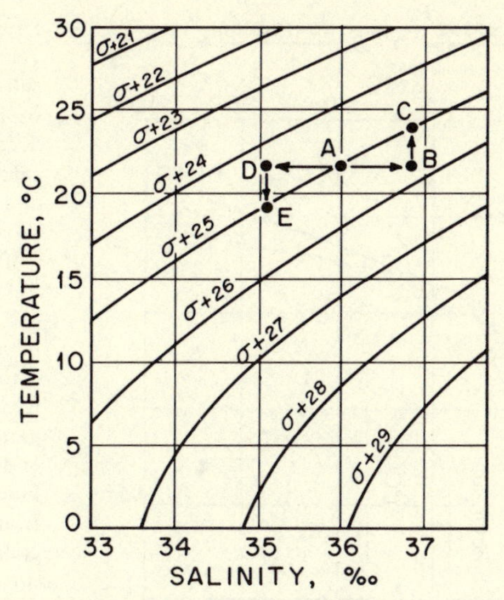

Figure 6.8 Temperature–salinity diagram: $\delta = 10^3 (\rho_w - 1)$, where ρ_w = water density. Density of sea water is a function of both temperature and salinity. Lines of equal density are shown. During glacial periods, when the removal of water to the continental ice sheets would have made oceanic salinity higher, foraminifera may have migrated to warmer water (generally upwards in the water column) to maintain a constant-density environment (illustrated schematically as A → B → C). In an interglacial, the opposite situation prevailed and the response of foraminifera may have been to move downward in the water column to a cooler zone below (A → D → E).

$CaCO_3$ which has been secreted in isotopic equilibrium with the upper mixed water layer, foram tests from the sea floor are significantly enriched with ^{18}O compared to their living counterparts (Duplessy *et al.* 1981). This is apparently due to calcification of the tests at depths (>300 m) considerably below the upper mixed layer, during the process of gametogenesis (reproduction). Gametogenic calcification may account for ~20% of foram test weight in samples from the sea floor and, because calcium carbonate has been extracted from water which is much cooler than that nearer the surface, the overall $\delta^{18}O$ values indicate a mean temperature significantly lower than the near-surface temperature (Fig. 6.9). Obviously, the rate at which the organism descends through the water column and the relative extent of gametogenic calcification will greatly influence the final isotopic composition of the test calcite, and make the difficult problem of isotopic paleotemperature estimates even more complex.

A final problem affecting paleotemperature calculations from the isotopic composition of test carbonate concerns the effect of dissolution on species composition in the thanatocoenoses. This is a pervasive factor which has implications not only for isotopic studies, but for all paleoclimatic studies based on faunal assemblages. Because of its importance, it is discussed in more detail below (Sec. 6.6.).

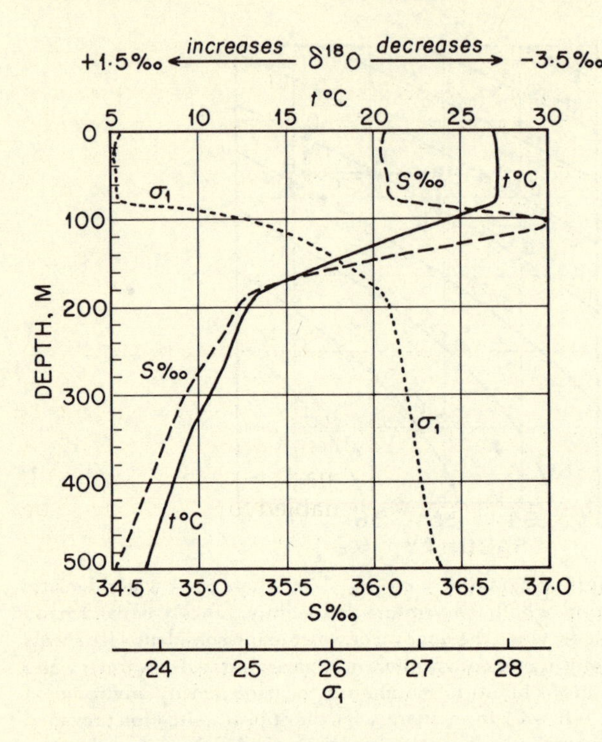

Figure 6.9 A typical vertical temperature, salinity, and density profile from the tropical oceans. As the amount of calcium carbonate secreted at depths below the upper mixed layer (generally ~100 m) increases, so the ^{18}O content of the test carbonate increases. Isotopic temperature estimates from foram tests which have undergone gametogenic calcification at depth are considerably lower than those obtained from living foraminifera collected in the upper mixed layer (~0.2‰ per degree Celsius).

6.3.2 *Oxygen isotope stratigraphy*

Oxygen isotope analyses have now been carried out on cores from most of the important areas of calcareous sedimentation throughout the world and, in many cases, studies have been made of both planktonic and benthic species (Shackleton 1977a). The overwhelming conclusion from such studies (after making due allowance for variations in sedimentation rates, vital effects, and other complicating factors mentioned in the previous section) is that similar isotopic ($\delta^{18}O$) variations are recorded in all areas. It thus seems extremely probable that the primary $\delta^{18}O$ signal being recorded is the result of ice volume changes on the continents and concomitant changes in the isotopic composition of the oceans (see Sec. 6.3.1). Because the mixing time of the oceans is relatively short (~10^3 years) this global-scale phenomenon results in essentially synchronous isotopic variations in the sedimentary record (though bioturbation, mixing by burrowing organisms in the upper sediments, tends to smooth out fine details in the record). These synchronous variations provide unique stratigraphic horizons, or markers, in the sediments, enabling correlations to be made between cores which may be thousands of kilometers apart. However, before these markers could be used for absolute time control, an independently dated timescale had to be established. This has mainly resulted from paleomagnetic analysis of ocean sediments supplemented by radio-isotopic studies, such as radiocarbon and uranium-series dating methods (see Chs 3 & 4). The general approach has been to locate the first major magnetic reversal in the sediment and assign to this an age of 730 000 years (i.e. to consider this as representing the Brunhes/Matuyama epoch boundary). A constant sedimentation rate between this level and the surface is then assumed and a tentative chronology is thereby established. Radioisotopic dates may be used to check on this chronostratigraphic framework and perhaps to adjust the assumed sedimentation rate.

With the recognition of consistent stable isotope signals in the sedimentary records of many different areas, isotope stages were defined (Emiliani 1955, 1966). Warmer periods (interglacials and interstadials) were assigned odd numbers (the present interglacial being number 1) and colder (glacial) periods were assigned even numbers (Fig. 6.10). A comparison of terrestrial chronostratigraphic markers of known age with equivalent horizons in the ocean sediments enabled further checks to be made on the isotopic chronology. For example, the Barbados high sea-level stands (dated by uranium-series analyses of raised corals at 82 000, 103 000, and 125 000 years BP; Mesolella *et al.* 1969) should correspond to isotopically light carbonate values in foraminiferal tests if interpolated timescales are correct. Such correspondence has been convincingly demonstrated by Shackleton and Opdyke (1973) and Shackleton and Matthews (1977), indicating that the chronology of the last interglacial–glacial cycle, at least, is fairly well established (Fig. 6.11). Five stages are recognized in the isotopic record of the last ~130 000 years, with stage 5 being subdivided into several substages, bearing letter designations (5a to 5c). Stages 5a, 5c, and 5e were periods of reduced terrestrial ice volume and/or higher temperatures, with substage 5e being the peak of the last interglacial (Shackleton 1969). Stages 5b and 5d were periods of cooler temperature and/or terrestrial ice growth, but on a smaller scale than occurred in stage 4. Interestingly, the change in $\delta^{18}O$

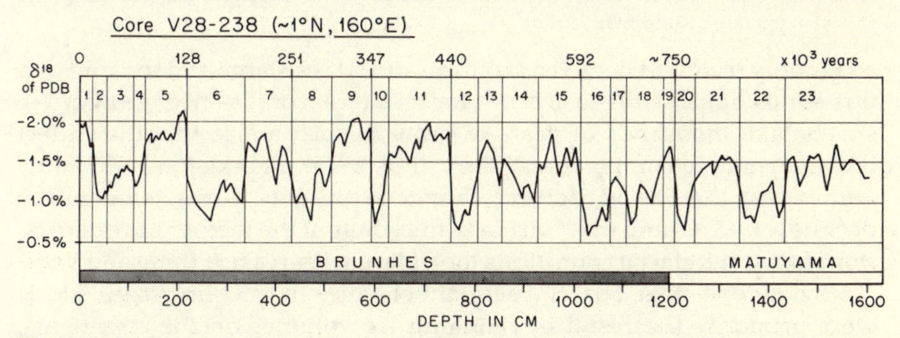

Figure 6.10 Oxygen isotope and paleomagnetic record of the last 1.6 million years in core V28-238 from the equatorial Pacific (~1°N, 160°E). Isotope stages are shown in the upper part of the diagram (periods with relatively low ice volume are shaded). Isotopic values are from measurements on *Globigerinoides sacculifer* (after Shackleton & Opdyke 1976).

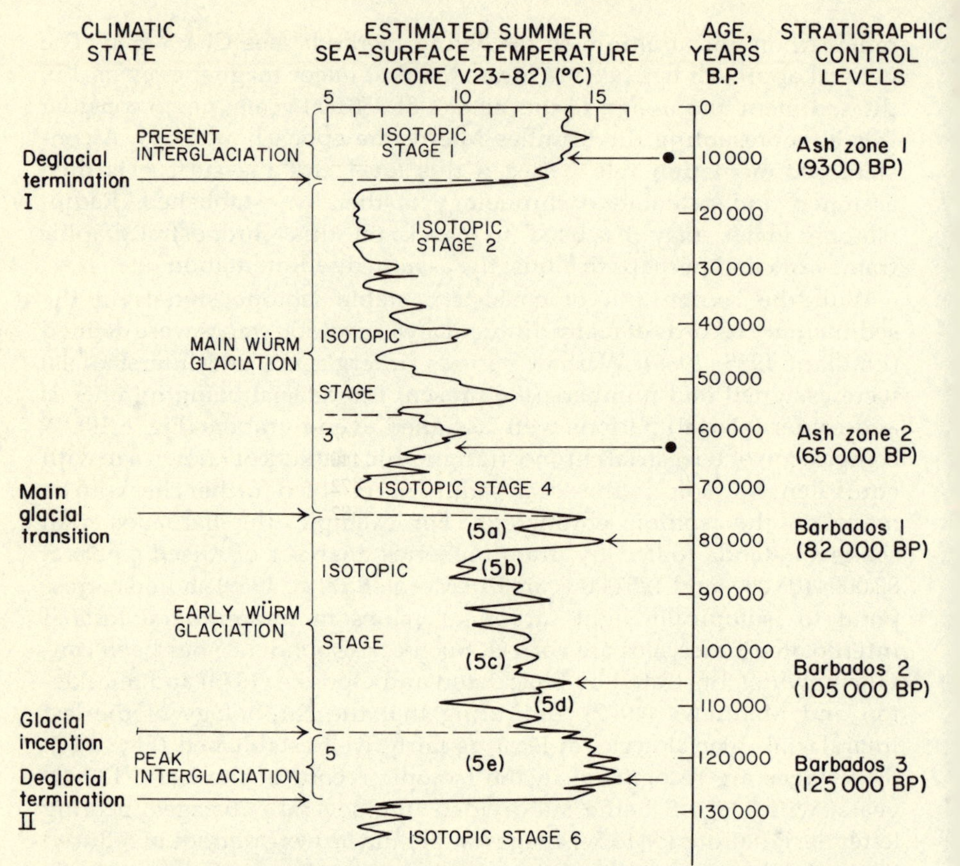

Figure 6.11 Summer sea-surface temperature reconstructions for the North Atlantic Ocean based on foraminiferal assemblage paleotemperature estimates, using core V23-82 from ~53°N 22°W (Sancetta *et al.* 1973a,b). Chronological controls used in other cores are shown on the right (tephra layers and Barbados sea-level stands). Isotopic stages are after Shackleton and Opdyke (1973). Generalized climatic conditions and major changes are shown at left (after Ruddiman 1977b).

commonly recorded between stages 5e and 5d is so large and so rapid that it is almost impossible to account for it in terms of ice-sheet growth. Ice sheets take thousands of years to grow to such a size that they affect oceanic isotopic composition (Barry *et al.* 1975). Shackleton (1969) thus argues that the 5e–5d isotopic change represents a true temperature decline (of ~3 °C) and that "virtually the whole of the temperature decline from fully interglacial conditions took place within a few thousand years . . . by isotope stage 5d." Subsequent changes in $\delta^{18}O$ (in stages 4 to 1) were primarily the result of changing ice volumes on the continents, according to this argument.

Isotope stages prior to the last interglacial have fewer cross-checks with terrestrial records or with radio-isotopic analyses. However, a

Table 6.2 Estimated ages of oxygen isotope stage boundaries and terminations.

Boundary	Termination†	Estimated ages (×10³ years)		
		A‡	*B*§	*C*‖
1–2	I	13	11	11
2–3		32	29	27
3–4		64	61	58
4–5		75	73	72
5–6	II	128	127	128
6–7		195	190	188
7–8	III	251	247	244
8–9		297	276	279
9–10	IV	347	336	334
10–11		367	352	347
11–12	V	440	453	421
12–13		472	480	475
13–14		502	500	505
14–15		542	551	517
15–16	VI	592	619	579
16–17		627	649	608
17–18		647	662	671
18–19		688	712	724
Brunhes/Matuyama boundary		700	728	
Jaramillo (top)			908	
Jaramillo (bottom)			983	
Olduvai (top)			1640	
Olduvai (bottom)			1820	

† Terminations from Broeker and Van Donk (1970). They defined terminations on the basis of their interpretation of the saw-toothed character of the oxygen isotope record.

‡ Estimates from Shackleton and Opdyke (1973) by linear interpolation in core V28-238, using a mean sedimentation rate of 1.7 cm per thousand years.

§ Estimates from Hays *et al.* (1976), Kominz *et al.* (1979), and Pisias and Moore (1981), based on the assumption that variations in the tilt of the Earth's axis (obliquity) have resulted in variations of global ice volume, and that the phase shift between the Earth's tilt and the 41 000 year component of the isotopic record has remained fixed with time. See Section 6.8 for discussion.

‖ Estimations from Morley and Hays (1981) based on adjustments to maintain a constant phase relationship between variations in oxygen isotope ratios and changes in obliquity and precession.

comparison of isotopic records in sediment cores from the equatorial Pacific, Atlantic, and Caribbean (each independently dated, but with different sedimentation rates) has enabled a reliable timescale for the entire Brunhes epoch to be established (Emiliani & Shackleton 1974, Shackleton & Opdyke 1973, 1976). This chronology (Table 6.2) is considered to be sufficiently reliable to be used independently to date other stratigraphic events. For example, biostratigraphic markers, such as the level at which a particular species became extinct, have been assigned ages based on their occurrence within a particular stable isotope stage

(Berggren *et al.* 1980). They may therefore be used as chronostratigraphic markers in their own right, independent of both radio-isotopic and stable isotope analyses on the sedimentary record in question. For example, extinction of the radiolarian *Stylatractus universus* has been found by stable isotope stratigraphy to have occurred throughout the Pacific and Atlantic Oceans at 425 000 ± 5000 years BP (Hays & Shackleton 1976, Moreley & Shackleton 1978). Similarly, the coccolith *Pseudomiliania lacunosa* became globally extinct in the middle of isotope stage 12, at ~458 000 years BP and the coccolith *Emiliania huxleyi* made its first appearance at ~268 000 years BP, late in isotope stage 8 (Thierstein *et al.* 1977). Stable isotope stratigraphy has also enabled volcanic ash horizons to be accurately pin-pointed in time (e.g. the 65 000 year BP ash layer in the North Atlantic), so that they can also be used as independent chronostratigraphic markers (e.g. Ninkovitch & Shackleton 1975, Ruddiman 1977b). The isotopic record can be viewed (in terms of its major low-frequency component) as being made up of periods of gradual increase in $\delta^{18}O$ separated by shorter, relatively abrupt episodes when $\delta^{18}O$ values increase. In a sense, the curve is "saw-toothed" in character, the slow increase in $\delta^{18}O$ apparently resulting from the gradual build-up of ice on the continents, followed by a period of rapid deglaciation when isotopically light water is returned to the oceans (Broecker & van Donk 1970). To formalize such a concept, Broecker and van Donk refer to the sharp decreases in $\delta^{18}O$ as a termination, signifying the end of a glacial period, the most recent deglaciation being Termination I. Estimated ages of other terminations are included in Table 6.2.

Because the isotopic record provides an integrated summary of global ice volume changes, it has been argued that isotope stages should be used as standard reference units for both marine and terrestrial deposits (Shackleton & Opdyke 1973). On the land, continuous stratigraphic sequences spanning the last 150 000 years or more are rare; more often, deposits are discontinuous both in time and space. In the oceans, the sedimentary record has been less disturbed so there are cogent reasons for using marine stratigraphic divisions to clarify and help understand the terrestrial record (Kukla 1977, cf. Miller *et al.* 1979). However, it should be emphasized that the isotopic signals in ocean cores contain both a temperature and an ice-volume component, which may not be synchronous. Furthermore, the ice-volume signal is an index of *global* ice volume and says nothing about ice extent in any one geographical area. This makes the application of a marine isotope stratigraphic frame of reference to many areas of questionable utility since the local stratigraphy may bear little relationship to the marine isotopic record. For example, northern Ellesmere Island and eastern Baffin Island were relatively ice-free during isotope stage 2 and may have experienced glacial advances and retreats out of phase with the larger continental ice sheets, variations of which dominate the oceanic isotope record (Andrews *et al.* 1974, England & Bradley 1978).

6.4 Relative abundance studies

The possibility of reconstructing paleoclimates by using the relative abundance of a particular species, or species assemblage, in ocean sediment cores was first proposed by Schott (1935). Schott recognized that variations in the number of *Globoratalia menardii* (a foraminifera characteristic of subtropical and equatorial waters) are indicative of alternating cold and warm intervals in the past. It was not until the 1950s and 1960s, however, that the availability of relatively long undisturbed cores, and improved dating techniques, enabled others to capitalize on Schott's work, and to develop his ideas further. Ericson *et al.* (1964), for example, confirmed that the *Globor. menardii* complex (a group of related species or subspecies) was diagnostic of mild subtropical–equatorial waters. In addition, they recognized other diagnostic species for cool mid-latitude waters (*Globor. hirsuta*) and relatively cold, northerly water bodies (*Globigerina pachyderma, Globig. inflata* and *Globig. bulloides;* cf. Fig. 6.6). In deep-ocean cores, 99% of the size fraction <74 μm is composed of the tests of planktonic foraminifera, so a simple index of the productivity of each of these species compared to the total foraminiferal population is to express the number of each diagnostic species as a percentage of the weight of the <74 μm fraction. Using this crude, but relatively quick, method, four major glacial episodes were recognized during the course of the last 2 million years (Ericson & Wollin 1968).

A more comprehensive examination of faunal composition as a paleoclimatic index was made by Lidz (1966), who examined the relationship between all species of foraminifera identified in a Caribbean core and Emiliani's (1966) isotope record from the same core. In this way he was able to identify those species that were well correlated (+ and −) with the isotope (paleoglaciation) record (concluding, incidentally, that the *Globor. menardii* complex of Ericson *et al.* (1964) was not a reliable temperature indicator). By plotting the ratio of selected warm-water to cool-water species, extremely good correlations with the isotope record could be obtained (Fig. 6.12). A similar approach was taken by Ruddiman (1971), who derived ratios from all warm- or cold-water species ("total fauna" paleoclimatic analysis). By plotting the percentage excess of warm- over cool-water species, qualitative paleotemperature estimates could be made which showed good correlations with oxygen isotope paleoglaciation curves (Fig. 6.13). Although somewhat of an improvement over the individual species approach, Ruddiman recognized that the technique was still relatively simplistic in considering all species as

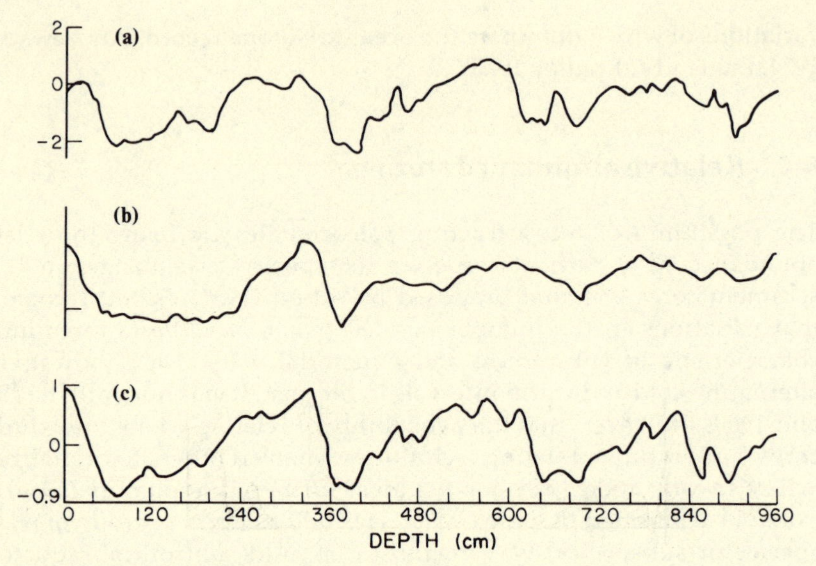

Figure 6.12 Ratios of the relative abundance of diagnostic stenothermic (temperature-sensitive) planktonic foraminiferal species in Caribbean core P6304-8, compared to oxygen isotope data. (a) Ratios of *Pulleniatina obliquiloculata* and *Sphaeroidinella dehiscens* ("warm water" preference) to *Globoratalia inflata* ("cool water" preference). (b) Ratios of *Globoratalia menardii menardii* ("warm water" preference) to *Globigerinoides ruber* ("cool water" preference). (c) $^{18}O/^{16}O$ ratio in the foram *Globigerinoides trilobus sacculifer* from the same core (Emiliani 1966). The time period spans approximately the last 350 000 years (after Lidz 1966).

equally "warm" or "cold" when gradations in their individual tolerances obviously exist.

In the early 1970s, major advances in paleoclimatic and paleo-oceanographic reconstructions were made by a number of workers. Multivariate statistical analyses of modern and fossil data were used to quantify former marine conditions ("marine climates") in an objective manner (Imbrie & Kipp 1971, Hecht 1973, Berger & Gardner 1975, Williams & Johnson 1975). The general approach in all of these studies is to calibrate the species composition of modern (core-top) samples in terms of modern environmental parameters (such as sea-surface temperatures in February and August). This is achieved by developing empirical equations which relate the two data sets together. These equations are then applied to down-core faunal variations to reconstruct past environmental conditions (Fig. 6.14). The equations are termed transfer functions. Mathematically, the procedure can be simply expressed as follows:

$$T_m = XF_m \quad \text{and} \quad T_p = XF_p,$$

where T_m and T_p are modern and paleotemperature estimates, respectively, F_m and F_p are modern and fossil faunal assemblages, respectively, and X is a transfer coefficient (or set of coefficients).

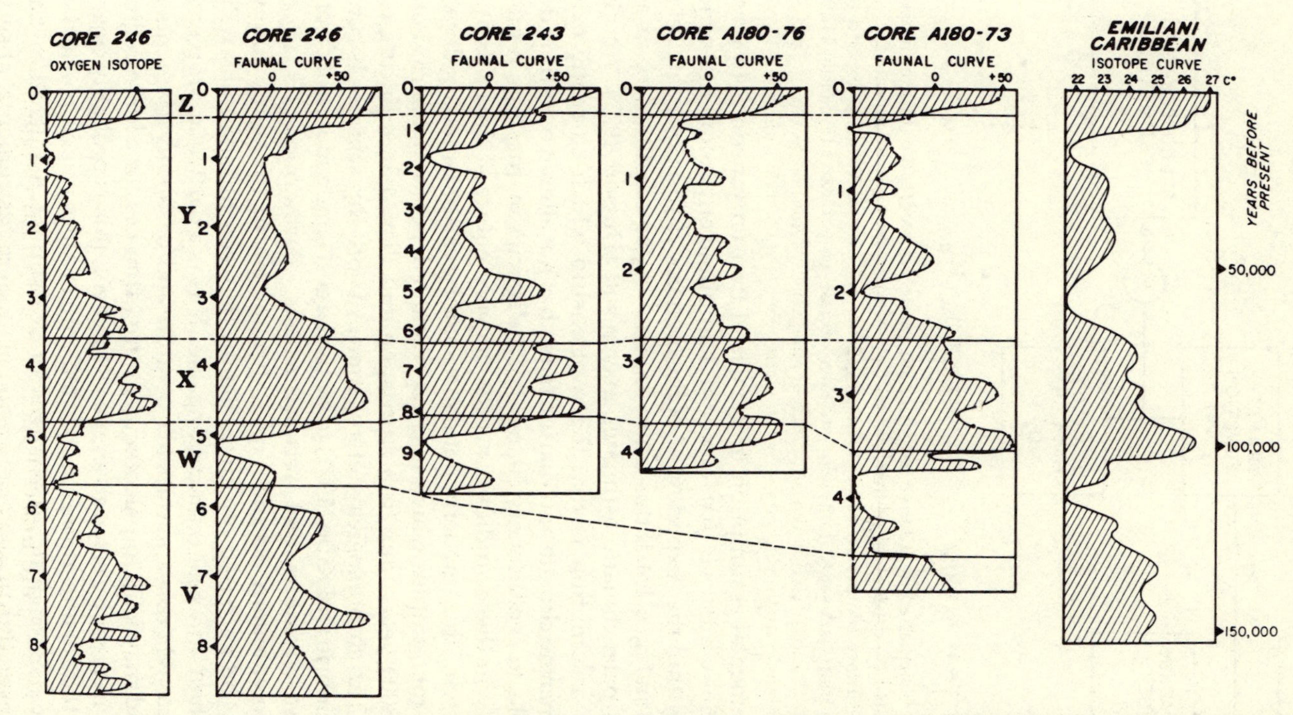

Figure 6.13 "Total fauna" (foraminiferal) records from four shallow equatorial cores, with oxygen isotope data from Emiliani (1955, 1966). In the faunal curves, a value of zero represents an equal number of "warm" and "cold" species. Positive values indicate a higher proportion of "warm" species and negative values a higher proportion of "cold" species. Horizontal lines suggest synchronous stratigraphic levels. (after Ruddiman 1971).

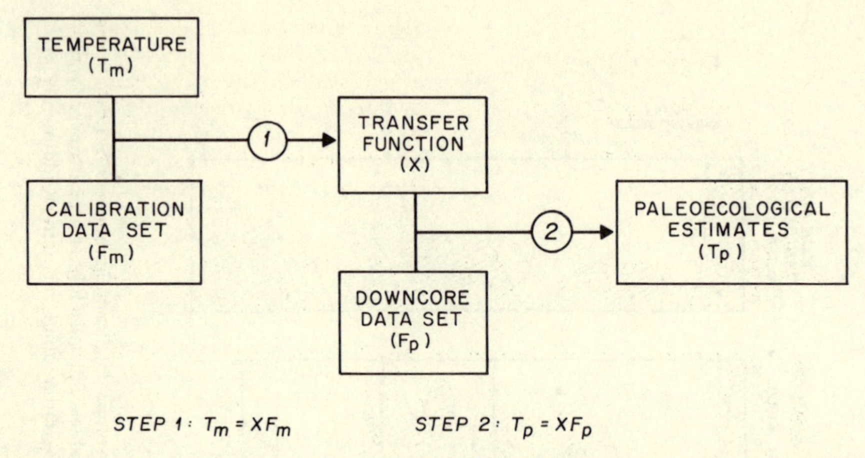

Figure 6.14 Schematic quantitative paleoclimatic model. In step 1, the transfer function (X) is calculated by calibration of the modern (core-top) foraminiferal data set (F_m) with modern sea-surface temperatures (T_m). In step 2 the transfer function is applied to a down-core (fossil) data set (F_p) to yield estimates of past temperatures (T_p) (after Hutson 1977).

A fundamental assumption in the use of transfer functions to reconstruct marine climates is that former biological and environmental conditions are within the "experience" of the modern (calibration) data set (as illustrated in Fig. 6.15). If this is not so, a no-analog condition exists and erroneous paleoclimatic estimates may result (Hutson 1977). Another important assumption is that the relationship which currently exists between marine climate and marine fauna has not altered over time due, for example, to evolutionary changes of the species in question. A major uncertainty in these studies concerns the very nature of the calibration data set itself. The "modern" faunal assemblages are generally derived from core-top samples which may represent a depositional period of as much as 8000 years, due to bioturbation and disturbance during core recovery. The chronological heterogeneity of core-top samples is considered by Imbrie and Kipp (1971) to be the largest single source of error in their paleo-environmental reconstructions. Furthermore, it is not unusual for modern oceanographic parameters to be poorly known, commonly being based on interpolation between observations which are both short and geographically sparse. This is a particular problem in remote areas where sea-surface temperature and/or salinity gradients are strong, and may lead to paleotemperature estimates for certain regions which are in error by several degrees. However, this is probably close to the magnitude of uncertainty associated with modern values, particularly in areas where significant changes of sea-surface temperature have occurred, even during the brief period of modern instrumental observations (e.g. Wahl & Bryson 1975). Under such circumstances the selection

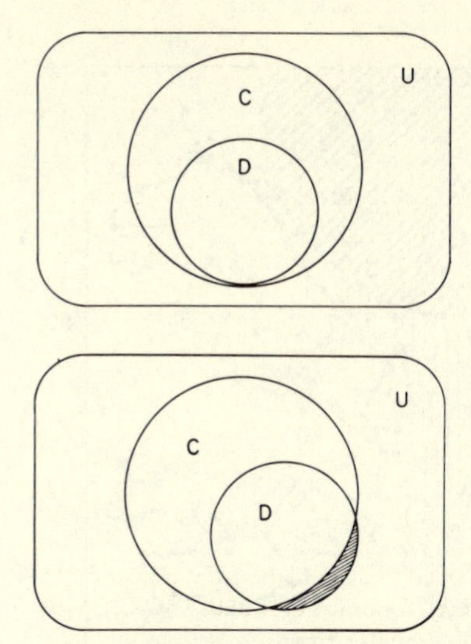

Figure 6.15 Venn diagrams illustrating conditions which are ideal (left) and non-ideal (right) for calibrating a transfer function. In an ideal situation, the calibration data set, *C*, encompasses the range of all biological and environmental conditions which exist in the down-core data set, *D*. In a non-ideal situation the calibration data set, *C*, does not reflect all the biological and environmental conditions which are represented within the down-core data set, *D* and a no-analog condition results (shaded area). *U* is the universe of all biological and environmental conditions both today and in the past (after Hutson 1977).

of a modern calibration value to equate with the core-top faunal assemblage is somewhat problematical, though by no means a problem unique to marine data.

Of all the multivariate approaches to the quantification of former marine climates, the methodology of Imbrie and Kipp (1971) has been most widely applied. In their original study, an attempt was made to reconstruct sea-surface temperature variations at a core site (V12-122) ~150 km south of Haiti. To achieve this, the species composition of core-top samples from 61 sites in the Atlantic Ocean (and part of the Indian Ocean) were used as the basic "modern fauna" data set. As a first step, Imbrie and Kipp reduced the number of independent variables in this data set by the use of principal components analysis. Principal components analysis is an objective way of combining the original variables into linear combinations (eigenvectors) which effectively describe the principal patterns of variation in a few primary components, leaving the less coherent aspects ("noise") for the last few components (Sachs *et al.* 1977). Thus, Imbrie and Kipp were able to condense much of the spatial variation of species abundance in 61 core-top samples from the Atlantic Ocean into five principal components or assemblages, which accounted for over 91% of variance in the original data set (cf. Sec. 10.2.4). By mapping the relative contribution of each component to the variance of each core-top sample, it was clear that four of these assemblages were related to temperature variations near the sea surface and could be simply described as tropical, subtropical, subpolar, and polar assemblages (Fig. 6.16). A fifth assemblage was more related to oceanic circulation around the subtropical high-pressure cells and was termed the gyre margin assemblage.

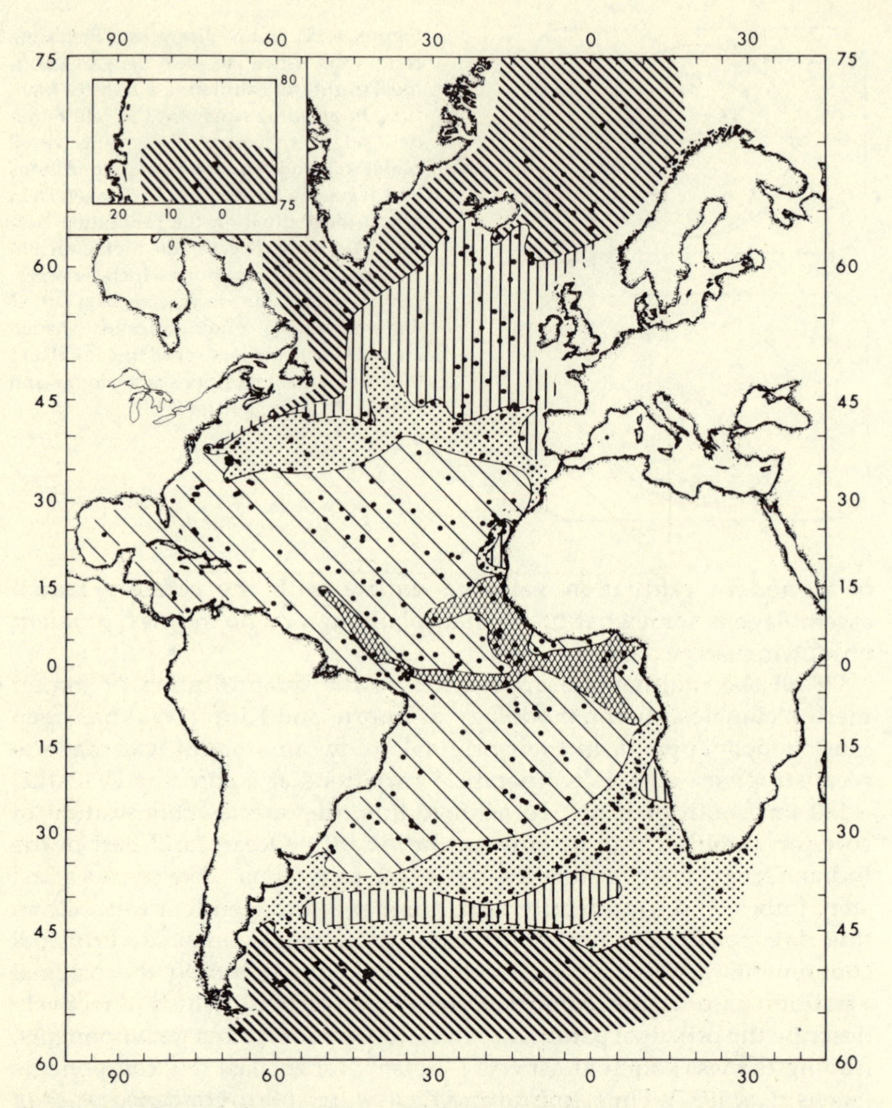

Figure 6.16 Core-top foraminifera assemblages, defined by factor analysis. Each zone represents an area dominated by a particular assemblage (polar, subpolar, transitional, subtropical, or gyre margin) (Molfino *et al.* 1982).

The next step was to utilize the relative weightings of each assemblage (factor scores) at each site to predict sea-surface temperatures. A stepwise multiple regression procedure was used, with temperature as the dependent variable and the factor scores as independent variables (predictors). In this way, an equation was derived which parsimoniously described sea-surface temperature in terms of the relative importance of the factor

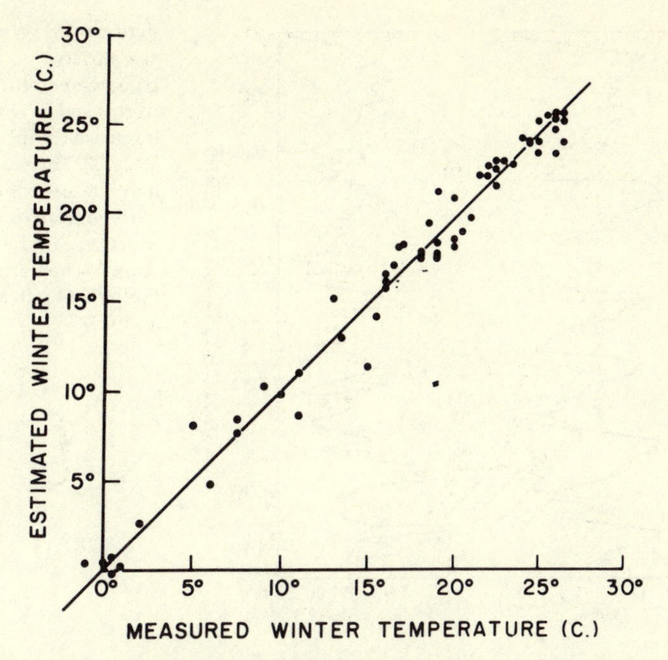

Figure 6.17 Winter sea-surface temperatures based on modern instrumental observations (including interpolated values) versus those estimated from faunal assemblages in 61 core-top samples using factor analysis and transfer function methods (after Imbrie & Kipp 1971).

scores at each site. In the case of winter temperatures, for example, the following calibration equation was derived:

$$T_w = 23.6A + 10.4B + 2.7C + 3.7D + 2.0K,$$

where *A*, *B*, *C*, and *D* refer to the four major assemblages (tropical, subtropical, subpolar, and polar) and *K* is a constant.† This equation explained 91% of variance in the winter sea-surface temperature observations (Fig. 6.17).

At this stage, the modern faunal data set had been calibrated in terms of sea-surface temperatures. It was then necessary to transform the fossil (down-core) faunal variations from core V12-122 into relative weightings of the major faunal assemblages already defined. Finally, these values were entered into the calibration equation to produce paleotemperature estimates. These are shown in Figure 6.18 together with $\delta^{18}O$ measurements on the foraminifera *Globigerinoides ruber* from the same core (Imbrie *et al.* 1973). A sequence of cooler episodes can be seen separated by warmer periods when temperatures approached modern (core-top)

†Gyre margin assemblage was not considered in this analysis.

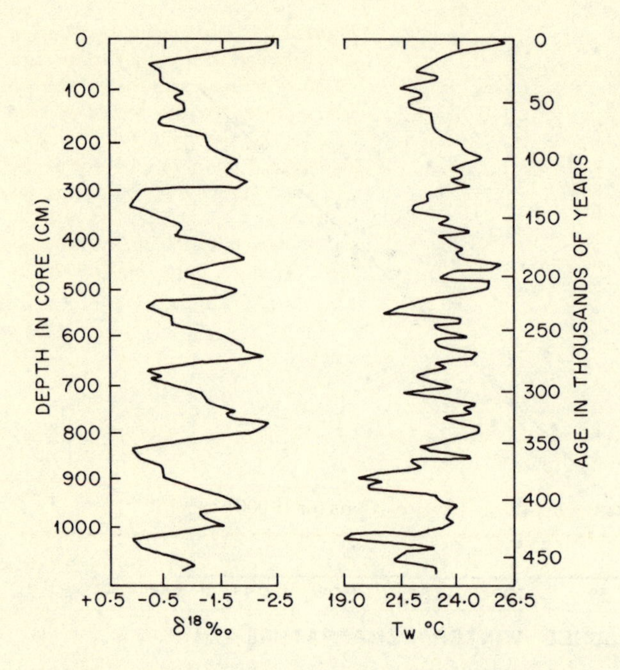

Figure 6.18 Winter sea-surface paleotemperature estimates (right) and $\delta^{18}O$ values (left) based on Caribbean core V12-122. The sea-surface temperature estimates are derived from transfer functions, in the manner shown schematically in Figure 6.14 (after Imbrie *et al.* 1973).

values. Interestingly, the cooler periods generally coincide with high $\delta^{18}O$ values (and vice versa). However, the changes in water temperatures can only account for a small fraction (~20%) of the isotopic change and, in fact, provide support for the view that global ice-volume changes are mainly manifested in the $\delta^{18}O$ record (see Sec. 6.3.1). From this reconstruction of Caribbean Sea paleotemperatures, it appears that sea-surface temperatures in this region have been predominantly cooler than today over the duration of the core record (~450 000 years), with winter temperatures as much as ~7.5 °C lower than today around 440 000 years BP. The anomalous nature of modern marine climate in this region is particularly well illustrated by a histogram of winter sea-surface temperatures integrated over the entire core (Fig. 6.19) (Imbrie *et al.* 1973).

One of the most rewarding and interesting applications of Imbrie and Kipp's methodology has been to use the technique to provide a synoptic view of paleo-oceanographic conditions in the past. By applying transfer functions to samples from a particular time horizon in many different cores, it is possible to reconstruct and map marine climates as they were at that time. This was one of the major objectives of the CLIMAP project, which focused attention on marine conditions at 18 000 years BP (CLIMAP Project Members 1976). The date of 18 000 years BP was selected as the time of the last maximum continental glaciation, defined by maximum $\delta^{18}O$ values during isotope stage 2 (Shackleton & Opdyke 1973). Using transfer functions derived for each of the major world oceans, February and August sea-surface temperatures have been recon-

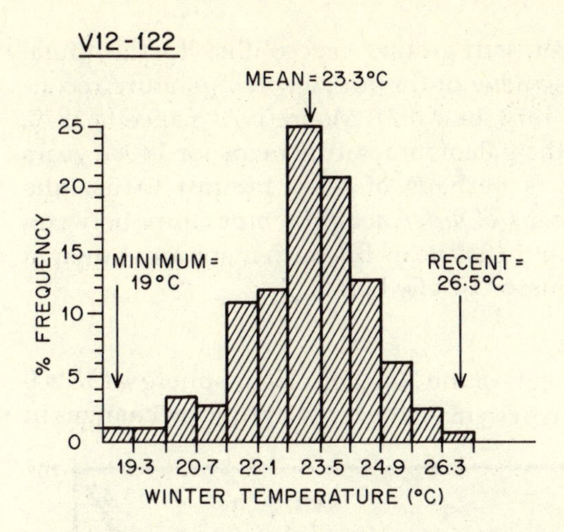

Figure 6.19 Frequency distribution of Caribbean winter sea-surface temperature estimates for the last 450 000 years (based on the record of core V12-122, shown in Fig. 6.18). Modern sea-surface temperatures are at the extreme upper range of conditions experienced over the last 450 000 years, more than 3 °C above the long-term average (after Imbrie *et al.* 1973).

Table 6.3 Sea-surface paleotemperature reconstructions for 18 000 years BP.

Area	Principal faunal groups used	Major reference
North Atlantic	Foraminifera	Kipp (1976)
		McIntyre *et al.* (1976)
South Atlantic	Radiolaria	Morley & Hays (1979)
Norwegian and Greenland Seas	Foraminifera	T. Kellogg (1975, 1980)
Caribbean and equatorial Atlantic	Foraminifera	Prell *et al.* (1976)
Western equatorial Atlantic	Foraminifera	Bé *et al.* (1976)
Eastern equatorial Atlantic	Foraminifera	Gardner & Hays (1976)
Indian Ocean	Foraminifera	Hutson (1978), Prell & Hutson (1979), Prell *et al.* (1980)
Antarctic Ocean	Radiolaria	Lozano & Hays (1976), Hays (1978)
Pacific Ocean		
South	Foraminifera	Luz (1977)
North and South	Coccoliths	Geitzenauer *et al.* (1976)
North	Diatoms	Sancetta (1979, 1983)
North and South	Radiolaria	Moore (1978)
North and South	All four groups (synthesis)	Moore *et al.* (1980)
World Ocean (summary)	Foraminifera Radiolaria Coccoliths	CLIMAP Project Members (1976)

structed for this period (Table 6.3). Most studies relied mainly on foraminiferal assemblage data, but in areas where siliceous fossils predominate (e.g. in the South Atlantic and Antarctic Oceans) the technique has been applied to Radiolarian assemblages (Lozano & Hays 1976, Morley & Hays 1979). In the Pacific Ocean, where preservational characteristics of carbonate and siliceous fossils vary significantly from one area to another, it has been found advantageous to develop transfer functions based on four major microfossil groups (coccoliths, foraminifera, Radiolaria, and diatoms) to achieve optimum paleotemperature reconstructions (Geitzenauer *et al.* 1976, Luz 1977, Moore 1978, Sancetta 1979, Moore *et al.* 1980). Although the paleotemperature maps for 18 000 years BP are of interest alone, it is perhaps of most interest to use the reconstructions to produce maps of *differences* in temperature between modern conditions and those at 18 000 years BP. Such maps are shown in Figures 6.20–6.26 and are discussed briefly below.

6.4.1 North Atlantic Ocean

Situated between major ice sheets of the Northern Hemisphere (at 18 000 years BP), the North Atlantic experienced the most significant changes in

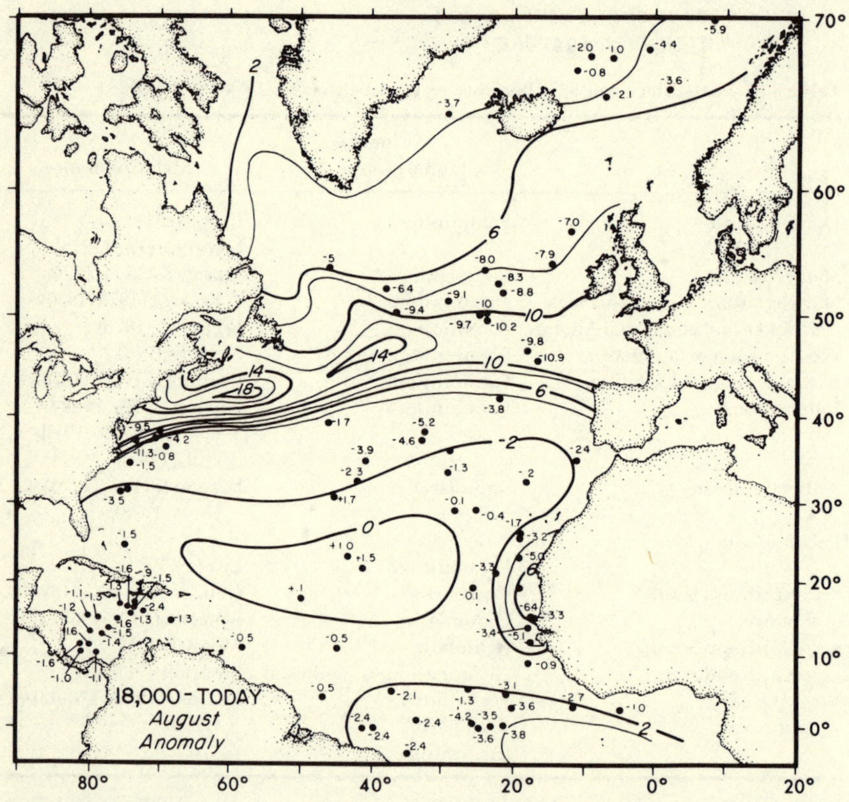

Figure 6.20 Difference between modern and 18 000 years BP August sea-surface temperatures. Contour interval is 2 °C. Cores used in paleotemperature study shown as dots. Paleotemperatures were derived from faunal assemblage transfer functions. Figure derived by subtracting map of modern temperatures from map of paleotemperatures (both containing interpolated data). Hence the estimates of paleotemperature increase since 18 000 years BP contain large values in some areas, even though core data may not have been available from those areas (e.g. the western North Atlantic between 42 and 50°N) (from McIntyre *et al.* 1976).

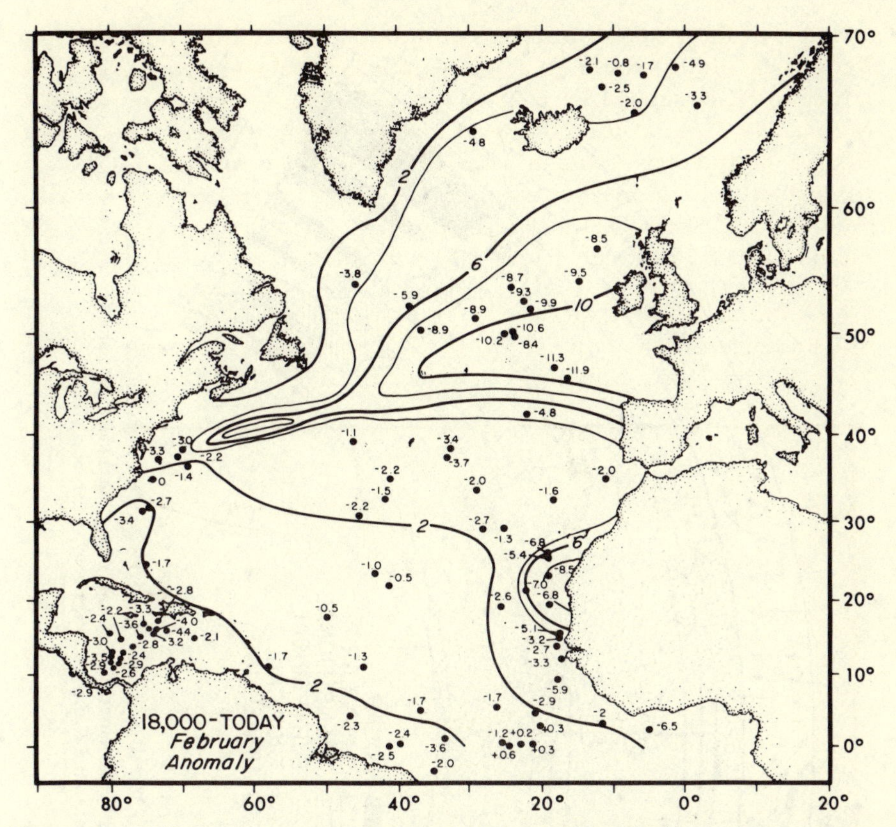

Figure 6.21 Difference between modern and 18 000 years BP February sea-surface temperatures, derived as explained in Figure 6.20. Contour interval is 2 °C (from McIntyre *et al.* 1976).

temperature of all oceanic areas (Figs 6.20 & 6.21). At 18 000 years BP August sea-surface temperatures were more than 10 °C cooler in a broad zone from 40–45°N in the west to 45–50°N in the east. This reflects a marked southward movement of polar and subpolar water at the time. In February, temperatures were less depressed off the North American coast (ΔT = 3–5 °C), but in the eastern North Atlantic temperatures were 6–12 °C cooler than today in a triangular area stretching from Scandinavia to Portugal. In both seasons, strong upwelling off the coast of north-western Africa (presumably related to stronger Trade winds, cf. Sec. 6.7) resulted in significantly cooler temperatures in that area (ΔT = 5–8 °C). Relatively minor temperature differences were apparent over most of the subtropical North Atlantic to the west, though February sea-surface temperatures in the Caribbean were 2–4 °C cooler than today at 18 000 years BP.

Considered together, the two maps suggest that the North Atlantic is

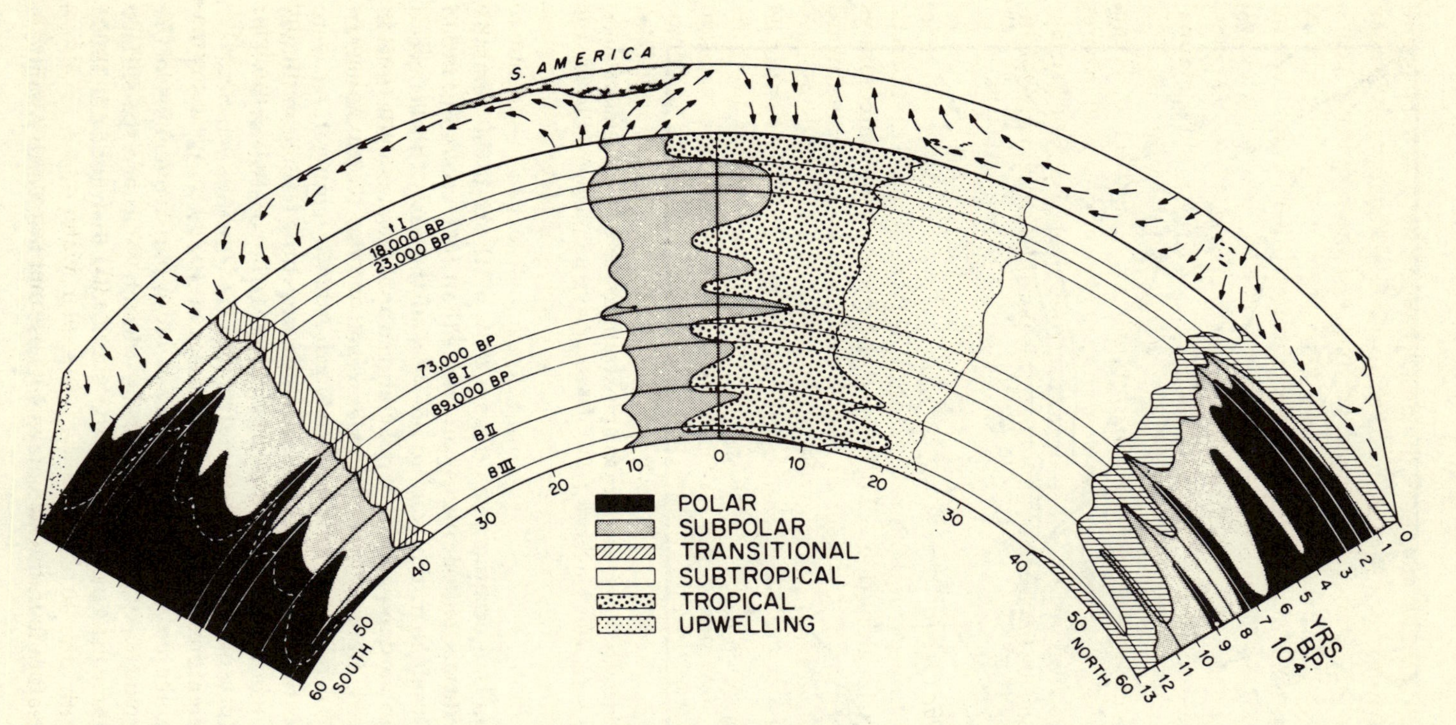

Figure 6.22 Variations of Atlantic surface water masses along the 20°W meridian, from 60°S (left) to 60°N (right) over the last 130 000 years (vertical axis). Principal ocean currents along the 20°W meridian are shown schematically at top. Water mass variations shown cover the last interglacial–glacial cycle. Note the large latitudinal variations in water masses in the North Atlantic (from McIntyre *et al.* 1975).

made up of two zones, an area of dynamic change from ~40 to 50°N and a relatively stable zone to the south. Such a condition has apparently been characteristic of a much longer period than just the last 18 000 years. McIntyre *et al.* (1975) have reconstructed major water-mass boundaries along a meridional north–south transect through the Atlantic Ocean (at 20°W) for the past 130 000 years (i.e. an interglacial–glacial cycle) and their study clearly indicates that the North Atlantic has been the most variable zone of both hemispheres (Fig. 6.22). At 50°N, faunal assemblages have varied from being predominantly subtropical at 125 000 years BP (the last interglacial) to polar from ~35 000 to 15 000 years BP. Furthermore, changes of a similar nature have been observed in cores reaching back over 600 000 years, indicating that seven complete glacial–interglacial cycles have occurred during this period (Ruddiman & McIntyre 1976).

6.4.2 *Pacific Ocean*

August sea-surface temperature differences were maximized in subarctic and equatorial regions (Fig. 6.23). In the area around Japan, temperatures were as much as 8 °C cooler at 18 000 years BP due to a southward displacement of the warm Kuroshio current at that time and its replacement by subarctic water (Oyashio current). Less pronounced temperature depressions occurred in the Gulf of Alaska and southward along the California coast (ΔT = 2–4 °C), a fact which stands in marked contrast to conditions in the eastern Atlantic Basin. In equatorial regions, temperatures were cooler by 2–4 °C in a broad band, perhaps related to greater advection of cool waters into North and South Equatorial currents due to intensified trade winds at the time. In the South Pacific, cooler waters adjacent to the coast of South America and west of New Zealand are noteworthy (ΔT = 2–4 °C). It is also of interest to note that, in addition to areas of major cooling, large parts of the Pacific Basin appear to have been *warmer* at 18 000 years BP than at present. In particular, core regions of subtropical high-pressure centers are predicted to have been 1–2 °C warmer than modern values. Temperatures along the eastern coast of Australia were also warmer (by up to 4 °C), perhaps due to enhanced equatorial flow from the stronger Equatorial currents.

Similar patterns of difference are observed in many areas in the February sea-surface temperature reconstructions, with maximum temperature changes in the area east and north-east of Japan (ΔT = 6–8 °C). Cooler temperatures at this season could have resulted in sea-ice formation over an extensive area (Sancetta 1983). In the Southern Hemisphere, cooling was relatively minor at 18 000 years BP, except in the Peruvian current off the western coast of South America (ΔT = 2–4 °C) and in the extreme south due to an expanded subpolar water mass. Again, a noticeable feature is the extensive area of warmer sea-surface tempera-

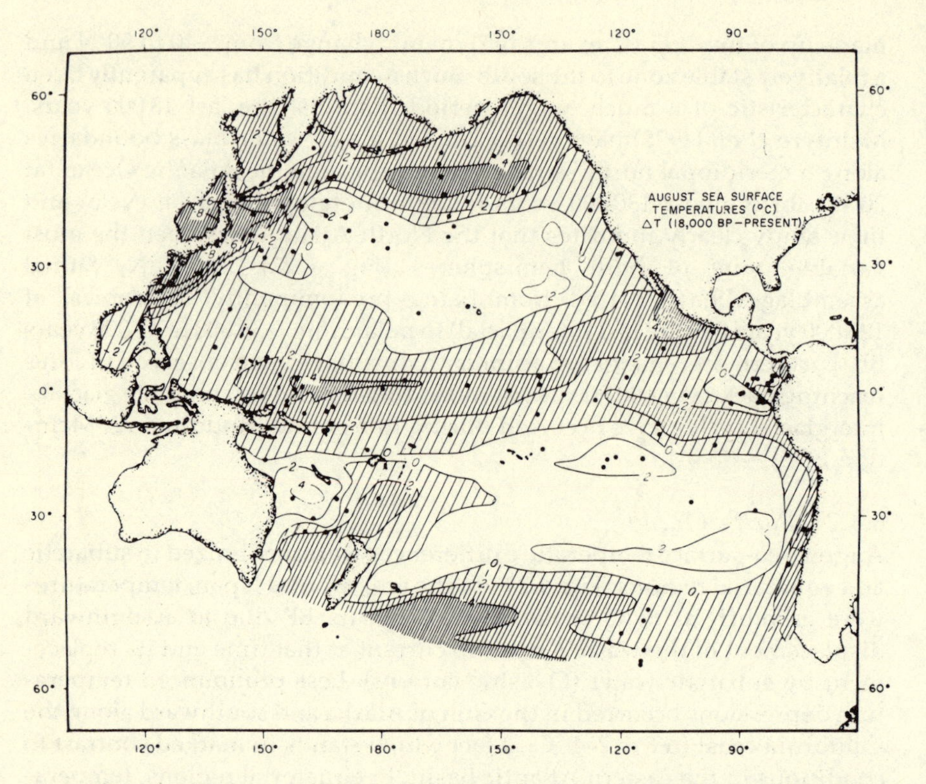

Figure 6.23 Differences between modern and 18 000 years BP August sea-surface temperatures in the Pacific Ocean. Dots show locations of cores used in paleotemperature estimates. Dark shading indicates 18 000 years BP temperatures >4 °C cooler; intermediate shading 2–4 °C cooler; light shading 0–2 °C cooler than modern values. Areas which were warmer than modern conditions at 18 000 years BP are unshaded (from Moore *et al.* 1980).

tures at 18 000 years BP centered over the subtropical high-pressure cells. It is particularly interesting to note the large extent of this positive anomaly in the Southern Hemisphere, associated with the poleward movement of the subtropical high-pressure center at this time of year. A corresponding southward shift in the Northern Hemisphere positive anomaly field is also apparent in the February maps compared to those for August. Such a pattern suggests a more intense Hadley cell circulation at 18 000 years BP, with well developed subtropical high-pressure centers. In these areas, adiabatic warming and clear skies would favor warmer sea-surface temperatures and, on the subtropical margins, trade winds and gyre margin ocean currents would be strengthened. All these factors fit together quite coherently in the light of the reconstructed paleotemperatures and provide important clues for terrestrial chronologies on adjacent land areas.

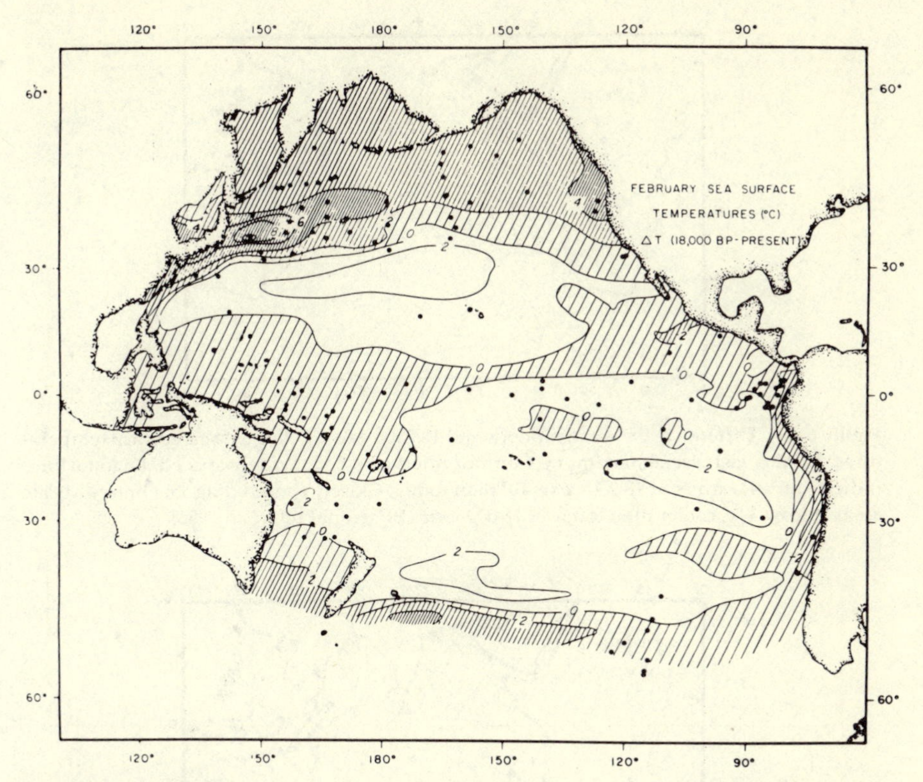

Figure 6.24 Difference between modern and 18 000 years BP February sea-surface temperatures in the Pacific Ocean. Dots show location of cores used in paleotemperature estimates. Shading as in Figure 6.23 (from Moore *et al.* 1980).

6.4.3 *Indian Ocean*

August sea-surface temperature anomalies reveal relatively minor differences between 18 000 years BP and today (Fig. 6.25). Apart from areas associated with the eastern and western boundary currents, off the western coast of Australia and off south-eastern Africa (the Agulhas Current) most areas at 18 000 years BP were within 1 or 2 °C of modern values. It is interesting that temperatures in the Arabian Sea were ~1 °C warmer at 18 000 years BP, suggesting a weaker South West Monsoon flow at that time, resulting in less upwelling of cool water (Prell *et al.* 1980).

February maps reveal larger temperature differences, particularly in the area centered on 40°S, where northward movement of the Antarctic Convergence zone and associated subpolar water caused temperatures to be lower by 4–6 °C at 18 000 BP (Fig. 6.26). Compared to the other ocean basins, however, temperature changes in the Indian Ocean were relatively small (overall cooling of only ~1.8 °C) and large areas off the coast

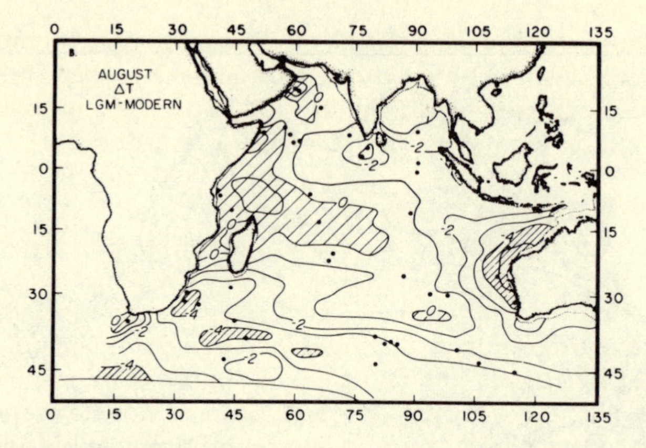

Figure 6.25 Difference between modern and 18 000 years BP August sea-surface temperatures. LGM = last glacial maximum. Contour interval is 1 °C. Widely spaced diagonal lines indicate areas warmer at 18 000 years BP than today. Closely spaced diagonal lines indicate areas at least 4 °C cooler than today at 18 000 years BP (from Prell *et al.* 1980).

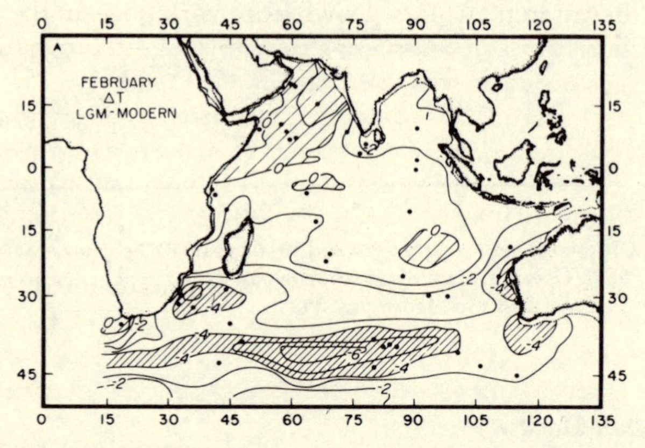

Figure 6.26 Difference between modern and 18 000 years BP (LGM) February sea-surface temperatures. Contour interval is 1 °C. Shading as in Figure 6.25 (from Prell *et al.* 1980).

of eastern Africa and in the Arabian Sea may have been slightly warmer at 18 000 years BP than they have been in recent years.

6.5 Morphological variations

Numerous studies have shown that morphological differences between individuals of the same species (phenotypes) may result from varying environmental conditions. Variance resulting from habitat variation (ecophenotypic variation) is particularly common in the Foraminiferida and this characteristic has important implications for paleo-oceanographic reconstructions (Kennett 1976). Four main types of morphological difference have been noted, though not all have yet been fully utilized in paleo-oceanographic work. The important parameters are as follows: coiling directions, size, shape, and surface structure.

6.5.1 *Coiling direction variations*

Perhaps the best known type of phenotypic variation in foraminifera is the difference in coiling directions of *Globoratalia truncatulinoides* (Ericson *et al.* 1954). In the North Atlantic, three geographically discrete provinces can be recognized, distinguished by a dominance of one or the other coiling direction. Unfortunately, a simple relationship between coiling direction and sea-surface temperature is not clear and this has relegated the use of *Globor. truncatulinoides* primarily to biostratigraphic correlations in North Atlantic sediments. However, in the South Pacific Ocean, a clearer relationship between water temperature and coiling direction is observed (Kennett 1970, Parker & Berger 1971). Dextral forms predominate in tropical and subtropical waters, sinistral forms in temperate and sub-Antarctic waters. Down-core variations in the percentage of individuals exhibiting coiling in a particular direction may be used as an index of surface water temperatures.

Another species with environmentally sensitive coiling preferences is *Neogloboquadrina pachyderma*, which occurs in dextral forms in temperate and subtropical oceans and sinistral forms in polar and subpolar oceans (Ericson 1959, Bandy 1960). In fact, the left-coiling form is virtually the only polar foraminiferal species. The boundary between dominance by one form or another (i.e. the 50% isopleth) approximates 7 °C in surface waters of the North Atlantic and 10 °C in the North Pacific (Ingle 1973). Similar patterns are observed in the South Pacific where abrupt change to sinistral forms occur south of 52°S, around the oceanic Antarctic convergence zone.

6.5.2 *Size variations*

Although size variations of fossil populations are subject to the uncertainty that dissolution may have removed part of the population record (Sec. 6.6), it has been noted that the variations in test sizes of certain species reflect changes in water temperature and salinity. This is well illustrated in the Indian Ocean, where populations of the foram *Orbulina universa* show a gradual decrease in average size from subtropical waters southwards, with an abrupt change in size around 25–30°S, at the subtropical convergence oceanic front (Bé *et al.* 1973). Test diameter is strongly related to surface water temperature and inversely related to surface water density (salinity). By assuming a mean shell size of 450 μm for the subtropical convergence zone, Bé and Duplessy (1976) were able to demonstrate a northward shift of 10° latitude in this zone during glacial

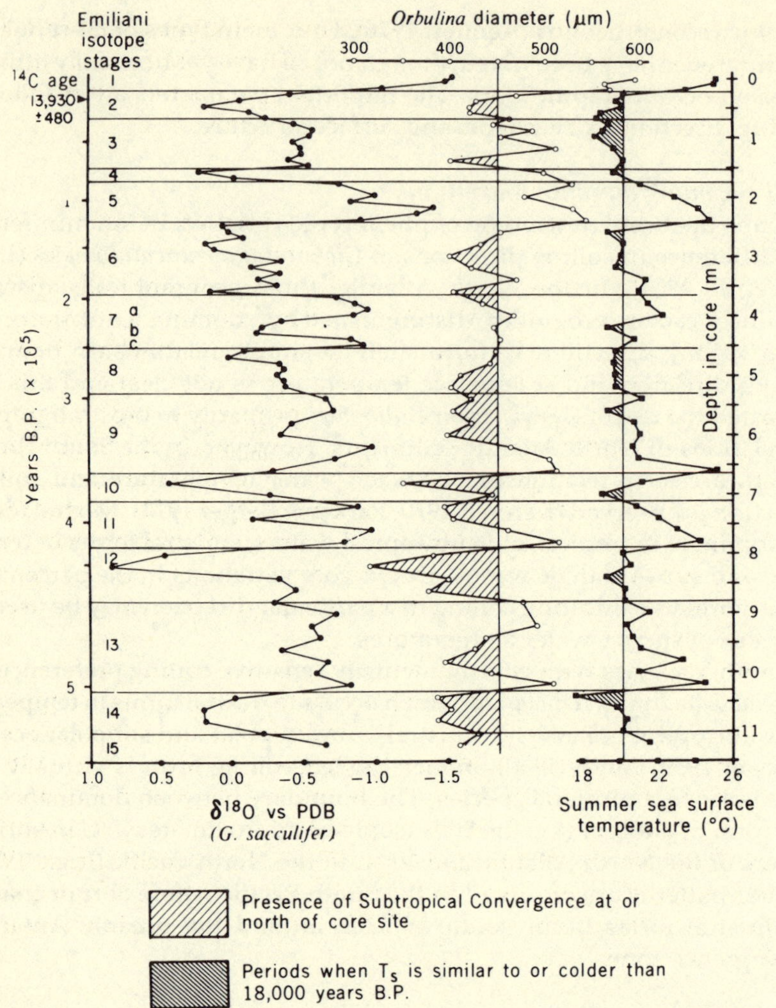

Figure 6.27 Variations in shell diameter of *Orbulina universa* in core RC17-69 (off the coast of South Africa) in relation to reconstructed summer sea-surface temperatures (T_s) and $\delta^{18}O$ measured in *Globigerina sacculifer*. Shell diameter is strongly related to both temperature and low $\delta^{18}O$ values (low global ice volume). Shell diameters less than 450 μm (shaded) indicate the subtropical convergence zone at or north of the core site (i.e. more extensive subpolar water). Values of T_s below those at 18 000 years BP are also shaded (after Bé & Duplessy 1976).

times. Size variations of *O. universa* strongly confirmed paleotemperature estimates based on faunal assemblage transfer functions (Fig. 6.27).

Similar studies of test size variations in *Globigerina bulloides* from the southern Indian Ocean have enabled regression equations to be produced, relating test size to surface water temperature (tests increase in size as water temperature decreases; Malmgren & Kennett 1976). These equations can then be used to assess former surface water temperatures by analyzing test-size variations in cores from the area. The resulting paleotemperature estimates are strongly confirmed by parallel studies of isotopic variations, foraminiferal assemblage abundances, coiling direction changes, etc. (Malmgren & Kennett 1978a,b).

6.5.3 *Shape variations*

Studies of *Globoratalia truncatulinoides* in surface sediments from both the North Atlantic and the South Pacific reveal distinctly different morphological types along a gradient from polar to tropical areas (Kennett 1968). Northward from Antarctic waters there is a gradual increase in the average test height, and a decrease in convexity of the dorsal side until it becomes flattened, or even concave, in tropical water. Using simple ratios of width to height in individual specimens, a statistically significant inverse relationship can be demonstrated between surface water temperatures and mean ratio. Individuals change from compressed forms in cold water to highly conical forms in tropical waters. This has been confirmed in an elegant study by Healy-Williams and Willams (1981), who use the amplitudes of Fourier series harmonics to characterize particular shape components of tests of *Globor. truncatulinoides*. The harmonic representing "conical shape" is highly correlated with surface water temperature.

6.5.4 *Surface structure variations*

Detailed studies of foram tests have demonstrated that distinct structural differences exist in individuals living in different environmental conditions. For example, Wiles (1967) noted that the number of pores in samples of *Globigerina eggeri* varied with oxygen isotopic concentration and faunal composition of the sediment. High pore counts reflect interglacial conditions and vice versa. Similarly, in *Orbulina universa*, variations in pore diameter and pore concentration ("test porosity") show latitudinal dependence in the Indian Ocean, with porosity increasing towards the warmer waters to the north (Bé *et al.* 1973). Multivariate analysis of several morphological parameters of *O. universa*, in samples from the Indian Ocean, confirm that test porosity (percentage of pore space per unit area) is a fundamental phenotypic variable with a major morphological discontinuity at the boundary of equatorial and central water masses (~20°S), corresponding to the 24 °C August sea-surface isotherm today (Hecht *et al.* 1976). As glacial maximum (18 000 years BP) reconstructions of sea-surface temperatures do not show a significant change in the location of the 24 °C isotherm, independent verification of this could be obtained by examining morphological characteristics of *O. universa*.

6.6 Dissolution of deep-sea carbonates

Throughout the deep ocean basins of the world, a major factor affecting preservation of carbonate tests is the rate of dissolution at depth. The oceans are predominantly undersaturated with respect to calcium carbonate at all depths below the upper mixed layer (the zone above the thermocline; Olausson 1965, 1967). After death of the organisms, deposition of the test on the ocean floor leads to dissolution in the undersaturated water (Adelsack & Berger 1977). Pteropod tests (composed of calcium carbonate in the form of aragonite) are most susceptible to solution and are the first to disappear; pteropods are thus only found in relatively shallow waters where undersaturation is not pronounced (Berner 1977). At greater depth, dissolution of tests made of calcite (e.g. foraminifera and coccoliths) becomes apparent. The level at which calcite dissolution is at a maximum is the lysocline (Berger 1970, 1975), generally encountered at 2500–4000 m depth in the oceans (Fig. 6.28). In the

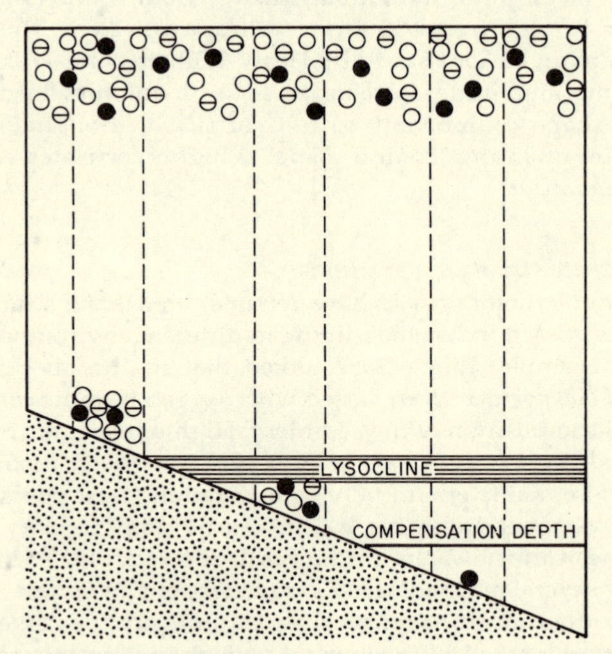

Figure 6.28 Schematic diagram illustrating selective dissolution of planktonic foraminiferal species at depth, due to undersaturation of the water with respect to calcium carbonate. Dark circles represent the resistant species *Globoratalia tumida*. Open circles represent *Globigerinoides ruber*, which is dissolved relatively easily. *Globigerina bulloides* (open circle with a line) is intermediate in resistance. Dissolution alters the species composition of the sediment so it may not be representative of species in the overlying water column. At depths below the compensation depth only the occasional *Globor. tumida* may survive. Changes in the depth of the lysocline and compensation depth through time may offset the sediment species composition, due to differential dissolution (after Bé 1977).

Atlantic, there is evidence that this corresponds to the boundary between North Atlantic Deep Water and the deeper Antarctic Bottom Water (Berger 1968). Below this level, calcite dissolution rates increase markedly, until at extreme depths the water is so corrosive to calcite that virtually no tests survive to be deposited. The depths at which the dissolution rate equals the rate of supply of carbonate tests from the overlying water column is the calcite compensation depth (CCD) (Berger 1970). This can be envisaged as analogous to a snowline on land; deep ocean basins below the compensation depth are devoid of carbonate sediments, and higher levels are increasingly blanketed by microfossil tests (Berger 1971). Because the calcite compensation depth is a function of both the rate of supply of carbonate tests and the dissolution rate, its actual depth varies from one area to another (Fig. 6.29), though generally it is <4000 m (Berger & Winterer 1974). Since vast areas of the ocean floor are below 4000 m, particularly in the Pacific Basin, this phenomenon greatly restricts the area in which foraminiferal studies can be usefully carried out (cf. Fig. 6.2). Even in less deep areas of the ocean, sediments accumulating below the lysocline are subject to significant dissolution. Most importantly, dissolution does not affect all species uniformly; selective removal of the more fragile, thin-walled species may significantly alter the original assemblages (biocoenoses), leaving behind thanatocoenoses which are unrepresentative of productivity in the overlying water column. Assemblages may be enriched with resistant species, which tend to be deep-dwelling, secreting their relatively thick tests in water which is significantly cooler than that near the ocean surface (Ruddiman & Heezen 1967, Berger 1968). Similarly, in populations of a particular species, the thicker walled, more robust individuals, which are preferentially preserved, tend to build their shells in deeper, colder water and are therefore isotopically heavier than their more fragile counterparts (Hecht & Savin 1970, 1972, Berger 1971).

Studies of the relative abundances of different foraminiferal species, in cores from various depths, have demonstrated these effects well (Fig. 6.30) and enabled species to be ranked according to their relative susceptibility to solution. Similar studies of coccoliths indicate corresponding problems, with structurally solid, cold-water forms preferentially preserved in thanatocoenoses (Berger 1973a).

Berger (1973b) suggests that the partially dissolved assemblages be designated taphocoenoses, to distinguish them clearly from assemblages more representative of the original biocoenoses. Clearly, paleoclimatic reconstructions based on taphocoenoses require very careful interpretation. This is particularly so if the rate or dissolution has changed over time as suggested by a number of studies (Chen 1968, Broecker 1971, W. Berger 1971, 1973b, 1977, Thompson & Saito 1974, Ku & Oba 1978). There is evidence that dissolution rates increased during interglacial times in the

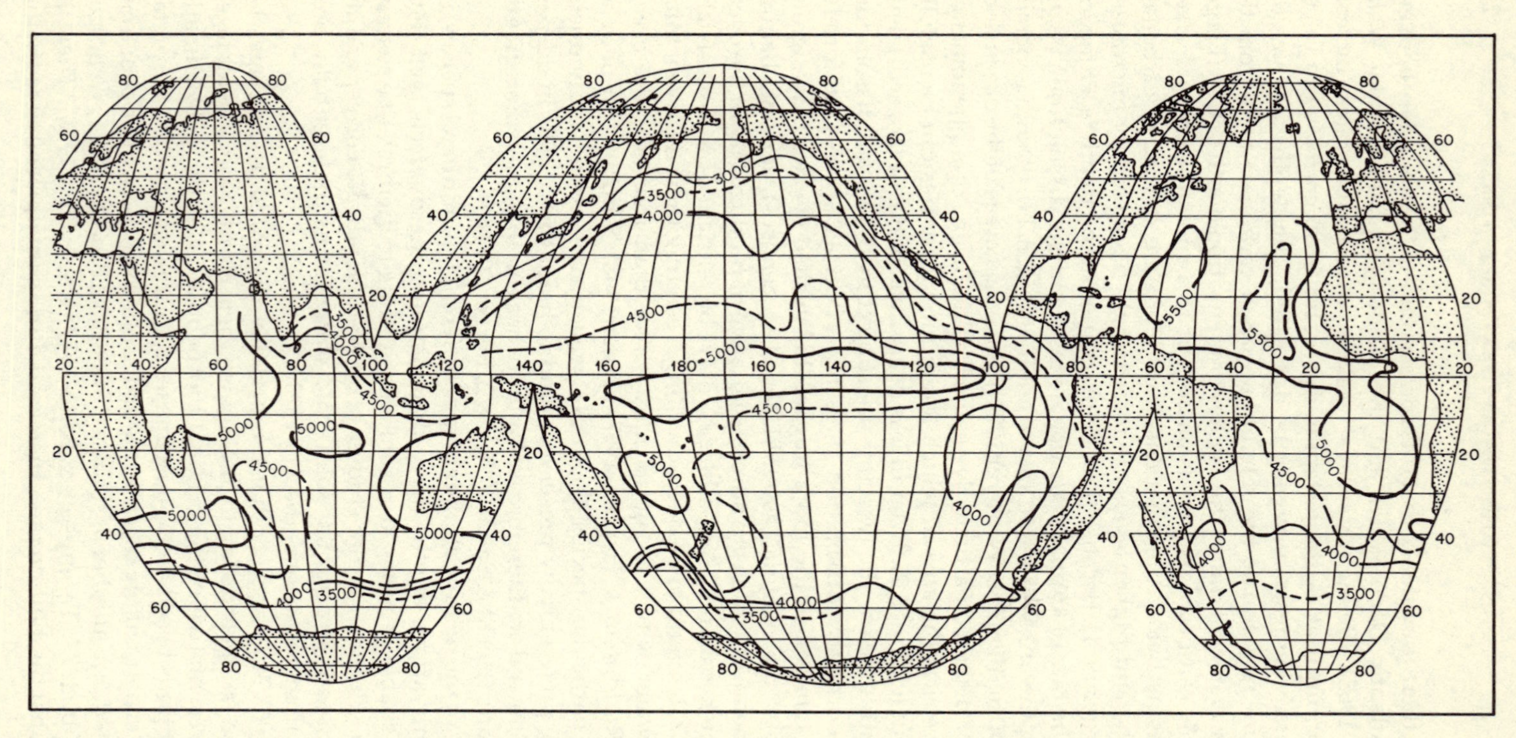

Figure 6.29 Distribution of calcium carbonate compensation depth. Depths in thousands of meters (after Berger & Winterer 1974).

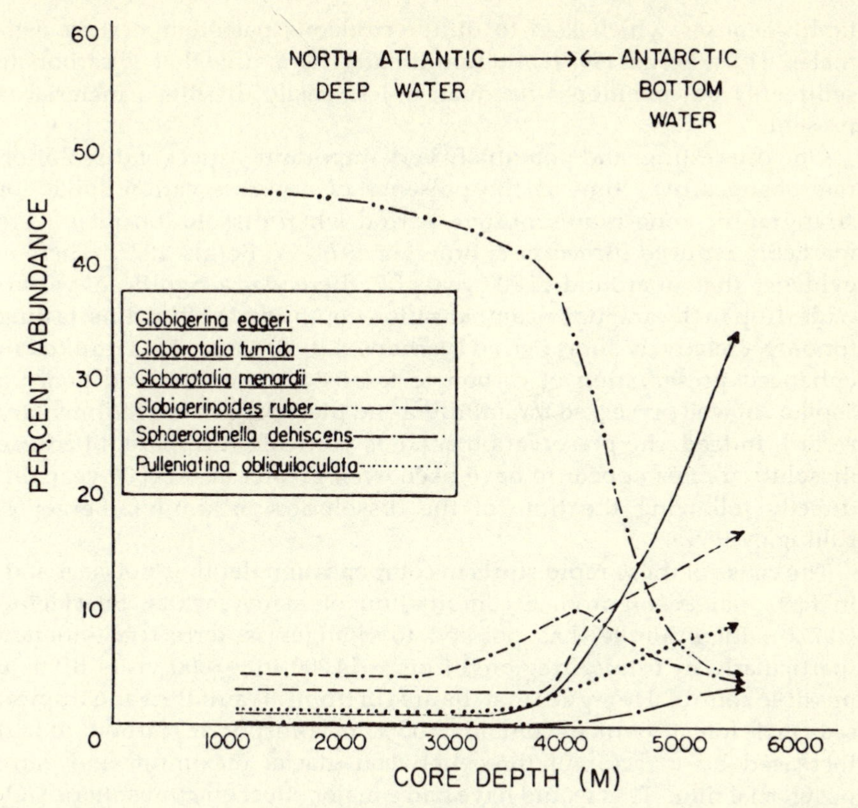

Figure 6.30 Changes in percentage abundance of several diagnostic planktonic foraminifera in equatorial Atlantic core tops with increasing water depths due to differential dissolution. The more corrosive Antarctica Bottom Water dissolves poorly resistant species (such as *Globigerinoides ruber*) so that more resistant species (such as *Globorotalia tumida*) increase in relative abundance (after Ruddiman & Heezen 1967).

tropical Pacific and Indian Oceans, resulting in the removal of many less resistant species, and a relative concentration of individuals with a cold-water aspect.† Conversely, in glacial times, dissolution rates were reduced, giving rise to assemblages of both solution-susceptible and solution-resistant forms. In short, glacial–interglacial changes may be characterized by corresponding dissolution cycles in these areas (Berger 1973b). Such effects would result in erroneous isotopic paleotemperature estimates since interglacial-age samples would have a higher abundance of cold-water individuals compared to glacial-age samples, thereby reducing the apparent glacial–interglacial temperature range (Berger 1971, Emiliani 1977, Berger & Killingley 1977). Similarly, in foraminiferal assemblage studies (Sec. 6.4), dissolution cycles may result in

†In the equatorial Atlantic and Gulf of Mexico, dissolution seems to have increased in glacial times (Gardner 1975, Luz & Shackleton 1975).

taphocoenoses which lead to quite erroneous paleotemperature estimates. Thus, Berger (1971) and Ruddiman (1977a) urge that all carbonate sediments be considered residual, unless easily dissolved material is present.

One interesting, and potentially very important, aspect of dissolution rate changes over time is the presence of a "preservation spike" or stratigraphic zone representing a period when dissolution rates were markedly reduced (Broecker & Broecker 1974, W. Berger 1977). There is evidence that, at around 14 000 years BP, there was a significant worldwide drop in the aragonite compensation depth and the lysocline, lasting for only a relatively short period (perhaps <1000 years). This resulted in enhanced preservation of carbonate fossils at that time and hence a "spike" of well preserved foraminifera and pteropods in the sedimentary record. Indeed, the preservation spike is particularly apparent because dissolution rates appear to have been even greater at ~12 000 years BP directly following the time of the dissolution minimum (Berger & Killingley 1977).

The cause of these rapid shifts in compensation depth is not clear and, in fact, may result from a combination of many factors. Shackleton (1977b), for example, has pointed to changes in terrestrial biomass (particularly in tropical regions) from ~14 000 to ~8000 years BP as a possible control. He argues that the area of tropical rainforest and tropical seasonal forest (which contain >50% of biospheric carbon today) increased by a factor of three between glacial maximum and early postglacial time. This would have had a major effect on atmospheric CO_2 levels. Expansion of the biosphere would have led to removal of CO_2 from the atmosphere and oceans, leading eventually to a lowering of the supersaturation zone and improved preservation of carbonate marine fossils. However, it is difficult to account for the seemingly rapid change from improved preservation to increased dissolution between 14 000 and 12 000 years BP by this mechanism alone. Other factors were probably involved. For example, redeposition of carbonate from the continental shelves as sea-level rose could have increased oceanic alkalinity and thereby reduced dissolution (W. Berger 1977). However, as the sea level rose a low salinity upper water layer may have formed (from continental ice-sheet ablation), creating a lid on the ocean and preventing vertical mixing (Worthington 1968). Continued biological activity in the oceans would have led to the accumulation of CO_2 and hence increased dissolution (Berger *et al.* 1977). Perhaps it is this signal which is observed following the dissolution minimum. Such a hypothesis has intriguing implications; if an extensive meltwater layer did exist during deglaciation (and if this resulted in a build-up of CO_2 in the subsurface waters), when ocean mixing was eventually restored an increase in atmospheric CO_2 concentrations would have ensued, resulting in an enhanced greenhouse effect. This may have been the sequence of events which resulted in the warm Hypsithermal period following deglaciation (W. Berger 1977, Berger *et al.* 1977).

The presence of a relatively shallow meltwater layer over a large part of the world ocean, as opposed to just marginal seas such as the Gulf of Mexico (Kennett & Shackleton 1975, Emiliani *et al.* 1975) has other implications of equal importance. Solar radiation receipts would have increased the temperature of the surface (thereby adding a thermal density stratification to the existing salinity stratification) and this would have inhibited mixing even more. Such a sequence of events would have enabled significantly higher temperatures to exist at the ocean surface than would have been the case without a meltwater layer, and hence may have facilitated the rapid demise of ice sheets during the deglaciation (Adam 1975). Such a mechanism would have been reinforced by the higher summer solar radiation receipts at this time, resulting from differing Earth–Sun orbital relationships (Lockwood 1978). However, it is difficult to imagine how a vast meltwater "lid" on the open ocean could be maintained for any significant length of time without being eliminated by mixing. Certainly the effect, if it occurred, is unlikely to have been important beyond the North Atlantic Ocean and could not have been a global phenomenon (cf. Sec. 6.9; Jones & Ruddiman 1982).

6.7 Paleoclimatic information from inorganic material in ocean cores

Weathering and erosion processes in different climatic zones may result in characteristic inorganic products. When these are carried to the oceans (by wind, rivers, or floating ice) and deposited in offshore sediments, they convey information about the climate of adjacent continental regions, or about the oceanic and/or atmospheric circulation, at the time of deposition (McManus 1970, Kolla *et al.* 1979). For example, numerous studies of inorganic material in cores from off the coast of West Africa have enabled climatic fluctuations of the adjacent land mass to be deduced. In this area today, vast quantities of silt and clay-sized particles (>25 million tons per year) are transported from the Sahara desert westwards across the Atlantic by the north-east trade winds (Chester & Johnson 1971). As the dust settles and/or is washed out of the atmosphere, it forms a significant proportion of the total sediment accumulation in the deep ocean (Windom 1975). By examining variations of the wind-blown fraction of ocean cores, an important index of trade wind intensity and continental aridity can be obtained. During late Quaternary glacial epochs, a higher proportion of terrigenous material accumulated in the equatorial and tropical Atlantic, off West Africa, than in interglacial periods. This has been measured directly (Diester-Haas 1976) and also indirectly by measuring the proportion of calcium carbonate in the cores (Hays & Perruzza 1972). Terrestrial detritus dilutes the relatively constant pelagic carbonate influx so that calcium carbonate content shows an inverse relationship with terrestrial material. Hence times of carbonate abundance indicate low terrestrial influx, i.e. low trade wind intensity and/or a decrease in aridity onshore. Conversely, carbonate minima signify periods of higher trade wind intensity and/or a larger arid zone. More detailed studies have helped clarify some of these uncertainties. By examining the ratios of quartz to mica (i.e. a hydraulically heavy to a hydraulically light material) Diester-Haas (1976) was able to resolve variations in transport energy (wind strength); stronger trade winds were indicated in glacial periods than in interglacials (cf. Parkin & Shackleton 1973, Parkin 1974). This conclusion was supported by a study of the number of characteristically red–yellow stained "desert quartz" grains which are a diagnostic feature of arid lands upwind from the core site. During the Holocene, cores north of 20°N showed a high percentage of desert quartz grains, with less during glacial times. South of 20°N, the reverse situation prevailed (Fig. 6.31). There was thus a southward shift

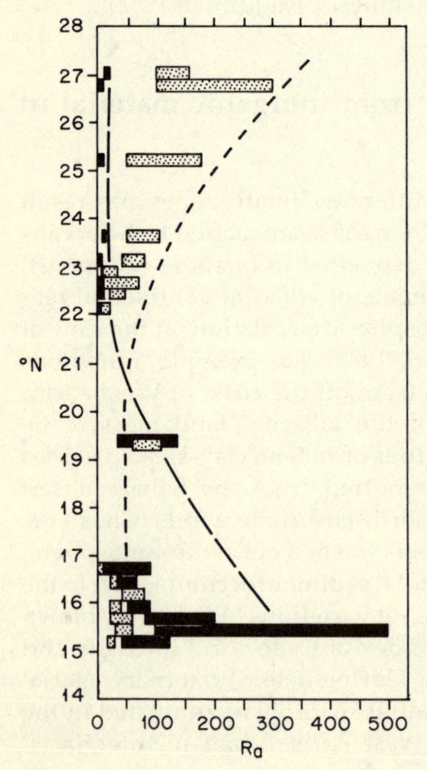

Figure 6.31 Ratios of red quartz to white quartz grains, R_Q, in Holocene and Würm (Wisconsin) sections of cores from off the coast of West Africa (15–27°N). The cores of Würm age show high levels of red (desert) quartz grains at more southerly latitudes, reflecting an increase in desert extent and/or a decline in white (alluvial) quartz grains due to reduced river flow. Offshore wind strength may also have been stronger at that time (after Diester-Haas 1976).

of the desert belt in glacial periods, and a drastic reduction of fluvial material from the Senegal River. Further support for this scenario is provided by studies of biogenic detritus in ocean cores. The concentration of freshwater diatoms and opal phytoliths (minute silica bodies derived from epidermal cells of land plants, particularly grasses) increases during glacial periods in cores south of 20°N off the coast of West Africa. It is suggested that this results from deflation of lacustrine sediments by stronger trade winds in relatively dry glacial times, following more humid interglacial periods when (grassland) vegetation was extensive and lakes were more common (Parmenter & Folger 1974).

The use of clay mineral assemblages to reconstruct paleoclimates has been proposed (McManus 1970) and indeed the global distribution of clay minerals in ocean sediments gives some support for this idea. For example, on a broad scale (making due allowance for local geological factors) chlorite has a distinct polar distribution and is clearly indicative of weathering processes at high latitudes. By contrast, kaolinite and gibbsite are primarily equatorial and tropical in distribution. Other clay minerals (e.g. illite and montmorillonite) are less diagnostic of climate (Griffin *et al.* 1968, Rateev *et al.* 1969). Large-scale mapping of clay mineral distribution at discrete intervals in the past (cf. Kolla *et al.* 1979) could capitalize on this source of paleoclimatic data, but as yet little work along these lines has been published.

Studies of clay minerals down core (i.e. continuous variations with depth) are not common, and where undertaken, have generally not provided much paleoclimatic information (e.g. Bowles 1975). One exception is the study of ocean cores from near the mouth of the Amazon (Damuth & Fairbridge 1970). These cores contain high proportions of undecomposed feldspars in sections of late Wisconsin age and a clay mineral assemblage typical of more arid conditions (less gibssite and kaolinite than in Holocene times). This led Damuth and Fairbridge to conclude that the Amazon Basin was considerably more arid than today, with the vast equatorial forests largely replaced by extensive grasslands. There is now considerable biogeographic evidence to support this hypothesis (e.g. Vuilleumier 1971, Colinvaux 1978), indicating that greater equatorial aridity in glacial times was not only characteristic of the African continent (see Ch. 8).

Another, perhaps less obvious, paleoclimatic indicator is the surface texture and micro-relief of quartz grains, seen under a scanning electron microscope (Krinsley & Donahue 1968, Krinsley & Margolis 1969). Quartz grains subjected to different environmental conditions have diagnostic surface textures which can be distinguished by trained observers. Thus, Krinsley and Newman (1965) were able to recognize the initial occurrence of ice-rafted debris in an ocean core from the souther Indian Ocean by examining the surface texture of quartz grains down the core. Their study lent support to faunal evidence that climatic deterioration in the Southern Hemisphere began well before the start of the Pleistocene (Ericson *et al.* 1963). More recent work indicates that surface texture of sand grains may be a useful paleo-environmental indicator even in ocean sediments as old as the lower Cretaceous (Krinsley 1978).

Ice-rafted debris has also been used to identify paleocirculation patterns in the North Atlantic and the timing of glacial episodes, which result in increased deposition rates of coarse detritus. By mapping the distribution and concentration of ice-rafted detritus in the North Atlantic at discrete intervals of time, Ruddiman (1977b,c) was able to recognize a major change in the pattern of ice-rafting deposition at ~75 000 years BP (Fig. 6.32). Prior to this period (for ~50 000 years) deposition had been greatest near Greenland and Newfoundland, more or less as occurs today. However, from ~75 000 to ~11 000 years BP the major depositional axis was displaced 1500 km to the south-east along a WSW–ENE axis from 46 to 51°N; during this period it appears that the warm North Atlantic Drift water was replaced by a strong cyclonic (counterclockwise) drift of subpolar water. This allowed debris-laden ice to travel much further towards the Equator before meeting water warm enough to cause it to melt. The marked change in depositional pattern and debris accumulation rate is in accord with isotopic evidence from the area, which also indicates that the most significant change in North Atlantic ocean-surface temperatures occurred around ~75 000 years BP (Sancetta *et al.* 1973a,b).

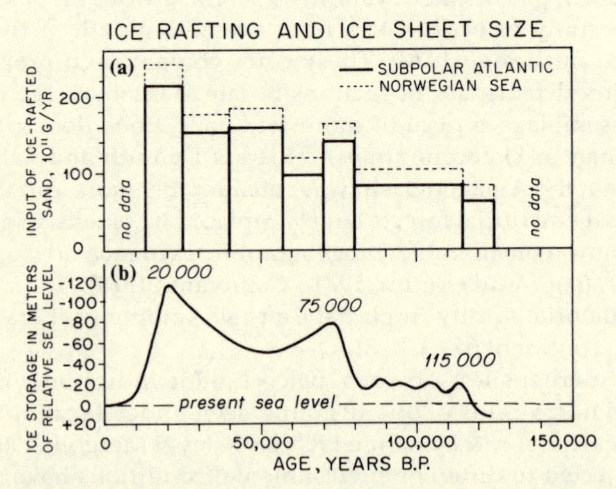

Figure 6.32 (a) Total absolute input rate of ice-rafted sand to the subpolar Atlantic and Norwegian Sea over the last interglacial–glacial cycle (~125 000 years). (b) Estimated changes in eustatic sea level based on an assumed relationship between $\delta^{18}O$ and global ice volume. Note the increase in ice-rafted debris at times of global ice volume increases (after Ruddiman 1977a).

6.8 Causes of glaciation and deglaciation: the oceanic evidence

The availability of continuous paleoclimatic records from the ocean floor, spanning several hundred thousand years, has enabled hypotheses of the causes of climatic change to be tested, and has facilitated the development of new models. One of the most important hypotheses is that propounded by Milankovitch (1941), who argued that glaciations in the past were principally a function of variations in the Earth's orbital parameters, and the resulting redistribution of solar radiation reaching the Earth (see Sec. 2.6 for a complete discussion). Emiliani (1955, 1966) was the first to note that $\delta^{18}O$ maxima in Caribbean and equatorial Atlantic cores closely matched summer isolation minima at 65°N, which was the latitude that Milankovitch had considered critical for the growth of continental ice sheets. Subsequently, Broecker and van Donk (1970) suggested revisions of Emiliani's timescale, but still concluded that insolation changes were a primary factor in continental glaciation. In addition, dates of coral terrace formation, indicative of a former higher sea level (lower global ice volume) were shown to be closely related to times of insolation maxima, again supporting the ideas of Milankovitch (Broeker *et al.* 1968, Mesolella *et al.* 1969, Veeh & Chappell 1970).

Although the orbital parameters are now known quite precisely, and the resulting insolation receipts on the upper atmosphere can be calculated (A. Berger 1978), three factors inhibit a rigorous test of the hypothesis:

(a) The orbital variations cause a redistribution of solar radiation, both temporally (seasonally) and spatially. It is commonly assumed that the critical location for continental ice-sheet growth is around 65°N and that summer insolation variations are of paramount importance. However, whether this is true or not is open to debate. Kukla (1975) has suggested that more critical periods of the year are the transition seasons, particularly the fall (September, October) when the establishment and persistence of snow cover is critical. This question assumes importance because calculated insolation values through time differ, depending on the season in question. Furthermore, the latitude for which insolation calculations are being made greatly affects the resultant curve. Although the reasons for selecting 65°N are reasonable, equally cogent arguments can be made for not selecting any one zone, but rather for calculating changes in insolation *gradient* between the Equator and the poles, which is, after all, the driving force of the general circulation (Young & Bradley 1984).

(b) An important signal which has been inspected for a relationship between orbital perturbations and climatic change is the marine core $\delta^{18}O$ record, which reflects changes in continental ice volume (principally in the Northern Hemisphere). Clearly, there will be a lag between the timing of any forcing factor and the timing of maximum glaciation (maximum $\delta^{18}O$ values in the ocean) since it takes many thousands of years for ice sheets to develop to such a size that they begin to have an effect on sea level and on the isotopic composition of the world ocean (Barry *et al.* 1975). However, the duration of this lag is not obvious, and may not be constant for all glaciations. Indeed, the entire mechanism or sequence of events leading to continental-scale glaciation is poorly understood, though several models have been proposed (e.g. Ruddiman & McIntyre 1979, 1981a, Imbrie & Imbrie 1980, Oerlemans 1981, Pollard 1982). Whether a given set of insolation conditions will inevitably lead to a particular cryospheric response is uncertain, and may critically depend on numerous other factors (e.g. volcanic dust loading of the atmosphere at the time and/or the preceding boundary conditions). There is even some doubt as to whether the same set of insolation conditions would result in an identical atmospheric response (Lorenz 1968, 1976). Hence, if the timing of glaciations, indicated in ocean cores, does not precisely match Milankovitch's predicted insolation minima, or does not maintain some constant lag relationship to those minima, would this necessarily negate the idea of a fundamental pulse to glaciations, induced by orbital perturbations?

(c) Testing of Milankovitch's hypothesis requires a very precisely dated proxy record extending back hundreds of thousands of years in time. Without such precision, apparent periodic components or phase relationships within the data would be meaningless. It is this fundamental chronological problem which has plagued the numerous efforts made to try and resolve the problem.

The first rigorous attempt to assess the evidence for orbital changes in paleoclimatic data was made by Hays *et al.* (1976) using two ocean core records from the southern Indian Ocean (43 and 46°S). Using oxygen isotope stratigraphy and extinction levels of certain species of coccolith and radiolaria, they constructed a chronology for the cores by analogy with isotope stage boundaries of "known" age on other cores (see Sec. 6.3.1). Three parameters were studied: $\delta^{18}O$ values in the foraminifera *Globigerina bulloides* (an index of global, but primarily Northern Hemisphere, ice volume); summer sea-surface temperature (T_s) derived from radiolaria-based transfer functions (an index of sub-Antarctic temperatures); and abundance variations of the radiolaria *Cycladophora davisiana* (considered to be an index of Antarctic surface water structure). Using the ~450 000 year record available, Hays *et al.* (1976) showed that much of the variance in these proxy records was concentrated at frequencies corresponding closely to those expected from an orbital forcing function. Specifically, spectral peaks were found at periods of ~100 000 years, 40 000–43 000 years, and 19 500–24 000 years. Such periodicities closely match spectral peaks in orbital data (at ~105 000, ~41 000, and 19 000–23 000 years, associated with variations in eccentricity, obliquity, and precession, respectively). Furthermore, not only are the proxy and orbital series closely matched in the frequency domain, but an examination of the time domain of each periodic component shows fairly consistent phase relationships (back to 300 000 years) between orbital parameters and the "resultant" climatic signal. Thus, in the 40 000 year frequency band, the three oceanographic parameters maintain a constant phase relationship with each other (with $\delta^{18}O$ lagging T_s by 2000 years) as well as a consistent and logical relationship with orbital (obliquity) variations. Following times of low obliquity, $\delta^{18}O$ values are high, values of T_s are low, and *C. davisiana* is abundant, with the three proxy records lagging behind obliquity minima by 9000, 8000, and 7000 years, respectively). Such results are very improbable by chance and provide strong evidence that changes in the Earth's orbital geometry played an important role in causing the observed variations over the past 300 000–400 000 years. In fact, it has even been argued that the ~41 000 year obliquity period is so closely matched with the proxy climatic data over the last ~300 000 years that where the relationship begins to disintegrate (generally in the lower, older section of a core) it is the chronology adopted which is in error. Hence, improvements in the chronology may be possible by assuming that the constant phase relationship between, say, $\delta^{18}O$ values and obliquity minima was maintained throughout the period of the core record (Hays *et al.* 1976). In this way, the conventional oxygen isotope chronology proposed by Shackleton and Opdyke (1973, 1976) has been revised by "tuning" the isotope record to fit the obliquity signal (Table 6.2). Whether such manipulations are justifiable is open to debate, but some support for the approach is given by the fact that it results in an age for the Brunhes/Matuyama geomagnetic epoch boundary very close (within 0.3%) to independent radio-isotopic estimates for this event (Kominz *et al.* 1979, Pisias & Moore 1981). Close approximations to radio-isotopic ages of the Jaramillo and Olduvai events have also been achieved by this technique. However, Morley and Hays (1981) have found inconsistencies in this chronological model when applied to other cores, leading them to compute a further revision of the oxygen isotope stage boundary chronology using a similar tuning approach (Table 6.2). No doubt, there will be yet more efforts to fine tune the timescale before a universally accepted chronology emerges.

6.9 Mechanisms of glaciation and deglaciation: the oceanic evidence

Variations of isotopic content, faunal and floral abundance, and inorganic components in ocean cores provide a comprehensive measure of oceanographic changes during the waxing and waning of continental ice sheets. These variations have provided strong evidence that orbital perturbations are of primary importance in the timing of glaciation and deglaciation, but the climatic mechanisms involved remain unclear. We may accept the role of orbital forcing, but exactly how this was translated into ice-sheet growth and decay is enigmatic. However, oceanic data can provide further insight into such mechanisms; indeed, some would argue that the oceans themselves may be the key to the whole problem, regardless of orbital perturbations (e.g. Weyl 1968, Worthington 1968, Newell 1974).

The most comprehensive hypotheses of ice-sheet growth and decay in relation to oceanic conditions have been developed by Ruddiman and McIntyre (1981a). According to their glaciation hypothesis, ice-sheet growth is favored when Northern Hemisphere summer insolation levels are low (due to orbital factors) but oceanic temperatures at high latitudes are warm, providing an abundant moisture source adjacent to the relatively cool continents. Strong thermal contrasts at the continental margin help steer depressions towards the developing ice sheets, thereby increasing the local accumulation rate.

Strong evidence for this scenario is provided by ocean cores from the subpolar North Atlantic; oxygen isotope analyses of benthic foraminifera (providing a record of continental ice volume) can be directly compared with sea-surface temperature estimates derived from planktonic foraminifera assemblages (Fig. 6.33). Comparison of such records at both the 5c – 5d isotope substage transition and the 5/4 isotope stage boundary indicates that ocean temperatures remained high during most of the period of ice growth. The main fall in surface temperatures lags 1000–5000 years behind ice growth (Ruddiman & McIntyre 1979, Ruddiman *et al.* 1980a). Once ice build-up proceeds beyond a certain point, isostatic adjustments and positive feedback mechanisms perpetuate ice growth, even though nearby ocean temperatures have cooled considerably (Fig. 6.34a).

Ocean temperatures may also have played a key part in the mechanism of deglaciation. The last two glacial – interglacial transitions have coincided with both high summer insolation totals and low winter insolation totals (due to orbital variations, principally the 23 000 year precessional periodicity). Both factors are considered to be crucial to the deglaciation process, but insufficient by themselves to bring about the destruction of continental ice sheets. However, as a result of positive feedback pro-

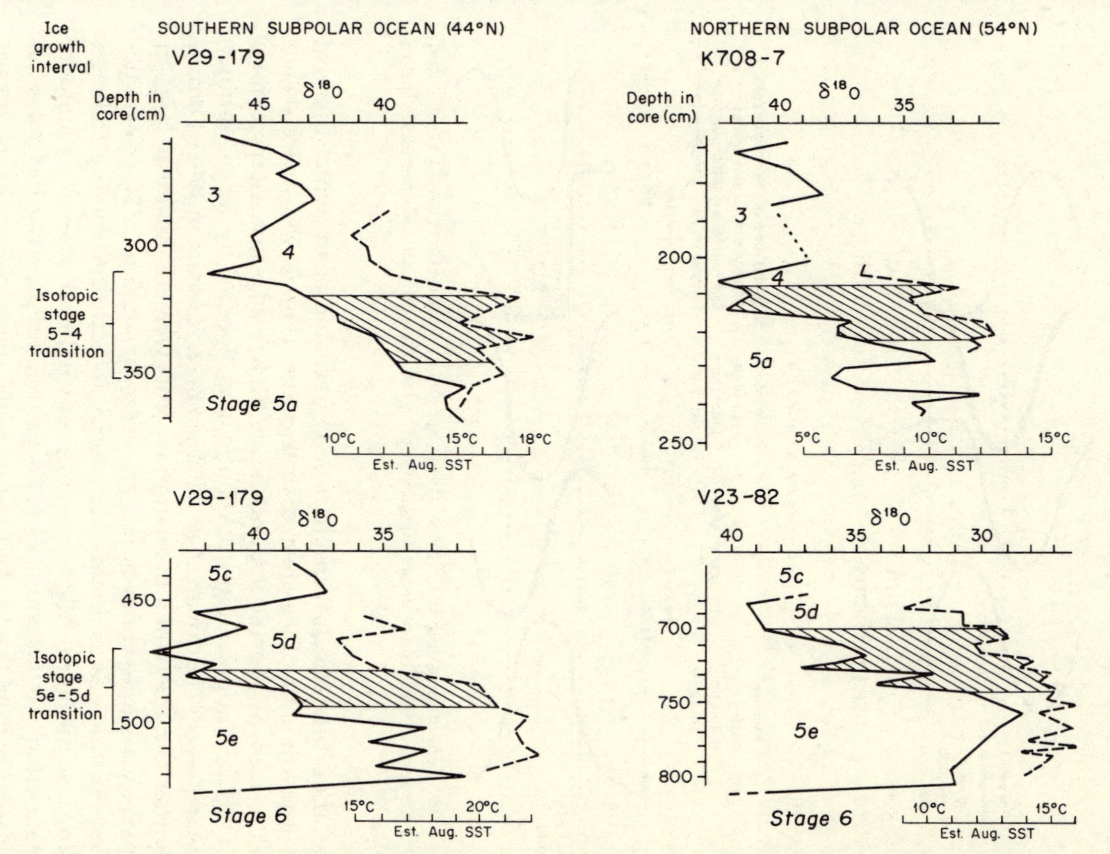

Figure 6.33 Estimated August sea-surface temperatures (SST) based on planktonic foraminiferal assemblages, and oxygen isotope ratios ($\delta^{18}O$) based on benthic foraminiferal analyses. Two records of the transitions from isotopic stages 5c–5d and 5–4 are shown. The subpolar ocean surface remained warm during most of each ice growth interval (represented by the $\delta^{18}O$ records, with *higher* $\delta^{18}O$ values indicating ice growth). The major period of warm-ocean lag is stippled (after Ruddiman & McIntyre 1979).

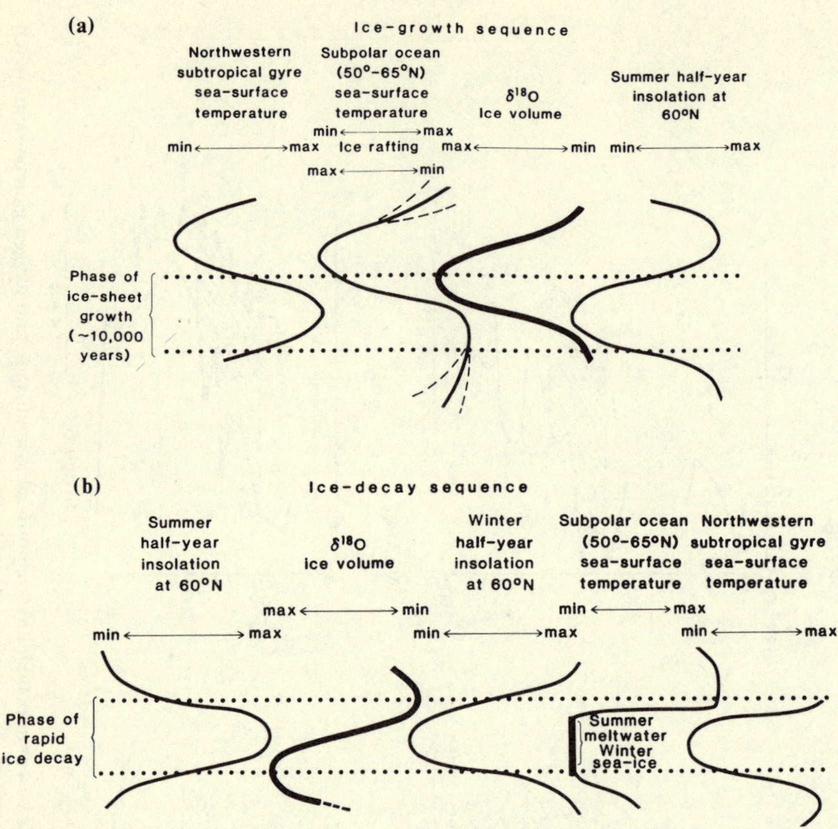

Figure 6.34 Schematic diagrams representing changes in North Atlantic ocean associated with (a) ice growth and (b) ice decay. See text for explanation (after Ruddiman & McIntyre 1981a).

cesses, the orbital effects were amplified by oceanic factors which thereby accelerated the rate of ice-sheet disintegration (Fig. 6.34b). As ice sheets began to melt, as a result of significantly higher solar radiation receipts, much of the subpolar North Atlantic was flooded by a low salinity meltwater layer (Ruddiman *et al.* 1980b, Jones & Ruddiman 1982). During the winter months (when insolation receipts were anomalously low) an extensive sea-ice cover developed at least as far south as 50°N, essentially cutting off this area as a moisture source. In addition, calving icebergs drifting into the North Atlantic significantly reduced sea-surface temperatures, perhaps as far south as 40°N, further suppressing the potential moisture flux from the North Atlantic to the decaying ice sheets. As sea level rose, due to increased ice-volume losses (on a global scale), so the rate of iceberg calving would have increased, further accentuating this feedback loop (though isostatic rebound exceeding the rate of eustatic sea-level rise would tend to negate this effect). The model is particularly

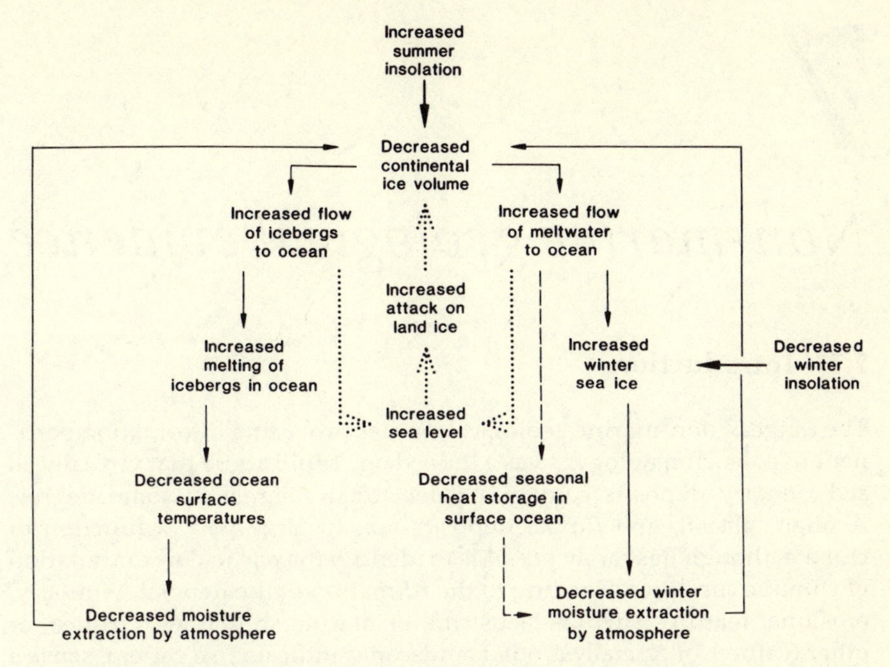

Figure 6.35 Schematic model of oceanic feedback loops that hasten ice disintegration. Increased summer insolation results from (Milankovitch) Earth-orbital changes, primarily the 23 000 year precessional cycle. Oceanic feedbacks amplify the initial perturbation (forcing) (after Ruddiman & McIntyre 1981a).

attractive since it describes a mechanism of self-regulation in ice-sheet growth and decay; the larger the size of the ice sheets at the time of (orbitally forced) ice decay, the stronger are the feedback loops operating to amplify the rates of ice disintegration (Fig. 6.35).

Oceanic evidence for such a sequence of events is provided by the ocean-core record from the 2/1 and 6/5 isotope stage boundaries (i.e. Terminations I and II). At both levels, there is an abundance of ice-rafted debris in cores from the subpolar North Atlantic, indicating a marked increase in iceberg production. At the same time, sea-surface temperatures in the zone 35–45°N fell to minimum levels, associated with increased iceberg and sea-ice outflow from the North Atlantic. Most remarkably, over the entire subpolar North Atlantic, coccolith and planktonic foraminifera concentrations fell by several orders of magnitude, as a result of a low salinity (stably stratified) meltwater layer inhibiting productivity (Ruddiman & McIntyre 1981a,b,c). This coincidence of exceptionally low floral and faunal productivity, at a time of increased ice-rafted debris and cooler sea-surface temperatures at lower latitudes, provides strong physical support for the deglaciation model (Fig. 6.34b).

7

Non-marine geological evidence

7.1 Introduction

The range of non-marine geological studies providing information pertinent to paleoclimatology is vast; indeed one could argue that virtually all sedimentary deposits convey a paleoclimatic signal to some degree. Aeolian, glacial, and fluvial deposits are, in large part, a function of climate, though it is rarely possible to identify the particular combination of climatic conditions leading to the formation of the deposit. Similarly, erosional features such as lacustrine or marine shorelines, cirques, or other features of glacially eroded landscapes indicate in a general sense a particular type of climate, but quantitative paleoclimatic reconstructions are rarely possible based on this kind of information (Flint 1976). More often than not, the only climatic inference to be drawn from such evidence is limited to the observation that, formerly, conditions were wetter and/or cooler, or warmer and/or drier; even then it may be impossible to provide an adequate date on the feature! Glacial striae, for example, indicate former abrasion by ice, but assigning a date to the event may be impossible . . . a definitive event but of dubious antiquity (e.g. Blake 1977).

In this chapter no attempt will be made to review all of the possible paleoclimate-related geological and geomorphological phenomena. Instead, discussion will be limited to the smaller number of approaches which have either provided quantitative paleoclimatic data or helped establish the chronology of paleoclimatic events.

7.2 Periglacial features

The use of fossil periglacial phenomena as an index of former climatic conditions is limited by two basic problems. First, dating periglacial features directly is often difficult, if not impossible; generally they are dated by reference to the deposits within which they are found, thereby obtaining only a maximum age for the features. Secondly, although regions of modern periglacial activity can be circumscribed by particular isotherms, the occurrence of similar activity in the past can only indicate an upper limit to temperatures at the time, not a lower limit (R. B. G. Williams, 1975). Thus, in general terms, permafrost today only occurs in areas where the mean annual air temperature is <−2 °C and it is virtually ubiquitous north of the −6 to −8 °C isotherm in the Northern Hemisphere (Ives 1974). Evidence of more extensive permafrost in the past, however, only demonstrates that temperatures were below these levels, but provides little information on *how much* lower. Mapping the distribution of relict periglacial features may indicate how far the southernmost boundary of the permafrost zone was displaced, but within this zone only the limiting *maximum* paleotemperature estimates are possible. Nevertheless, periglacial features are of particular interest because they provide information about the periods of extreme temperature depression during past glacial episodes. They also provide information about areas close to the ice-sheet margins, for which there are few other sources of proxy paleoclimate data.

It has already been mentioned that permafrost only occurs in areas with mean annual temperatures below a certain level, but permafrost itself may leave no morphological evidence of its former existence. Paleoclimatic inferences can only be based on features which develop in regions of permafrost and disturb the sediments in a characteristic manner. In this way fossil or relict features can be identified and their distribution mapped (Fig. 7.1). The most useful and easily identified features include fossil ice wedges, pingos, sorted polygons, stone stripes, and periglacial involutions (Washburn 1979a). The problem is to identify those climatic factors which are necessary for the formation of the features in question; commonly this can only be done in general terms (Table 7.1). Ice wedges, for example, result from thermal contraction at subfreezing temperatures. Winter temperatures of −15 to −20 °C or less are required before active frost cracking occurs, but the exact requirements depend on the material being considered. Cracking and ice-wedge formation will occur at higher temperatures in silts and fine-grained material than in gravels, where mean annual temperatures of −12 °C may be necessary. Furthermore, the amount of snowfall is a significant factor since snow will insulate the ground surface from the effects of severe cold. This has been demonstrated in many areas where today active ice-wedge formation does not normally occur; where snow is artificially removed (e.g. from roads or airport runways) frost cracks and ice wedges will develop. Paleoclimatic reconstructions based on such phenomena are thus subject to a certain amount of uncertainty, and similar problems have to be faced when dealing with other types of periglacial features. Nevertheless, it is possible to make conservative estimates of temperature change based on the former distribution of different types of periglacial features (Fig. 7.2). Accuracy is really limited by our under-

Figure 7.1 Distribution of patterned ground of Devensian (=Wisconsin) age in Britain (after Washburn 1979b).

Figure 7.2 Temperature increases in Europe since the last glacial maximum, based on periglacial features. Increases are minimum estimates. Maximum ice limits are shown as heavy lines (after Washburn 1979a).

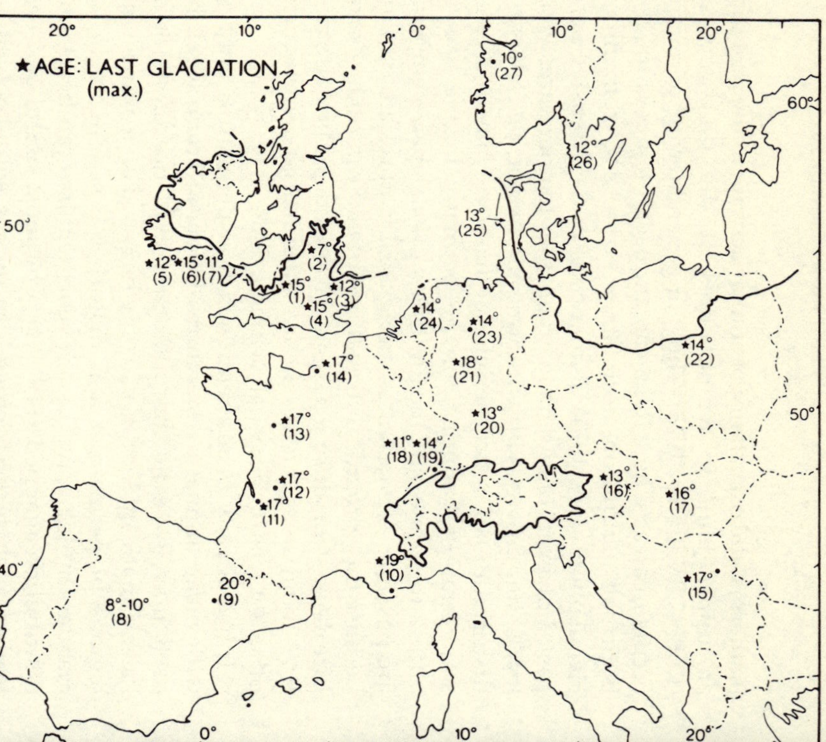

Table 7.1 Climatic threshold values for the distribution of periglacial geomorphic features (after Karte and Liedtke 1981)

	Climatic threshold values[‡]	
Periglacial geomorphic features[†]	*MAT (°C)*[§]	*MAP (mm)*[‖]
1 Periglacial geomorphic features whose formation requires permafrost		
1.1 Features connected with continuous permafrost Ice-wedge polygons	<−4 to <−8 °C	>50–500 mm
	Other climatic indication: rapid temperature drops in early winter	
Sand-wedge polygons	<−12 to <−20 °C	<100 mm
Closed system pingos	<−5 °C	
1.2 Features connected with discontinuous permafrost		
Open system pingos	<−1 °C	
1.3 Features which occur in connection with continuous, discontinuous, and sporadic permafrost		
Depergelation forms ("thermokarst" forms, active layer failures, detachment failures, ground ice slumps, permafrost depressions, alas, "baydjarakhs," "dujodas," alas thermokarst valleys, beaded drainage, thaw lakes, oriented lakes, thermo-erosional niches, thermo-abrasional niches, degradation polygons, thermokarst mounds)	<−1 °C Other important indication: high ground-ice content	
Seasonal frost mounds (frost blisters, hydrolaccoliths, bugor)	<−1 to <−3 °C	
Palsas	<0 to <−3 °C	
Rock glaciers	<+2 to <0 °C	<1200 mm
	Other climatic indications: continental climates with high incoming radiation, sublimation, evaporation and little snowfall	
2 Features whose formation requires intense seasonally frozen ground but which also occur in connection with permafrost		
2.1 Seasonal frost-crack polygons (ground wedges)	<0 to <−4 °C	
	Other climatic indication: mean temperature of coldest month <−8 °C	
2.2 Frost mounds (thufurs)		
Tundra hummocks (high latitude occurrences)	<−10 °C	
Earth hummocks (high latitude occurrences)	<−6 °C	
Earth hummocks (high altitude occurrences)	<+3 °C	
2.3 Non-sorted circles (mud boils, mud circles)	<−2 °C	>400–800 mm
2.4 Sorted circles and stripes (ϕ>1 m)[¶]	<−4 °C	
2.5 Sorted circles and stripes (ϕ>1 m)[¶]	<+3 °C	

Table 7.1 continued

	Periglacial geomorphic features[†]	Climatic threshold values[‡] MAT (°C)[§]	MAP (mm)[‖]
2.6	Gelisolifluction microforms (lobes, steps, ploughing blocks)	<−2 °C	
2.7	Nivation and cryoplanation features (nivation hollows, cryoplanation terraces, frost-riven cliffs)	<−1 °C	
3	Features which are linked to diurnally frozen ground and needle ice but which also occur in connection with seasonally frozen ground and permafrost		
3.1	Miniature polygons		
3.2	Miniature sorted forms and stripes	<+1 °C	
3.3	Microhummocks		

[†] For a description of these features, see Washburn (1979b).
[‡] The thermal threshold values represent upper limits for development of features.
[§] Mean annual air temperature.
[‖] Mean annual precipitation.
[¶] $\phi = -\log_2 d$ where d = sediment diameter (Krumbein's sediment size scale).

standing of the climatic controls on similar contemporary features. From this preliminary map, it would appear that mean annual temperatures in Europe were *at least* 14–17 °C below recent averages during the maximum phase of the last (Wisconsin/Würm/Weichsel) glaciation (Washburn 1979b). Although many features used to compile the map are not well dated, and often are simply considered to reflect conditions during the maximum stage of the last glaciation, it is worth considering the point made by Dylik (1975) that maximum temperature depression was generally *not* coincident with the maximum extent of ice (maximum "glaciation"). Dylik considers the extent of glacier ice to be more of an index of snowfall (i.e. of cold and humid conditions) than simply of low temperatures. Periglacial features may thus achieve their maximum development during periods of minimum temperature prior to the maximum extent of major ice sheets; this may explain the large discrepancy between botanically derived paleotemperatures for the last glacial maximum (generally indicating mean annual temperatures only 3– 6°C below those of today) compared with paleotemperature estimates derived from periglacial phenomena.

Where a variety of periglacial features of varying ages can be identified in a limited area, it may be possible to reconstruct paleotemperature through time. This has been attempted by Maarleveld (1976) using observations of relict periglacial features in the Netherlands (Fig. 7.3). Maarleveld associated each type of feature with particular temperature constraints; pingo remnants, for example, indicated maximum mean

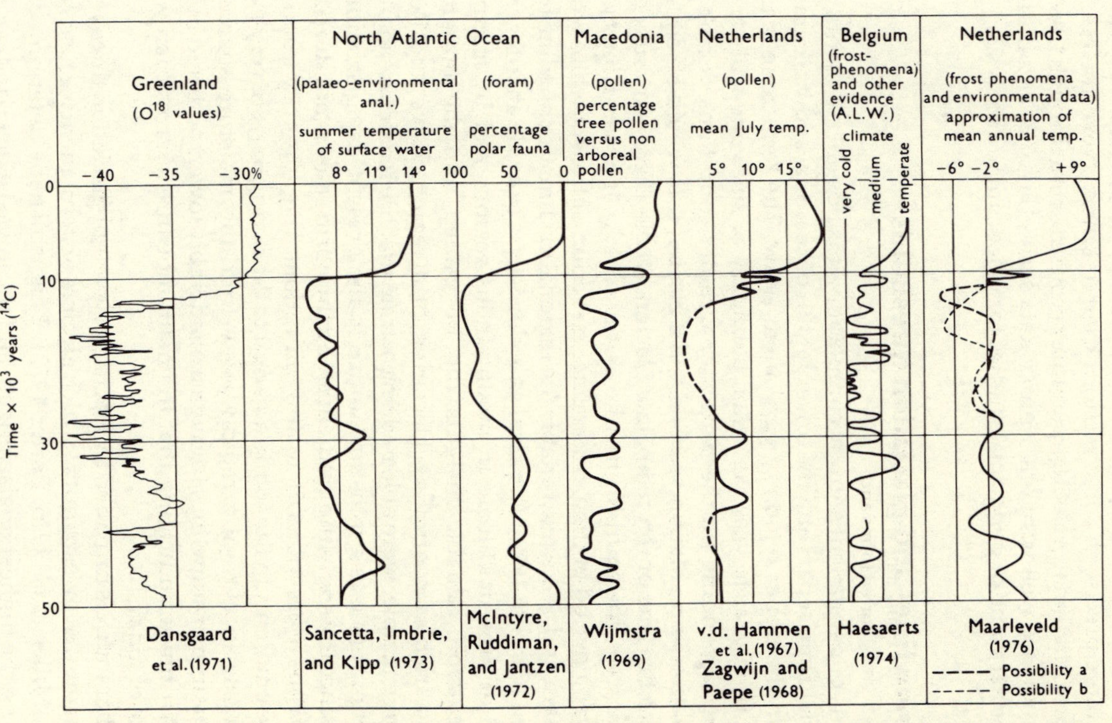

Figure 7.3 Paleotemperature reconstruction for Europe based on periglacial features (right column) compared to other long proxy data records (after Maarleveld 1976).

annual temperatures of −2 °C, whereas "extensive coarse snow meltwater deposits" were indicative of a range in mean annual temperature of −5 to −7 °C (see Fig. 7.3 caption). Unfortunately, Maarleveld does not identify which sections of his graph are based on estimates of maximum temperature and which are based on a defined temperature range, so the graph may be more precise in some sections than in others. Nevertheless, as a first approximation to paleotemperature reconstruction through time the results compare well with other proxy data series (Fig. 7.3) and indicate the potential value of periglacial studies for paleoclimatic analysis.

7.3 Snowlines and glaciation thresholds

In regions of permanent snow accumulation, it is possible to identify an altitudinal zone separating the lower region of seasonal snow accumulation from the upper region of permanent snow. The term zone is used because from year to year the actual boundary or snowline will vary in elevation, depending on the particular weather conditions during the accumulation and ablation seasons. On a glacier, this would be equivalent to the firn line or (on temperate glaciers where there is no superimposed ice zone) the equilibrium line altitude (ELA). If observations were made over a period of time, the *average* elevation of the snowline would be apparent and this would enable the regional or climatic snowline to be identified (Østrem 1974). Regional snowlines thus provide an integrated measure of present climate in mountainous regions and by mapping their elevation valuable insights can be gained into the important climatic variables controlling glaciation of a region. Péwé and Reger (1972), for example, were able to demonstrate that the Arctic Ocean plays no significant role as a moisture source in the present-day glaciation of Alaska, since the snowline gradient along the north coast is not steep and the snowline does not fall to low elevations (Fig. 7.4). By contrast, snowline gradients in the south, adjacent to the Bering Sea, are very steep and increase in elevation rapidly away from the moisture source. A similar pattern of snowlines (though considerably lower) occurred in late Wisconsin times, indicating that the main moisture sources then were like those of today.

A similar index is provided by glaciation levels or glaciation thresholds which define the lowest limit at which glaciers or permanent ice fields can develop (Miller *et al.* 1975, Porter 1977). This is usually determined by identifying the highest unglacierized, and the lowest glacierized, mountain summits in a region and averaging the two elevations. Both snowlines and glaciation thresholds can be used in paleoclimatic reconstructions if similar features can be mapped for periods in the past (Osmaston 1975), though obviously they only provide information about (extreme)

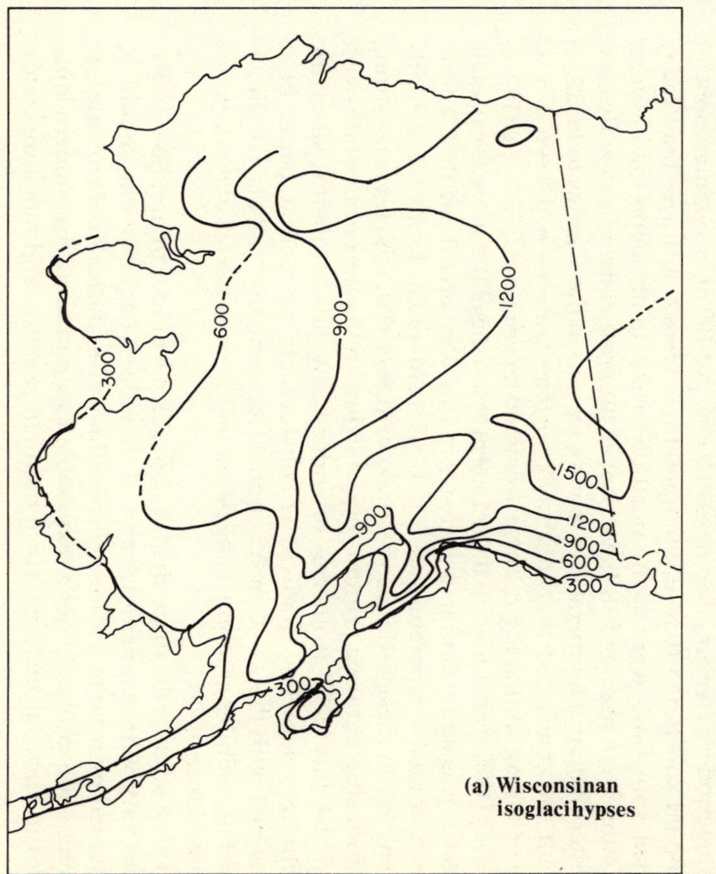

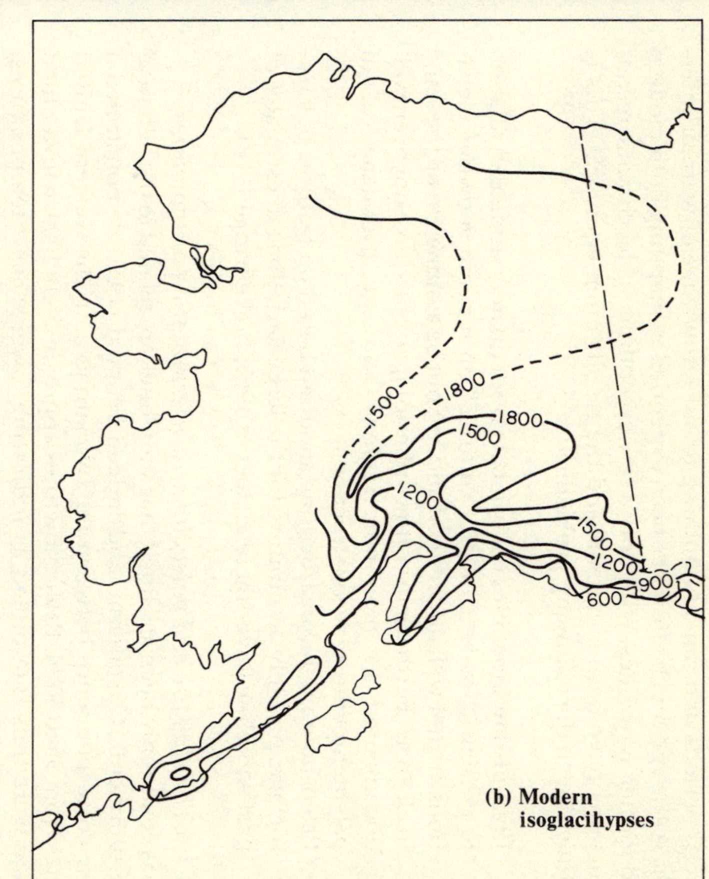

Figure 7.4 Modern and Wisconsin snowlines in Alaska. Snowlines today show the same pattern as glacial age snowlines, suggesting that no major change in moisture sources has occurred. The Gulf of Alaska, not the Beaufort Sea, was thus the most important source of moisture for glaciation (after Péwé & Reger 1972).

glacial periods and can add nothing to our knowledge of warmer intervals. Although much effort has been expended in mapping both modern and former snowlines, with only a few exceptions, paleoclimatic reconstructions have been simplistic and the results often equivocal. This is mainly due to the following problems:

(a) Present-day snowlines have not been adequately studied in relation to present climate; climatic "controls" on modern snowline elevations are not well understood and can not be assumed to be the same in all areas. Furthermore, atmospheric lapse rates have not been well documented in mountain regions and are problematical in paleotemperature reconstructions.
(b) Paleosnowline reconstructions are often based on features of varying age, possibly accounting for the large variations in estimates of past snowline lowering (e.g. Reeves 1965, Brakenridge 1978).

7.3.1 The climatic and paleoclimatic interpretation of snowlines

It has commonly been assumed that snowlines are related to the height of the summer 0 °C isotherm, and indeed Leopold (1951) demonstrated a close correspondence between the two surfaces, in the western United States from 35 to 50°N. Paleosnowlines approximately 1000 m lower than today were thus interpreted as indicating lower *summer* temperatures. Using modern free air lapse rates of 0.6 °C per 100 m in summer months, Leopold concluded that July temperatures were 6 °C lower than today when snowlines were at the position of maximum depression. Other months were assumed to have cooled proportionately less, with midwinter (January) temperatures having remained the same as today. As a result, he concluded that mean *annual* temperatures had been 4–5 °C lower. Using similar logic, but an assumed lapse rate of 0.75 °C per 100 m, Reeves (1965) concluded that the 1300 m lowering of the snowline in New Mexico was equivalent to a decrease in July temperature of 10 °C and in mean annual temperature of 5.1 °C. Considerable doubt has recently been cast on these estimates by Brakenridge (1978), who finds no strong relationship between present-day snowline in the American south-west and the July 0 °C isotherm, or indeed any July isotherm, since the latitudinal snowline gradient is considerably steeper. A better fit is obtained with the −6 °C mean annual isotherm; the "full glacial" snowline has a similar gradient, suggesting that there was a fall in *mean annual* temperature of 7 °C.

These studies illustrate the basic problems of identifying how snowlines vary with temperature and, if they do, what lapse rate should be assumed in order to use former snowlines as an indicator of temperature change. It is probably not a reasonable assumption to use modern lapse rates in such calculations, since both temperature and moisture condi-

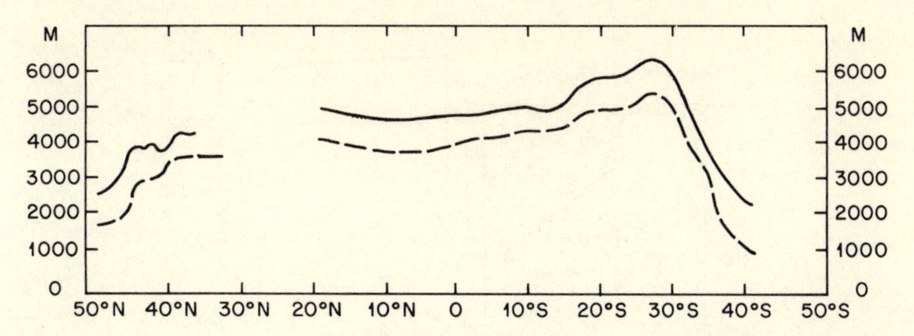

Figure 7.5 Modern (———) and "Pleistocene" (----) (or "last glaciation") snowlines along a transect from the northern Rocky Mountains through Central America to Chile (after Heuberger 1974 and Hastenrath 1967).

tions in the past would have been different from today, perhaps resulting in lower lapse rates in many areas. However, this entire approach is extremely simplistic and neglects other important factors. Snowline is not only a function of lapse rate but also depends on the variation of accumulation with elevation (accumulation gradient) and the variation of albedo with temperature (since temperature influences the frequency of snowfall versus rainfall events) (L. D. Williams 1975).

The importance of precipitation in controlling snowline elevation and spatial variation has been noted for the Cordilleran mountains of both North and South America. In the Cascade Range of Washington, for example, 86% of the variance in glaciation thresholds is explained by accumulation season precipitation (though mean annual temperature is highly correlated with precipitation) (Porter 1977). By assessing the amount of temperature change during the late Wisconsin (Fraser) glaciation from palynological evidence, Porter estimated that winter precipitation was 20–30% less than today, accompanied by ablation season temperatures 5.5 ± 1.5 °C lower than at present.

In the Andes, snowlines are highest (>6000 m) at the latitude of maximum aridity (Fig. 7.5). "Pleistocene" snowlines were lower by 650–1500 m, the depression being greatest in the hyperarid regions of southern Peru and northern Chile, and least in the equatorial zone. This suggests a general temperature decrease over the entire region, but substantially higher precipitation amounts in the desert zone (Hastenrath 1967). Further insight is provided by east – west transects across the mountain barrier which indicate a reversal of modern snowline gradients between 28 and 32°S (Fig. 7.6). This is related to the prevailing circulation around subtropical high-pressure cells centered at ~30°S; prevailing winds are easterly to the north and westerly (onshore) to the south. During glacial periods the lower snowline gradients generally parallel modern snowlines, but at 28°S this is not the case. "Pleistocene" snowlines show a reversal of gradient, suggesting an equatorward shift of ~5°

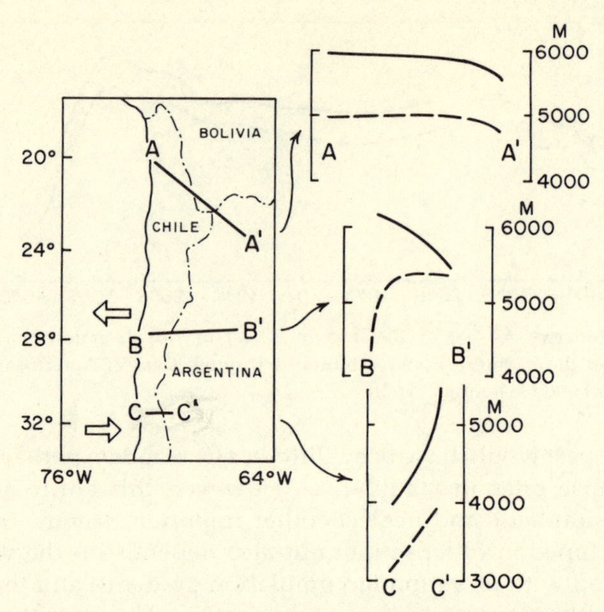

Figure 7.6 Modern (———) and Pleistocene (----) snowline elevations along west–east transects in the South American Andes. Principal zonal wind components today are shown by arrows at diagram on left. At the northern and southern locations Pleistocene snowlines were uniformly lower throughout the transect. In the central location (~28°S) a reversal of snowline gradient is apparent between Pleistocene and modern times. This resulted from a shift in the subtropical high-pressure cell and associated wind fields, changing from predominantly onshore flow in Pleistocene time to offshore today (after Hastenrath 1971).

in the boundary between temperate latitude westerlies and tropical easterlies in the lower troposphere. Where westerly wind regimes became established, a marked eastward rise of Pleistocene snowline resulted (Fig. 7.6). Similar reversals of firn line gradient, from modern times to glacial periods, have also been noted in parts of East Africa (Hamilton & Perrott 1979).

From this brief survey it is apparent that snowlines in different regions are controlled by different climatic parameters and that these must first be identified before paleosnowlines can be usefully used in paleoclimatic reconstructions. However, even if modern snowline controls are understood, there is no unique solution to the paleoclimatic interpretation of former snowlines since both precipitation amounts and temperatures are likely to have changed. Furthermore, such changes would have been accompanied by variations in cloudiness (and hence direct and diffuse radiation amounts), relative humidity, wind speed, etc. Only if independent estimates of such parameters are available can precise interpretations be made of former snowlines. However, some assessment of the *relative* importance of different climatic variables to snowline lowering can be made by the use of an energy balance model which takes into

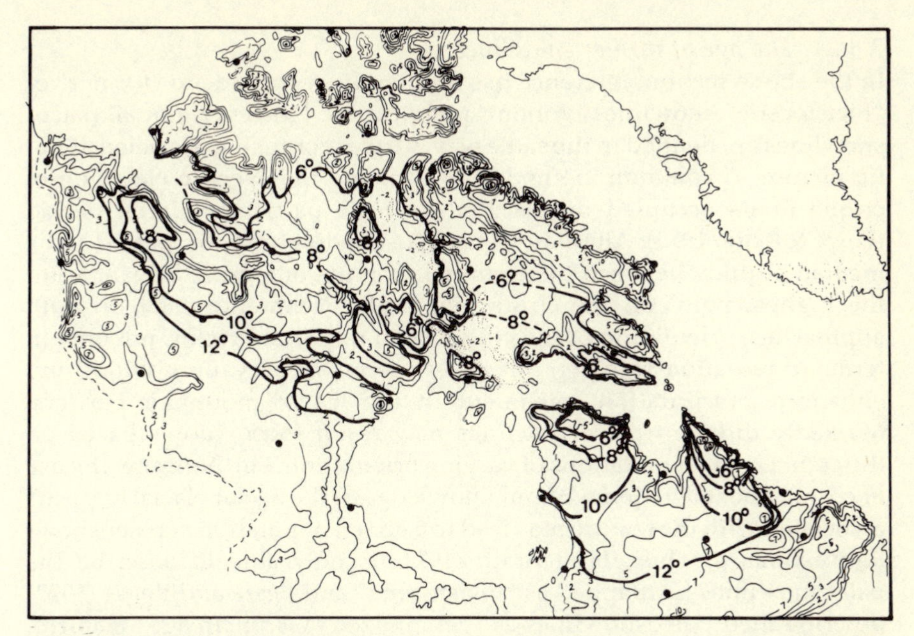

Figure 7.7 Boundaries of perennial snow cover predicted by an energy balance model for various amounts of uniform spring and summer temperature decrease, assuming "normal" (1931–70) snow accumulation and Earth-orbital parameters of 116 000 years BP for 31 March. For example, with a spring–summer lowering of 6–8 °C the perennial snow cover would have been confined mainly to the Arctic Islands and the northern tip of Labrador–Ungava; with a 10–12°C temperature lowering extensive areas of Keewatin and Labrador would have been glacierized (from L. D. Williams 1979).

account many of the relevant variables and the interactions between them. For example, such a model has been applied to the question of what climatic conditions were necessary to bring about extensive glacierization of northern Canada (L. D. Williams 1979). By calculating the regional snowline for varying climatic conditions it was possible to determine where perennial snow cover is most likely to have developed in the past (i.e. which areas are most susceptible to glacierization) and the extent of glacierization brought about by different changes in climatic conditions. Interestingly, Williams' model indicates that a substantial fall in summer temperatures (10–12 °C) is necessary to extensively glacierize Keewatin and Labrador, though smaller changes in temperature are sufficient to glacierize Baffin Island, to the north (Fig. 7.7). This confirms the view that Baffin Island is particularly sensitive to climatic fluctuations, and is likely to have been a primary site for ice-sheet initiation in the past (and, probably, in the future also) (Tarr 1897, Bradley & Miller 1972, Andrews *et al.* 1972). It is also of interest that the model results indicate little impact on the areal extent of glacierization with increased snowfall; temperature changes seem to be of most significance for regional snowline lowering and glacierization (Williams, 1979).

7.3.2 The age of former snowlines

In the above section, reference has frequently been made to "former" or "Pleistocene" snowlines without qualification. However, not all paleo-snowlines are defined in the same way, further confusing the paleoclimatic picture. A common method is to estimate the average elevation of cirque floors occupied by glaciers during a particular glacial period (Péwé & Reger 1972). Alternatively, paleosnowlines may be located at the median altitude between the terminal moraine of a given advance and the highest point on the cirque headwall (Richmond 1965). In both approaches, orientation of the cirque is an important factor, not only in terms of radiation receipts, but also in terms of prevailing winds and enhanced precipitation catchment in the lee of mountain barriers. Markedly different paleosnowlines may result from studies based on different cirque populations of varying orientation. Furthermore, the use of cirque floor elevation without knowledge of the age of glacial deposits associated with the feature may lead to a mixed population of paleosnowline estimates. Thus, Hastenrath (1971) is only able to describe the paleosnowlines he mapped as "Pleistocene" and Péwé and Reger (1972) describe their paleosnowlines as "generalized Wisconsin age" features. Such factors may explain the large variations in snowline depression commonly reported; in New Mexico, for example, estimates of regional snowline lowering vary from 1000 to 1500 m (Brakenridge 1978) and even larger variations are noted in other studies (e.g. Reeves 1965). This imprecision in dating, and uncertainty over climatic controls on snowline elevation, severely limits the value of most snowline studies for paleoclimatic reconstruction. However, careful study of modern conditions and of well dated glacial deposits can provide important insights into paleoclimatic conditions of mountain regions, as shown by Porter (1977). Indeed, there would appear to be considerable potential in a re-evaluation of this subject for many mountainous parts of the world.

7.4 Mountain glacier fluctuations

Glacier fluctuations result from changes in the mass balance of glaciers; increases in net accumulation lead to glacier thickening, mass transfer and advance of the glacier snout; increases in net ablation lead to glacier thinning and recession at the glacier front. A glacial advance therefore corresponds to a positive mass balance brought about by a climatic fluctuation which favors accumulation over ablation (e.g. Fig. 7.8). However, there are many combinations of climatic conditions which might correspond to such a *net* change in mass balance, so that evidence of a formerly more extensive ice position does not provide an unequivocal picture of climate at the time. However, if a check on one important

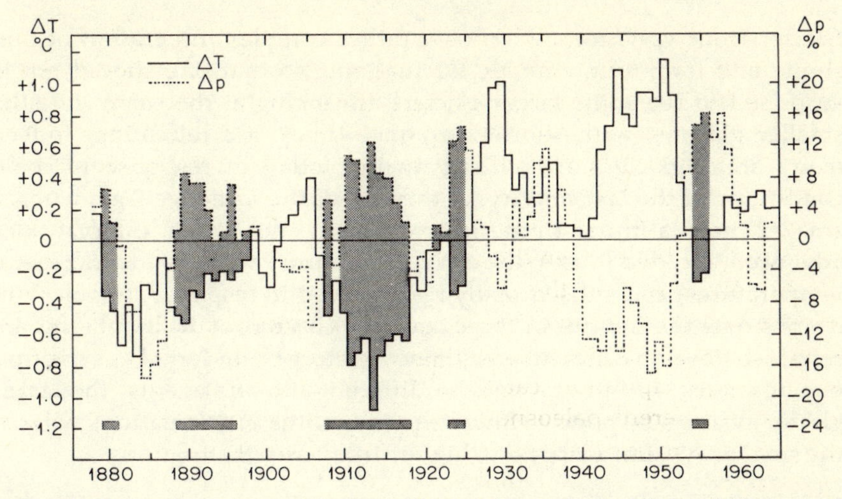

Figure 7.8 Mean departures of summer (June, July, and August) precipitation and temperature (from 1851–1950 averages) at seven stations above 2000 m in the Swiss Alps. Precipitation data shown as a percentage departure from the long-term mean; temperatures in degrees Celsius. Shaded areas indicate times of both below average temperatures and above average precipitation which correspond to times of principal glacier advances in the area (bars at bottom) (after Hoinkes 1968).

variable (such as temperature) is available from an independent source of paleoclimatic data, it may be possible to calculate, or model, the overall change in climate which resulted in the advance or retreat of the glacier terminus (e.g. Allison & Kruss 1977).

Changes in mass balance are not transformed immediately into changes in glacier front positions. There may be a period of down-wasting, during which the glacier loses mass but does not recede. For example, the lower Khumbu Glacier, in the Mount Everest massif, thinned by ~70 m between 1930 and 1956, but the glacier snout remained stationary (Müller 1958). Even when down-wasting is not a significant factor, glacier front positions will lag behind climatic fluctuations. Different glaciers have different response times to mass balance variations. An increase in net mass results in a kinematic wave moving down the glacier at a rate several times faster than the normal rate of ice flow. The time it takes for this wave to reach the glacier snout is the response time of the glacier and depends on a number of factors including the glacier length, basal slope, ice thickness and temperature, and overall geometry of the glacier itself (Nye 1965). The South Cascade Glacier (Washington), for example, has a response time of only about 25–30 years, whereas large ice sheets have response times on the order of millennia.

There is some evidence that an excess of ablation over accumulation may cause a more rapid response, with ice recession lagging very little behind the climatic fluctuation which caused the mass loss (Karlén 1980).

Glacier front variations are thus a rather complex integration of both short- and long-term climatic fluctuations, so that one should not be surprised to see some larger glaciers advancing at the same time that smaller glaciers, with shorter response times, are retreating. Indeed, many arctic glaciers are still advancing today in response to cooler conditions of the last century, at the end of the Little Ice Age, whereas smaller mid-latitude alpine glaciers over the same interval have advanced, receded (due to the early 20th century world-wide increase in temperatures) and subsequently re-advanced in response to cooler conditions over the last two or three decades. Glaciers of different sizes and response times in different areas may therefore be undergoing synchronous advances, but in response to different climatic events; the largest glacier systems respond to low-frequency climatic fluctuations whereas the smaller systems respond to higher-frequency fluctuations.

7.4.1 *Evidence of glacier fluctuations*

The complexity of climatic conditions and differing glacier response times makes glacier fluctuations, as viewed through the misty window of the paleoclimatic records, a rather complicated source of paleoclimatic data. This is compounded by the difficulties inherent in dating former glacier front positions in environments which generally have very little organic material for dating, and where weathering rates are extremely slow.

A record of changes in glacier front positions is generally derived from moraines produced during glacial advances; periods of glacier recession, and the magnitude of the recession, are much harder to identify in the field, though the record of down-valley lake sediments may assist in the interpretation of former glacier front positions (Karlén 1976). The problem with the record of moraines is that they are commonly incomplete, with recent advances (which were often the most extensive) obliterating evidence of earlier, less extensive advances. However, recent work by Röthlisberger (1976) and Schneebeli (1976) has demonstrated that detailed stratigraphic studies of glacial deposits may reveal buried soils and weathered profiles indicative of former subaerial surfaces subsequently buried by more recent morainic debris.

By far the greatest difficulty in the use of glacier front positions as a paleoclimatic index is the problem of dating the glacial deposits. Radiocarbon dates on organic material in soils which have developed on moraines may be obtained, but they only provide a minimum age on the glacial advance. There may be a delay of many hundreds of years between the time a moraine becomes relatively stable, and the time it takes for a mature soil profile to develop. If a soil has been buried by debris from a subsequent glacial advance, a date on the soil would provide only a maximum age on the later glacial episode because the organic material in the soil may be hundreds of years (at least) older than the subsequent ice advance (Griffey & Matthews 1978, Matthews 1980). For older glaciations in volcanic areas, tephrochronology (Sec. 4.4) or the dating of interstratified lava flows may assist in the interpretation of glacial events (e.g. Löffler 1976, Porter 1979).

Lichenometry is commonly used to date moraines (see Sec. 4.6) but the technique is uncertain and probably no more reliable than ±20% before 1000 years BP; in most cases, lack of a well dated calibration curve, or the derivation of a curve based on observations in an environment totally unlike that of the glacial valley or cirque in question, makes the uncertainties even larger. Consequently, the chronology of glacier fluctuations is quite uncertain in most parts of the world.

7.4.2 *The record of glacier front positions*

In spite of the difficulties of dating glacial deposits, studies of glacier fluctuations have been conducted in virtually all mountainous parts of the world (e.g. Field 1975). Because of the difficulties of dating, most work has focused on postglacial (Holocene) glacier fluctuations. Early work in the Rocky Mountains led Matthes (1940, 1942) to suggest that many alpine glaciers disappeared during a mid-Holocene warm and dry period (the Altithermal; Antevs 1948) only to be regenerated during subsequent cooler and/or wetter periods ("neoglaciations;" Porter & Denton 1967). In other areas, there is evidence for glacial advances at intervals throughout the mid to late Holocene. For example, in a review of the numerous studies on Scandinavian glacier fluctuations, Karlén (1982) was able to identify periods of glacier expansion at 7500, 6300, 5600, 5100, 4800, 4500, 3000, 2200, 1900, 1400, 1050, 600, and 430 years BP. Over the same period, numerous glacial advances also occurred in the European Alps, though not always at the same time as those in Scandinavia (e.g. Patzelt 1974, Röthlisberger *et al.* 1980). Some smaller European alpine glaciers may also have wasted away during relatively brief (century-long?) warm intervals in the mid-Holocene, but such events are difficult to resolve within the coarse framework of alpine glacier advances that have been dated.

In many areas of the world, the most recent Neoglacial episode (which occurred between the 15th and early 19th centuries) resulted in the most extensive Holocene glacial advances. This period is commonly referred to as the "Little Ice Age" and is particularly well documented in western Europe (Lamb 1963, Le Roy Ladurie 1971). In the European Alps, glacier advances can be traced through historical records, paintings, and sketches, and this has greatly facilitated the interpretation of glacial deposits in the field. Detailed reconstructions of glacier front positions over the past 300–400 years have been constructed for the Grindelwald Glacier, Switzerland, using a variety of such sources, and periods of both

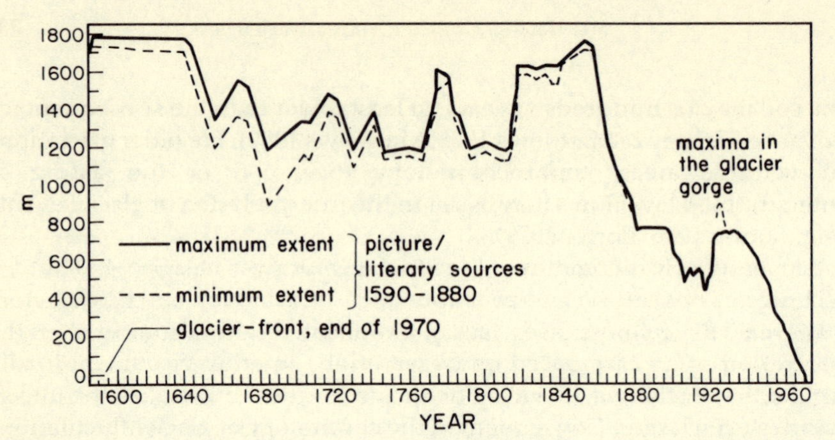

Figure 7.9 Fluctuations of the lower Grindelwald Glacier front between 1590 and 1970 relative to the 1970 terminal position. The major recession since ~1860 (with minor interruptions around 1880 and 1920) is clearly seen. Moraines are indicated by bold curves (after Messerli *et al.* 1978).

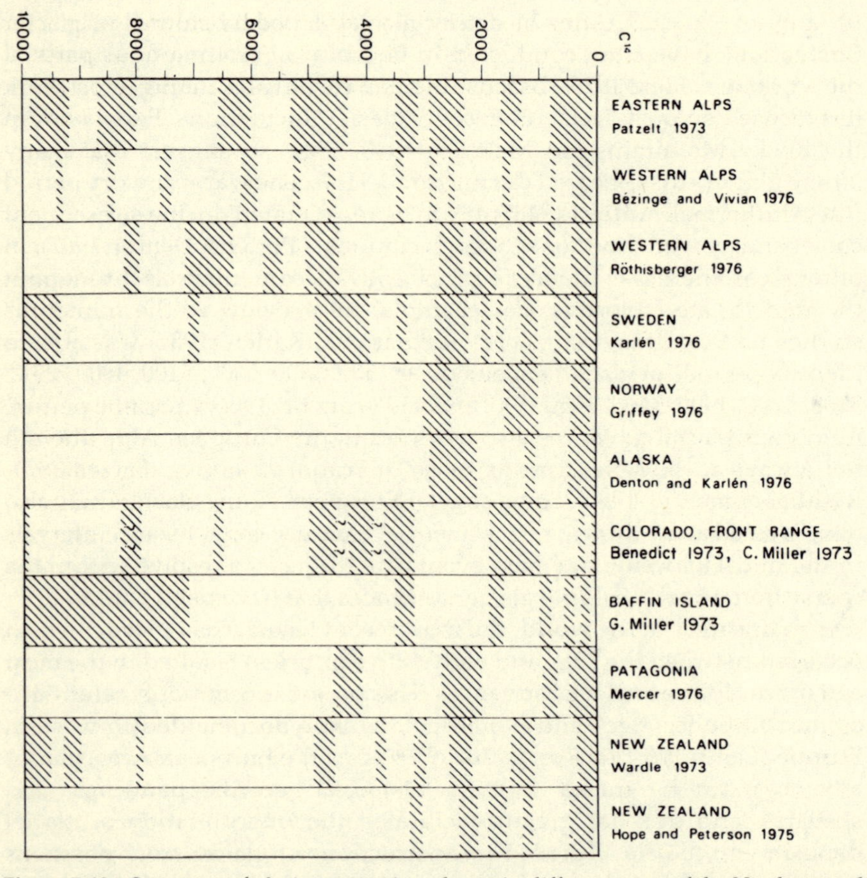

Figure 7.10 Summary of glacier expansion phases in different areas of the Northern and Southern Hemispheres. The data suggest a general lack of synchronism except possibly within the last 500 years and around 2500 years BP. This may be due to climatic fluctuations which are regional, not hemispheric or global in extent, or to poor dating or an incomplete data set (after Grove 1979).

glacier advance and recession have been determined (Fig. 7.9; Messerli *et al.* 1978). Such a detailed record is rare and would have been impossible with only geomorphological evidence to go on.

Because the Little Ice Age advances were most extensive in many areas, the record of earlier glacial events was often destroyed, or buried. This has made world-wide correlations very difficult, and the present evidence of globally synchronous glacial episodes is somewhat equivocal. Certainly there were many periods of alpine glacier expansion throughout the Holocene (Fig. 7.10) and these varied in magnitude from one area to another. However, on the basis of current evidence, there is little support for claims of world-wide synchrony in glacier fluctuations (Grove 1979, Williams & Wigley 1983) and even less support for the notion of a 2500 year periodicity in mountain glacier fluctuations (as proposed by Denton & Karlén 1973a). Until more detailed studies are carried out (along the lines of those by Patzelt 1974 and Röthlisberger 1976 in the European Alps), perhaps supplemented by palynological and lake sediment studies (e.g. Andrews *et al.* 1975, Karlén 1976), it seems probable that the record of glacier fluctuations in many areas will remain incomplete.

7.5 Lake-level fluctuation

Throughout the arid and semi-arid part of the world, there is commonly no runoff (surface water discharge) to the oceans. Instead, surface drainage may be essentially non-existent (in areic regions) or it may terminate in interior land-locked basins where water loss is almost entirely due to evaporation (endoreic regions). In these basins of inland drainage (Fig. 7.11) changes in the hydrological balance, as a result of climatic fluctuations, may have dramatic effects on water storage. During times of positive water budgets, lakes may develop and expand over large areas, only to recede and dry up during times of negative water balance. Studies of lake-level variations can thus provide important insights into paleoclimatic conditions, particularly in arid and semi-arid areas. In modern lake basins, periods of positive water balance are generally identified by abandoned wave-cut shorelines and beach deposits (Fig. 7.12) or perched deltas from tributary rivers and streams, and exposed lacustrine sediments at elevations above the present lake shoreline (e.g. Morrison 1965, Butzer *et al.* 1972, Bowler 1976). Periods of negative water balance (relative to today) are identifiable in lake sediment cores or by paleosols developed on exposed lake sediments. A study of the stratigraphy, geochemistry, and microfossil content of lake sediments may be particularly valuable in deciphering lake history (e.g. Bradbury *et al.* 1981).

Using such lines of evidence, lake-level fluctuations have been studied

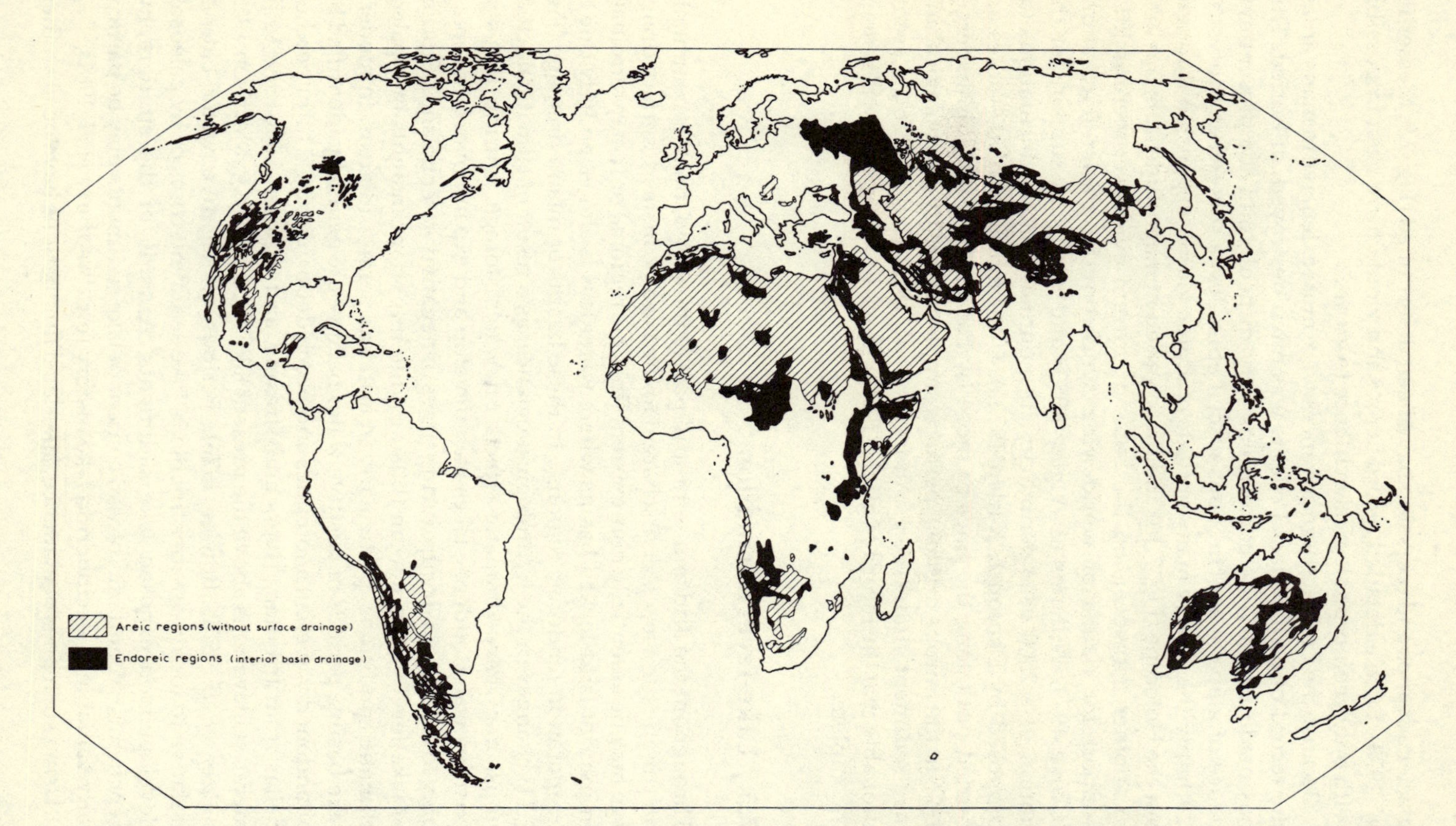

Figure 7.11 Areas of endoreic and areic drainage. Areic regions have no permanent surface drainage; endoreic areas are basins of inland drainage (from Cook & Warren 1977, after de Martonne & Aufrère 1928).

Figure 7.12 Wave-formed lake–shore terraces, Lake Chewaucan, Oregon. The highest shorelines (~80 m above present day Lake Albert) were formed during the last lake cycle (~15 000 years BP) (photograph kindly provided by W. McCoy).

in dozens of closed basins throughout the world (for a list of principal works, see Street & Grove 1979 Appendices 1 & 2). The majority of these studies are stratigraphic and provide little more than a qualitative estimate of climatic conditions. Periods of higher lake levels are commonly described as pluvials, but the question of whether such conditions result from increased precipitation, or lower temperatures and more *effective* precipitation (via reduced evapotranspiration) is controversial (cf. Brakenridge 1978 and Wells 1979). In an attempt to resolve the controversy, a number of studies have attempted to use the geomorphological evidence, together with empirically derived equations relating climatic parameters today, to make quantitative estimates of paleoclimatic conditions associated with particular lake stages in the past. These can be considered in two general categories: hydrological balance models and hydrological–energy balance models.

7.5.1 *Hydrological balance models*

In a closed basin, variations in lake level are a function of water volume which is, in turn, a reflection of the balance of water supply and water loss:

$$\frac{dV}{dt} = \frac{d(P + R + U)}{dt} - \frac{d(E + O)}{dt},$$

where V is water volume in the lake; P is precipitation over the lake; R is runoff from the tributary basin, into the lake; U is underground (subsurface) inflow to the lake; E is evaporation from the lake; and O is subsurface outflow from the lake. For any particular lake stage, if the hydrological balance is considered to be at equilibrium, such that

$$\frac{dV}{dt} = 0,$$

then, $P + R + U = E + O$. Generally, subsurface inflow and outflow are considered to be negligible and are omitted from the equation, even though in some cases they may be substantial; in the case of the Great Salt Lake, Utah, for example, subsurface inflow has been variously estimated at 3–15% of total lake input (Arnow 1980). In most cases, however, the subsurface components are unknown even for modern lake levels and trying to estimate them for paleolakes would be extremely speculative. Assigning values of zero to these components, the hydrological balance equation for a particular basin is thus

$$A_L P_L + A_T(P_T k) = A_L E_L,$$

where A_L is the lake area; A_T is the area of the tributary basin from which water drains to the lake; P_T is mean precipitation per unit area over the tributary basin; k is a coefficient of runoff (hence $P_T k$ equals the runoff per unit area, R_T, from the tributary basin); P_L is mean precipitation per unit area, over the lake; and E_L is mean evaporation per unit area, from the lake. Since only A_L and A_T are known for any given lake stage, the equation may be re-arranged thus:

$$\frac{A_L}{A_T} = \frac{P_T k}{E_L - P_L}.$$

A solution of the equation therefore requires a knowledge of precipitation over the lake and adjacent catchment basin, runoff from the surrounding basin, and the amount of water which evaporates from the lake.

Table 7.2 Factors affecting rates of evaporation and runoff.

Evaporation	Runoff
Temperature (daily means and seasonal range)	Ground temperature
Cloudiness and solar radiation receipts	Vegetation cover and type
Wind speed	Soil type (infiltration capacity)
Humidity (vapor pressure gradient)	Precipitation frequency and seasonal distribution
Depth of water in lake and basin morphology (water volume)	Precipitation intensity (event magnitude and duration)
Duration of ice cover	Precipitation type (rain, snow, etc.)
Salinity of lake water	Slope gradients; stream size and number

All of these parameters are a function of many other variables which are also unknown, making a unique solution to the equation extremely difficult to say the least. To illustrate the uncertainties involved, Table 7.2 lists some of the principal factors affecting evaporation and runoff. Of major importance to *both* parameters is temperature, and this has enabled some limits to be placed on estimates of former runoff and evaporation values when reasonably good paleotemperature estimates are available. If paleotemperatures are known, empirically derived equations relating runoff, evaporation, and temperature (e.g. Figs 7.13 & 7.14) can be used to solve the hydrological balance equation. However, these empirical relationships are often limited in the range of values considered, and constrained by inadequate or even non-existent data. Consider, for example, the relationship between lake evaporation and temperature. Most empirical relationships are based on standard measurements of evaporation from metal pans 1.2 m in diameter, at different temperatures: lake evaporation is assumed to be less than pan evaporation by a factor of 0.7, based on empirical studies by Kohler *et al.* (1966). However, evaporation rates depend on a number of factors which have not been constant through time (Table 7.2), e.g. lake volume and salinity variations. In large lakes, such as Lakes Superior and Ontario, there is a very poor correlation

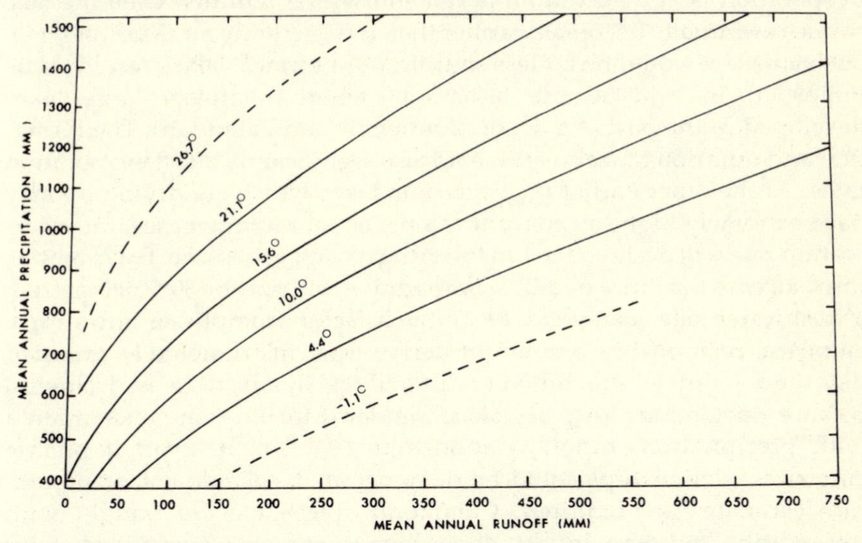

Figure 7.13 Relationship between mean annual runoff and mean annual precipitation for areas with different mean annual temperatures (in degrees Celsius). Temperatures are weighted by dividing the sum of the products of monthly precipitation and temperature by the annual precipitation. The quotient gives a mean annual temperature in which the temperature of each month is weighted in accordance with the precipitation during that month. A weighted mean *annual* temperature greater than the mean which would normally be computed indicates that precipitation is concentrated in warm months (and vice versa) (after Langbein *et al.* 1949).

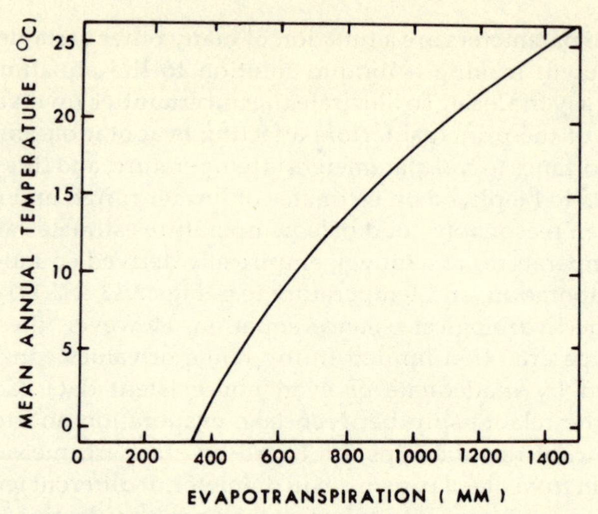

Figure 7.14 Relationship between mean annual temperature and evapotranspiration loss in humid areas, based on data from the eastern United States (after Langbein *et al.* 1949).

between monthly temperature and evaporation because much energy is used in raising the water temperature at depth (i.e. in heat storage). Evaporation is at a maximum in fall and winter months when the lake surface eventually becomes warmer than the overlying air (Morton 1967). Such an effect would be far less significant in tropical lakes, but could be important in mid-latitude situations where relatively large lakes developed in the past (e.g. Lakes Bonneville and Lahontan). The formation and duration of ice cover would also significantly affect evaporation rates. At the other end of the spectrum, lakes which are drying up may have extremely high salt concentrations; as salinity increases, the evaporation rate will decline due to a lowering of vapor pressure. For example, in a lake with a salinity of 200‰, evaporation will only be 80% of that from a freshwater lake (Langbein 1961). Such factors complicate any simple empirical relationship one might derive from instrumentally recorded data and point to the inherent difficulties involved in hydrological balance calculations for paleolakes. Similar difficulties are encountered with precipitation – runoff relationships (Table 7.2). Even if precise empirical relationships could be demonstrated, reliable paleotemperature estimates are required. Commonly, these too are fraught with uncertainty, and may, in fact, depend implicitly on assumptions about paleoprecipitation amounts. For example, paleotemperatures derived from studies of snowline depression depend on the assumption that precipitation amounts are similar to those of today. Any increase in precipitation would require a smaller fall in temperature to produce the same amount of snowline depression. Hence, the use of paleotemperature estimates based on snowline depression (e.g. Leopold 1951, Braken-

Table 7.3 Paleoprecipitation estimates from selected western US hydrological balance studies.

Study area	Author(s)	Paleotemperature change assumed (°C)		Paleoprecipitation/ modern precipitation
		July	Annual	
Lake Estancia, New Mexico	Leopold (1951)	−9	−4.5	1.5
Lake Estancia, New Mexico	Brakenridge (1978)	−8	>−7.5	1.0
Lake Estancia, New Mexico	Galloway (1970)	−10	−10.5	0.86
Spring Valley, Nevada	Snyder & Langbein (1962)	−7	−3.5	1.67
Various, in Nevada	Mifflin & Wheat (1979)		−2.8	1.68

ridge 1978) leads to suspiciously circular reasoning. Without accurate paleotemperatures, quite divergent conclusions may ensue. In Table 7.3, for example, three different studies of Paleolake Estancia, New Mexico, are summarized. Although each used slightly different approaches and empirical relationships, the fundamental difference in their final paleoprecipitation estimates (ranging from 80 to 150% of today's values) lies in the different paleotemperatures assumed. The larger the change in temperature assumed, the smaller is the required increase in precipitation (cf. Benson 1981). Given a sufficiently large decrease in temperature, values of precipitation even smaller than today can be shown to balance the hydrological budget at times of relatively high lake levels (e.g. Galloway 1970).

It is unlikely that controversies over paleoprecipitation estimates will be resolved until (a) more detailed studies of modern relationships between evaporation and temperature, precipitation, and runoff are undertaken, to provide more reliable empirical equations, and (b) better (independent) paleotemperature estimates are available. Paleotemperatures calculated from the extent of amino-acid epimerization in the shells of freshwater gastropods of known age may help to resolve this issue (McCoy 1981, 1982).

7.5.2 *Hydrological–energy balance models*

An alternative to the conventional hydrological balance models described above has been proposed by Kutzbach (1980) who applied the method to Paleolake Chad in North Africa. Kutzbach utilizes the climatonomic approach of Lettau (1969) by considering the hydrological balance of a lake basin in terms of energy fluxes at the surface. In simple terms, a positive hydrological balance results when there is insufficient energy available to evaporate precipitation falling on the basin. Instead of calculating paleoprecipitation amounts from estimates of runoff and evaporation (via paleotemperature estimates) a hydrological–energy balance model utilizes estimates of net radiation and sensible and latent heat fluxes over the lake and tributary basin. Modern values of these components are used, based on measurements in locations which are thought to characterize the paleo-environments of the basin being studied. Paleotemperature estimates are thus implicit in this "analog" approach; in the Paleolake Chad study, for example, the changes in vegetation which were assumed for 5000–10 000 years BP correspond to an area-weighted fall in mean annual temperature of 1.5 °C, by analogy with areas of similar vegetation today. Precipitation was estimated to have been almost double modern values (~650 mm versus 350 mm today), a result which is similar to previous estimates of precipitation for the area at that time, based on a variety of paleo-environmental data.

Kutzbach's study exemplifies a new avenue of research in paleolake studies and, no doubt, the approach will be applied to many other lake systems in due course. Whether it represents a major improvement over conventional hydrological studies is debatable, since it involves at least as many assumptions, and may involve a good deal more (e.g. Benson 1981). Nevertheless, as an "independent" approach, it provides a valuable alternative model.

7.5.3 Regional patterns of lake-level fluctuations

The large number of relatively well dated stratigraphic studies of lake-level fluctuations throughout the world has enabled maps of relative lake levels to be prepared for selected time intervals (Street & Grove 1976, 1979). Although individual errors in dating lake levels no doubt occur, mapping relative lake levels for discrete time periods has the advantage that regionally coherent patterns may be discerned, even if isolated "anomalies" occur. Thus, Street and Grove (1979) were able to map relative lake levels in much of the arid and semi-arid world for eight time intervals (24 000–23 000, 21 000–20 000, 18 000–17 000, 15 000–14 000, 12 000–11 000, 9000–8000, 6000–5000, and 1000–0 years BP). Taking the total range of lake-level fluctuations as 100%, from the level of complete desiccation to overflow (or known maximum level), three categories were defined. *Low levels* were when lakes were at no more than 15% of their maximum range; *intermediate levels* were when lake levels fluctuated between 15 and 70% of their range; *high levels* were when lake levels exceeded 70% of the total altitudinal range. Maps were then prepared to show the spatial distribution of low, intermediate, or high lake levels, so defined, at each time interval (e.g. Figs 7.15 & 7.16). These maps demonstrate remarkable coherence in the spatial and temporal patterns of lake-level fluctuations. During the glacial maximum (18 000–17 000 years BP) most of the evidence from intertropical Africa indicates that the area was relatively dry (Fig. 7.15); only in extratropical regions (the North

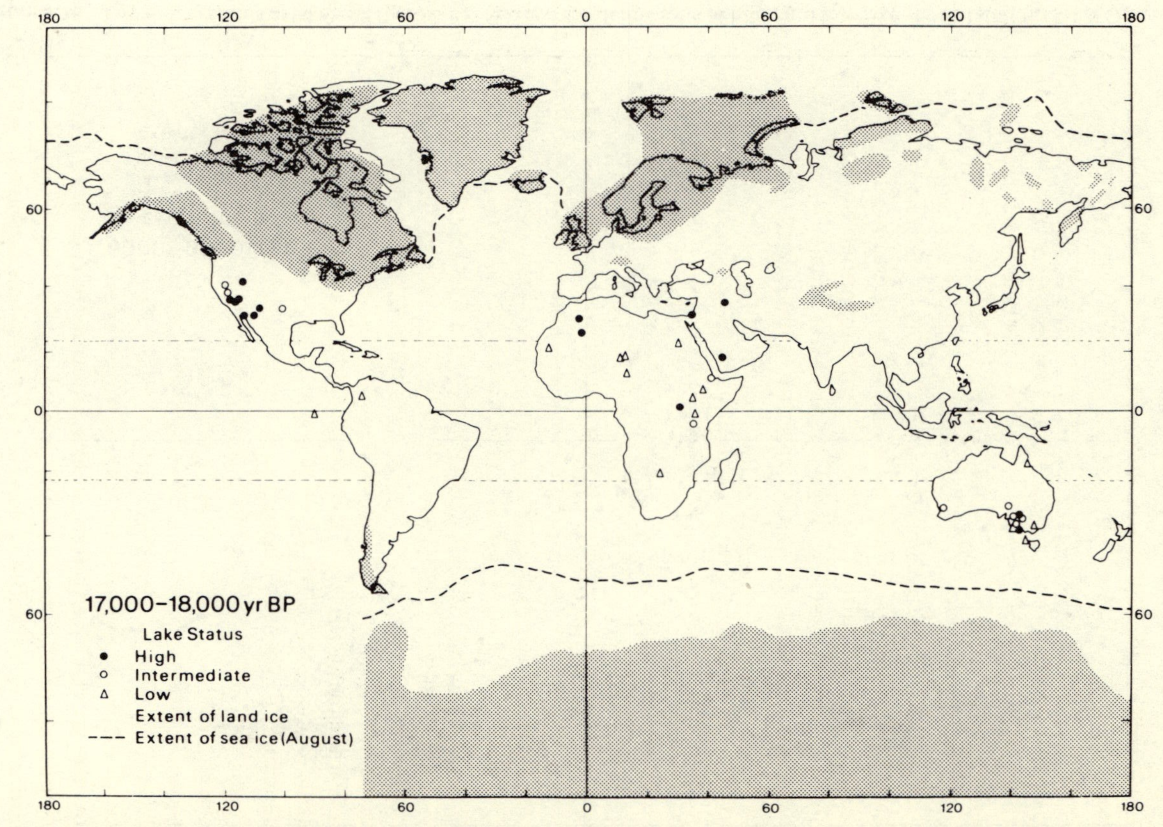

Figure 7.15 Lake-level status at 18 000–17 000 years BP. See text for definitions of high and low lake status (after Street & Grove 1979).

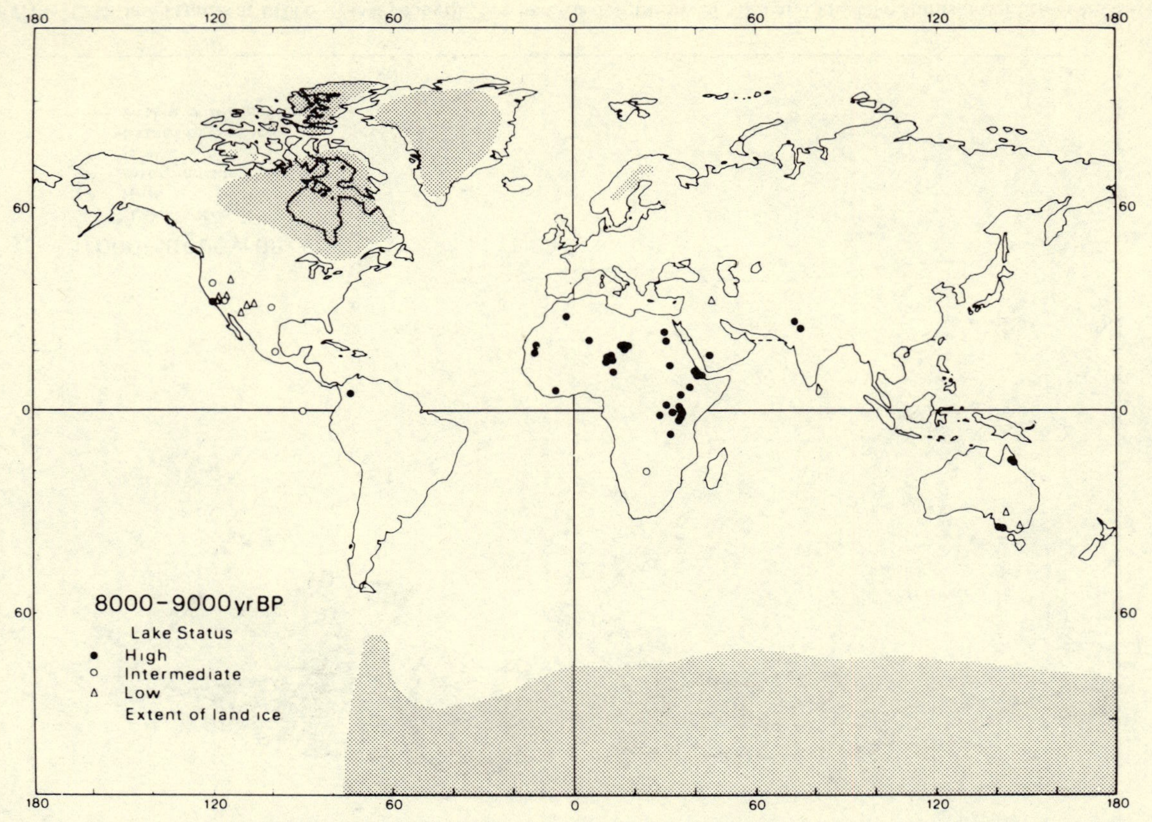

Figure 7.16 Lake-level status at 9000–8000 years BP. See text for definitions of high and low lake status (after Street & Grove 1979).

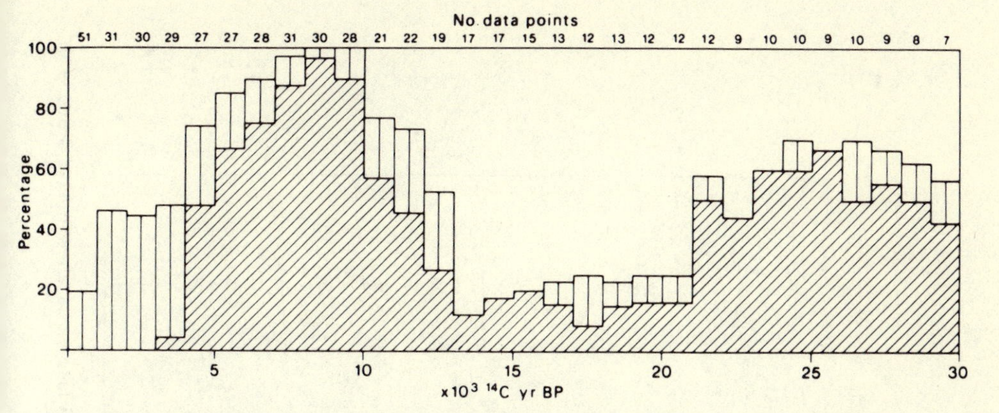

Figure 7.17 Histogram of intertropical African lake-level status for 1000 year time periods from 30 000 years BP to the present. Based on percentage of radiocarbon dates relating to high (shaded), low (blank), or intermediate (vertical ruling) lake levels (after Street & Grove 1979).

American Great Basin, in particular) is there an abundance of evidence for extensive lake stages at that time. Thus, the so-called "pluvial" climate of the western USA at the glacial maximum is not a viable model for tropical and equatorial regions (Butzer *et al.* 1972, Nicholson & Flohn 1980; cf. Sec. 8.3). During this arid phase, dune systems on the equatorward margins of the Sahara and Kalahari deserts greatly expanded (Sarntheim 1978); in the south-western Sahara, for example, dunes even blocked the much-reduced flow of the Senegal River in Mali (Michel 1973). Low lake levels persisted in these regions until ~12 000 years BP, when many lakes began to fill, commonly spilling over from their enclosed catchment basins and initiating new drainage systems. Maximum lake development throughout most of Africa and Australia peaked at 10 000–8000 years BP (Rognon & Williams 1977; Figs 7.16 & 7.17), though interestingly lake levels were generally low in mid-latitudes at this time. In subsaharan Africa, lake expansion was particularly spectacular, with Lake Chad expanding in area to a size comparable with the Caspian Sea today (Grove & Warren 1968, Rognon 1976, Street & Grove 1976). The high lake phase continued until 4500–4000 years BP, interrupted in most of Africa, at least, by an episode of increased aridity and somewhat lower lake levels (though still *relatively* high) around 7000 years BP (Nicholson & Flohn 1980). Profound ecological changes accompanied these periods of more effective precipitation, enabling human occupancy and cultural activities to take place in Saharan Africa at a scale which is almost inconceivable today (Sec. 8.3.2). Since 4500 years BP there has been a relatively steady decrease in lake size leading to a situation over the last 1000 years in which lakes over most of the Tropics and lower

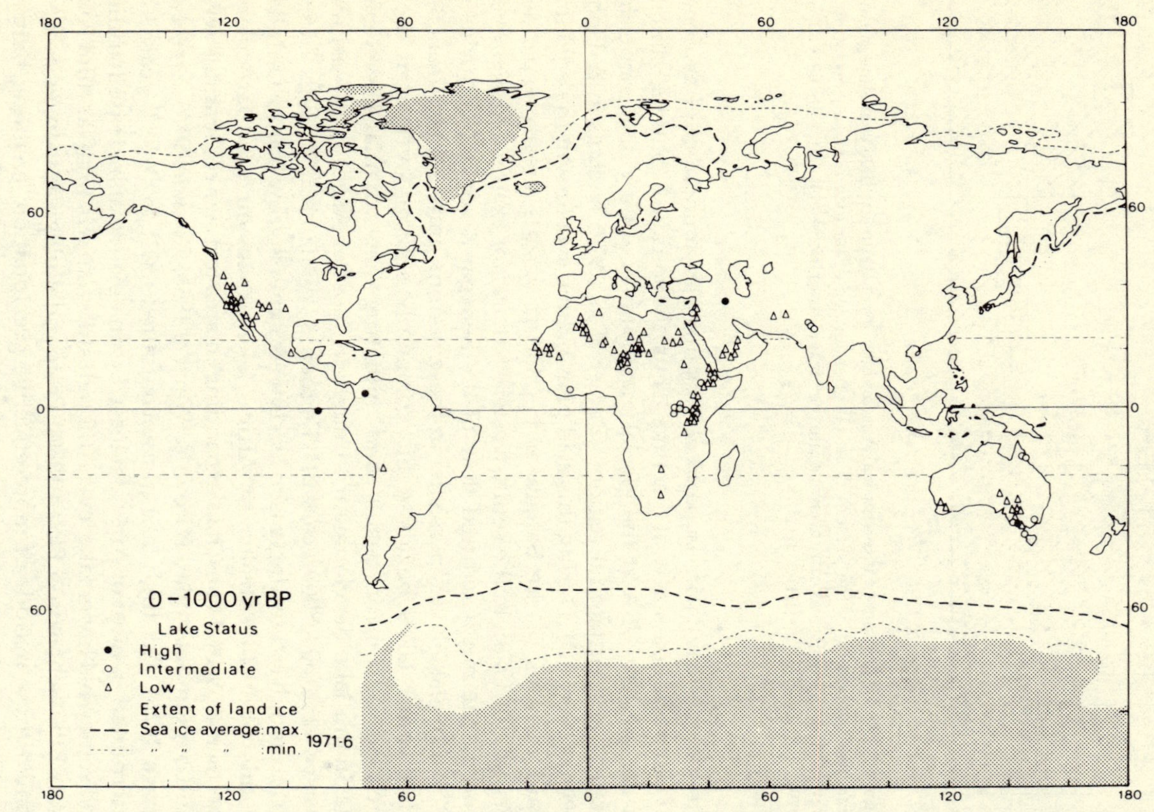

Figure 7.18 Lake-level status from 1000 years BP to the present. See text for definitions of high and low lake status (after Street & Grove 1979).

mid-latitudes are virtually all at low stages (Fig. 7.18). In many areas, it appears that lake levels have rarely been lower during the past 20 000 years.

In spite of the excellent spatial coherence displayed in the maps shown, caution must be used in interpreting lake-level data in this way. Once a lake has desiccated, there is no way of knowing just how much drier conditions may have been during the period of desiccation. Dry periods are thus likely to be underestimated. Comparison of lake basins of vastly different sizes can lead to erroneous conclusions. Small volume lakes respond much more rapidly to hydrological variations than large deep-water lakes and are likely to record higher frequency climatic variations than large volume lakes. A good analogy to this would be the problems encountered in trying to correlate fluctuations of a small alpine glacier with those of a large ice sheet; clearly the response times of the two systems are different by perhaps an order of magnitude. The problem is not quite as profound in lake systems, however, since mass turnover rates in lakes are rarely longer than a few decades even for very large lakes (Langbein 1961). Thus, providing a coarse enough time interval is used, and major low-frequency components of hydrological changes are being considered, it should be possible to make broad regional comparisons. When regional patterns show little spatial coherence for a particular interval, it may be that the climate was fluctuating fairly rapidly over a wide range so that no major low-frequency signal dominates the record. Finally, it should be noted that the 15 and 70% category boundaries (for low and high stages) used by Street and Grove (1976, 1979) may represent quite different surface area states, depending on the morphology of the basin. Consider, for example a lake which overflows from a deep narrow basin to a broad shallow plain at some level; the increase in lake depth represents a vastly greater change in surface area (and hence in evaporation from the surface) than a fall in lake level of comparable magnitude. When evaporation from the expanded lake area balances inflow, a new equilibrium is reached. Thus lake surface area is the critical variable controlling lake depth, and this in turn is a function of the basin morphology. Unfortunately, there are not enough reliable data on lake area changes to make a global-scale study feasible at present.

In spite of these caveats, lake level data from Africa, together with palynological, geomorphological, and archeological data, have enabled a fairly detailed picture of paleoclimatic fluctuations over the last 20 000 years to be obtained. This led Nicholson and Flohn (1980) to speculate on what the major circulation features over the continent were like at different periods in the past. Figures 7.19 and 7.20 show the principal differences in circulation during the main arid phase (20 000–12 000 years BP) and the subsequent period of high lake levels (10 000–8000 years BP)

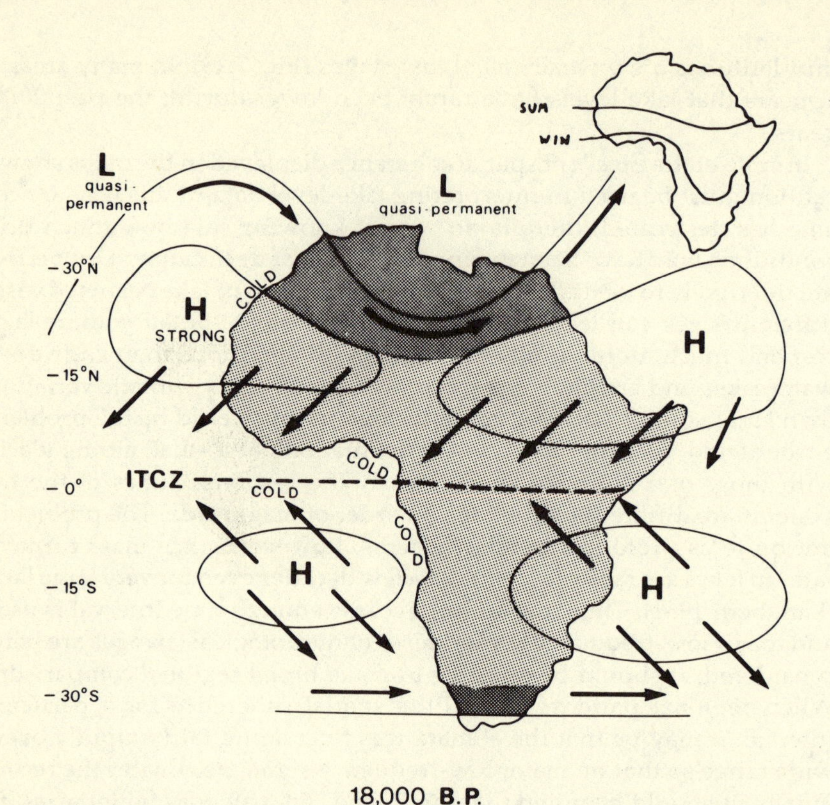

Figure 7.19 Proposed atmospheric circulation scheme ~18 000 years BP (and prevailing circulation pattern from 20 000 to 12 000 years BP) based on geological and palynological data. Dark shading = areas more humid than today; light shading = areas drier than today. Inset (top right) shows present position of intertropical convergence zone (ITCZ) in summer and winter months (after Nicholson & Flohn 1980).

as envisaged by Nicholson and Flohn (1980). Major changes in position of the subtropical high pressure centers are evident in their reconstructions; at 20 000–12 000 years BP a much stronger Hadley cell circulation would have resulted in increased subsidence in the subtropical high pressure cells and intensified upwelling of cooler equatorial water, thereby reducing oceanic evaporation rates in those regions. The seasonal migration of the intertropical convergence zone (ITCZ) would have been greatly reduced, preventing moisture-bearing winds from the Gulf of Guinea reaching southern Saharan regions. Displaced westerly flow (due to a strong baroclinic zone along the ice-sheet margin over northern Europe) would have brought relatively frequent depressions, and hence relatively moist conditions, to North Africa. By contrast, from 10 000 to 8000 years BP subtropical high pressure zones may have been displaced poleward as the ice sheet over Scandinavia diminished in size and

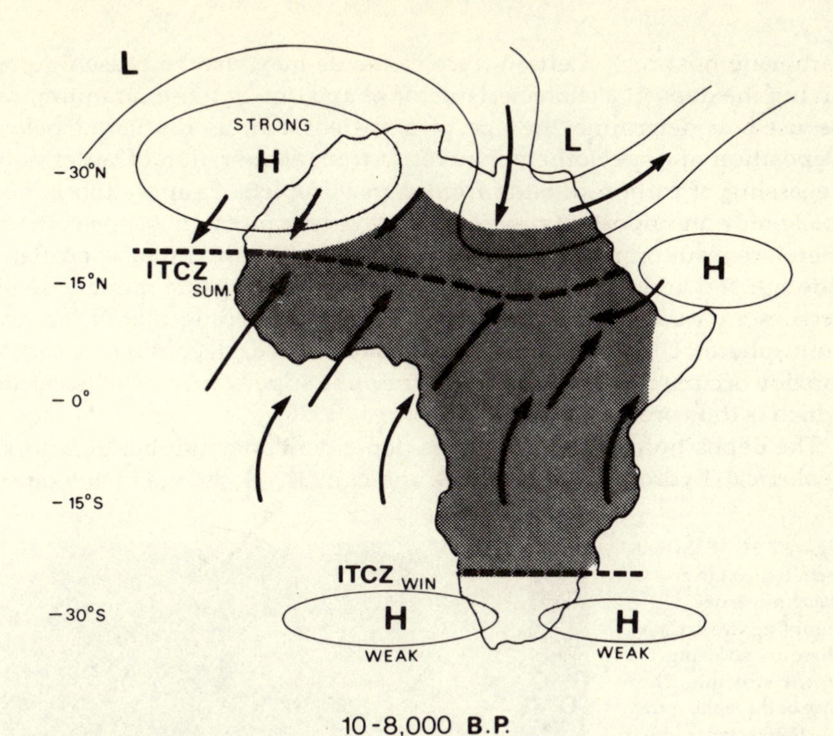

Figure 7.20 Proposed atmospheric circulation scheme ~10 000–8000 years BP. Dark shading = areas more humid than today; light shading = areas drier than today. $ITCZ_{winter}$ refers only to the position over southern Africa (after Nicholson & Flohn 1980.

Equator–Pole temperature gradients (in both hemispheres) were reduced. An increase in *interhemispheric* temperature differences could have resulted in a northward displacement of the ITCZ and increased moisture flux to the continent (facilitated by a warmer equatorial ocean). Evaporation rates from the ocean may have increased by as much as 50% in areas where upwelling of cool water was no longer occurring. Finally, the interaction of upper level troughs with low level tropical disturbances may have led to increased cyclogenesis and a significant contribution to Sahara rainfall totals from the resultant "Sudano-Saharan depressions" (Flohn 1975, Nicholson & Flohn 1980).

7.6 Speleothems

Speleothems are mineral formations occurring in limestone caves, most commonly as stalagmites and stalactites, or slab-like deposits known as flowstones. They are primarily composed of calcium carbonate, precipitated from ground water which has percolated through the adjacent carbonate host rock. Certain trace elements may also be present (often giving the deposit a characteristic color) and one of these, uranium, can be used to determine the age of a speleothem, as discussed below. Deposition of a speleothem may result from evaporation of water or by degassing of carbon dioxide from water droplets. Evaporation is normally only an important process near cave entrances; most speleothems therefore result from the degassing process. Water which has percolated through soil and been in contact with decaying organic matter usually accrues a partial pressure of carbon dioxide exceeding that of the cave atmosphere. Thus, when water enters the cave, degassing of carbon dioxide occurs, causing the water to become supersaturated with calcite, which is thus precipitated (Atkinson *et al.* 1978).

The deposition of speleothems is dependent on a number of factors – geological, hydrological, chemical, and climatic. A change in any one of

Figure 7.21 Polished section through a stalagmite from Tumbling Creek Cave, Missouri, showing growth structure. The base of the stalagmite is ~140 000 years old. It grew continuously until ~100 000 years ago when there was a hiatus in growth, marked by the distinct dark layer. Growth resumed ~64 000 years BP and continued until about 20 000 years BP. Well-defined growth layers are visible, particularly in the oldest section (photograph kindly provided by H. Schwarcz).

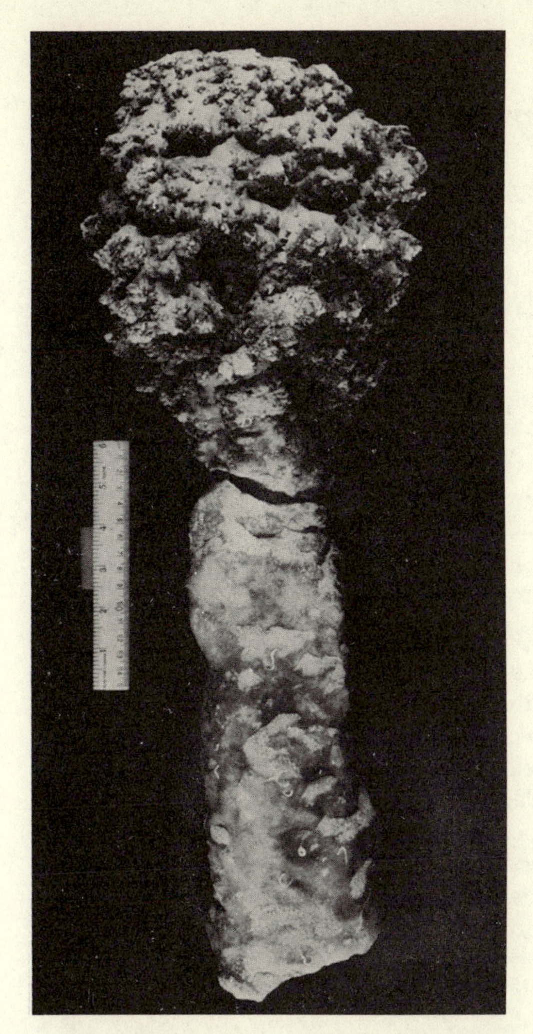

Figure 7.22 Coral overgrowth on a speleothem which formed below present sea level during glacial times when sea levels were lower. Subsequent eustatic sea-level rise resulted in the coral overgrowth (from Harmon *et al.* 1978b).

these factors could cause water percolation to cease, terminating speleothem growth at a particular drip site. However, cessation of speleothem growth over a large geographical area is more likely to be due to a climatic factor than anything else, so dating periods of speleothem growth can provide useful paleoclimatic information (Harmon *et al.* 1977). Uninterrupted speleothem growth is recognizable in a polished section (Fig. 7.21) as a series of very fine growth layers; major hiatuses in deposition are usually marked by erosional surfaces, desiccation and chalkification, dirt bands, and sometimes color changes. In speleothems deposited close to sea level, a rise in relative sea level may result in an overgrowth of marine aragonite on the deposit (Fig. 7.22).

7.6.1 Paleoclimatic information from periods of speleothem growth

Speleothems grow in sheltered environments which often have escaped the radical surface alterations resulting from glaciation. For example,

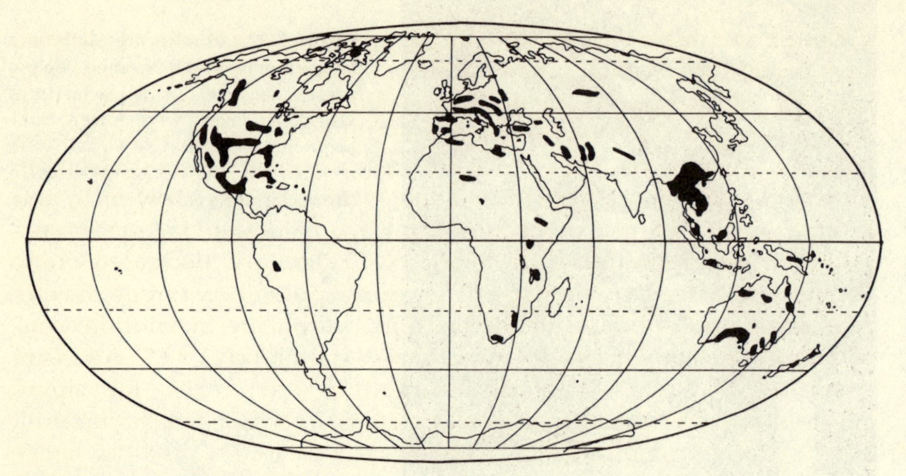

Figure 7.23 Distribution of karst in the world, showing the potential sources of paleoclimatic information from speleothems.

speleothems dated at up to 350 000 years BP are found beneath the present-day Columbia Icefield of Alberta, whereas the geomorphology of the surface has been repeatedly altered by glacial events. Speleothems may thus provide a long, often continuous record, of past environmental conditions. Furthermore, the extensive distribution on karst landscapes (Fig. 7.23) means that studies can be undertaken on a world-wide basis.

Paleoclimatic studies have focused on the timing of speleothem growth periods, their isotopic composition (of both the minerals and fluid inclusions), and their relationship to sea-level fluctuations. These are discussed separately in the following sections.

7.6.2 Dating of speleothems and the significance of depositional intervals

Speleothems are most commonly dated by uranium-series disequilibrium methods (generally ^{230}Th/^{234}U) described in Chapter 3. Isotopes of uranium leached from the carbonate bedrock are co-precipitated as uranyl carbonate with the calcite of the speleothems. Normally, the precipitating solution contains no ^{230}Th because thorium ions are either adsorbed onto clay minerals or remain in place as insoluble hydrolysates (Harmon *et al.* 1975). Thus, providing the speleothem contains no clay or other insoluble detritus which are carriers of detrital thorium, the activity ratio of ^{234}U to its decay product ^{230}Th will give the sample age (Harmon *et al.* 1975). The method is useful over the time range 350 000–10 000 years BP. A number of precautions are taken to ensure a reliable age estimate, most notably that any samples containing more than 1 percent of acid-insoluble detritus are rejected. Also, any indication that recrystallization has occurred (suggesting that the sample may not have remained a closed system) would cause it to be rejected.

Dating the onset and termination of speleothem growth, in samples from a wide area, enables regional chronologies to be built up and may indicate large-scale (climatically related) controls on periods of speleothem growth. This is most successful when dating samples from subalpine or subarctic sites which are presently marginal for speleothem growth. During glacial periods, colder conditions, less snow-melt, and more extensive permafrost would result in a marked reduction, even a cessation, of groundwater percolation. Furthermore, decreased biotic activity would lead to a decrease in the partial pressure of carbon dioxide in the soil atmosphere, and therefore less carbonate in solution; consequently speleothem growth might cease (Harmon *et al.* 1977, Atkinson *et al.* 1978). Figure 7.24 shows uranium-series dates for alpine speleothems from western Canada, collected at sites which currently have a surface mean annual temperature close to 0 °C and are hence marginal for speleothem growth today. The dates appear to cluster into four distinct groups which are assumed to represent interglacials, when conditions for speleothem growth were most favorable. These periods of relatively warm climate occurred from ~320 000 to 285 000 years BP, from ~235 000 to 185 000 years BP, from 150 000 to 90 000 years BP, and from ~15 000 years BP to the present (Harmon *et al.* 1977). In addition, a brief interval, possibly an interstadial, is identified at ~60 000 years BP. The earliest period may, in fact, have begun >350 000 years BP, and represents a long warm interval of massive spleothem deposition (Harmon 1976).

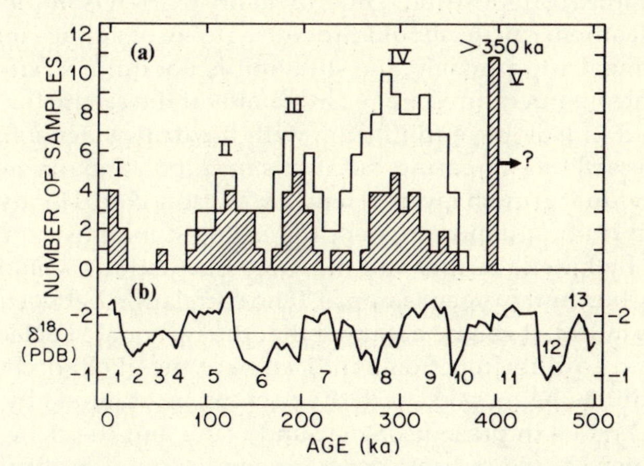

Figure 7.24 (a) Histogram of uranium series dates on speleothems from the Canadian Rocky Mountains. Line (A) is a histogram constructed by considering error bars (±1 S.D.) on each date; line (B) is a histogram based on dates without error bars. Arrow at upper right indicates minimum age estimates due to limitation of dating method. (b) $\delta^{18}O$ values in ocean core V28-238 (from Shackleton & Opdyke 1973) considered to represent glacial–interglacial changes in global ice volume (from less negative to more negative values). The speleothem dates cluster into groups which correspond temporally with global interglacial conditions (after Harmon *et al.* 1977).

These dates compare favorably with periods of speleothem growth observed in caves in England from 17 000 years BP to the present, between ~140 000 and 90 000 years BP and >170 000 years BP (Atkinson *et al.* 1978); as in the North American studies, a brief depositional interval at ~60 000 years BP has also been noted. It is interesting that these periods of speleothem growth correspond reasonably well with warm intervals noted in the $\delta^{18}O$ record of foraminifera in ocean cores (Fig. 7.24) and also with dates on corals which grew during interglacial periods of high relative sea-level stands (at 10 000 years BP to the present, between 145 000 and 85 000 years BP, between 235 000 and 190 000 years BP, and between 350 000 and 300 000 years BP; Harmon *et al.* 1977). More dates on speleothems from selected localities are needed to clarify the record further.

7.6.3 *Isotopic variations in speleothems*

In addition to using periods of speleothem growth as a rather crude index of paleoclimatic conditions, attempts have also been made to use oxygen isotope variations along the speleothem growth axis as an indicator of paleotemperatures. When air and water movement in a cave is relatively slow, a thermal equilibrium is established between the bedrock temperature and that of the air in the cave, approximating the mean annual surface temperature. During deposition of calcite from seepage (drip) water, a fractionation of oxygen isotopes occurs which is dependent on the temperature of deposition. Thus, in theory oxygen isotopic variations in the speleothem calcite should provide a proxy of surface temperature through time. Unfortunately, the situation is not quite so simple! First, isotopic paleotemperatures are recorded only if the calcite (or aragonite) is deposited in isotopic equilibrium with the drip-water solution. This can be assessed by comparing variations in carbon and oxygen isotopes along individual growth layers (Hendy & Wilson 1968, Hendy 1970). If a non-equilibrium situation existed, the isotopic composition would be controlled by kinetic factors and the same fluctuations would be found for both carbon and oxygen isotopes. If no correlation between these two isotopes is found, it can be assumed that the carbonate speleothem was deposited in equilibrium. Some indication of the likelihood of equilibrium conditions being present in the past can be obtained by analyzing the $^{18}O/^{16}O$ ratios in present-day ground water and in calcite deposited from it, which should indicate deposition in isotopic equilibrium.

Interpretation of $^{18}O/^{16}O$ variations in speleothems are not easy because a number of climatic factors other than cave temperature can influence observed $\delta^{18}O$ values. First, with a *decrease* in cave temperature, the fractionation factor between calcite and water *increases*, causing an increase in the calcite $\delta^{18}O$ values. However, as air temperature at the surface *decreases*, so the $^{18}O/^{16}O$ of precipitation, and thus the $\delta^{18}O$ value of drip water tends to *decrease*. Finally, during glacial periods, the growth of ^{18}O-depleted continental ice sheets results in an *increase* of $\delta^{18}O$ values of oceanic water and hence also of precipitation (see Secs 5.3.1–5.3.3). Thus, for a given climatic shift it is difficult to assess, *a priori*, in which direction the $\delta^{18}O$ value of the calcite will change (Thompson *et al.* 1976, Harmon *et al.* 1978a); indeed the calcite $\delta^{18}O$ – temperature relationship may not even be constant through time. This has led to diametrically opposed interpretations of $\delta^{18}O$ variations in speleothems. Duplessy *et al.* (1970b, 1971) for example, assumed that measured $\delta^{18}O$ variations were the result of variations in the ^{18}O content of precipitation; hence lower $\delta^{18}O$ values were interpreted as indicating colder conditions. This was disputed by Emiliani (1972) who observed that the speleothem $\delta^{18}O$ record, as interpreted by Duplessy *et al.*, was the inverse of paleotemperatures derived from oceanic foraminifera. He therefore concluded that the speleothem $\delta^{18}O$ variations were not controlled by variations in the $\delta^{18}O$ of precipitation, but due to the dominant effect of temperature-dependent fractionation. This has subsequently been confirmed in at least four localities, by the analysis of the isotopic composition of drip water, trapped as tiny liquid inclusions as the speleothem grew (Schwarcz *et al.* 1976). These inclusions vary in abundance, and when present in large amounts ($>1\%$ by weight) give speleothems a milky appearance; by isolating and analyzing the liquid at successive levels along the growth axis of a speleothem it is possible to assess, directly, whether isotopic variations of precipitation have occurred (Thompson *et al.* 1976, Harmon *et al.* 1979). Because it is possible that the inclusion water may have continued to exchange oxygen isotopes with the surrounding calcite following its entrapment, it is of no value to measure oxygen isotopes directly. A measure of the oxygen isotope fractionation between calcite and inclusion water would probably give a temperature close to present-day ambient temperature levels. Instead, the deuterium–hydrogen (D/H) ratio is measured since there is no hydrogen in the calcite with which hydrogen in the water might have exchanged. It is assumed that the relationship between δD and $\delta^{18}O$ in drip water approximates that noted in meteoric water by Dansgaard (1964), viz.

$$\delta D = 8\delta^{18}O + 10.$$

A similar relationship has also been noted in deep Antarctic ice (Epstein *et al.* 1970), suggesting that such a relationship has prevailed throughout the last glacial period, at least (Schwarcz *et al.* 1976). In this somewhat circuitous manner it is possible to estimate the *former* $\delta^{18}O$ values of meteoric water over very long periods of time and, by comparing these with $\delta^{18}O$ values of the surrounding calcite, to estimate paleotemperatures. Thompson *et al.* (1976) have carried out such studies on speleothem

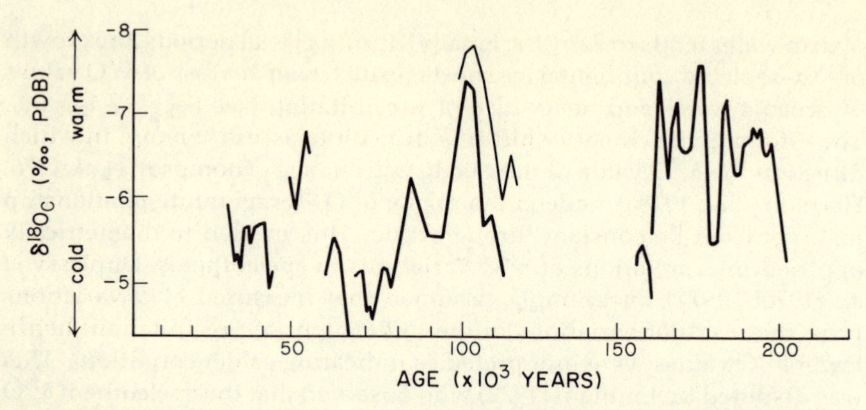

Figure 7.25 Variations of $\delta^{18}O$ in calcite from West Virginia speleothems. Paleotemperature interpretation is given at left. Ages determined by uranium-series dating methods (after Thompson *et al.* 1976).

calcite and inclusion waters from caves in West Virginia USA. Their results show that the oxygen isotopic composition of inclusion water at this site has changed very little over time, supporting the view of Emiliani (1972) that changes in $\delta^{18}O$ have been largely controlled by variations in calcite–water fractionation factors (i.e. temperature changes at the site). Thus, in this case at least, the $\delta^{18}O$ values of speleothem calcite increase with falling temperatures. Using this interpretation, the $\delta^{18}O$ data (Fig. 7.25) indicates that West Virginia has experienced three major warm episodes in the last 200 000 years – at <10 000 years BP, at 110 000–100 000 years BP, and at 175 000 ± 10 000 years BP. Cold intervals appear to have occurred prior to 200 000 years BP, ~180 000 years BP, from 165 000 to 110 000 years BP, and from 95 000 to 15 000 years BP, the last-mentioned perhaps interrupted by a warmer period at ~50 000 years BP. These records, although incomplete, do show some similarities with other isotopic records from sites in Alberta, Iowa, Kentucky, and Bermuda (Harmon *et al.* 1978b) and are in reasonable agreement with the oceanic foraminiferal records shown in Figure 7.25. However, it must be noted that when all the various calcite $\delta^{18}O$ records of the last 250 000 years are compared, there is little correspondence between them! Presumably this reflects the fact that no one factor has influenced isotopic fractionation to the same extent through time and at all sites, posing a real problem for the interpretation of these records.

The key to resolving all the possible interpretations which could be attached to calcite $\delta^{18}O$ variations through time lies with deuterium values in fluid inclusions and the relationship between δD and $\delta^{18}O$. If this relationship has not remained constant, paleotemperature estimates, even in a relative sense, will not be possible. However, if this problem can be resolved, speleothems offer great potential for more detailed studies which could provide resolution of paleotemperatures on time-scales as small as decades (e.g. Wilson *et al.* 1979). There is also the potential for mapping δD values of meteoric precipitation (as measured in speleothem inclusions) for different periods in the past, to assess if regional variations in δD did occur, and, if so, whether these can be related to changes in atmospheric circulation patterns (Harmon *et al.* 1979).

7.6.4 *Speleothems as indicators of sea-level variations*

Speleothem growth in carbonate island locations close to sea level can provide an extremely valuable indicator of former sea-level position, and hence ice volume changes on land. As speleothems must form above sea level in air-filled caves, the occurrence of speleothems in locations presently below sea level provides an upper limit to sea level at the time of formation. Similarly, speleothems exposed today which have overgrowths of marine aragonite (Fig. 7.22) indicate unequivocally that sea level was formerly higher. Uranium-series dating of the speleothems and coral deposits enables a picture of relative sea-level variations through time to be built up. Thus, Harmon *et al.* (1978b) were able to conclude, from studies of Bermuda speleothems, that interglacial conditions (high sea-level stands) occurred at around 120 000 and 97 000 years BP; between these events, a lower sea-level stand (−8 m) occurred at ~114 000 years BP. Minimum estimates could be placed on relative sea-level positions at other times (generally −8 m, or below) but only further (submarine) sampling could reveal in detail the precise magnitude of former sea-level depressions.

8 Non-marine biological evidence

8.1 Introduction

The study of non-marine biological material as a proxy of climate spans a wide range of subdisciplines, of which two are so large that separate chapters must be devoted to them (see Ch. 9, Pollen Analysis, and Ch. 10, Dendroclimatology). Here, those topics which have less spatial and/or temporal coverage, or which are relatively new areas of research, will be discussed.

8.2 Former vegetation distribution from plant macrofossils

It is not uncommon for plant macrofossils to be found far beyond the range of the particular species today. Where climatic controls on present-day plant distributions are known, their former distribution may be interpreted paleoclimatically, from dated macrofossils. Fluctuations of three major biogeographical boundaries have been studied in considerable detail using macrofossils: the arctic treeline, the alpine treeline, and the lower or "dryness" treeline of semi-arid and arid regions. In each case, the precise definition of treeline poses considerable problems since there is rarely a clearly demarcated boundary. Commonly, there is a gradual transition from mature dense forest through more open, discontinuous woodland to isolated trees or groups of trees, which may include dwarf or krummholz (deformed) forms, particularly in the alpine case (LaMarche & Mooney 1972). Topoclimatic factors are particularly important in determining the precise limit of trees. It is not necessary to dwell on this at length here, but sufficient to note that the location of the modern treeline is itself often problematical, and may make the interpretation of macrofossils somewhat difficult. For further discussion of the problem, see Larsen (1974) and Wardle (1974).

8.2.1 Arctic treeline fluctuations

Macrofossil evidence of formerly more extensive boreal forests has been found throughout the Northern Hemisphere, in tundra regions of Alaska, northern Canada, and the USSR (e.g. Miroshnikov 1958, Tikhomirov 1961, McCulloch & Hopkins 1966, Ritchie & Hare 1971). In addition, paleopodsols (relict forest soils) and charcoal layers (relating to forest fire episodes) have been found in many tundra areas of Keewatin, North West Territories, Canada (Bryson *et al.* 1965, Sorenson & Knox 1974). In most areas the macrofossil evidence is episodic in nature, made up of radiocarbon dates on isolated tree stumps located north of the modern treeline. Only in Keewatin have a sufficiently large number of dates (also on organic material in paleopodsols, and on charcoal layers) been obtained to construct a time series of the forest/tundra boundary during the latter part of the Holocene (Fig. 8.1; Sorenson 1977). According to these data, the northern treeline was 250 km or more north of the modern treeline between 6000 and 3500 years BP. Less extensive northward migrations occurred around 2700–2200 years BP and 1600–1000 years BP. By contrast, the presence of arctic brown paleosols (relict tundra soils) buried beneath more recent podsols south of the modern treeline, suggests that the treeline was at least 80 km further south around 2900, 1800, and 800 years BP (Fig. 8.1).

What paleoclimatic significance can be ascribed to such fluctuations?

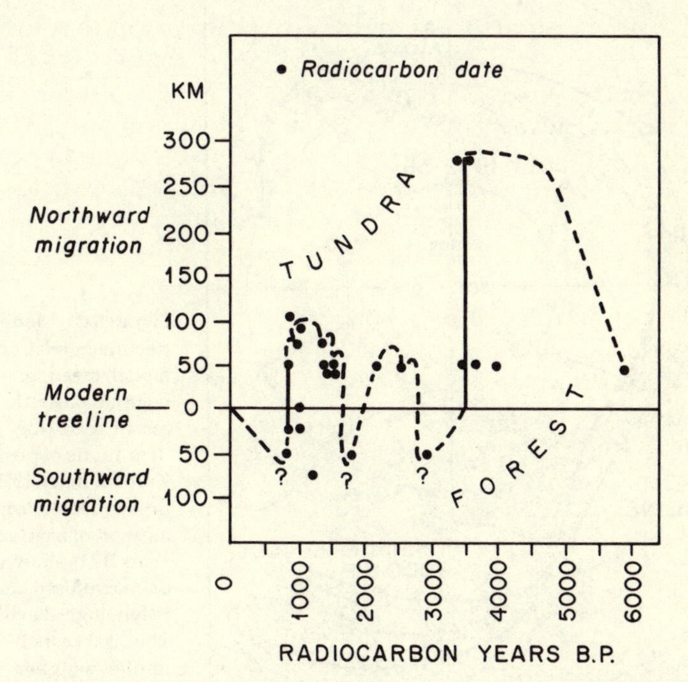

Figure 8.1 A reconstruction of Holocene treeline fluctuations in southwestern Keewatin, North West Territories, Canada. Treeline position is based on radiocarbon-dated tree macrofossils *in situ* north of the present treeline, and on dates on buried forest and tundra soils north and south of the modern treeline (after Sorenson & Knox 1974).

Several authors have noted the correspondence of northern treelines with isotherms of summer or July mean temperatures (Larsen 1974), so a northward migration of the ecotone may indicate warmer summer conditions. A tentative calibration of treeline migration in terms of July temperatures was made by Nichols (1967). Nichols assumed that, when the forest limit moved northward 250 km, July temperatures at the modern treeline were similar to locations 250 km south of the treeline today. In this way, July paleotemperatures were reconstructed for Keewatin, using both paleosol, macrofossil, and palynological evidence (see Fig. 9.14). Modern treeline is also closely related spatially to the mean or modal position of the arctic front in summer over North America and the median front position over northern Eurasia (Fig. 8.2; Bryson 1966, Krebs & Barry 1970). Whether this is a causative factor in the location of the northern forest border or whether the vegetation boundary itself largely determines the climatic differences noted across the vegetation boundary is difficult to assess. If air mass boundaries are a determinant of the forest border, then mapping paleoforest limits may provide an important insight into the dynamic climatology of the past (e.g. Ritchie & Hare

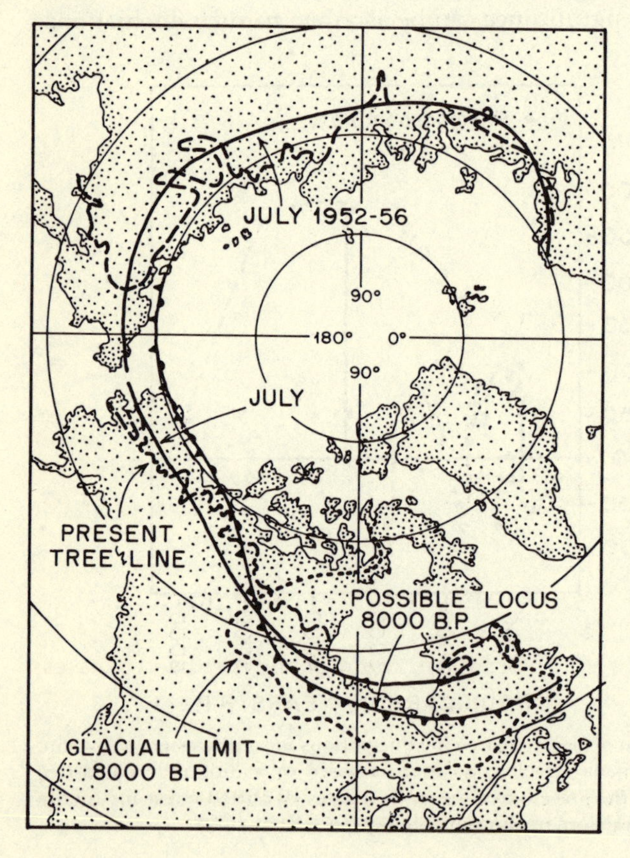

Figure 8.2 Modern treeline in relation to modal, mean, or median position of Arctic Front in recent years (for definitions of position, see Krebs & Barry 1970 and Bryson 1966). Proposed location of front at 8000 years BP is shown, based on macrofossil and palynological evidence. The 8000 years BP position implies a higher amplitude upper level westerly flow pattern at that time (after Ritchie & Hare 1971).

1971). Unfortunately, a number of factors make such interpretations difficult. Northward treeline migration during a climatic amelioration is more rapid than southward tree line migration in response to a climatic deterioration. Once established, trees may survive periods of adverse climate, and the treeline will only slowly "recede" as the trees which die are not replaced (cf. alpine treelines; LaMarche & Mooney 1967). Thus, in some areas today, trees at the northern forest boundary are out of phase with contemporary climatic conditions and are either not reproducing at all, or are doing so only vegetatively (Elliott 1979). Trees at the forest limit in Keewatin, for example, are relicts of a former warm period (Nichols 1976), a rather problematical fact when trying to assess the climatic significance of the modern treeline. This situation could continue for hundreds of years until the trees either die or are destroyed by forest fires. Such events in the paleo-environmental record could easily be misinterpreted, so we must recognize that treeline fluctuations are unlikely to provide information on high-frequency climatic fluctuations. In particular, climatic deteriorations on the order of several hundred years may not appear in the treeline macrofossil record, and climatic ameliorations are more likely to be accurately recorded than climatic deteriorations.

8.2.2 *Alpine treeline fluctuations*

In a survey of upper treelines in different climatic zones, C. Troll remarked: "It is absolutely clear that upper timberlines in different parts of the world cannot be climatically equivalent, not even in a relatively small mountain system such as the Alps or Tatra mountains" (Troll 1973 p. A6). In the paleoclimatic interpretation of macrofossils from above the modern treeline, we must therefore recognize that treeline variations in one area may result from different climatic factors than in another area, and that controls on tree growth may be complex. In the arid sub-Tropics, treeline is influenced by both temperature and the availability of moisture. In the humid Tropics, where seasonal temperature differences are extremely small, the transition to a treeless zone is generally quite abrupt, apparently related to a critical temperature threshold. In mid to high latitudes (particularly in the Northern Hemisphere) treelines are more diffuse, often reflecting strong topoclimatic controls. Climatic studies generally point to summer or July temperature as the major controlling factor in these latitudes (Wardle 1974). Evidence of higher treeline in the past is thus generally interpreted as indicating warmer summer temperatures, a lapse rate of 0.6–0.7 °C per 100 m commonly being used to assess the magnitude of temperature change. For example, Karlén (1976) estimated that summer temperatures in northern Sweden have varied by only ~1.5 °C over the last 9000 years, based on treeline variations of <200 m (Fig. 8.3). However, Karlén recognizes numerous problems of interpretation, among them the following:

(a) The incomplete nature of the macrofossil record; the highest trees may not have been found, or indeed may not have been preserved. Furthermore, in some areas, mountain summits may not extend far enough beyond the modern treeline to give a maximum estimate on former treeline extent (e.g. LaMarche 1973). Thus, paleotreeline evidence should be considered as providing a minimum paleotemperature estimate only.
(b) The present altitude of the treeline is often not precisely determined and may not be in equilibrium with modern climate (Ives 1978); a recent history of fire, over-grazing, avalanches, gales, or insect infestation may have resulted in a treeline well below the potential maximum for modern climatic conditions (cf. Griggs 1938).
(c) Trees take many years to become established during periods of favorable climate and may not have reached their highest position until after the temperature maximum; again this factor would tend to make any paleotemperature estimates based on treeline fluctuations minimum only.
(d) Treeline may have been affected by regional isostatic uplift following deglaciation of the region; such an effect must be taken into account when assessing the treeline record, particularly from the early Holocene (Fig. 8.3).

It is also worth noting that the time of treeline establishment is of more significance than the time of tree death (which may not have been climatically related). Denton and Karlén (1977), for example, have dated wood from trees growing above the modern treeline that were killed by explosive volcanic eruptions around 1800 and 1150 years BP. Ideally then, one should attempt to date the innermost wood fraction either by radiocarbon analysis, or dendrochronology, or both (LaMarche &

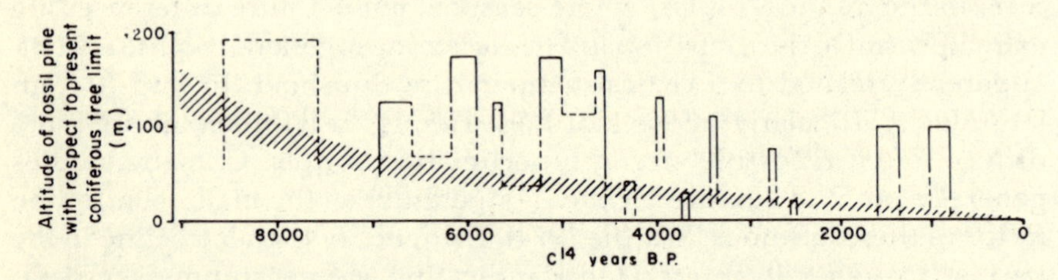

Figure 8.3 Variations in the upper limit of pine in northern Sweden during the Holocene, based on macrofossils. Horizontal lines represent minimum estimates on both the former altitude of the treeline above the present tree limit and the minimum lengths of time that higher treelines were present. Shaded zone indicates the approximate amount of rebound which has occurred in the area over the 9000 years. Thus pine macrofossils dated at ~8000 years BP are ~200 m above the present treeline, but, taking into account 100 m of postglacial uplift, the *relative* treeline was only ~100 m higher than today.

Table 8.1 Alpine treeline variations: selected estimates.

Location	Elevation above present (m)	Radiocarbon date (years BP)	Reference
San Juan Mts, Colorado	60–70	280 ± 80	Carrara & Andrews (1975)
White Mts, California	≥150	~5800–3100	LaMarche & Mooney (1967), LaMarche (1973)
Mt Washington, Nevada	100	~4100–1950	LaMarche & Mooney (1972)
Alaska/Yukon border	55–76	~1900–1150	Denton & Karlén (1977)
(Skolai Pass area)	10–60	~3600–3000	Denton & Karlén (1977)
	30–40	~5250	Denton & Karlén (1977)
Scotland	100–150	4630–4100	Pears (1969) and Lamb (1964)
Alberta	(?Higher)	4500–3600	Jungérius (1969)
Western Sweden	up to 200	1000–800	Lundquist (1959)
		1600–1350	Karlén (1976)
		~2400	
		~2700	
		~3300	
		~3600	
		~3900	
		~4250	
		5200–4500	
		~5600	
		6900–5900	
		8500–7600	

Mooney 1967, 1972). Often, the inner wood has weathered away, or decayed, making this impossible and providing only a minimum estimate on the timing of alpine treeline advances. Karlén (1976) considers that higher treelines in the past probably indicate mean temperatures above contemporary values for periods of 50–100 years, in order for young seedlings to become established and for the treeline as a whole to "advance." Once established, trees can survive periods of adverse climate, perhaps equally as long; thus the treeline record is both low frequency in resolution and biased towards recording warmer intervals of >50 years in duration.

Bearing all these factors in mind, it is perhaps not surprising to find that the alpine treeline record, considered on a world-wide basis, is extremely complex, showing little evidence of a coherent spatial or temporal pattern. There is, indeed, evidence that treelines were higher during the mid-Holocene, perhaps reflecting a globally extensive warmer interval (e.g. LaMarche & Mooney 1967, Pears 1969) but, in one area or another, alpine treelines have been higher than today more or less throughout the last 8500 years (Table 8.1). This suggests a number of possible explanations:

(a) Treeline fluctuations reflect different climatic factors in different areas, as Troll (1973) suggested.
(b) Climatic fluctuations are regional, not global, in extent.
(c) Climate is not the only factor involved in treeline fluctuations.
(d) The record, so far, is inadequate to draw any broad-scale conclusions.

In all probability there is probably some element of truth in each of these conclusions, but we must await more detailed studies, such as those of LaMarche (1973) and Karlén (1976), before a final assessment can be made.

8.2.3 *Lower treeline fluctuations*

Throughout the arid and semi-arid regions of the south-western United States, there is an altitudinal zonation of vegetation with xerophytic desert scrub (commonly sagebrush, evergreen creosote bush, and evergreen blackbrush) at lower elevations, grading into mesophytic woodland (juniper, pinyon pine, and live oak) at successively higher elevations (Fig. 8.4). The precise elevation of the woodland/desert scrub boundary varies more or less with latitude, being lowest in the Chihuahuan Desert of Mexico and highest in the interior Great Basin of Nevada. This elevational gradient, decreasing to the south, is related to distance from the source of summer moisture (the Gulf of Mexico and the tropical Pacific); the interior Great Basin is both furthest from these source regions of tropical maritime air and isolated from temperate Pacific moisture sources by the mountain ranges to the west (Wells 1979). The lower treeline elevational gradient thus strongly reflects the importance of moisture for tree growth and for this reason it is sometimes referred to as the "dryness treeline."

Fluctuations of the lower treeline have been greatly facilitated by the analysis of fossil packrat (*Neotoma*) middens from caves throughout the south-western United States (Wells & Jorgensen 1964, Wells & Berger 1967). Packrats forage incessantly within a very limited (<100 m) range of their dens which are constructed of plant material from the surrounding site. Because of the rats' propensity for collecting items at random, and not simply food stocks, the dens or middens effectively provide a remarkably complete inventory of the local flora (Wells 1976). Middens are cemented together into hard, fibrous masses by a dark brown, varnish-like coating of dried *Neotoma* urine (sometimes known as amberat). The amberat cements the deposit to rocky crevices in caves and prevents its destruction by fungi and bacteria (Fig. 8.5). Because the cave sites are so dry, *Neotoma* middens may remain preserved for tens of thousands of years; in fact over 130 macrofossil records from *Neotoma* middens have so far been dated, ranging in age from the late Holocene to >40 000 years BP. Middens are constructed over relatively short intervals,

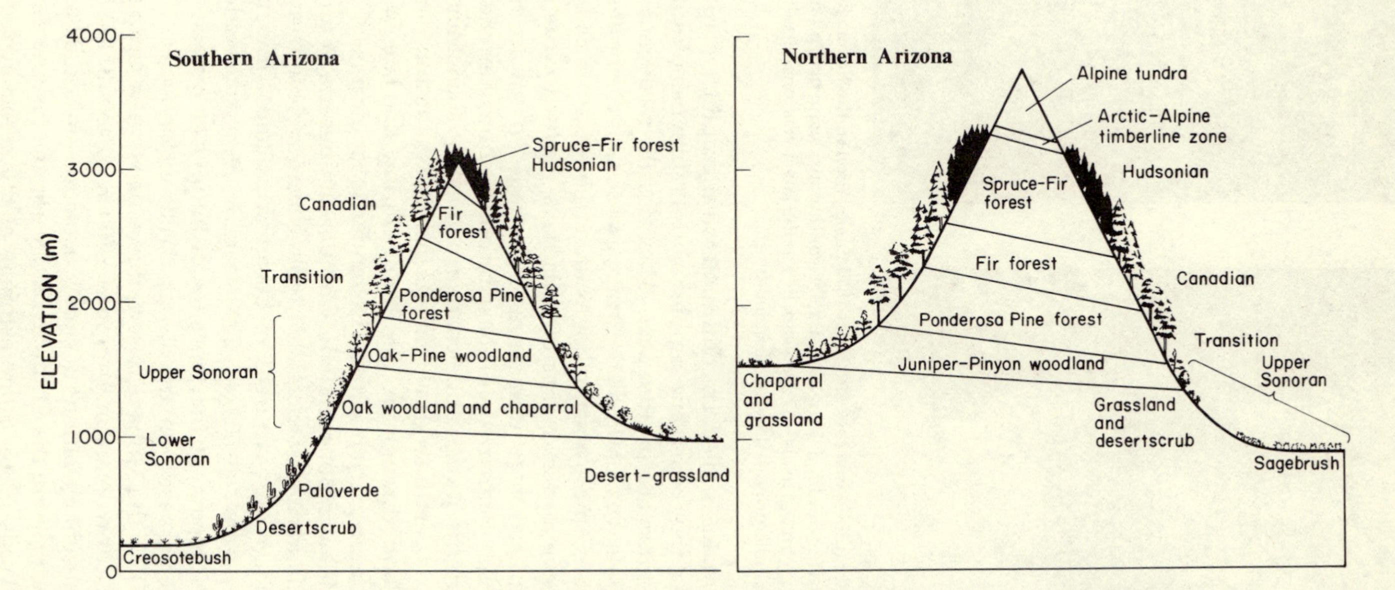

Figure 8.4 Transect through mountains of the south-western USA to show vegetation change with elevation. During glacial times, vegetation boundaries were generally lower due to increased effective moisture (resulting from lower temperatures and/or higher precipitation).

Figure 8.5 Packrat midden in a rock shelter at 1810 m elevation in the Eleana Range, Nye County, Nevada. This midden provides a near-continuous sequence of macrofossil assemblages from the full-glacial (~17 000 years BP) to the latest Wisconsin (~10 600 years BP) (photographs kindly provided by W. G. Spaulding).

until the rock crevice is filled, so continuous stratigraphic records are not available; rather, they represent samples of vegetation near the site, from discrete time intervals in the past. Macrofossils recovered include branches, twigs, leaves, bark, seeds, fruits, grasses, invertebrates such as snails and beetles, and even the bones of vertebrate animals (Fig. 8.6). Such prolific inventories of macrofossils have enabled quite detailed pictures of the local vegetation around the midden site to be reconstructed. Regional comparison of these reconstructions has enabled the broad-scale patterns of vegetation change to be established. Most significantly, the results demonstrate a dramatic increase in the area of pinyon–juniper woodland throughout the South during the late Wisconsin period of maximum glaciation (Van Devender & Spaulding 1979). In the Great Basin, Mohave, Sonoran, and Chihuahuan deserts of today (Fig. 8.7) such woodlands are restricted to higher elevations, often on isolated peaks surrounded by vast areas of desert scrubland. However, over periods ranging from >40 000 to ~8000 years BP, *Neotoma* middens from hyperarid sites document the presence of juniper or pinyon–juniper woodlands over a vastly enlarged area (Wells & Jorgensen 1964, Van Devender 1977). In the Mohave desert, this expansion of juniper woodlands involved a downward migration of ~600 m (17 450–10 100 years BP) compared to present elevations of such woodlands on similar substrates (Wells & Berger 1967). In the Chihuahuan Desert, the treeline depression was ~800 m (Wells 1966). Some estimate of the associated change in

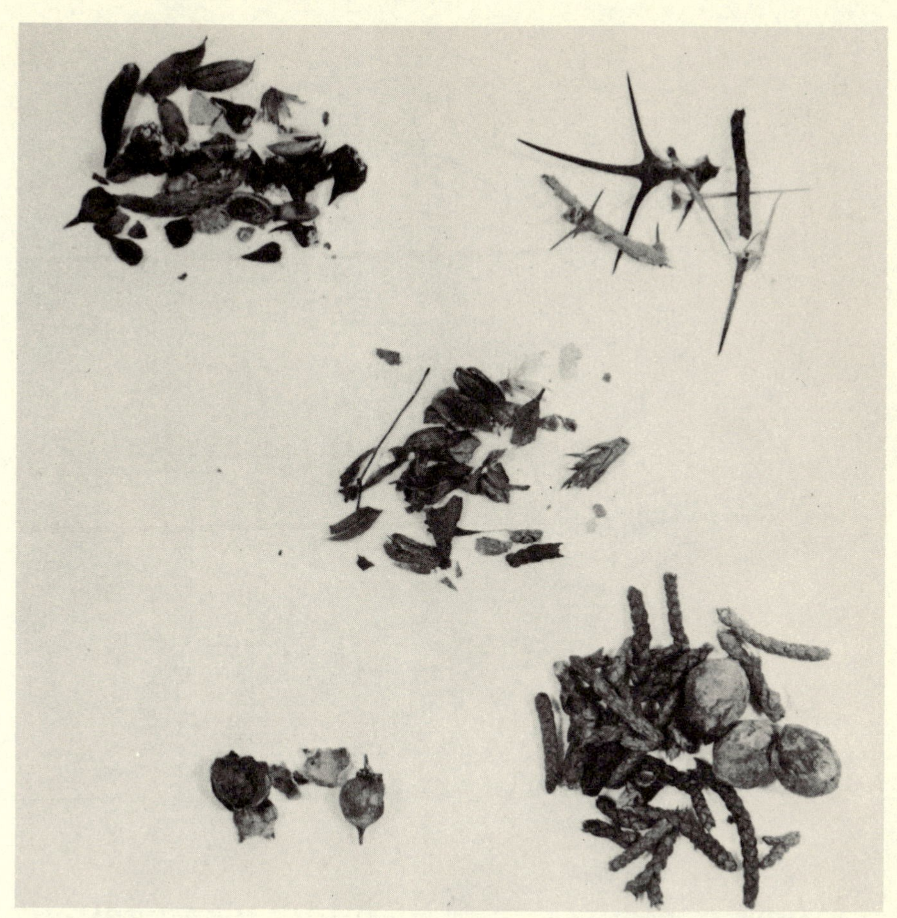

Figure 8.6 Dissected contents of a packrat midden from the Sheep Range, Nevada. This material has a ^{14}C date of 24 400 ± 760 years; clearly seen is the remarkable state of preservation of the material (material kindly provided by W. G. Spaulding).

precipitation can be obtained by comparing modern precipitation data from stations in the various vegetation zones today (Table 8.2). According to this analogy, precipitation in the midden areas may have been almost twice as high as today during late Wisconsin time. This assumes, however, that the cooler temperatures at higher elevations today (which increase the effectiveness of precipitation via reduced evaporation rates) were similar to those at lower elevations in the past, during the cooler glacial stage. It is possible, of course, that precipitation at the time was no greater than today, but due to lower temperatures and reduced evaporation rates the limited amount of precipitation which did occur was sufficient for juniper and pinyon to survive at lower elevations (Brakenridge 1978, McCoy 1981; see Sec. 4.2.1.4). However, such an argument neglects the persuasive evidence of a *steeper* latitudinal/elevational grad-

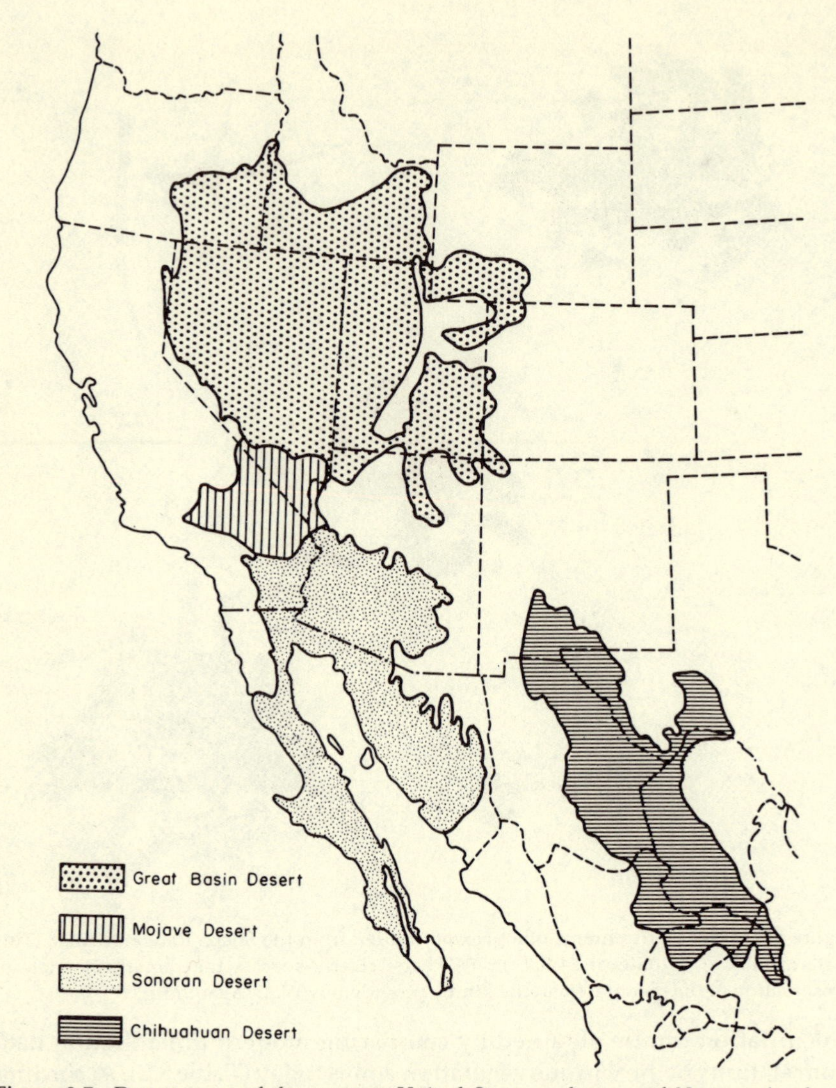

Figure 8.7 Desert areas of the western United States and range of *Neotoma* (packrat). Changing altitudinal vegetation zonation within desert areas is recorded in fossil packrat middens.

ient for the lower treeline in late Wisconsin times, i.e. trees migrated *further* into desert lowlands in the southern Chihuahuan Desert of Mexico (and, in fact, the forests contained more mesophytic species) than further north and north-west. If lower temperatures alone were responsible, one would have to argue that temperatures in Mexico decreased *more* than in areas to the north, and this seems unlikely. A more reasonable explanation is that precipitation was indeed greater and that the present precipitation gradient, increasing to the south and south-

Table 8.2 Precipitation data from stations in various vegetation zones of the Great Basin and Chihuahuan deserts.

Vegetation zone	Number of stations	Range of annual precipitation (cm)	Average annual precipitation (cm)
1. *Great Basin* (after Wells & Jorgensen 1964)			
Pinyon–juniper	20	26.4–42.4	33.3
Coleogyne	9	11.4–32.3	20.6
Larrea	20	6.4–22.1	12.4
2. *Chihuahuan Desert* (after Wells 1966)			
Lower pinyon–juniper–oak woodlands		42–50	45
Desert shrub		23–31	27

Note: Effectiveness of precipitation reinforced by differences in mean temperature between the different vegetation zones, the *Larrea* zone being warmest, and the Pinyon–juniper zone being coolest.

west, was enhanced in late Wisconsin times (Wells 1979). In the Chihuahuan Desert, at least, the term "pluvial" thus seems appropriate. It is quite possible, of course, that both temperature and precipitation factors played a role in explaining the observed lower treeline fluctuations. In the Great Basin, cooler temperatures with little or no increased precipitation may have occurred, whereas to the south and south-west increased low-level moisture (particularly in summer) may have accompanied somewhat lower temperatures. As a result the *effective* precipitation would have been enhanced throughout the desert areas of the western and south-western USA, and of northern Mexico, primarily as a result of cooler temperatures in the northern deserts, but due mainly to more precipitation further south.

8.3 Modern biological distributions

In many parts of the world, the modern distribution of plants and animals provides important clues to past environmental conditions. For example, many biogeographers have argued that the distribution of plant and animal species observed today in Africa and South America can only be satisfactorily explained by hypothesizing environmental conditions quite different from the present day at some time in the recent past (e.g. Moreau 1963, Vuilleumier 1971). These large-scale environmental changes are said to have resulted from markedly different climatic conditions, most probably during periods of continental glaciation at higher latitudes. The precise timing of such changes, is, however, somewhat speculative. The biological evidence alone provides little help in this matter since rates of speciation are themselves controversial (Endler 1977). Only by reference to other, independent lines of evidence (generally geomorphological or palynological) can reasonable estimates be made regarding when environmental changes took place in the past. To illustrate the application of this approach to paleoclimatic reconstruction, two areas will be briefly considered: tropical South America and tropical Africa.

8.3.1 Tropical South America

Although it is commonly assumed that the large numbers of species of flora and fauna in tropical South America are the result of long periods of stable climate, a number of biogeographical studies have cast doubt on that assumption. In a comprehensive study of different species of birds in and around the Amazon Basin, Haffer (1969, 1974) identified a number of geographical regions which he considered to have been refuges for groups of birds during drier periods in the past. At such times, the extensive tropical forests of today were reduced to discrete forest enclaves separated by savanna vegetation. Forest-dwelling species, which were isolated in this way, differentiated (developed new species) independently from members of the same species which had been separated into *other* forest enclaves. When wetter conditions returned and the forests re-occupied the savanna region, forest-dwelling species also expanded their ranges, coming into contact with other population groups in the intervening areas. In these areas of "secondary contact," hybridization of species took place so that the discrete morphological characteristics of species which had evolved within the forest refugia were no longer obvious.

According to proponents of the refuge hypothesis, the results of these changes can be seen today in contemporary biogeographical distribution patterns. Within the extensive tropical forests, zones of relatively high species diversity (i.e. zones containing extreme concentrations of different plant and animal species) can be identified. These are sometimes referred to as centers of endemism (Brown & Ab'Saber 1979). Within these zones, individual species may exhibit very uniform morphological characteristics (Vanzolini & Williams 1970). Such regions are considered by many to be the former forest refuges which served as "survival centers" for forest-dwellers during drier intervals. Between these centers of endemism, contact areas or "suture zones" are found, characterized by far fewer species than in the refuges and by more diverse morphological characteristics in the population of a particular species.

The refuge hypothesis is controversial. On the one hand there is considerable biogeographical evidence that there are indeed certain regions where species diversity is extraordinarily high. Such regions are generally identified by first mapping the ranges of individual species, then superimposing the distributions, and selecting those areas which exhibit very high levels of species diversity (Haffer 1982). In this way, studies of rainforest trees, butterflies, and lizards have been undertaken, and all reveal geographically similar core areas to those suggested by Haffer (1974) on the basis of his detailed studies of tropical birds (Fig. 8.8; see Vanzolini & Williams 1970, Vanzolini 1973, Brown *et al.* 1974, Prance 1974, 1982, Brown 1982). It is interesting to note that there is also linguistic and ethnographic evidence which points to the existence of similarly distributed forest refuges in prehistoric Amazonia (Migliazza 1982, Meggers 1982). The general coincidence of all these regions is quite impressive, considering the range of evidence involved. However, it could be argued that the distribution patterns observed do not reflect former refuges at all, but merely reflect modern ecological units which have evolved together in response to contemporary edaphic and climatic conditions, the uniqueness of which may or may not be immediately obvious (Endler 1982). Similarly, zones of "secondary contact" simply reflect significant environmental gradients (Benson 1982). However, Brown and Ab'Saber (1979) and Brown (1982) have noted that "independent" geoecological evidence (from soils, geomorphology, vegetation, and paleoclimatic models) can be used to delimit areas which are likely to have been refuges during drier intervals; these correspond quite clearly to those areas identified as refuges on the basis of species distribution, thereby confirming the refuge hypothesis. Clearly, the arguments can only be satisfactorily resolved by well dated stratigraphic evidence demonstrating that certain areas contained savanna *at the same time* as other areas (i.e. the postulated refuges) were under forest cover (Livingstone 1982). So far, such evidence is lacking from the Amazon Basin itself (though there is undated evidence of savanna in one area which is currently under tropical forest; Absy & Hammen 1976). There is, however, considerable palynological and geomorphological data from adjacent regions which point to a former period of cooler and drier climate in tropical South America during the late Wisconsin glaciation of higher latitudes (~21 000–13 000 years BP). At that time, Andean treelines were lower by 1200–1500 m and savanna was more extensive in the tropical forest regions of northern South America (Wijmstra & Hammen 1966, Hammen 1963, 1974, Salgado-Labouriau & Schubert 1976). This interpretation may explain the presence of extensive aeolian deposits within the equatorial forest zone (Tricart 1974), as well as offshore sediments which point to more extensive semi-arid conditions in the Amazon Basin in the past (Damuth & Fairbridge 1970). However, it still does not provide information on the location of forest enclaves at such times. Whether there was a significant reduction in forest cover in postglacial time is not yet known, though somewhat drier periods since the mid-Holocene have been proposed by Absy (1982). Only more

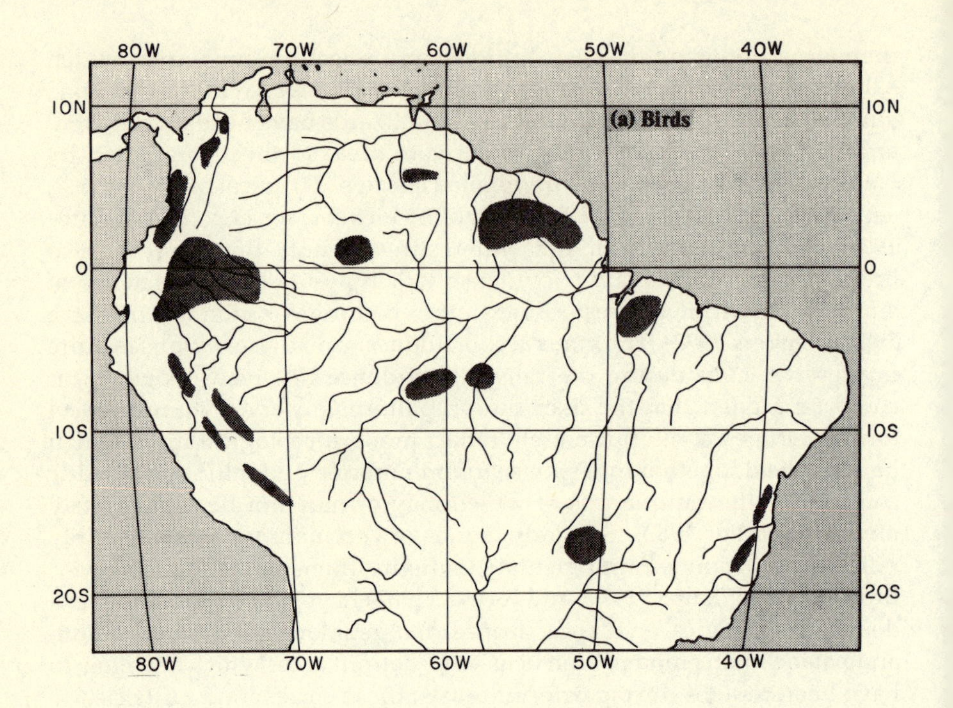

80W
70W
60W
50W
40W
10N
0
10S
20S
(a) Birds

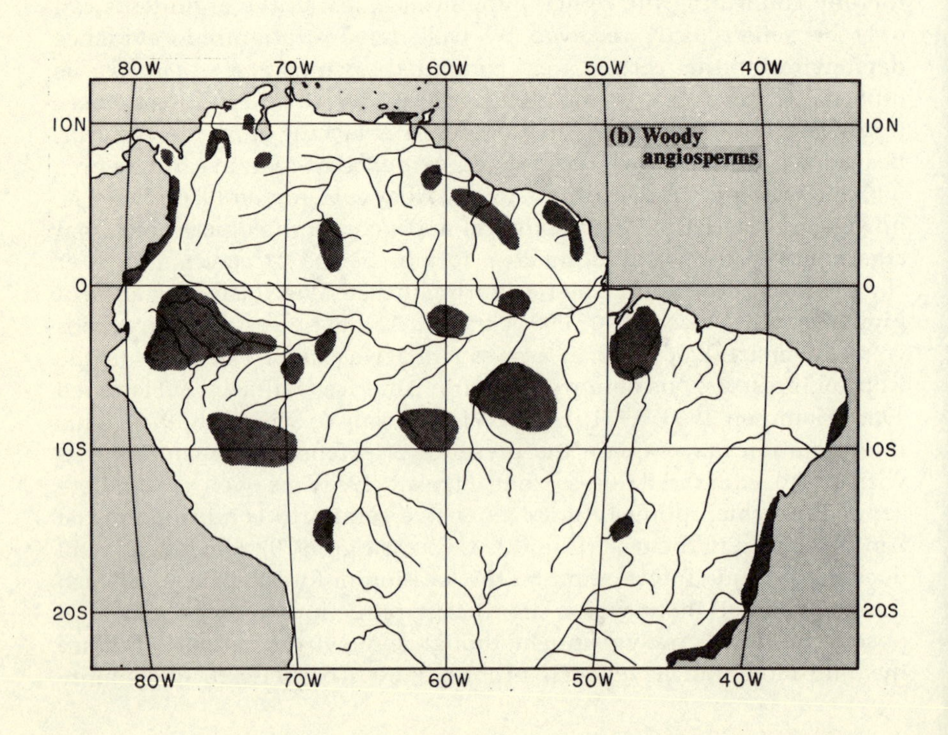

80W
70W
60W
50W
40W
10N
0
10S
20S
(b) Woody angiosperms

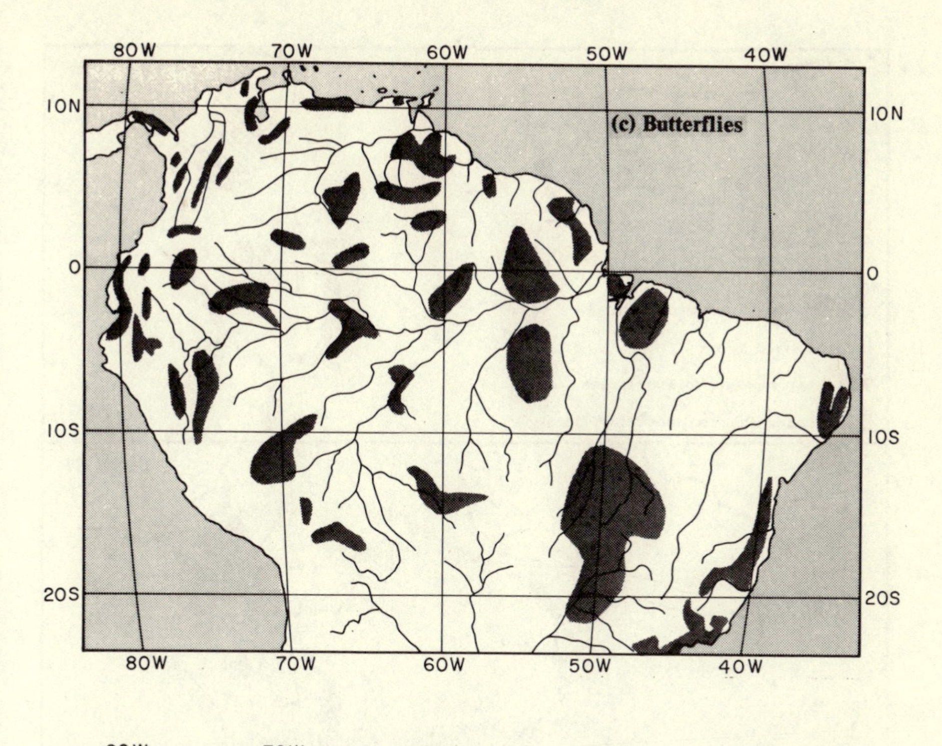

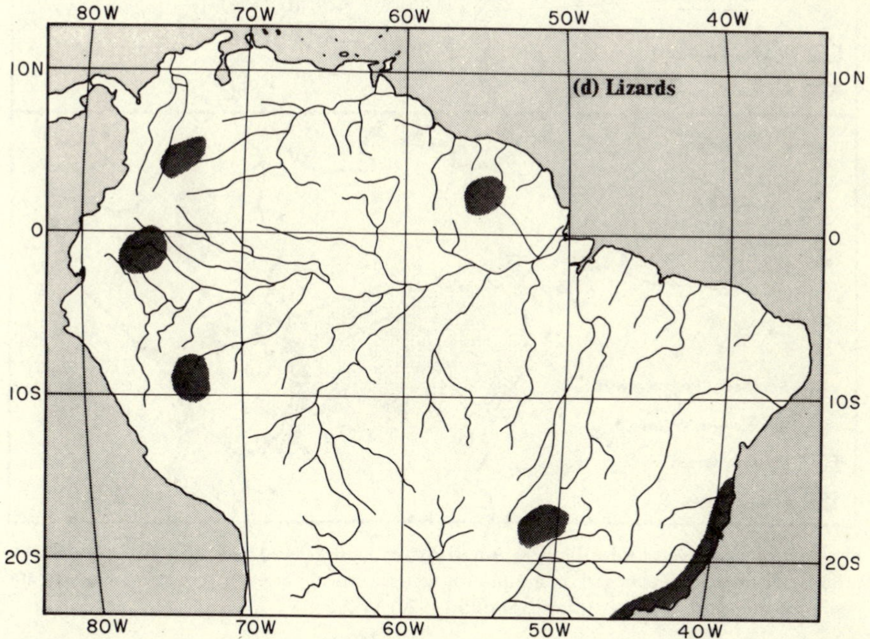

Figure 8.8 Proposed refuge areas for certain species of (a) bird, (b) woody angiosperms, (c) butterflies, and (d) lizards in the Amazon Basin during dry climatic phases of the Pleistocene, based on modern biogeographical distributions (after Haffer 1974, Prance 1982, Vanzolini & Williams 1970 and Brown 1982).

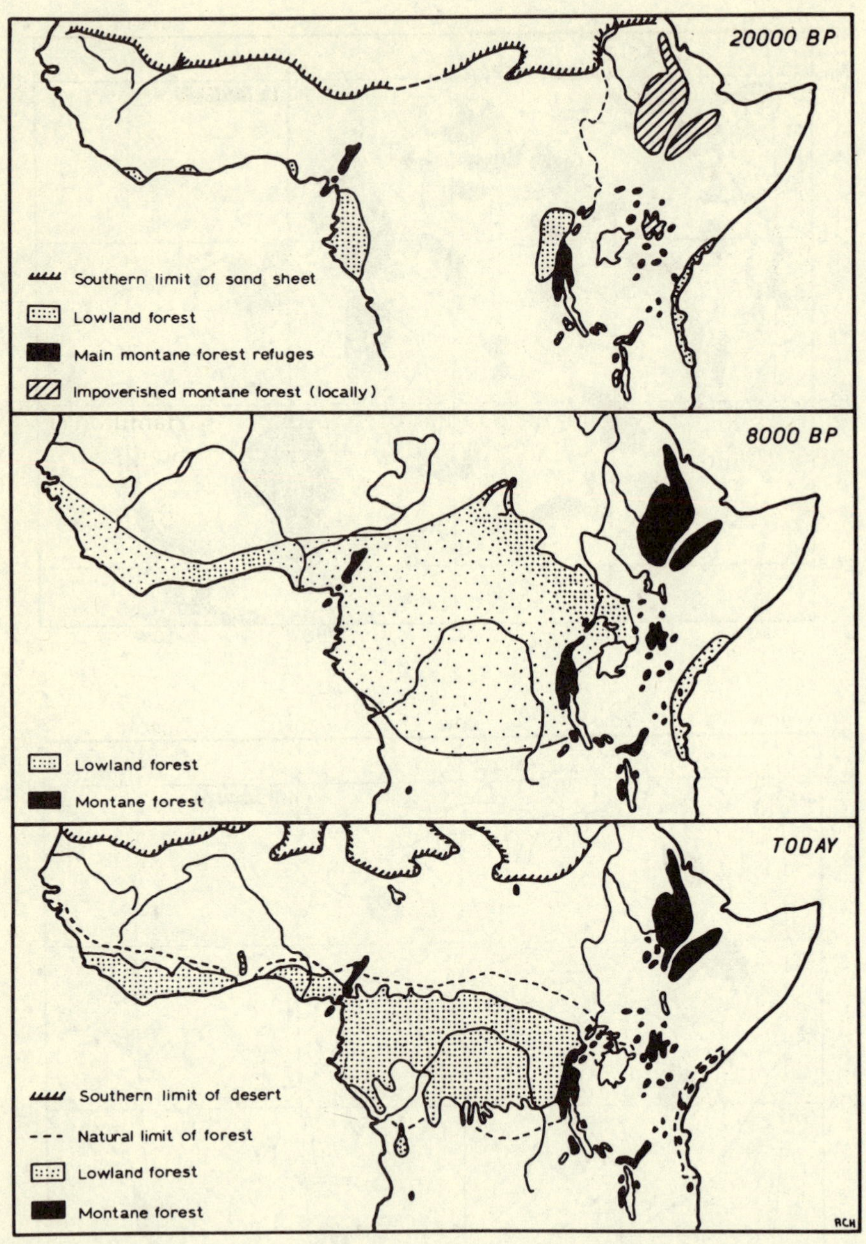

Figure 8.9 Proposed forest distribution in Africa at 20 000 and 8000 years BP compared to today. Reconstructions based on modern biogeographical patterns, plus palynological and geomorphological evidence (after Hamilton 1976).

detailed studies in different areas will reveal the extent to which rainforest was dispersed to isolated regions, as the refuge hypothesis predicts, and most importantly *when* such changes took place. On the basis of current evidence, it is possible that the biogeographical evidence relates to a protracted period of lower effective rainfall during the late Wisconsin glaciation of higher latitudes (cf. Colinvaux 1972, Newell *et al.* 1975, M. A. J. Williams 1975, Flenley 1978). However, the definitive stratigraphic evidence to support or refute the refuge hypothesis has yet to be found.

8.3.2 Tropical Africa

The biogeographical evidence for significantly different climate in tropical Africa at some time in the past has been reviewed by Hamilton (1976) and Livingstone (1975). Like the studies conducted in South America, there is much biogeographical evidence that the extensive tropical forests of the Congo Basin and adjacent coastal regions in the Gulf of Guinea were formerly much more limited in extent (e.g. Moreau 1963, Grubb 1982). However, the same problems of interpretation exist as previously discussed for South America (Livingstone 1982). Using independent geological and palynological evidence (cf. Chs 7 & 9), Hamilton has attempted to characterize the extremes of recent forest extent, from a minimum around 20 000 years BP (a situation which continued until an abrupt change occurred around 12 000 years BP) to a maximum forest cover around 8000 years BP (Fig. 8.9). At ~8000 years BP, lowland forest vegetation covered an area more than 15 times larger than at the 20 000 years BP minimum. At the same time as lowland forest vegetation was expanding, climatic conditions in subsaharan Africa became less arid, enabling the semi-arid savanna vegetation belt to extend farther northward (cf. Sec. 7.5.3). As a result, the range of large herbivores (such as giraffe, elephant, hippopotamus, and gazelle) was also more extensive and today the bleached bones of these animals provide a mute reminder of the remarkably different climatic conditions which existed there in relatively recent times. Accompanying the animal migration into subsaharan Africa were aboriginal hunters who recorded their way of life on magnificent rock paintings and carvings (pictographs; Lhote 1959, Monod 1963, Lajoux 1963). These are found today, hundreds of kilometers from the nearest permanent settlements (Fig. 8.10).

Of particular significance during the period of more extensive savanna and tropical forests were the much more extensive riverine and lacustrine environments, which effectively provided water connections across the entire sub-Saharan region, from the Nile to Senegal (Beadle 1974; see Ch. 7). Fauna of the Lake Chad Basin, for example, provide unequivocal evidence for recent connections, not only with the Niger and Congo basins, but also with the Nile drainage system over 1000 km to the east

Figure 8.10 Pictographs from caves in Tin Tazarift, Tassili n'Ajjer, central Sahara. Scene shows fishermen in a large boat, and was drawn in an area which today is totally arid. Estimated age is mid-Holocene ("Bovidienne" period) (photograph by P. Colombel, kindly provided by C. Roubet).

(Fig. 8.11). Even today, relict populations of animals and plants are found isolated in topographically favorable environments, far removed from their nearest adjacent populations. For example, the Eurasian green frog (*Rana ridibunda*) has been found in the streams of the Ahaggar mountains at least 1000 km from adjacent population groups. Furthermore, and perhaps most remarkably, a Nile crocodile (*Crocodilus niloticus*) was found in a pool in the Tassili-N-Ajjer Mountains, separated by vast stretches of desert from major population centers to the east (Seurat 1934, Beadle 1974). Such disjunct species bring climatic changes to life. When did these vastly different environmental conditions prevail? Again, the chronology of events based on biogeographical evidence is crude and equivocal but can be supplemented by numerous studies of geomorphological features, lake-level fluctuations, and pollen (e.g. van Zinderen Bakker & Coetzee 1972, Street & Grove 1976, Livingstone 1975, Flenley 1978). These studies provide evidence that tropical Africa is probably more arid today than it has been during most of the Holocene and was only more arid during the late Wisconsin glacial maximum (cf. Figs 7.15 & 7.18). However, the record is still far from complete in either time or space, so brief periods of relatively moist conditions (in the order of 10^2 years), which may have been significant events biogeographically, may

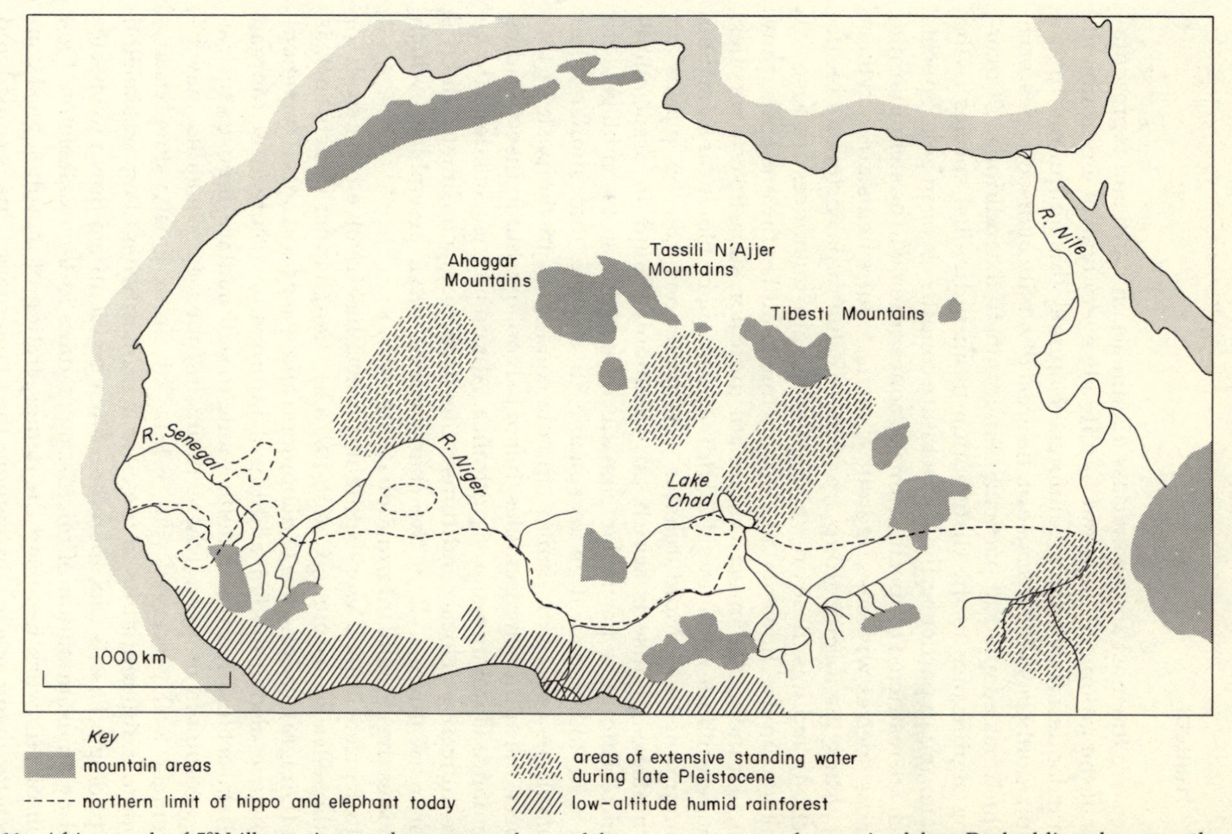

Figure 8.11 Africa north of 5°N illustrating modern geography, and former occurrence of extensive lakes. Dashed line shows northern limit of elephant and hippopotamus today.

be beyond the resolution of methods of paleo-environmental reconstruction currently in use.

8.4 Insects

Insects are the most abundant class of animals on Earth and representatives of the group can be found in virtually every type of environment, from polar desert to tropical rainforest. Naturally this ubiquitous distribution is only possible because of the great diversity of insect types, each of which has adapted to particular environmental conditions. Of overriding significance to the distribution of an individual species is the climate, and in particular the temperature conditions, of an area. Species which are restricted to specific climatic zones are said to be stenothermic, whereas species with less rigorous climatic requirements are eurythermic; clearly the former group are of most value in paleoclimatic reconstructions and it is these on which paleoclimatic inferences are based. It would be unwise, however, to place too much faith in the presence of any particular individual insect as a climatic indicator, since insects are often extremely mobile and inevitably individuals will be blown far from their optimum habitat. More reliable interpretations can be placed on assemblages of insects which are commonly found in associations characteristic of a particular climatic regime. Such assemblages are observed today and it is reasonable to assume that similar fossil assemblages represent similar climatic conditions in the past. In this respect, the approach resembles that of palynology, but in insect studies abundance is not of major significance. Abundance is considered to be more indicative of local conditions rather than the macroclimate which is of primary interest. It is the characteristic fossil assemblage which provides the climatic information (Coope 1967).

Most paleoclimatic work utilizing insects has involved the study of fossil beetles (Coleoptera; Coope 1977a,b), though other insects such as flies (Diptera), caddis flies (Trichoptera) and wasps and ants (Hymenoptera) have also provided additional information (Morgan & Morgan 1979). Insect fossils are commonly found in sedimentary deposits such as lake sediments or peat, where their chitinous exoskeletons may be extremely well preserved (Fig. 8.12). This is of great value because taxonomic differentiation of the class is primarily based on exoskeleton morphology. Fossils can, therefore, often be identified down to species level by an examination of microscale features in the exoskeleton. One result of this work has been the demonstration of morphological constancy for many species throughout the Quaternary. This is considered to be evidence that they have also exhibited physiological constancy, in other words, they have not altered their ecological requirements, at least

Figure 8.12 Fossil beetle (Chrysomelidae; Donaciinae) from 11 m depth (*c.* 10 000 years BP) in a core from Stafford, England (photograph kindly provided by A. V. Morgan).

over the last 2 million years or so. No direct evidence of this can be obtained, but the fact that fossil *assemblages* are often so similar to modern assemblages, in what are assumed to be similar environmental conditions, suggests that radical changes in physiological development have not occurred. This is a fundamental assumption in using fossil insects as paleoclimatic indices, since any change in their climatic tolerances would, of course, invalidate any conclusions that might be drawn from their presence. This problem is no different from that facing palynologists or marine microfaunal analysts, however, and indeed entomologists have considerably more evidence for genotypic stability in their fossils than can be provided in many other branches of biology.

From a paleoclimatic viewpoint, one of the most important attributes of insects is their ability to occupy new territory fairly rapidly following a climatic amelioration. They thus provide a much more sensitive index of climate variation than plants, which have much slower migration rates. Indeed, Coleoptera may occupy and abandon a new territory in response to a marked but brief warm interval, whereas there may be no evidence for such an event in the pollen record because of the lag in vegetation response time (Coope & Brophy 1972, Morgan 1973). In short, "this combination of sensitivity and rapidity of response to climatic changes, coupled with their demonstrated evolutionary stability, makes the Coleoptera one of the most climatically significant components of the whole terrestrial biota" (Coope 1977a).

8.4.1 Paleoclimatic reconstructions based on fossil Coleoptera

Most paleoclimatic work using insects has been carried out in Britain, where the temperate assemblages of Coleoptera today were replaced in the past by an alternation of boreal or polar assemblages during glacial and stadial events, and by more southern or subtropical assemblages during interglacials and interstadials (Coope 1975a, 1977b). A large number of sites have now been studied, ranging in age from interglacial to postglacial (Flandrian). Figure 8.13 illustrates the estimated average July temperature record of the last 120 000 years, since the last (Ipswichian = Sangamon) interglacial. July temperatures are assumed to be a major control on insect distribution since the northern limits of most thermophilous (warmth-loving) species more closely parallel July or summer season isotherms than isotherms of winter months (Morgan 1973). Nevertheless, some estimate of winter temperatures can be made by

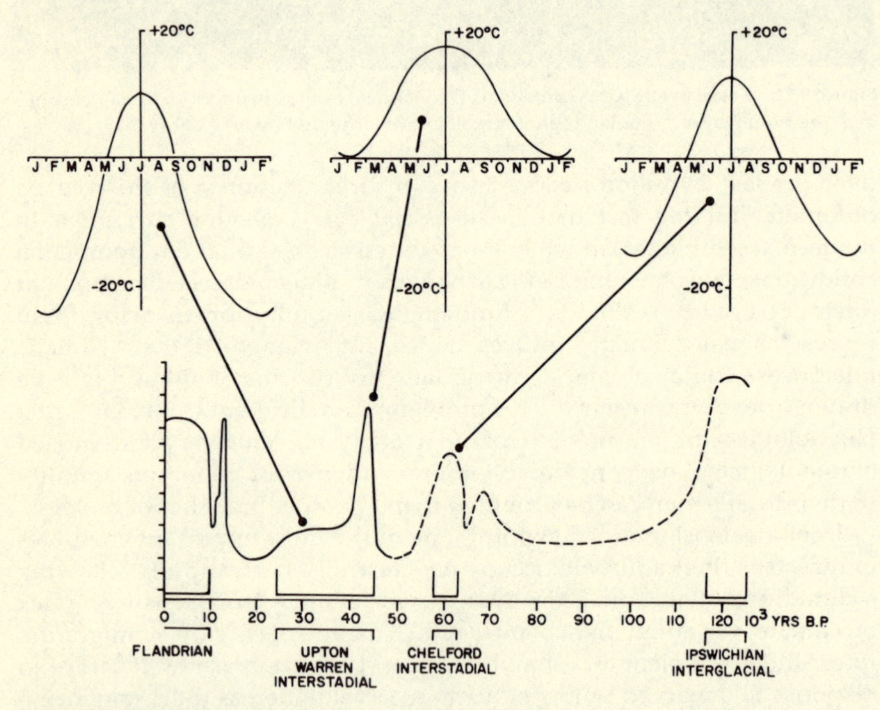

Figure 8.13 Reconstructed July paleotemperatures based on insect remains in areas of the southern and central British Isles since the last (Ipswichian) interglacial. Annual temperature ranges are also shown (after Coope 1977b). The period before ~50 000 years BP (dashed line) is very uncertain and there may have been a more gradual, monotonic decline in temperature from ~120 000 years BP to ~60 000 years BP.

considering the occurrence of species which are today characteristic of continental Eurasia. A species may be an arctic stenotherm (having a northern distribution) but it may live in continental areas where July temperatures are relatively high and winter temperatures extremely low. Bearing such factors in mind, and considering modern distributions of fossil species, it is possible to assess the annual temperature *range* at intervals in the past (Fig. 8.13; Coope 1977b).

Over the last 125 000 years, there appear to have been three distinct periods when temperatures in central England were at least as warm, or warmer, than they are at present: the Ipswichian interglacial, the Upton Warren interstadial, and the Lake Windermere interstadial. The last interglacial was, by definition, the warmest of these episodes, with Coleopteran assemblages characteristic of southern Europe today, present in lowland England; July temperatures are estimated to have been ~3 °C higher than today (Coope 1974). Between 50 000 and 25 000 years BP the climate of Great Britain appears to have fluctuated rapidly between temperate and cold continental conditions. This inference is based on the occurrence of climatically contrasting Coleopteran assemblages which follow one another quite abruptly in stratigraphic sequences. The period has been termed the Upton Warren interstadial complex, and includes one brief period (~43 000 years BP) when temperatures seem to have been warmer (by 1–2 °C) than the present day (Fig. 8.13). The duration of this interval is uncertain (indeed uncertain radiocarbon dates, close to the limit of the method, may account for the apparently rapid temperature fluctuations of this period) but it may have lasted for 1000–2000 years, followed by a gradual fall in temperature. This cooling was accompanied by more continental conditions, as evidenced by a beetle assemblage typical of parts of Eurasia today; average February temperatures of −20 °C and July temperatures of only +10 °C seem probable. In spite of periods of relative warmth during the "interstadial complex," central England was devoid of trees all the time and there is little *palynological* evidence for any climatic amelioration (Coope 1975b). Evidently the Coleoptera were sufficiently mobile that they could rapidly move northward as the climate improved, whereas certain plants could not migrate northward fast enough to become established in Britain before the climate once again deteriorated. A similar situation occurred at the end of the Devensian (Weichselian) cold phase when temperatures again rose abruptly, but for only a relatively short period (the Lake Windermere interstadial; Coope & Pennington 1977). At this time, an abrupt change from arctic to thermophilous beetle assemblages took place, with maximum warmth occurring around 12 500–13 000 years BP. By 12 000 years BP (when from pollen data it is apparent that birch began to colonize the north of England) the interstadial peak of warmth had passed and a significantly cooler episode was already beginning. The newly established birch forest declined and the thermophilous beetle assemblage was replaced by a northern assemblage typical of tundra regions today. By 9500 years BP this sequence had been entirely reversed and thermophilous species again rapidly replaced the arctic stenotherms which had been abundant only 500 years earlier (Osborne 1974, 1980). Again, the more mobile insects were in advance of the vegetation and provide a more accurate assessment of paleoclimatic conditions than could be obtained simply from palynological data.

It is perhaps appropriate to note that coleopteran and palynological data do not always appear to be out of phase; such situations are probably the exception rather than the rule. The Chelford interstadial (radiocarbon dated at ~60 000 years BP), for example, must have lasted long enough for trees to migrate northward into Britain following the cold, early Devensian period. Coleopteran assemblages are in complete accord with palynological evidence for a cool but quite continental climate at this time (Fig. 8.13); conditions in central England were similar to those in southern Finland today (Simpson & West 1958, Coope 1959, 1977b).

Paleoclimatic studies based on insects in other parts of the world are still relatively uncommon, though a number of sites have been studied in North America (Matthews 1968, 1974, 1975, Ashworth 1977, Morgan & Morgan 1979). However, it has not yet been possible to reconstruct long-term temperature variations in any detail, as Coope has done for Great Britain. This is primarily due to two factors:

(a) The insect fauna of North America is significantly larger than that of Europe and systematic relationships between the modern faunal elements are not as well known.
(b) The distribution and ecology of modern insects are also not well known in North America and many areas remain entomologically unexplored (Ashworth 1980, Morgan & Morgan 1981).

Consequently, it is more difficult in North America to identify paleoenvironmental conditions precisely by fossil insect faunal assemblages. As more studies of both modern and fossil assemblages are carried out, this situation should improve significantly.

Fossil insects would appear to offer tremendous potential for paleoclimatic reconstructions, particularly as they provide information on relatively brief climatic fluctuations, rarely resolved in other proxy data. They also shed much light on vegetation migration rates in response to climatic change. Comparisons of palynological and insect data are particularly valuable (e.g. Coope & Brophy 1972, Matthews 1974). Multivariate analyses of insect assemblages might also provide useful in stratigraphy by resolving the difficulties of correlating discontinuous deposits beyond the range of radiocarbon dating.

9

Pollen analysis

9.1 Introduction

Every year millions of tons of organic material are dispersed into the atmosphere by flowering plants and cryptogams in an effort to reproduce. The higher plants (angiosperms and gymnosperms) produce pollen grains containing the male genetic material; sexual reproductive success is assured only if this material reaches a female receptacle of the same plant species. The lower plants or cryptogams (plants without true flowers or seeds) produce spores which contain the necessary genetic material for the growth of an independent generation of plants. Pollen grains and spores are the basis of an important aspect of paleoclimatic reconstruction, generally referred to as pollen analysis, or palynology, the study of pollen and spores. Where pollen and spores have accumulated over time, a record of the past vegetation of an area may be preserved. In many cases, changes in the vegetation of an area may be due to changes of climate; hence, interpreting past vegetation through pollen analysis may lead to strong inferences about the former climate of an area. Of course, not all changes in the accumulation of pollen and spores are necessarily due to a change in climate; fire, insect infestation, plant successional changes, interference by man as well as changes in factors leading to the accumulation and preservation of the fossil material itself often makes interpretation of the pollen and spore record complex. The pollen analyst must therefore carefully evaluate the record to isolate whatever climatic signal it contains.

9.2 Basis of pollen analysis

Of fundamental significance to the discipline of pollen analysis are the facts that, in general, pollen grains and spores (a) are extremely resistant to decay and possess morphological characteristics which are specific to a particular genus or species of plant; (b) are produced in vast quantities and are distributed widely from their sources; (c) reflect the natural vegetation of the area around the preservation site.

As pollen grains have been studied far more than spores for reconstructing past climates (i.e. the emphasis has been on higher plants; Table 9.1) the following sections will focus on pollen grains. However, from a methodological viewpoint, most of the problems of studying pollen grains apply equally to the study of spores.

Table 9.1 Some important plant taxa in North American and European Quaternary palynology (c.f. Fig. 9.1).

Genus	Family	Common name
Abies		Fir
Acer		Maple
Alnus		Alder
Ambrosia		Ragweed
Artemisia		Wormwood/sage
Betula		Birch
Carpinus[†]		Ironwood
Carya		Hickory
	Chenopodiaceae	Goosefoot
Corylus		Hazel
	Cyperaceae	Sedges
Ephedra		Horsetail
Eucalyptus[‡]		Eucalyptus
Fagus		Beech
Fraxinus		Ash
	Gramineae	Grasses
Juglans		Walnut
Juniperus		Juniper
Larix		Larch
Liquidambar		Sweet gum
Lycopodium[‡]		Clubmoss
Nyssa		Tupelo
Ostrya		Hornbeam
Picea		Spruce
Pinus		Pine
Populus		Poplar
Pseudotsuga		Douglas fir
Quercus		Oak
Salix		Willow
Taxodium		Bald cypress
Tilia		Basswood/lime
Tsuga		Hemlock
Ulmus		Elm

[†] *Ostrya* and *Carpinus* pollen are indistinguishable and are generally considered together.
[‡] Exotic pollen added to samples for absolute pollen influx calculations (see Sec. 9.4).

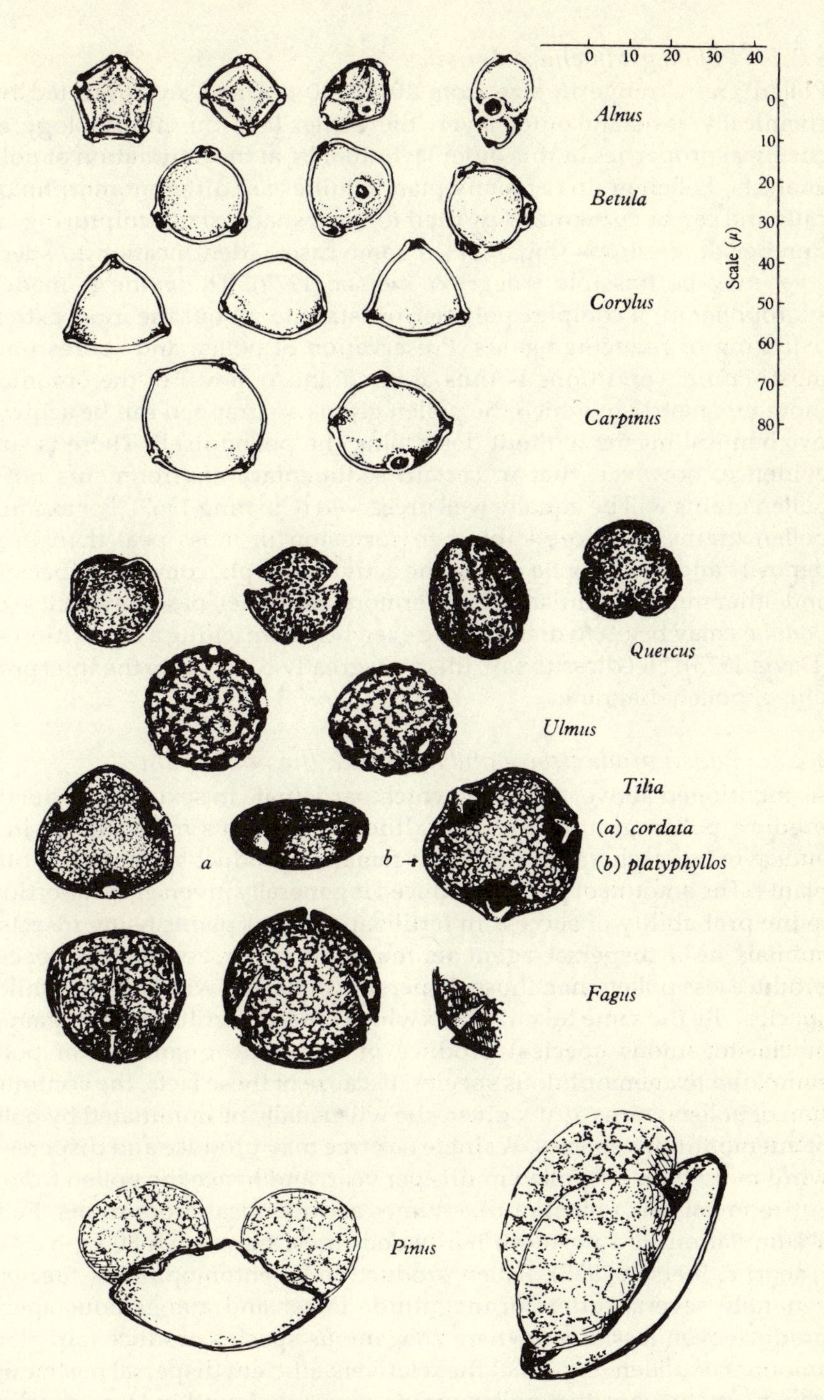

Figure 9.1 Some of the principal pollen types in British Holocene deposits, drawn to the same scale. Common names of plants are given in Table 9.1 (from Godwin 1956).

9.2.1 *Pollen grain characteristics*

Pollen grains range in size from 10 to 150 μm and are protected by a chemically resistant outer layer, the exine. It is the morphology and chemical properties of this outer layer that lie at the foundation of pollen analysis. Pollen grains of many plant families are different morphologically and can be recognized by their distinct shape, size, sculpturing, and number of apertures (Fig. 9.1). In some cases, identification to species level may be possible (Faegri & Iversen 1975). The exine is made of sporopollenin, a complex polymer resistant to all but the most extreme oxidizing or reducing agents. Preservation of pollen and spores under most natural conditions is thus assured and removal of the organic or inorganic matrix in which the pollen grains are trapped can be achieved by chemical means without destroying the pollen itself. There is some evidence, however, that in certain sedimentary environments not all pollen grains will be equally well preserved (Cushing 1967). For example, pollen grains are more subject to corrosion in moss peat than in silt deposits and this may be due to the activities of phycomycetes, bacteria, and other micro-organisms. Furthermore, the pollen of some species (e.g. *Populus*) may begin to disintegrate even before reaching a deposition site (Davis 1973). Needless to say, this may greatly complicate the interpretation of pollen diagrams.

9.2.2 *Pollen productivity and dispersal: the pollen rain*

As mentioned above, all plants which participate in sexual reproduction produce pollen grains, dispersing them by various mechanisms in an endeavor to reach and fertilize the female reproductive organs of other plants. The amount of pollen produced is generally inversely proportional to the probability of success in fertilization; thus plants using insects or animals as a dispersal agent (entomophilous or zoophilous species) produce less pollen than those dispersing pollen by wind (anemophilous species). By the same token, plants which are self-fertilizing (autogamous or cleistogamous species) produce only minute quantities of pollen compared to anemophilous species. Because of these facts, the accumulation of pollen grains at any given site will usually be dominated by pollen of anemophilous species. A single oak tree may produce and disperse by wind more than 10^8 pollen grains per year, and hence the pollen from an entire forest (the pollen rain) assumes astronomical proportions. Pollen accumulation in a northern hardwood forest may reach 80 kg ha^{-1} a^{-1} (Faegri & Iversen 1975). Pollen production by entomophilous species is generally several orders of magnitude lower and autogamous species produce even less. Even where zoogamous species produce fairly large amounts of pollen (e.g. *Tilia*) the relatively efficient dispersal mechanism (via insects) means that pollen grains are rarely found in large numbers, even in forests where *Tilia* is abundant (Janssen 1966).

The vast majority of pollen grains dispersed by wind are not carried more than 0.5 km beyond their source. Dispersal by wind is a function of grain size, the larger and heavier grains falling to the ground sooner than the smaller, lighter grains (Dyakowska 1936). Pollen grains of beech (*Fagus*) and larch (*Larix*), for example, are relatively heavy and settle out close to their source. Consequently, the occurrence of fossil beech or larch grains in a deposit would indicate the former growth of a species in the immediate vicinity of the site. Field measurements of pollen dispersal from artificial sources and from isolated stands of vegetation indicate that pollen produced by an individual plant is not identifiable above background levels (the regional pollen rain) beyond a few hundred meters. This is also indicated by theoretical dispersal models (Tauber 1965). Many investigators thus favor the analysis of sediments from fairly large lakes (>1 km^2) because they act as catchment basins for the regional pollen rain and are not unduly influenced by vegetation in the immediate vicinity of the sampling site. However, lake sediments have other problems, as will be seen in Section 9.3, below.

Pollen dispersal by wind is obviously not equal from day to day or from year to year, depending greatly on the prevailing synoptic conditions at the time of pollen production. In particular, the frequency and intensity of precipitation events is crucial because precipitation is particularly effective at cleansing the atmosphere of its particulate load (Schmidt 1967). Nevertheless, if the meteorological conditions are suitable, pollen may be dispersed over extremely long distances, even reaching mid-oceanic and polar locations (Erdtman 1969, Fredskild & Wagner 1974, Lichti-Federovich 1975). Because of these factors, even with equal pollen productivity from year to year, interannual deposition at any one site may vary dramatically; as most studies of pollen focus on low-frequency ($>10^2$ years) aspects of the record (indeed the poor resolution due to sampling problems and the dating of deposits makes this almost inevitable), the yearly variations in pollen rain are not significant. However, for studies of pollen in annually laminated sediments (varves and firn) interannual production patterns (e.g. Andersen 1980) and the frequency of individual synoptic types and their effect on pollen dispersal assume more significance.

Differences in pollen dispersal have also been observed between non-arboreal plants, which grow close to the ground where wind speeds are lower, and arboreal plants, which generally extend into a more turbulent zone where wind speeds are higher. Consequently, non-arboreal pollen (NAP) is rarely dispersed far from its source (generally <25 m).

The interpretation of shrub and herb pollen is also a difficult problem and has not been adequately studied. Because of their low growth forms shrubs and herbs produce pollen which is not carried great distances and generally reflects the local site conditions. This can be used to advantage if species which are particularly diagnostic of the prevailing environmental conditions can be identified (e.g. high frequencies of *Salix* or *Gramineae* pollen in an arboreal site may indicate an open canopy forest). However, shrub and herb pollen may be carried further afield in an open forest where wind speeds are higher than in a closed canopy forest so, again, careful interpretation is necessary. It has also been suggested that shrubs intercept pollen grains from the atmosphere and may selectively filter certain grains from the underlying surface (Tauber 1965). Thus, Tauber (1967) found an inverse relationship between *Salix* and *Fagus* pollen, the *Salix* filtering out the *Fagus* when the former is particularly abundant. Janssen (1966) found no such selective filtration but it seems likely that shrubs and herbs do play some part in at least smoothing out short-term variations in pollen accumulation. Because of all these problems, many investigators have emphasized the advantages of analyzing lake sediments in preference to deposits in bogs or small woodland clearings where pollen production at the site may confuse the interpretation of regional vegetation change.

Particular problems arise when a species may occur as an undercanopy shrub or as an upper canopy tree cover (e.g. *Corylus*); in full illumination pollen production by *Corylus* is far greater than when it forms an understorey, but such occurrences are often very difficult to distinguish in a fossil deposit (Andersen 1973). Consequently, *Corylus* pollen may sometimes be included in the arboreal pollen (AP) sum (Sec. 9.5) or sometimes it may be counted as a shrub (NAP).

9.2.3 *Pollen rain as a representation of vegetation composition*

Differences in pollen productivity and dispersal rates pose a significant problem for paleoclimatic reconstruction because the relative abundance of pollen grains in a deposit can not be directly interpreted in terms of species abundance in the area. It is necessary to know the relationship between plant frequency in an area and the total pollen rain from that plant species in order to use pollen data to calculate the actual composition of the surrounding vegetation. For example, a vegetation community composed of 10% pine, 35% maple, and 65% beech may be represented in a deposit by approximately equal amounts of pine, maple, and beech pollen, because of the differences in their pollen productivity and dispersal rates. The problem has been approached in two ways, both adopting the uniformitarian principle, "the present is the key to the past." These may be referred to as the adjustment method and the analog method.

9.2.3.1 The adjustment method. The first approach attempts to equate modern pollen rain data with statistics on the basal or crown area of a particular species (Davis 1963). If a simple relationship can be found, a direct method for converting fossil pollen rain statistics to former vegetation composition would be available. Davis (1963), following the work of Fagerlind (1952), proposed the use of a simple index, R, where, for tree species X,

$$R = \frac{\text{pollen of X as a percentage of AP counted}}{\text{percentage of land occupied by X}}.$$

Hence, for over-represented species, $R > 1$; for under-represented species, $R < 1$. If R values can be calculated from modern pollen rain in forests of known composition, they can then be used as adjustment factors in interpreting fossil pollen spectra.

Andersen (1970), in an exhaustive study of the problem, notes that R is not constant for any one species because it is dependent on the amount and type of other vegetation present and their relative pollen productivity. It is also very dependent on the size of the land area considered. He suggests the use of a relative pollen representation factor R_{REL}, which is the R value of one species divided by the R value of a reference species, such as *Fagus*; R_{REL} values are generally constant for each pair of species. From his studies, Andersen suggests that tree pollen spectra from within forests should be adjusted as shown in Table 9.2 to reflect vegetation composition accurately within 20–30 m of the sample site. Such an adjustment is shown in Figure 9.2 for a late Holocene pollen diagram from eastern Jutland, Denmark (Andersen 1973). Without correction factors, the diagram might have been interpreted as indicating the dominance of *Quercus* or *Betula* for most of the period (Fig. 9.2b). However, from Figure 9.2c it seems likely that the site was occupied by climax forest dominated by *Tilia* and *Fagus* at different times (stages II and IV), separated by periods of disturbance (stages I and III). The high frequency of *Betula* in stage III indicates destruction of the *Tilia* climax forest by fire, possibly as a result of human interference.

Several studies have examined the question of adjustment factors and

Table 9.2 Adjustment factors for pollen productivity (data from Denmark; after Andersen 1973).

Pinus, Quercus, Betula, Alnus	1 : 4
Carpinus	1 : 3
Ulmus, Picea	1 : 2
Fagus, Abies	1 : 1
Tilia, Fraxinus, Acer	2 : 1

Example: Pinus pollen is four times as frequent in arboreal pollen (AP) spectra as its frequency in the vegetation would suggest. Conversely, *Tilia, Fraxinus,* and *Acer* are under-represented in the arboreal pollen spectra. *Fagus* and *Abies* pollen as a percentage of AP accurately reflects their real proportion in the surrounding forest.

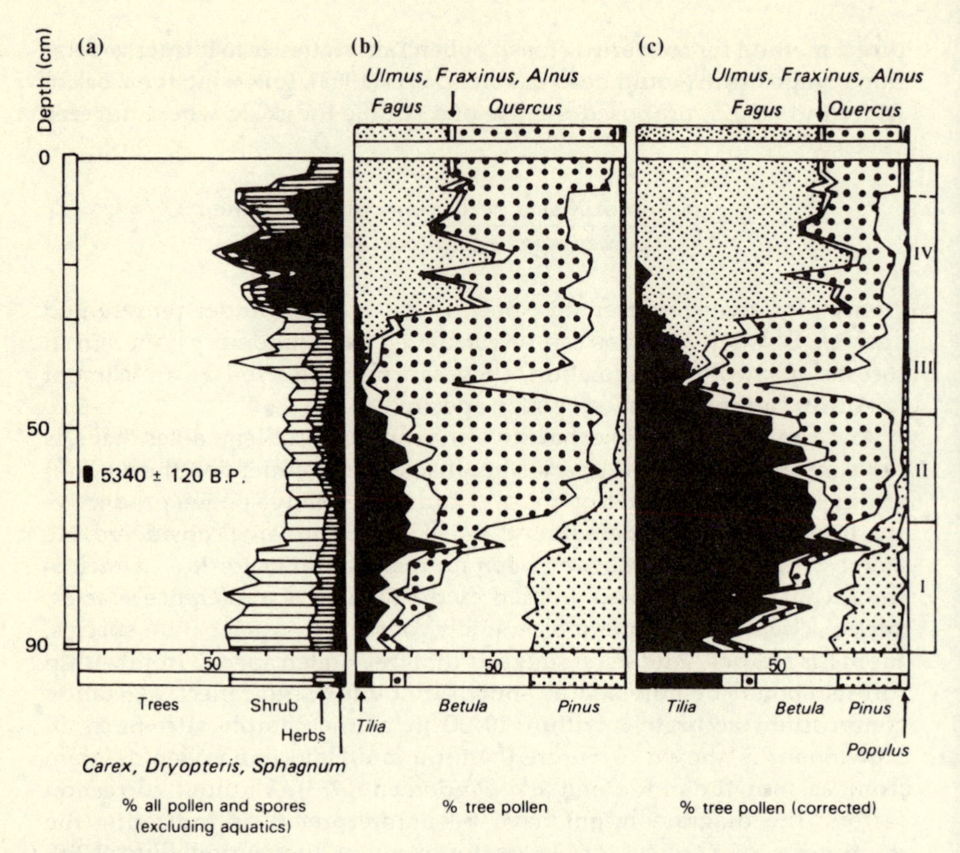

Figure 9.2 Pollen diagram from eastern Jutland, Denmark. (a) Total pollen and spore diagram; (b) tree pollen diagram; (c) corrected tree pollen diagram. Tree pollen counts were corrected (prior to the percentage calculations) by the factors shown in Table 9.2 (after Anderson 1973).

although the general picture for those species which are grossly over- or under-represented is clear, there is considerable variation in the actual adjustment factors calculated. For example, Livingstone (1968) found that the R_{REL} values calculated for a particular genus at different sites were very dependent on the actual percentage of the genus in the vegetation, the very factor on which the adjustment factors are supposed to shed light. *R* values, in some cases, varied by as much as an order of magnitude, which led Livingstone to conclude that the *R* values were so poorly understood that the uncertainty was probably greater than the total amount of postglacial change in the abundance of forest tree genera (in eastern Canada). Certainly, the adjustment factors are complex and differ from one environment to another; in more open sites, such as large lakes, or bogs, or in tundra environments, pollen representation will be significantly different than in densely forested sites (e.g. Maher 1963, Lichti-

Federovich & Ritchie 1968). However, much of the inconsistency in *R* values may be due to inadequate statistics on vegetation composition around the sample site and, indeed, inadequate knowledge about how large an area is an appropriate "source area" for pollen reaching the sample site (Janssen 1966, Livingstone 1968).

9.2.3.2 The analog method. The second approach to interpreting fossil pollen assemblages uses different modern vegetation communities as analogs of former vegetation cover; if pollen assemblages in the modern pollen rain resemble fossil pollen assemblages, the former vegetation is assumed to be similar to that of today. If similar pollen assemblages can not be found today, presumably no modern analog for the former vegetation cover exists (e.g. Ritchie 1976). However, there are a number of problems with this approach, notably the difficulty of sampling today a vast array of possible vegetation types to find a good analog and the fact that much "natural vegetation" in both the Old and New Worlds has been destroyed or greatly modified (Iversen 1949). Furthermore, it is likely that individual species in a vegetation community migrate into a new site at different rates so that initially, at least, the fossil pollen rain may reflect seral communities, whereas today the "natural" vegetation is closer to a climax state, and such transitory seral communities may no longer exist (Wright 1964, Birks 1981).

The analog method does possess great value in shedding light on the sensitivity of pollen rain as an index of vegetation cover. If modern vegetation communities cannot be distinguished in the associated pollen assemblage, there is little hope of using this approach to identify such communities in the fossil pollen rain. Lichti-Federovich and Ritchie (1968) have tackled this problem by analyzing modern pollen rain from over 100 lake surface samples along a north – south transect from southern Manitoba to Keewatin. This transect crossed grassland and broad-leaved deciduous forests to coniferous forests and tundra. Pollen spectra for each major vegetation (formation) zone were then examined to see if the regional vegetation types could be distinguished in the pollen rain (Fig. 9.3; see also Birks *et al.* 1975a). They conclude that there is a direct correlation between the pollen spectra and broad biogeographical zones, though some difficulties were apparent. Tundra was quite distinctive, characterized by low total pollen influx and a predominance of *Betula glandulosa*, *Pinus*, *Picea*, and *Alnus* in the AP rain. Forest tundra had high percentages of *Alnus* (20%) and conifer pollen (40%) but this was often hard to distinguish from the conifer forests proper, to the south. There, conifer pollen averaged 60% but subzones within the forest (open, closed boreal, or mixed forests) are not seen in the spectra. This lack of sensitivity may be due to the presence of "blind-spots" in the landscape (or "silent areas;" Davis 1967a) where certain species occupy part of the forest but

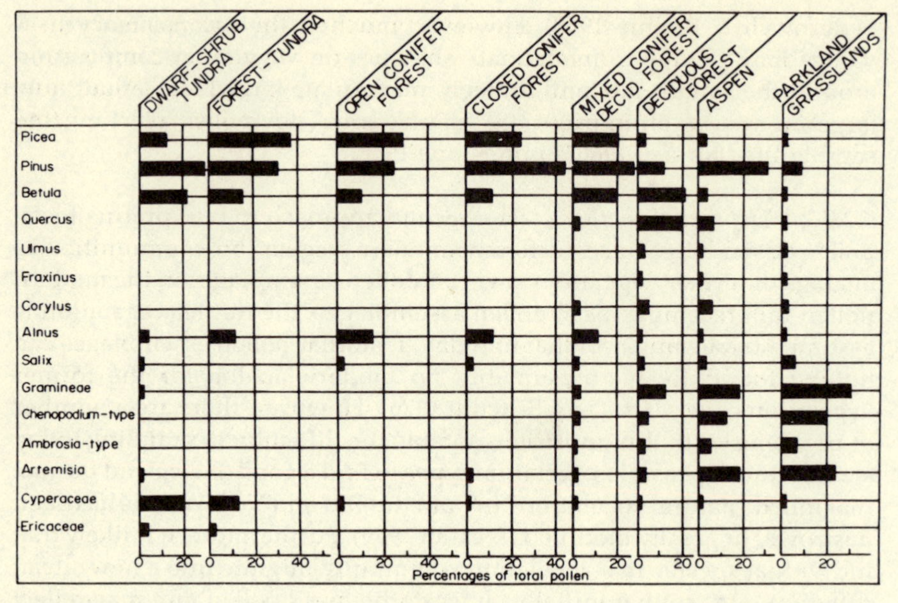

Figure 9.3 Average pollen percentages in surface lake muds from the main vegetational zones of western interior Canada. All frequencies expressed as percentages of total pollen. Each vegetation zone has a characteristic pollen assemblage (after Birks 1973).

do not contribute to the pollen accumulation because of poor productivity, dispersal, or preservation of their pollen (e.g. *Populus*). As the mixed deciduous – coniferous forests to the south are approached, more deciduous pollen is encountered (particularly ash, elm, and oak pollen), reaching a maximum in that zone. Similarly, herbaceous pollen increases to the south, reaching a maximum of over 60% herbaceous pollen in the grassland zone itself.

A similar approach was taken by Davis (1969) in interpreting a pollen diagram from New England. Pollen assemblages from surface lake sediments along a transect from the deciduous forests of southern Ontario to the tundra forest margin near James Bay, Quebec (Fig. 9.4a) were used to interpret a pollen record from southern New England (Fig. 9.4b). Modern-day analogs for former vegetation zones in New England were thus suggested (Fig. 9.4c).

Pollen rain in mountainous regions poses particular problems of interpretation, because vegetation communities may be confined to narrow zones along mountainsides (Maher 1963). In parts of South America, for example, vegetation may grade all the way from equatorial rainforest below 500 m to tundra-like *páramo* (or the drier *puna*) above 3500 m, over a horizontal distance of less than 100 km. Nevertheless, modern pollen rain studies demonstrate that very good discrimination is possible between different vegetation zones (Salgado-Labouriau 1979).

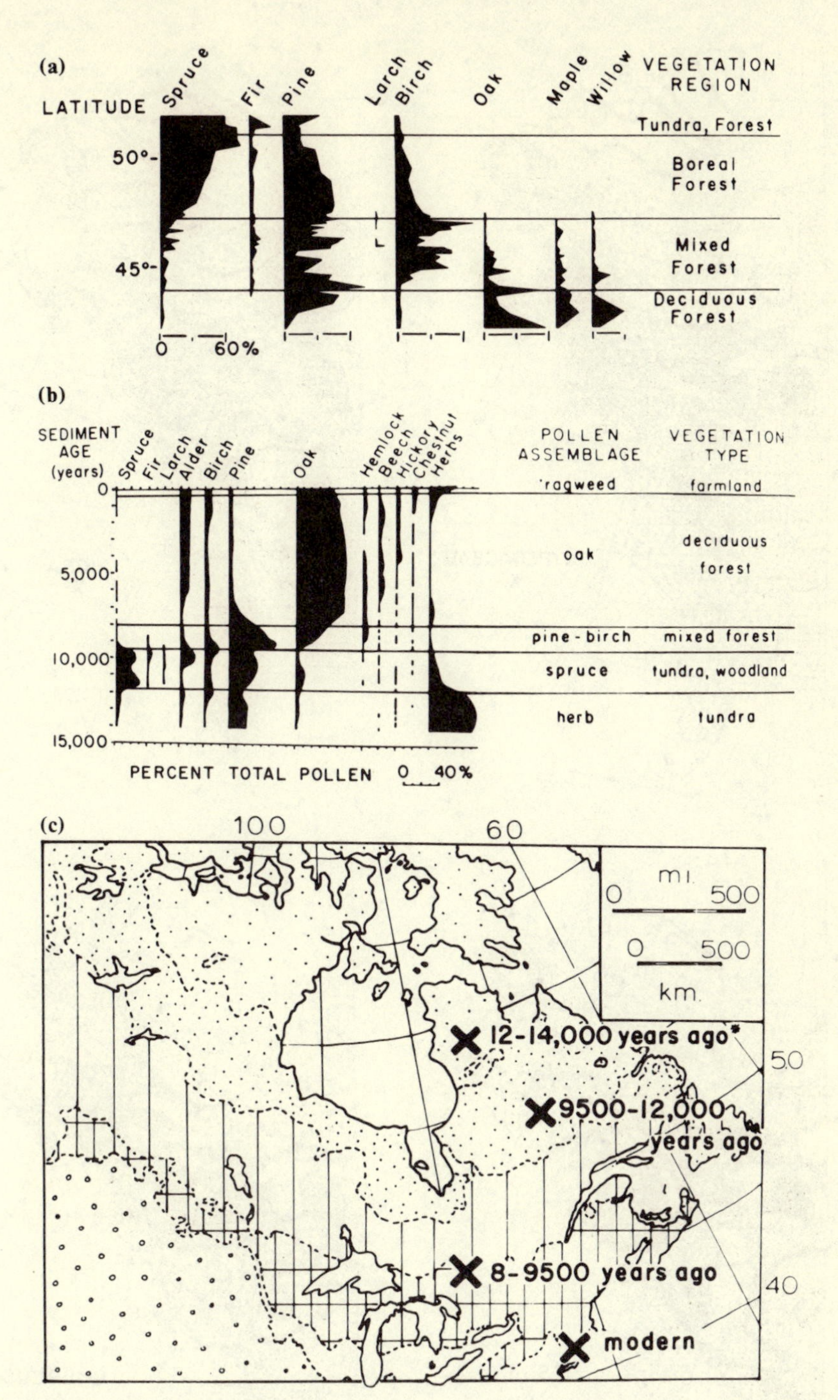

Figure 9.4 (a) Pollen assemblages in surface samples collected along a north–south transect through the main vegetation zones of Canada (shown in (c)). (b) Fossil pollen assemblages in a sediment core from southern New England. Principal pollen assemblages and the vegetation zones of which they are characteristic are shown at right. (c) Localities in Canada where surface pollen assemblages today (shown in (a)) resemble fossil pollen assemblages in the sedimentary record shown in (b). Age of analogous fossil pollen spectra is indicated for each locality (after Davis 1963).

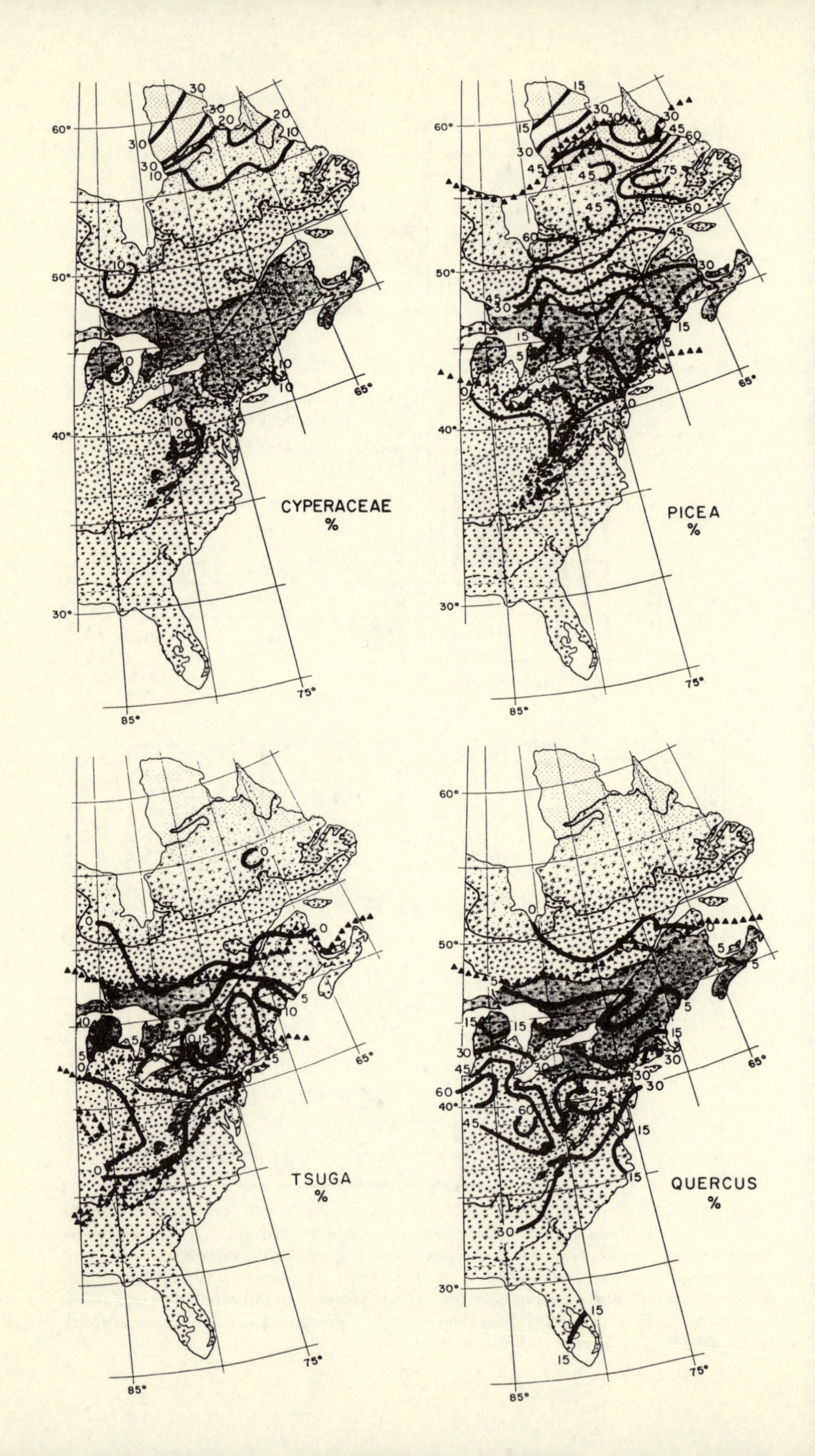
CYPERACEAE
%
PICEA
%
TSUGA
%
QUERCUS
%
60°
50°
40°
30°
85°
75°
65°

Pollen from lower vegetation zones is commonly carried upwards to higher elevations by daytime upslope winds. However, with increasing elevation, gradient winds predominate, so that there may be an upper limit to upslope pollen transport (Markgraf 1980). Pollen from high elevations is not dispersed very far beyond the high altitude vegetation zone (Hamilton & Perrott 1980). Characteristic pollen rain assemblages corresponding to different elevations can thus be identified and usd to reconstruct altitudinal changes in vegetation zones through time (e.g. Salgado-Labouriau *et al.* 1978). Such changes may be converted to paleoclimatic estimates if modern climatic controls on vegetation boundaries are known.

Extensive studies of modern pollen rain and modern vegetation have been made by R. B. Davis and Webb (1975), Webb and McAndrews (1976), and Webb *et al.* (1978) for eastern, central, and north-eastern North America, respectively. In these studies, the amount of pollen of a particular genus was expressed as a percentage of the total pollen accumulation at each site; isopolls (lines of equal percentage pollen representation) were thus mapped for each major genus and these can be superimposed on, and compared directly with, maps of the regional vegetation (Fig. 9.5). In general, the zero isopoll corresponds well with the range limit of the taxa and isopoll maxima coincide with the zone of maximum frequency of the taxa in the vegetation. A major exception to this is *Picea*, which reaches maximum relative values in the forest tundra zone and the zero isopoll lies well beyond the range of the genus (Fig. 9.5). Absolute pollen frequency, calculated as the number of grains per cubic centimeter of upper lake sediment, also reflected the general pattern of vegetation, but more extensive work on this question is required in view of the probable difference in time represented by 1 cm^3 of sediment from, for example, the tundra zone compared to an equivalent sample from the southern USA.

One of the most detailed and convincing demonstrations of the correspondence between modern vegetation and modern pollen is that of Webb (1974), who compared pollen content of the upper 2 cm of the 64 lake cores from throughout Michigan with detailed forest inventory records. Maps showing the percentage distribution of individual tree genera were compared with corresponding isopoll maps (Fig. 9.6). Clearly, the spatial distribution of pollen of each genus closely resembles the percentage cover of that genus in the state (cf. Prentice 1978). Furthermore, the forest composition (i.e. the covariation of individual

Figure 9.5 Isopolls of *Cyperaceae* (sedge), *Picea* (spruce), *Tsuga* (hemlock), and *Quercus* (oak) based on modern pollen samples, compared to present-day vegetation zones (after R. Davis & Webb 1975). The base map is divided into the following vegetation zones (from north to south): tundra, with forest tundra in the south; boreal forest; conifer–hardwood forest; deciduous forest; oak–hickory–pine forest; and southern mixed forest.

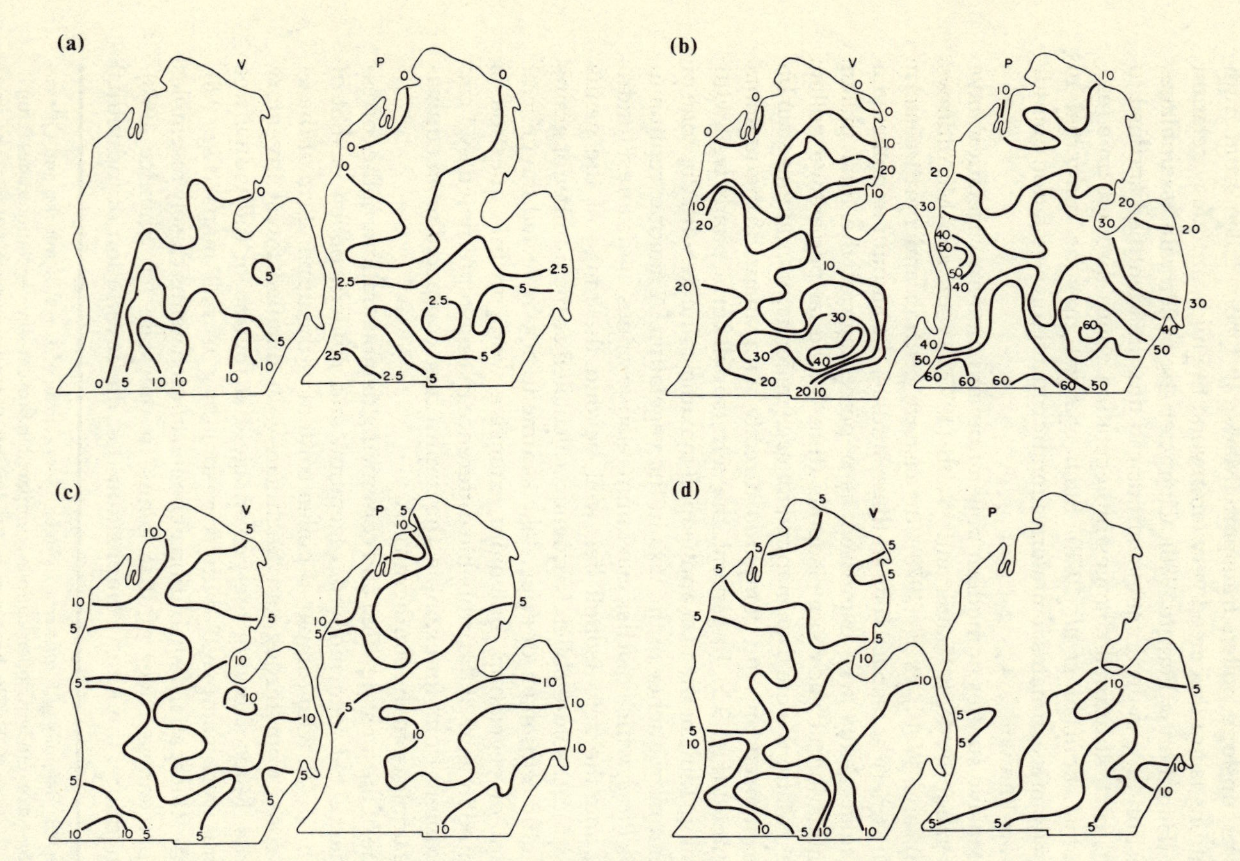

Figure 9.6 Maps of the percentages of (a) hickory, (b) oak, (c) elm, and (d) ash in the vegetation of Michigan (V) compared to the percentages of pollen (P) from the same trees in the modern pollen rain (based on arboreal pollen sum). Modern pollen data based on analysis of uppermost lake sediments.

genera) when summarized by principal components is also revealed by principal component analysis of the pollen rain data (Fig. 9.7).

Another approach to the analog method involves calculating pollen influx, the absolute quantity of pollen falling on the sample site in a given period of time (grains cm^{-2} a^{-1}). It has long been known that tundra, for example, produces far less pollen per unit area than coniferous forest, and this is reflected in the pollen rain of the two areas. If pollen influx values are characteristic of each vegetation formation, they can be used in the interpretation of the fossil record. The principal problem here is to obtain a sample which spans a known period of time in order to estimate influx per square centimeter per year. Ritchie and Lichti-Federovich (1967) used air sampling devices at a network of stations across northern Canada to estimate the pollen accumulation rate during a specific time interval. Their results gave values of only 5 grains cm^{-2} a^{-1} in an arctic block-field to an average of 44 grains cm^{-2} a^{-1} in a sedge-moss environment, 335 in dwarf shrub communities, ~1100 in forest tundra, and ~5000 in the northern coniferous forest zones. Although these values do not include pollen from rainout, they are similar to fossil pollen accumulation rates in Rogers Lake, Connecticut, during late and Postglacial times; pollen influx increased from 600–900 grains cm^{-2} a^{-1} to over 9000 grains cm^{-2} a^{-1} during a period which was interpreted as a transition

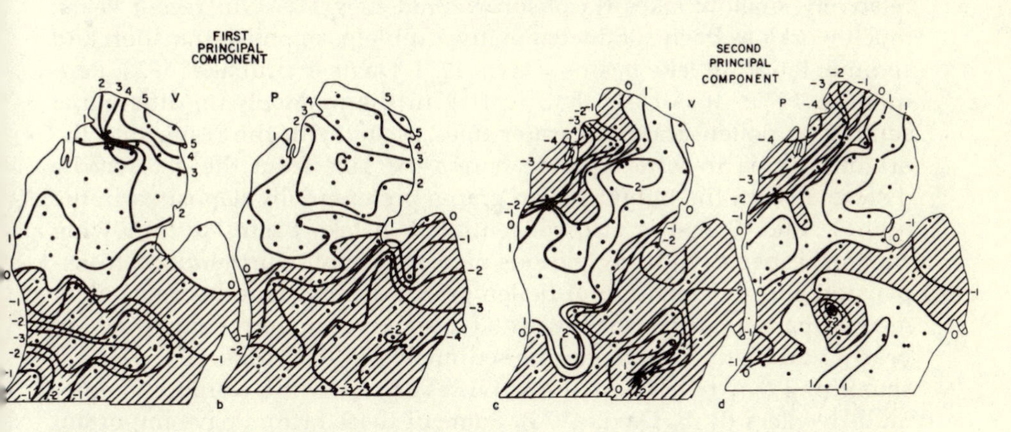

Figure 9.7 Map of the first two principal components (PC) of vegetation percentages (V) and pollen type percentages (P) in Michigan. The principal components of vegetation reflect major vegetation formations in the state. PC1 has a distribution reflecting the change from deciduous forest in the south to mixed coniferous–hardwood forest in the north. It accounts for 25% of variance in the original data set. PC2 depicts primary divisions within the two major vegetation formations, differentiating the northern hardwoods from the pine–birch–aspen forests in the north and the beech–maple and elm–ash–cottonwood forests from the oak–hickory forests in the south. It accounts for a further 35% of variance in the original data set. The first two principal components of pollen mirror the principal components of vegetation, indicating that the spatial pollen data may be used as a reliable indicator of vegetation distribution. Lakes shown by dots (after Webb 1974).

from tundra to coniferous forest (Davis & Deevey 1964). The modern pollen influx values provide an additional line of support for such an interpretation. This example is, of course, fairly straightforward, but on many occasions simply looking at relative (percentage) variations in the total pollen assemblage can lead to erroneous conclusions and pollen influx data can help to clarify the situation (see Sec. 9.5, below).

These studies clearly indicate that in spite of all the problems involved in pollen dispersal, preservation, and accumulation (Sec. 9.3), the broad geographical patterns of vegetation are closely mirrored by the composition and amount of the pollen rain. We may thus conclude that fossil pollen has great potential for the reconstruction of former vegetation cover and hence, perhaps, of paleoclimatic conditions.

9.3 Sources of fossil pollen

Pollen is an aeolian sediment that will accumulate on any undisturbed surface. Pollen falling on sites where organic or inorganic sediments are accumulating will be incorporated into the sediments and become part of the stratigraphic record. Pollen has thus been recovered from peat, lake sediments, alluvial deposits, marine sediments, ice, and firn. It is most commonly studied in peat deposits from bogs and marshes and from relatively shallow lakes (Jacobson & Bradshaw 1981). In recent years, much work has been conducted on the problems of pollen transport and sedimentation in lake basins (Davis 1973, Davis & Brubaker 1973, Pennington 1973). It is clear that, just as in the atmosphere, differential settling of pollen grains in water does occur, with the result that the original ratios in which pollen enters the lake from the air may be distorted, with the lighter pollen grains preferentially deposited in the littoral zone. Pollen is also concentrated in lake basins by inflowing streams, especially during periods of heavy runoff. Furthermore, resuspension and redeposition of pollen grains during periods of turbulent mixing, particularly in shallow water, also occur, thereby smoothing out yearly variations in pollen and sediment inputs to the lake. Further smoothing may result from the activities of burrowing worms and other mud-dwellers (R. B. Davis 1974). Some of these factors have important implications for studies of interannual variations of pollen in varves (e.g. Swain 1973, 1978) and indicate the importance of sampling lake sediments from the deepest parts of lake basins. For comparative purposes, lake basin samples should also be of similar morphometry. In spite of these difficulties, lake sediments are preferred by most palynologists for the reconstruction of past vegetation and climate.

Pollen has also been recovered from alluvium (Grichuk 1967, Solomon *et al.* 1982), archaeological sites (van Zeist 1967), wood rat middens (King & Van Devender 1977), and coprolites (fossilized fecal matter of animals;

Martin *et al.* 1961). Until recently the majority of pollen records spanned only the last 12 000 years or so, but there are now a number of studies of continuous pollen records back to Mid-Wisconsin times or even to the last interglacial (marine isotope stage 5e) (e.g. Tsukada 1966, Hammen *et al.* 1971, Kershaw 1978, Wijmstra 1978, Woillard 1978, Heusser & Shackleton 1979, Adam *et al.* 1981). The development of more sophisticated coring equipment will no doubt lead to many more long, continuous stratigraphic records.

9.4 Preparation of the samples

In order to isolate pollen grains and spores from the matrix of organic or inorganic sediment, rigorous chemical treatment by hydrochloric, sulfuric, and hydrofluoric acid is generally required, as well as acetolysis by a mixture of acetic anhydride and sulfuric acid. For the complete laboratory procedure, the reader is referred to Faegri and Iversen (1975) or Moore and Webb (1978). Removal of the matrix enables the remaining pollen grains and spores to be seen clearly when stained and mounted on slides for microscopic analysis (Fig. 9.8). Generally, the original core is sampled at intervals of a few centimeters and slides are prepared of pollen and spores at each level. These are then scanned and the number of

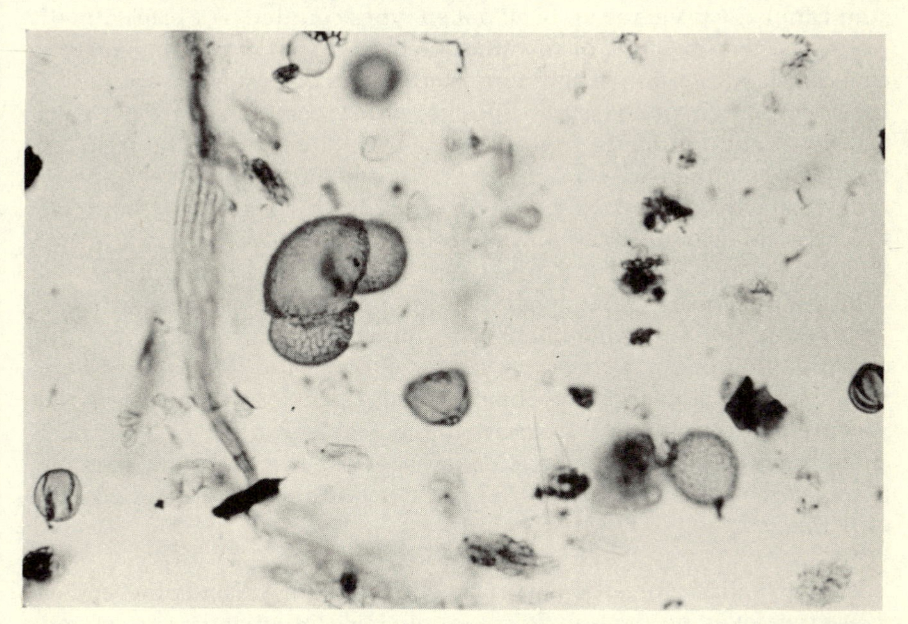

Figure 9.8 Photomicrograph of pollen grains (photograph kindly provided by W. Patterson).

different grains in each sample are noted. The total number of grains counted at each level would depend on the purpose of the study and the source of material being studied (Moore & Webb 1978), but usually at least 200 grains are counted.

For absolute pollen influx calculations (Sec. 9.5) it is necessary to know accurately how the pollen of the slide(s) counted related to the total pollen content of the sample from the level being considered. The most widely used method is to add a known quantity of exotic pollen or spores (e.g. *Eucalyptus* or *Lycopodium*) to the sample initially and then to count the number of these grains which occur on the final slide preparation. The ratio of exotic pollen counted to the number of exotic pollen added originally can be used to estimate the total pollen content of the original sample (Stockmarr 1971, Bonny 1972).

9.5 Pollen analysis of a site: the pollen diagram

Pollen data from a stratigraphic sequence are generally presented in the form of a pollen diagram composed of pollen spectra from each level sampled (Fig. 9.2). A pollen spectrum is the number of different pollen grains at a particular level. Generally, at least 200 grains are counted at each level and the number of grains of each genus are expressed as a percentage of the total pollen count (the pollen sum). Actually, the pollen sum is not always made up of all pollen types counted. For paleoclimatic purposes, the objective of the analysis is to depict regional vegetation change, so both arboreal and non-arboreal (shrub and herb) species are included in the pollen sum. Species which commonly grow in wet (lowland) environments around the sample site are usually excluded, though difficulties arise when a particular genus has different species (not easily distinguished by their pollen) which grow in both wet lowland and drier upland environments (e.g. *Picea mariana*, black spruce, and *Picea glauca*, white spruce, respectively; Wright & Patten 1963).

In the pollen diagram, changes in the percentage of one species are assumed to reflect similar changes in the vegetation composition (due consideration being given to the factors of over- and under-representation discussed above). The problem with this is that apparent changes in species occurrence may occur in the percentage data as a result of changing receipts of pollen from *other* species. The following example, from Faegri and Iversen (1975, p. 160) succinctly summarizes the difficulty:

> If we visualize a forest consisting of equal parts of oak and pine, and we use the pollen production figures quoted, we find that the corresponding spectrum will contain 15 percent oak, 85 percent pine. If beech is

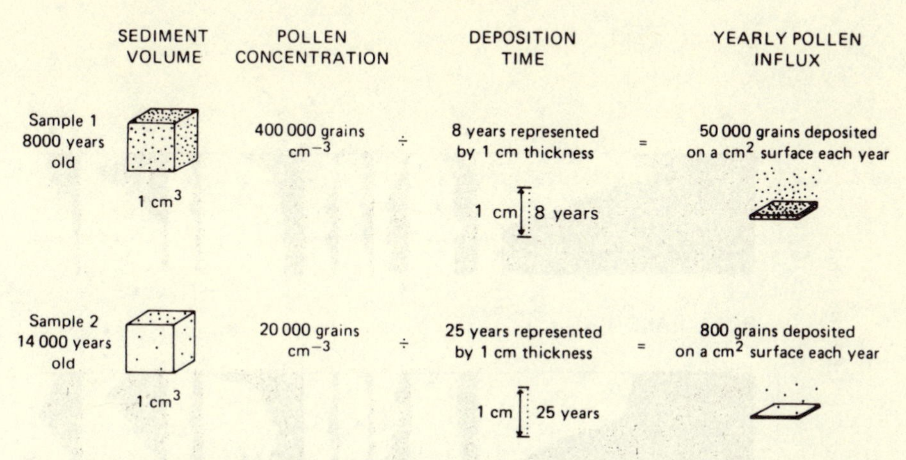

Figure 9.9 Examples of pollen influx calculations from measurements of pollen concentration and the rate of accumulation of the sediment matrix (after Davis 1963).

> substituted for pine (apart from the botanical improbability of that succession) the same quantity of oak will give 60 percent of the pollen as against 40 percent beech. If the beech is then replaced by a tree, e.g. *Acer* spp. or *Populus balsamifera*, which is scarcely, or not at all, registered in the spectra, we shall find almost 100 percent oak pollen, although the quantity of oak has not changed at all. It is necessary to take into account not only the curve under discussion, but the others as well.

To circumvent such problems, palynologists in recent years have attempted to calculate pollen influx, the number of grains accumulating on a unit of the sediment surface per unit time (Fig. 9.9). To do this, the sediment accumulation rate must be known. Two methods have been used to obtain sedimentation rates: radiocarbon dating and the identification of settlement horizons in lake sediments from North America. In the former case, ^{14}C dates are made at close intervals to establish a mean sedimentation rate. In the second case, the rise of *Ambrosia* (ragweed) pollen at the time of colonial settlement (when forests were being cleared and herbaceous plants were increasing rapidly in numbers) is clearly seen in North American lake sediments (Fig. 9.10); since the dates of settlement are known, sedimentation rates are readily calculated (Bassett & Terasmae 1962, McAndrews 1966, Davis *et al.* 1973). Clearly the latter method is only useful for obtaining modern pollen influx statistics; ^{14}C dating enables influx to be calculated throughout the Holocene, and beyond.†

†In some cases, it is not possible to obtain reliable ^{14}C dates to calculate sedimentation rates. In such cases, pollen concentration (grains cm^{-3} or grains gm^{-1}) can be used to correct percentage values (e.g. Salgado-Laboriau 1979).

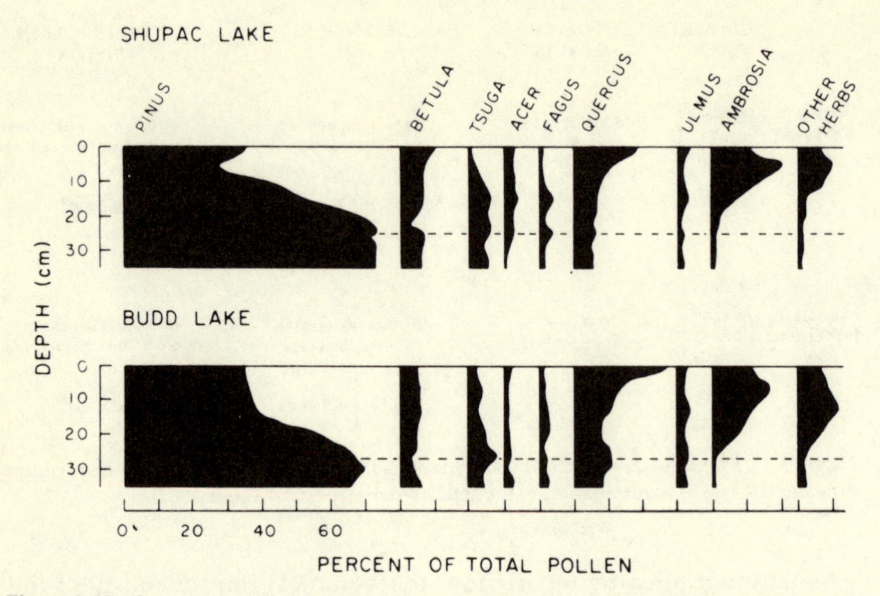

Figure 9.10 Summary pollen diagrams from two lakes in southern Michigan, illustrating the increase in *Ambrosia* (ragweed) pollen following colonial settlement. The ragweed pollen rise is a useful stratigraphic marker in North American palynological studies (after Webb 1973).

Pollen influx can often clarify a stratigraphic record, as illustrated by Davis (1967b) in a core from Rogers Lake, Connecticut. Figure 9.11a shows the pollen (percentage) diagram using total pollen from terrestrial plants as the pollen sum. Before 12 000 years BP *Pinus* was the most abundant tree pollen; subsequently *Pinus*, *Picea*, and *Quercus* increased, followed by a decline of *Quercus* to a minimum around 10 000 years BP. At this time, *Pinus*, *Picea*, and *Alnus* increased in frequency and by 9000 years BP *Pinus* had become the dominant pollen type. One interpretation of this sequence would be that gradual warming occurred after 12 000 years BP, enabling warm-loving genera such as *Quercus* to migrate into the area; a cold interval centered at 10 000 years BP led to a decline of oak and an increase in pine, spruce, and alder (Davis & Deevey 1964). The conundrum here is the high amount of pine early in the Postglacial period; elsewhere pine is absent at this time and is considered to be a slow migrant into recently deglaciated terrain.

The Rogers Lake record can be re-appraised by considering pollen accumulation rates. Figure 9.12 shows the sedimentation rates based on ^{14}C dates and the resultant estimates of pollen influx in grains per square centimeter per year. This enables the entire diagram to be replotted as shown in Figure 9.11b. From this figure, it is clear that before 12 000 years BP very little pollen at all was deposited. Arboreal pollen influx was extremely low, though on a percentage basis (Fig. 9.11a) it would appear

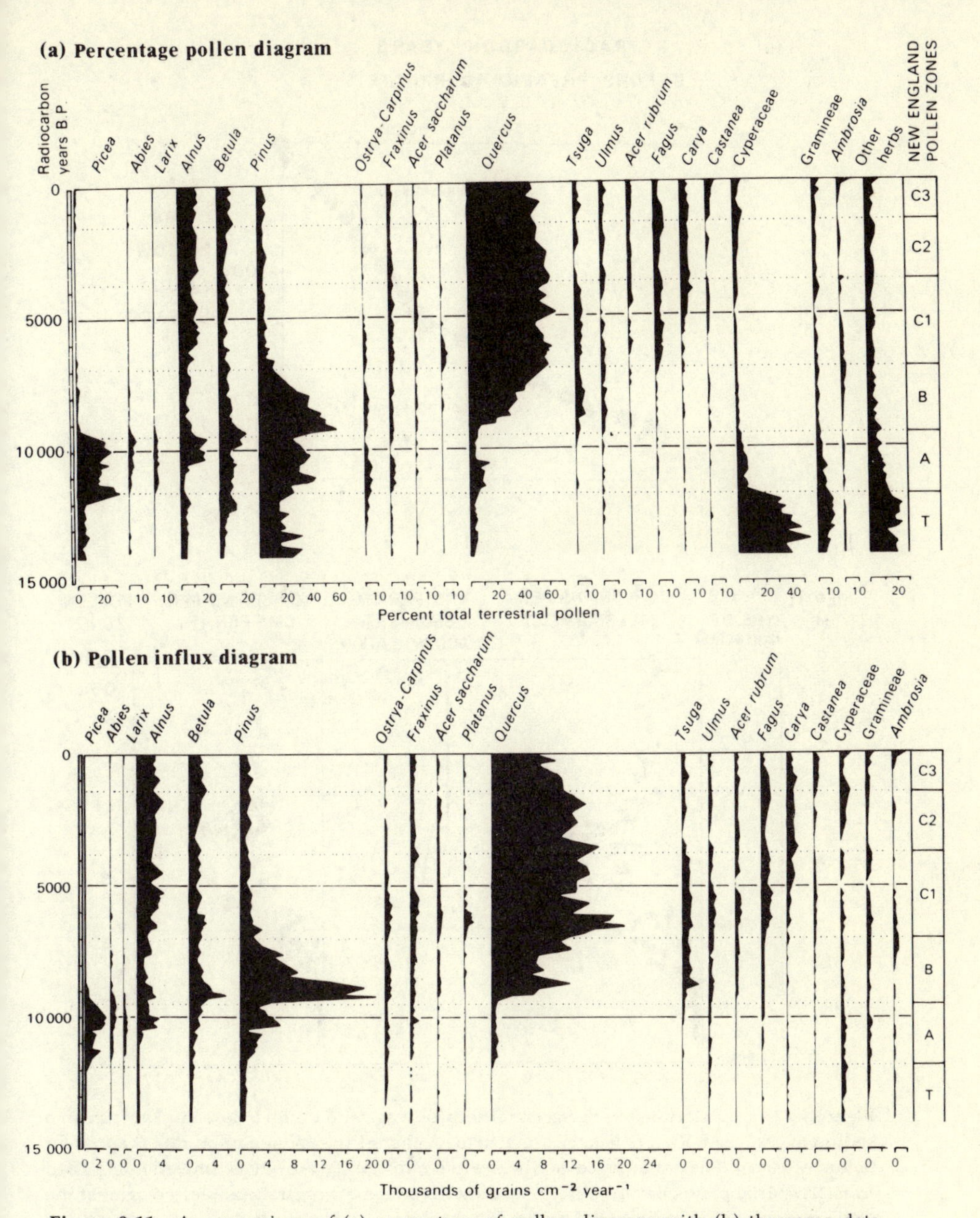

Figure 9.11 A comparison of (a) percentage of pollen diagram with (b) the same data expressed in terms of pollen influx (grains per square centimeter per year). Ordinate is linear for age, not depth. Percentages are calculated as percentages of total pollen from terrestrial plants. Accumulation rates calculated by dividing the estimated number of grains per sample by the estimated number of years necessary for their accumulation (see Fig. 9.9). Significant differences are apparent between the lower part of each diagram because the accumulation rate changes around the C1/B pollen zone boundary. This is explained in Figure 9.12. Two accumulation rates for each type are shown in (b) in samples from the zone of overlap of the two equations used to estimate the number of years per centimeter of sediment thickness (see Fig. 9.12). These are shown by the shaded and unshaded silhouettes (after Davis 1967b).

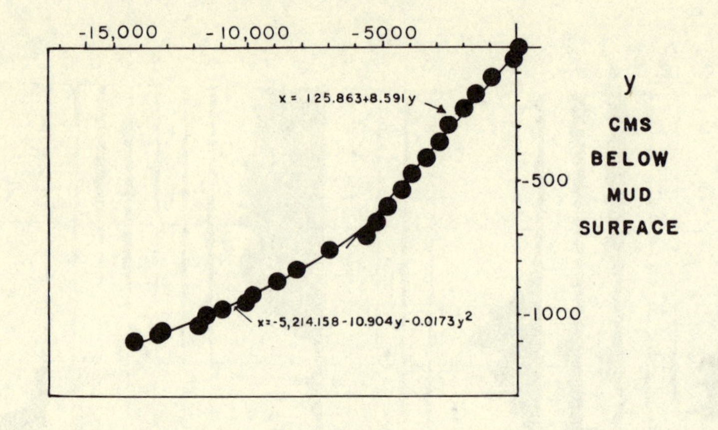

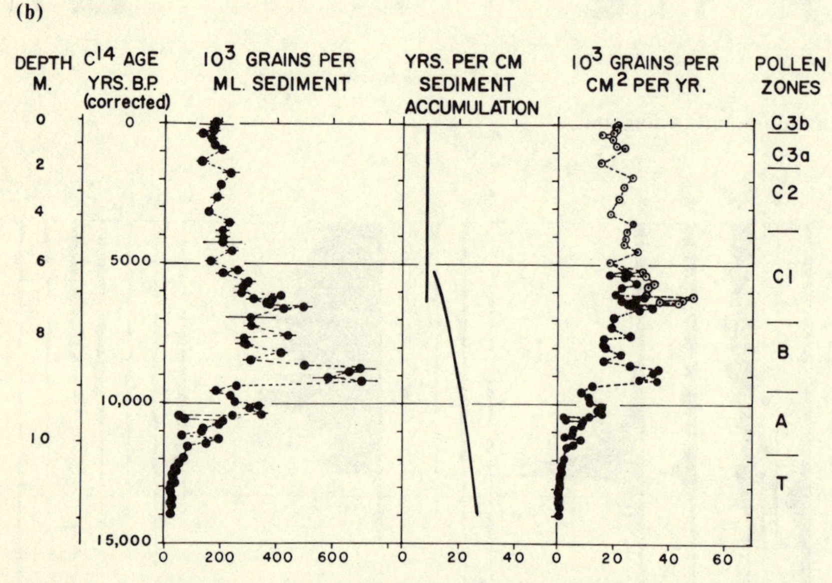

Figure 9.12 (a) Relationship between radiocarbon age and depth below mud surface in a sediment core from Rogers Lake, Connecticut. A change in sedimentation rate is apparent below ~700 cm. The equations describe the two rates. (b) To the left, estimated total pollen from terrestrial plants per milliliter of sediment (pollen concentration) plotted against the estimated sample age; horizontal bars = 95% confidence intervals. In the center column, the estimated number of years represented by a centimeter thickness of sediment (i.e. the derivatives of the curves shown in (a)) are plotted against the estimated age of the sediment. To the right are estimates of the rates of total pollen accumulation at various times in the past, derived by dividing the number of grains per cubic centimeter by the estimated number of years represented by the sample (pollen influx). Black dots represent estimates based on the derivative of the polynomial fitted to the *older* radiocarbon-dated samples; open dots represent estimate based on the derivative of the straight line fitted to the *younger* samples. Area of overlap gives rise to two possible estimates of pollen influx, shown on the right and in Figure 9.11b (after Davis 1967b).

that trees were a dominant feature of the vegetation at this time. In reality, the tree pollen falling at the site was probably wind-blown from forests to the south. Between 12 000 and 10 000 years BP coniferous trees increased in frequency, perhaps reflecting the development of open spruce woodland at this time. Note that the pollen of temperate deciduous trees remained more or less constant at this time; the apparent decline in oak around 10 000 years BP, seen in the percentage diagram, is not apparent, and it would seem that oak frequency remained steady until 9000 years BP when higher temperatures favored its expansion. Changes in oak shown in the percentage diagram are thus a reflection of increasing deposition rates for coniferous tree pollen ~10 000 years ago and not of a climatic oscillation, as previously suggested by Davis and Deevey (1964).

These differences in interpretation result from the fact that before ~8000 years BP total pollen influx varied considerably; 14 000 years ago influx averaged only ~1000 grains $cm^{-2}\ a^{-1}$, but by 9000 years BP it had risen to ~40 000 grains $cm^{-2}\ a^{-1}$. During the last 8000 years (i.e. once the area became dominated by forest) pollen deposition rates remained relatively steady at around 20 000–25 000 grains $cm^{-1}\ a^{-1}$. Consequently, the major differences in interpretation between percentage and influx diagrams occur in the period before 8000 years BP, when drastic changes in vegetation and pollen influx occurred as tundra was replaced by park tundra, woodland, and forest. It is in this period that absolute pollen influx values are of most assistance in clarifying the pollen stratigraphic record (Davis 1967b).

9.5.1 Zonation of the pollen diagram

Pollen diagrams contain a large amount of information on the covariance of different pollen types through time. In order to facilitate comparison between different sites (diagrams) it is common practice to subdivide the stratigraphic record into pollen zones; these are biostratigraphic units defined on the basis of the characteristic fossil pollen assemblage. Generally pollen zones contain a homogeneous assemblage of pollen and spores, but some investigators may recognize a zone which is characterized by abrupt changes. Needless to say, the definition of what constitutes a zone is a rather subjective decision and may not be agreed upon by different scientists. Human nature also compels us to look for correlations with zonations previously "identified," thereby reinforcing systems which may not justify such blind faith! To avoid these problems, a number of more objective, computer-based methods have been suggested, ranging from simple sequential correlation (Yarranton & Ritchie 1972) to informational analysis without any stratigraphic constraint (Dale & Walker 1970) to stratigraphically constrained, agglomerative, and divisive dissimilarity techniques (Gordon & Birks 1972). These techniques are capable of identifying both major and minor zone boundaries (i.e. of defining zones and subzones) and permit objective comparisons to be made between sites. Thus, if similar local pollen assemblage zones can be identified over a large geographical area, it may be possible to define pollen assemblage zones, reflecting *regional* vegetation changes, which may be of paleoclimatic significance (Gordon & Birks 1974, Birks & Berglund 1979). Objective computer-based zonation can also be applied to other variables in a sedimentary sequence (e.g. macrofossils, diatoms, sediment characteristics) to shed further light on the major climate-related features of the stratigraphic record (Birks 1978). For example, in a zonation of plant macrofossils and pollen data from Kirchner Marsh, Minnesota, Gordon and Birks (1972) found that

> In some instances the pollen and the macrofossil zonations correspond, in other instances they do not. As pollen tends to reflect the regional and upland vegetation and macrofossils tend to reflect the local or site vegetation, correspondence between the two suggests major vegetational changes both regionally and locally. Changes in the macrofossils that are seemingly independent of the pollen stratigraphy imply local changes only, whereas changes in the pollen stratigraphy that do not correspond to changes in the macrofossils may reflect forest successional and other non-climatically induced changes (Birks 1978 p. 118).

There would seem to be great potential for further work along these lines in the future.

9.6 Mapping vegetation change: isopolls and isochrones

Studies of modern pollen rain and modern vegetation, discussed in Section 9.2.3, above, indicate that there is a fairly good spatial correspondence between them. Maps of pollen data can reproduce broad-scale patterns of individual taxa over large areas (R. B. Davis & Webb 1975, Webb & McAndrews 1976). Studies such as these paved the way for synoptic mapping of vegetation distribution at discrete periods in the past. This approach was first suggested by Szafer (1935), who defined the term isopoll as lines of equal percentage representation of a particular pollen type in the pollen sum. Using isopolls, Szafer constructed maps showing the distribution of beech and spruce across East Germany and Poland at five intervals from late-glacial to late Holocene time. However, Szafer's time framework was speculative since, at that time, there was no means of accurately dating organic material. It was only with the extensive use of ^{14}C dating that identical stratigraphic horizons could be identified (by interpolation between dates) at sites over large geographical areas, thus facilitating the preparation of time-sequential maps, such as those compiled by Birks *et al.* (1975b), Birks and Saarnisto (1975), and Bernabo and Webb (1977). As an example, Figure 9.13 shows isopolls of spruce (*Picea*) and oak (*Quercus*) over north-eastern North America at intervals from 11 000 to 500 years BP. Between 11 000 and 9000 years BP spruce pollen decreased over a wide area in the north and was replaced by pine pollen. By 7000 years BP spruce pollen had essentially disappeared from the United States; if comparisons with modern pollen rain studies are valid (Fig. 9.4), we might place the southern boundary of the genus at this time just north of the USA/Canada border. During the same period, isopolls of oak show a rapid increase in the relative amount of oak pollen across most of the region. In this case, modern pollen rain studies (Fig. 9.5) would suggest that the northern boundary of oak was close to the USA/Canada border by 7000 years BP. By 4000 years BP the relative proportion of oak pollen had decreased, perhaps due to a slight reversal of the migration pattern. Spruce pollen percentages also decreased slightly in eastern Canada during this interval. During the period 4000–500 years BP spruce percentages have increased over most of Canada whereas oak isopolls indicate a general southward shift in the limits of the genus.

Isopoll maps can be re-interpreted to show the migration of a particular genus or ecotone through time by isochrones (equal time lines). Figure 9.14a, for example, shows the location of the 15% spruce isopoll at

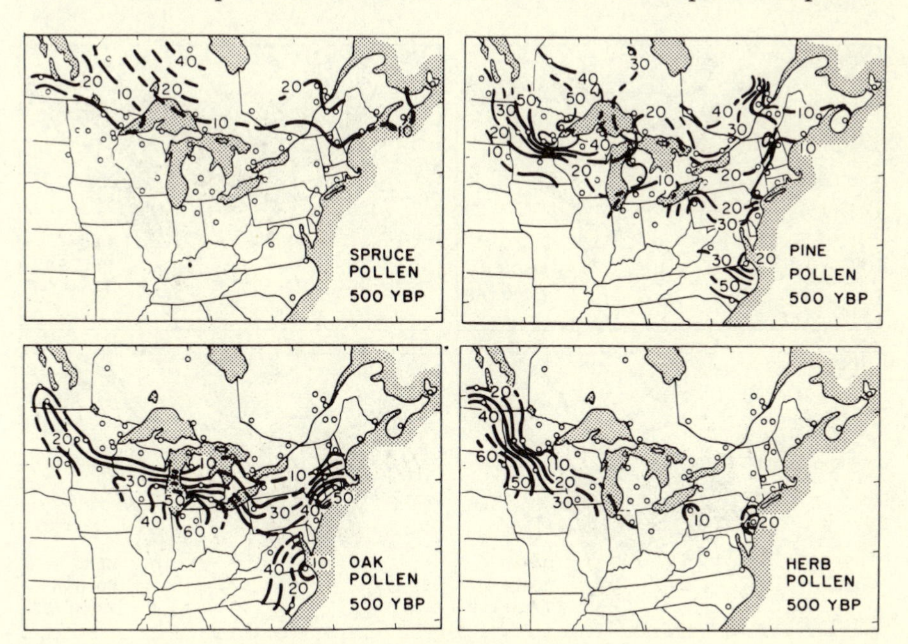

Figure 9.13–*caption on p. 315.*

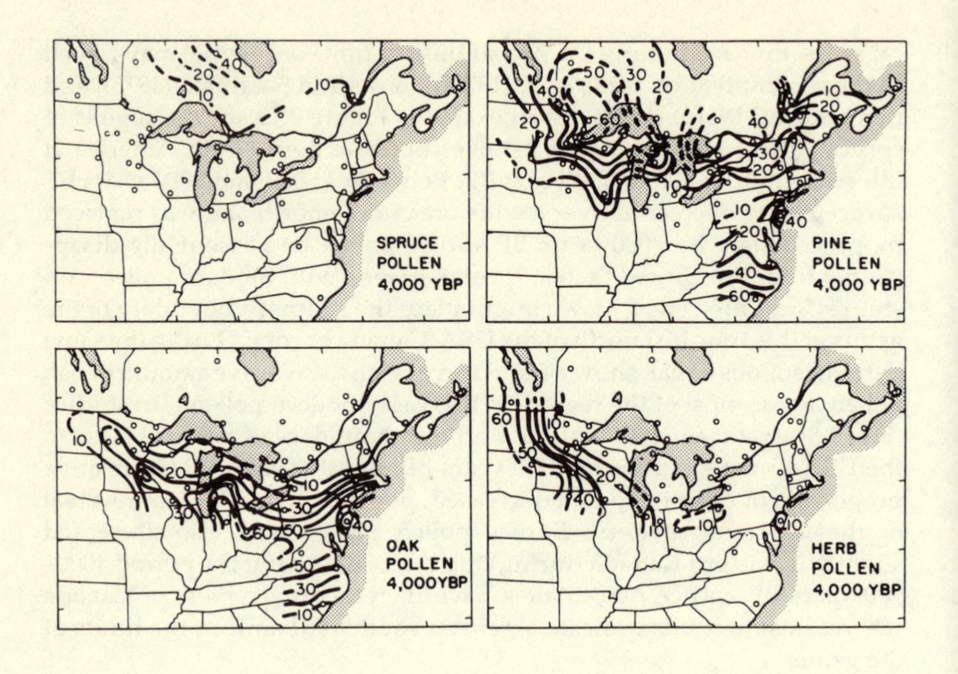

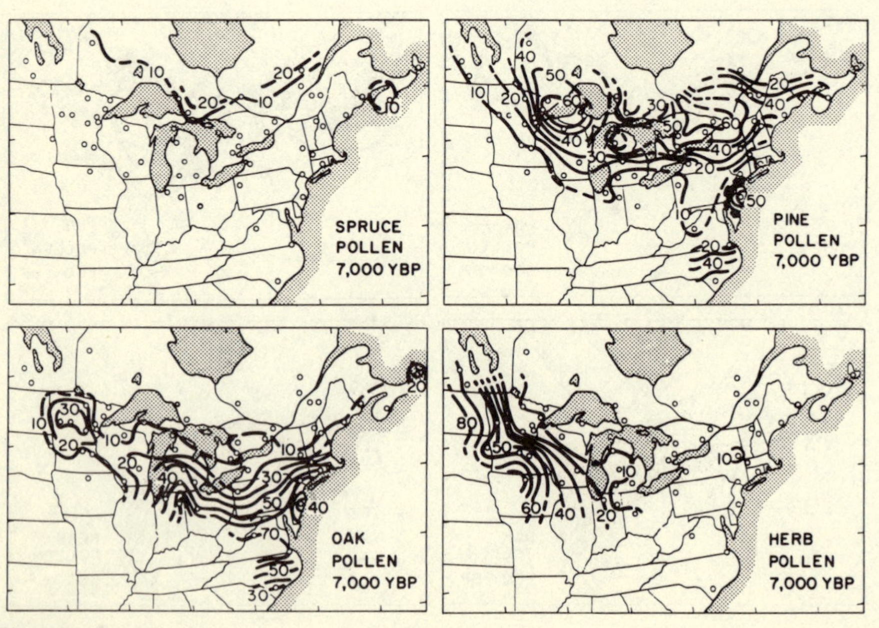

Figure 9.13–*continued*

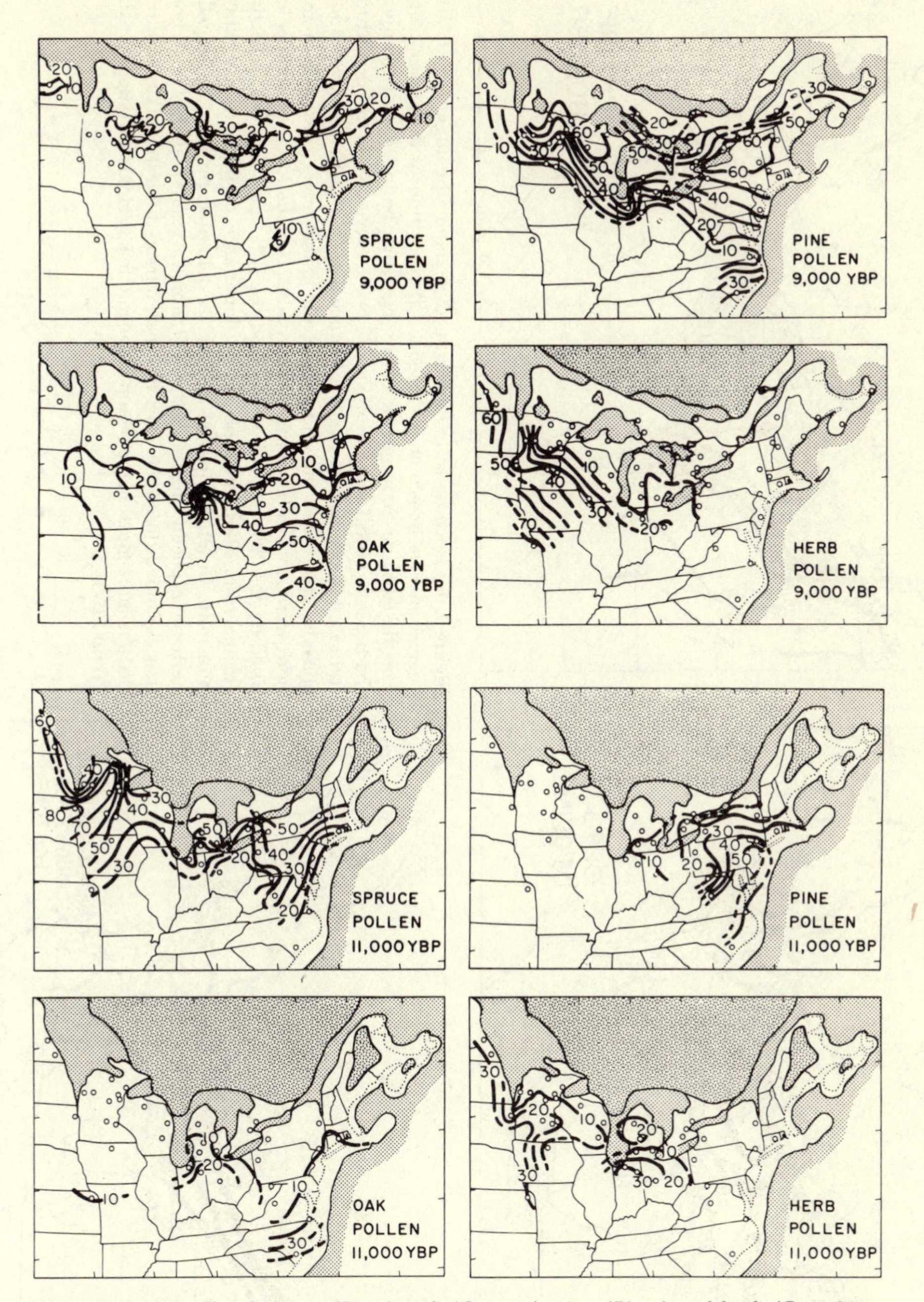

Figure 9.13 Isopolls of spruce (*Picea*), oak (*Quercus*), pine (*Pinus*) and herb (Graminae, Chenopodiineae, *Arteinisia*, *Ambrosia*, Compositae and *Plantago*) at intervals since 11 000 years BP. Circles indicate sample sites. Vegetation moved northward during this interval, following recession of the Laurentide Ice Sheet. Note the slight southward recession of spruce from 4000 to 500 years BP (after Bernabo & Webb 1977).

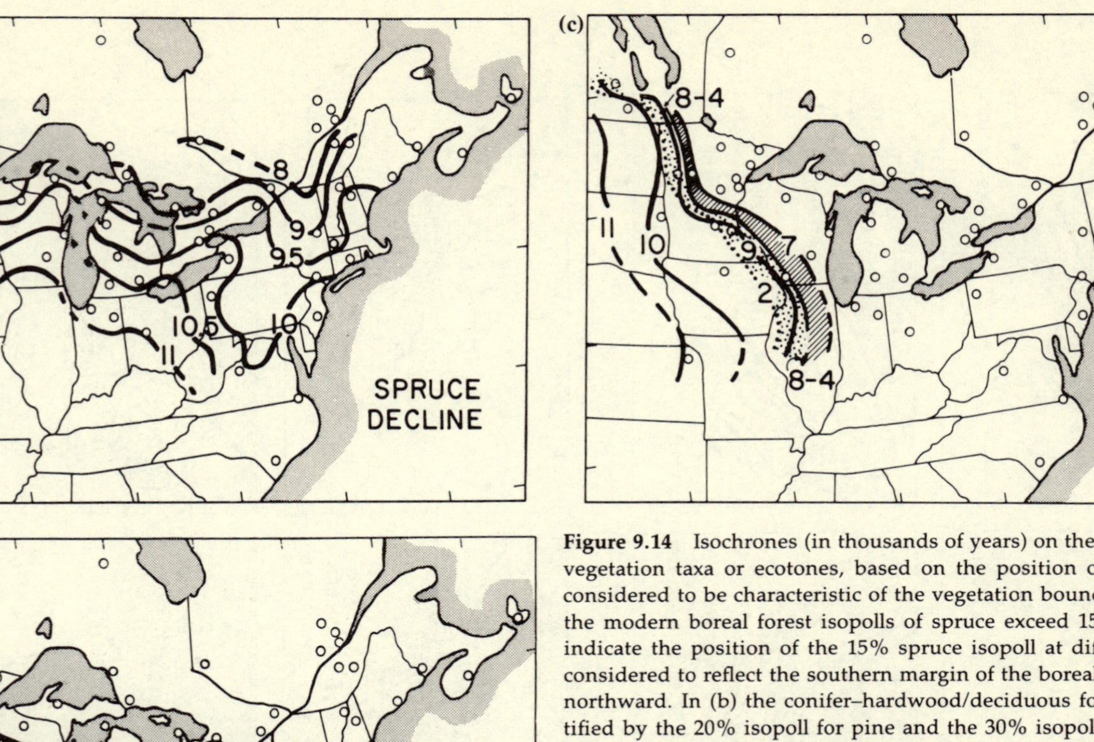

Figure 9.14 Isochrones (in thousands of years) on the migration of various vegetation taxa or ecotones, based on the position of diagnostic isopolls considered to be characteristic of the vegetation boundary. For example, in the modern boreal forest isopolls of spruce exceed 15%. In (a) isochrones indicate the position of the 15% spruce isopoll at different times and are considered to reflect the southern margin of the boreal forest as it migrated northward. In (b) the conifer–hardwood/deciduous forest ecotone is identified by the 20% isopoll for pine and the 30% isopoll for oak, reflecting a change from dominance of oak to dominance of pine (in a northward direction). In (c) the "prairie border" is delimited by the 30% isopoll for herbaceous pollen. Shading indicates the area across which the Prairie border first expanded (to ~7000 years BP) then retreated (westward) as conditions became more moist in the region in the mid to late Holocene (after Bernabo & Webb 1977).

intervals from 11 500 to 8000 years BP, a line thought to approximate the southern boundary of the late-glacial boreal forest. Similarly, Figure 9.14b shows the position of the conifer–hardwood/deciduous forest ecotone derived by analogy with modern pollen, which indicates that this boundary coincides with the 20–30% isopolls for oak and pine. Both maps indicate a rapid northward migration of forests in late-glacial times. A final map, Figure 9.14c, illustrates the position of the "prairie border," the grassland/forest ecotone of midwestern North America based on 30% isopolls for herb pollen. Following rapid eastward shift in the early Holocene, this boundary regressed westwards after ~7000 years BP, indicating that the period of minimum precipitation and maximum warmth in the area had already passed.

Isopoll and isochrone maps depicting former vegetation patterns may have value for paleoclimatic reconstructions in view of the apparent coincidence of major vegetation boundaries (ecotones) with air mass frequency boundaries (Bryson & Wendland 1967). However, it is likely that they contain little direct paleoclimatic information, because of the differing migration rates of different species. Lags in the response of vegetation to climatic fluctuations are undoubtedly complex and make paleoclimatic interpretations particularly vulnerable to misinterpretation (Davis 1978, Birks 1981).

9.7 Paleoclimatic reconstructions based on pollen analysis

The sections above have illustrated the problems associated with using fossil pollen to reconstruct past vegetation cover, and some of the methods which have been used circumvent these problems. Pollen analysis can be successfully applied to the reconstruction of past vegetation . . . but what does this tell us about past climate? In the vast majority of pollen studies the answer, unfortunately, is: very little. Usually investigators allude to the climatic implications of their work in a qualitative way only – the climate was wetter/drier or warmer/colder. Others may point to modern analogs of former vegetation cover and suggest, implicitly or explicitly, that the former climate at the study site was like the present climate in analogous locations. This is implied in the study of Davis (1969), shown in Figure 9.4, but never explicitly stated. The problem with this approach is that finding a precise analog is extremely difficult, and perhaps should not be expected, at least for late-glacial vegetation, when recently deglaciated terrain was being occupied for the first time in millennia, and different species were migrating in at different rates (Davis 1976). Indeed, there is increasing evidence that modern vegetation communities do not have a long history and are "simply temporary aggregations of species developed under certain historical

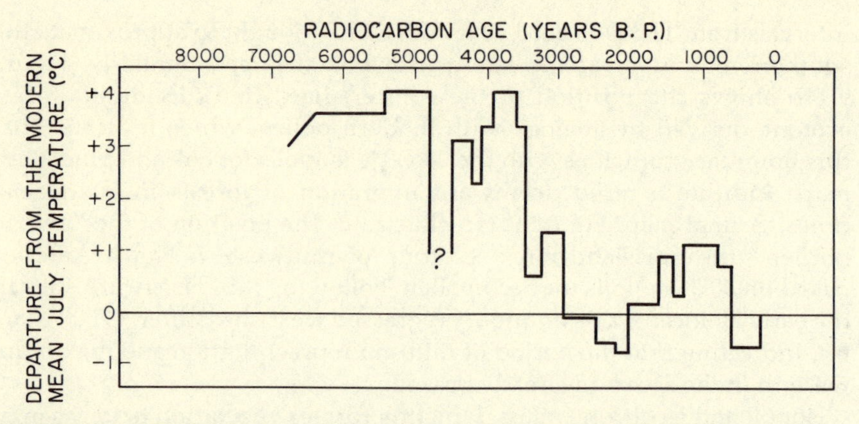

Figure 9.15 Estimated July paleotemperatures in Keewatin and Mackenzie Districts, North West Territories, Canada, based on palynological reconstructions of the former position of the northern treeline. Temperatures estimated from modern meteorological measurements at sites north or south of present treelines (after Nichols 1975).

and climatic factors" (Birks 1981). We will return to this point again later.

Some palynologists have been more adventurous in their interpretations of past climatic conditions, providing quantitative estimates. Nichols (1967, 1975), for example, has estimated mid to late Holocene temperature change north of the Canadian treeline by analogy with present temperatures in areas closer to, and within, the boreal forest (Fig. 9.15). By assuming that mean July temperatures are a major control on treeline location, Nichols estimates that mid-Holocene July temperatures were 3–4 °C warmer than in recent years; at that time the treeline was 200–300 km further north than today. Cooling began 3500–3000 years BP and the treeline has been within 100 km of its present position since then (cf. Sec. 8.2.1).

A similar type of study has been carried out by Rampton (1971) in south-western Yukon territory, Canada; Rampton examined pollen in a core from a lake presently about 600 m below the altitudinal treeline. Comparison of fossil pollen assemblages with modern pollen influx enabled contemporary analog sites to be identified. The present climate of these analog locations was then used to estimate mean annual and July temperatures and mean annual precipitation for various periods in the past (Fig. 9.16). During late Wisconsin times, for example, Rampton estimates that the mean July temperature was ~7 °C cooler than during the last 8700 years and that precipitation was only 25% of recent values. This study of altitudinal change in treeline suffers from the paucity of both climatic and pollen rain data for high altitudes. Thus, Rampton's alpine fell-field (blockfield) zone is compared to high arctic rather than alpine fell-fields today, and climatic estimates are thereby derived. It seems extremely unlikely that such a comparison is valid in view of the

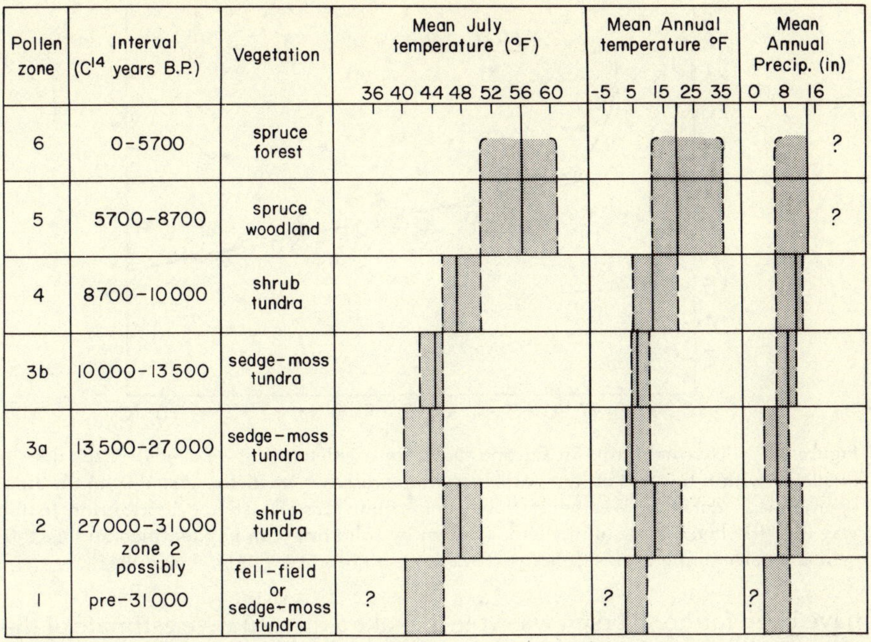

Pollen zone	Interval (C^{14} years B.P.)	Vegetation	Mean July temperature (°F) 36 40 44 48 52 56 60	Mean Annual temperature °F -5 5 15 25 35	Mean Annual Precip. (in) 0 8 16
6	0–5700	spruce forest			?
5	5700–8700	spruce woodland			?
4	8700–10000	shrub tundra			
3b	10000–13500	sedge-moss tundra			
3a	13500–27000	sedge-moss tundra			
2	27000–31000	shrub tundra			
1	zone 2 possibly pre-31000	fell-field or sedge-moss tundra	?	?	?

Figure 9.16 Paleoclimatic estimates for a location at ~62°N in Yukon Territory, Canada, based on fossil pollen data and analogous climatic conditions in locations with similar pollen rain today. Maximum, minimum, and "most probable" temperature and precipitation are shown for intervals over the last 30 000+ years (after Rampton 1971).

quite different annual energy budgets of arctic and alpine locations. Estimates of former precipitation amounts derived in this way are particularly unreliable. It may be that there is simply no modern analog to the sort of alpine fell-field vegetation which this area experienced more than 13 000 years ago (cf. Ritchie 1977).

A number of people have quantified paleoclimatic estimates by the use, not of total pollen assemblages, but of individual indicator species, plants which may not be abundant but which are thought to be limited by certain climatic conditions. The classic work of this type is that of Iversen (1944), who examined the present distributions of *Ilex aquifolium* (holly), *Hedera helix* (ivy), and *Viscum album* (mistletoe) in northern Europe. Iversen found that the distributions of these species today are closely related to mean temperatures in the warmest and coldest months of the year, though a single isotherm does not describe the species distribution (Fig. 9.17). He argues that the straight, slightly dipping part of the "thermal limit curve" is determined by summer warmth, the curved section by summer warmth, growing season length, and winter cold, and the steep vertical section by winter cold alone. By plotting present temperatures at sites where fossil occurrences of these plants

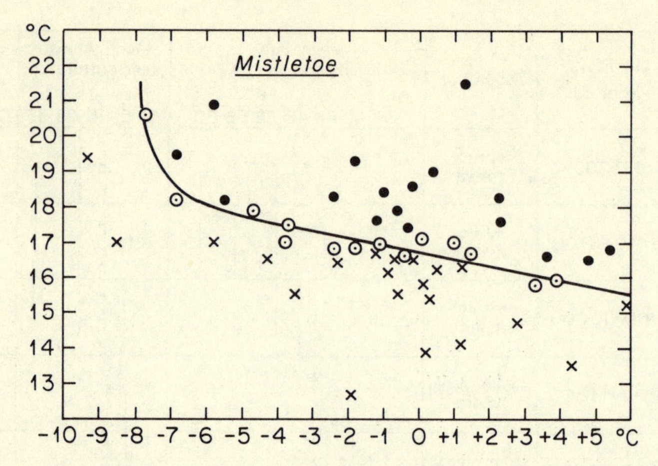

Figure 9.17 Thermal limits in Europe today for mistletoe (*Viscum album*). Black dots = mistletoe within the area of the weather station; crosses = mistletoe absent from the area; open circles = data from weather station near the boundary of mistletoe distribution. In this way the "thermal limits" of mistletoe (shown by solid line) can be identified and used to estimate paleotemperatures from fossil mistletoe occurrences (after Iversen 1944).

have been found, Iversen was able to make a quantitative estimate of the change in temperature at different periods in the past. For example, based on fossil occurrences of holly in Denmark and Germany during "the last interglacial," Iversen concluded that winter temperatures were warmer at that time (by 1–2 °C), suggesting a more oceanic climate than today. Similarly, data on the former distribution of mistletoe were used to conclude that in Atlantic and sub-boreal time (7450–2450 years BP) July temperatures were 1.5–2 °C warmer than in recent years. By the use of three thermal limit curves for holly, ivy, and mistletoe, Iversen then attempted to trace changes in both summer and winter temperatures in eastern Denmark from Atlantic time (7450–4450 years BP) to the present day (Fig. 9.18).

A similar approach has been taken by Churchill (1968) in a study of *Eucalyptus* in south-western Australia. Churchill demonstrates that various species of *Eucalyptus* are presently limited, not by temperature, but by the amount of rainfall in the wettest and driest months of the year (Fig. 9.19). This information is then used to provide estimates of rainfall change in the mid to late Holocene; however, in spite of his attempts to quantify the way in which precipitation amounts affect the distribution of *Eucalyptus* species, in the final analysis Churchill does not venture from the qualitative terms "wetter" and "drier."

One other notable study which utilizes the indicator-species approach is that of Grichuk (1969), who attempts to extend this concept of climatic limits to entire assemblages of vegetation. Grichuk reverses the process somewhat by selecting a fossil horizon and listing all the flora rep-

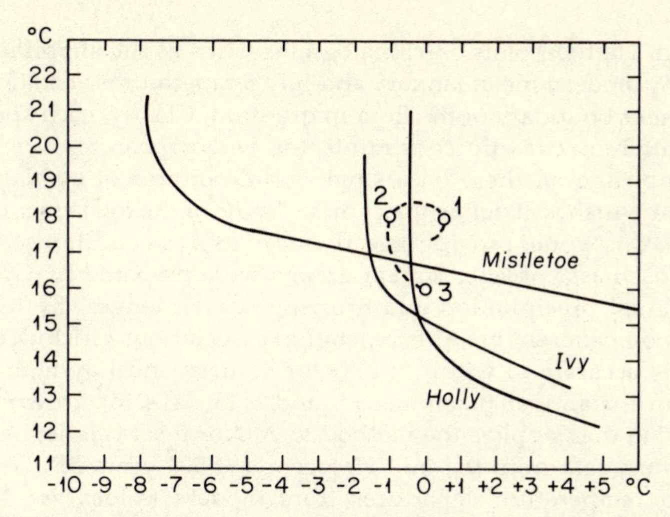

Figure 9.18 Thermal limit curves for mistletoe, holly, and ivy in Europe (solid lines) and paleotemperature estimates for eastern Jutland, Denmark, based on fossil occurrences there from (1) Atlantic time (~7000 years BP) through (2) sub-boreal time (~2750 years BP) to (3) the present day (after Iversen 1944).

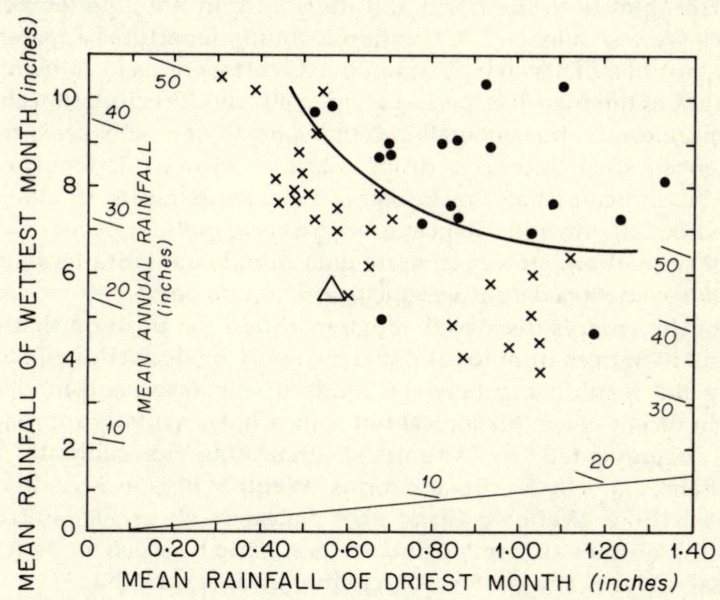

Figure 9.19 Moisture limits from *Eucalyptus diversicolor* within the 30–50 inch (762–1270 mm) annual rainfall regime. Average values of the driest month (abscissa) and wettest month (ordinate) are shown for each rainfall station. Black dots indicate *E. diversicolor* close to station; crosses indicate absence from station area. Solid line shows approximate moisture threshold for *E. diversicolor*; triangle shows modern precipitation conditions at site of fossil location, indicating considerably wetter conditions there in the past. With a few other fossil locations of similar age, a quantitative assessment of the precipitation changes could be made (after Churchill 1968).

resented. He then plots on climatograms (such as the simplified one in Fig. 9.20) modern mean January and July temperatures at stations along the present boundary of the flora in question. Clearly, each species will have different climatic constraints (or will appear to have) but by superimposing all these on a single graph an area of overlap may be apparent which will define the climatic limits of the total assemblage. By analogy, this would provide an estimate of former conditions at the time of the fossil assemblage. Similar graphs were prepared by Grichuk for total annual precipitation and growing season length. By testing his method on different modern vegetation associations Grichuk estimates that it is accurate to within ±1 °C for January and July temperatures, ±50 mm for annual precipitation, and ±15 days for growing season length. He then applies the method to 20 Northern Hemisphere pollen sites using data from the mid-Holocene (~5500 years BP). The reconstructed temperature departures from modern values are shown in Figure 9.21; data are presented for sites along a transect which approximates 35°E in Europe (17–46°E) and 95°W in North America (73–134°W). Clearly, temperature departures were largest in January at around 58°N, diminishing to both the north and the south. In July, the temperature changes were smaller (~2 °C), with maximum departures apparently at higher latitudes. This change of up to 2 °C corresponds to an increase in the length of the frost-free period of up to 40 days. Precipitation changes were more erratic but generally latitudes north of ~42°N seem to have experienced small increases (from +20 to +75 mm). Grichuk's study seems to indicate that the Equator–Pole temperature gradient was reduced during the mid-Holocene, with a correspondingly less vigorous general circulation. However, more data from lower latitudes would be needed to come to a definitive conclusion on this point.

All of the studies discussed above maintain the position that paleoclimatic inferences from fossil pollen can only be derived by first determining the relationship between modern vegetation and modern climate. In recent years this logical but somewhat circuitous approach has been circumvented by scientists attempting to calibrate pollen assemblages *directly* in climatic terms (Webb & Bryson 1972, Bryson & Kutzbach 1974, Webb & Clark 1977, Sachs *et al.* 1977). Multivariate statistical routines are used to find a relationship between modern pollen rain assemblages and contemporary climatic data such that

$$C_m = T_m P_m,$$

where C_m is the modern climatic data, P_m is the modern pollen rain, and T_m is a functional coefficient or set of coefficients ("transfer functions") based on modern climate and pollen data.

Former climatic conditions (C_f) are then derived by using the fossil

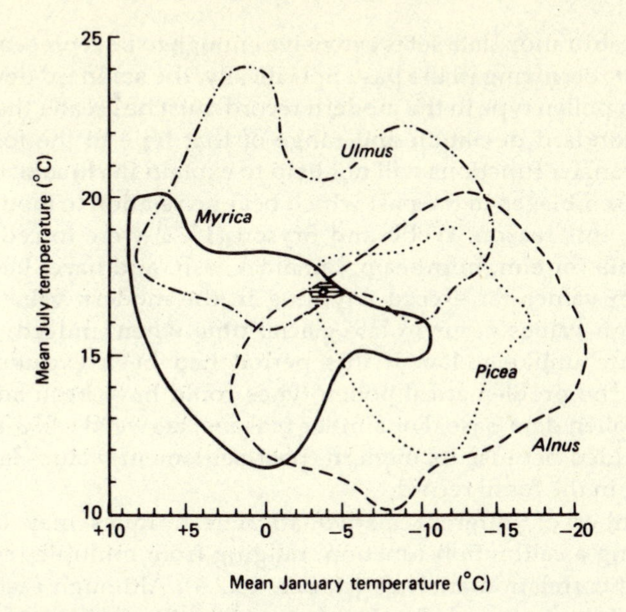

Figure 9.20 Thermal limits of *Ulmus* (elm), *Picea* (spruce), *Alnus* (alder), and *Myrica* (myrtle) in northern Europe. Each line represents the range of temperatures within which the species is known to survive today. The solid area where the lines overlap indicates the mean temperatures in July and January when a fossil sample containing the pollen of all four species was deposited (after Grichuk 1969).

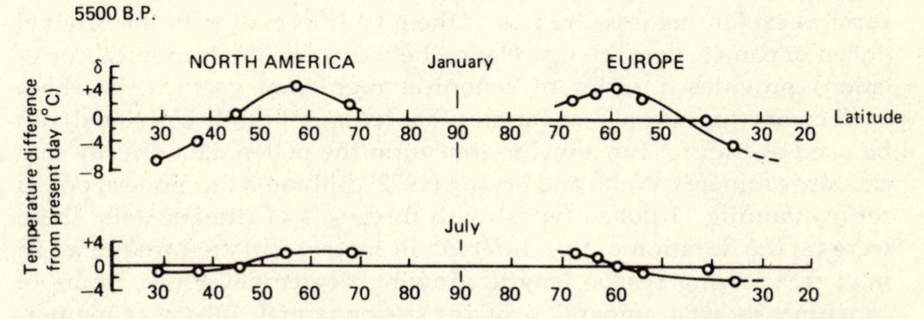

Figure 9.21 January and July temperature deviations from the present day at 5500 years BP. Two north–south transects are shown, one along the 90°W meridian (through North America) the other along the 35°E meridian (through eastern Europe). No data are available north of ~68°N in either transect (after Grichuk 1969).

pollen assemblage (P_f) and the modern transfer function (T_m). A basic assumption in this approach is that the functional relationship established between modern climate and modern pollen data was also operative in the past; in particular, it assumes that both modern vegetation and past vegetation represent an equilibrium with climate. Paleoclimatic reconstructions based on transfer functions are only reliable if the modern calibration data set is extensive enough to be representative of all conditions occurring in the past. Specifically, the standard deviation and range of a pollen type in the modern record must be greater than, or equal to, the standard deviation and range of that type in the fossil record. Hence, transfer functions will not help to explain in climatic terms fossil pollen assemblages in the past which bear no relation to modern experience. For this reason, Webb and Bryson (1972) were forced to discard pollen data for elm, hornbeam, tamarack, ash, and hazel because their maximum values far exceed anything in the modern values sampled. These high values occur in late glacial time when, indeed, conditions were quite unlike today. If this period had been excluded in their analysis, the problematical pollen types could have been added to the overall pollen data base. For similar reasons, ragweed (*Ambrosia*) pollen was excluded because its modern (post-settlement) values far exceeded anything in the fossil record.

A number of different mathematical techniques may be used in developing a calibration function, ranging from multiple regression to canonical correlation analysis (cf. Sec. 10.2.4). Although each approach has its advantages and disadvantages, they all yield generally similar results (Webb & Clark 1977). Webb and Bryson (1972) used canonical correlation analysis to identify relationships between pollen and climatic data. This technique finds an ordered set of mutually orthogonal patterns (canonical variates) among both the climatic and pollen data. Primary canonical variates express most of the variance in the data, successive variates explaining less and less of the total information in the original pollen or climate data set. Correlation between variates (canonical correlation) provides a series of canonical regression coefficients; these coefficients (for the pairs of variates that are significantly correlated) can be used as transfer functions to transform the pollen data directly into climatic estimates. Webb and Bryson (1972) calibrated the modern pollen set (containing 11 pollen types) with three sets of climatic data; these were (a) the duration of five different air masses; (b) their frequency in July; (c) growing season length, growing degree day totals, hours of sunshine, snowfall amounts, growing season rainfall, July mean temperature, and precipitation minus potential evapotranspiration (an index of moisture stress). The resulting canonical regression coefficients were then applied to the fossil pollen data from three midwestern US sites to produce the first comprehensive estimates of climatic conditions during late-glacial and postglacial time (Table 9.3 and Fig. 9.22). Webb and Bryson were thus able to describe the major change of climate around 11 300 years BP in terms of changing air mass frequencies; at this time, there was a marked increase in "Pacific south" air. At two sites, Kirchner Marsh and Disterhaft Farm, this drier air was associated with a decrease in cloudiness and less snowfall. Mean July temperatures rose by 2–3 °C,

Table 9.3 Paleoclimatic reconstructions based on palynological data from Kirchner Marsh, southern Minnesota (after Webb & Bryson 1972).

Time in Years BP	Dominant pollen types in zones	Air mass durations (months)			July air mass Frequencies† (%)					Standard climatic variables						
		S	W	N	mT	cT	Ps	Pn	A	Growing season rainfall (mm)	Snow (mm)	Sun (h)	P – PE‡ (mm)	July mean Temperature (°C)	Degree days below 18.3 °C	Growing season length (days)
2 000	Oak, pine, sedge	4.0	5.0	3.0	14	1	64	14	7	470	1250	2500	30	21.5	3900	137
4 000	Oak	5.5	4.5	2.0	16	2	64	12	6	440	1120	2550	−50	22.0	3900	138
6 000	Oak	4.0	6.0	2.0	14	2	69	11	4	420	1030	2600	−150	22.5	3950	140
	Herbs	3.5	7.0	1.5	11	2	75	10	2	410	1000	2650	−220	22.5	3950	139
8 000	Oak, elm, herbs	4.0	6.5	1.5	14	2	70	11	3	430	1080	2580	−130	22.0	3850	137
10 000	Pine, elm, oak, birch, alder	3.0	4.5	4.5	13	–	61	17	9	460	1250	2480	50	21.5	3800	134
12 000	Spruce, ash	4.0	3.0	5.0	6	–	40	32	21	490	1870	1850	350	19.0	3070	114
13 500	Spruce, herbs	4.0	–	8.0	7	–	20	42	32	520	2200	1470	480	18.0	2880	108

† mT = maritime tropical air; cT = continental tropical air; Ps = North Pacific (south) air; Pn = North Pacific (north) air. A = Arctic air.
‡ P – PE = precipitation minus potential evapotranspiration.

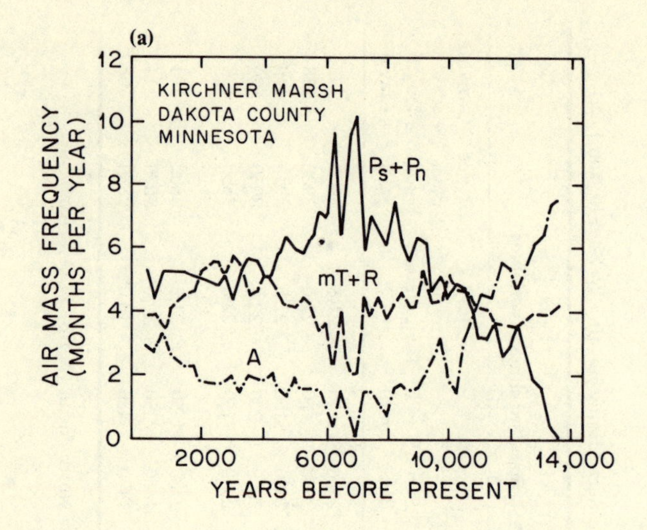

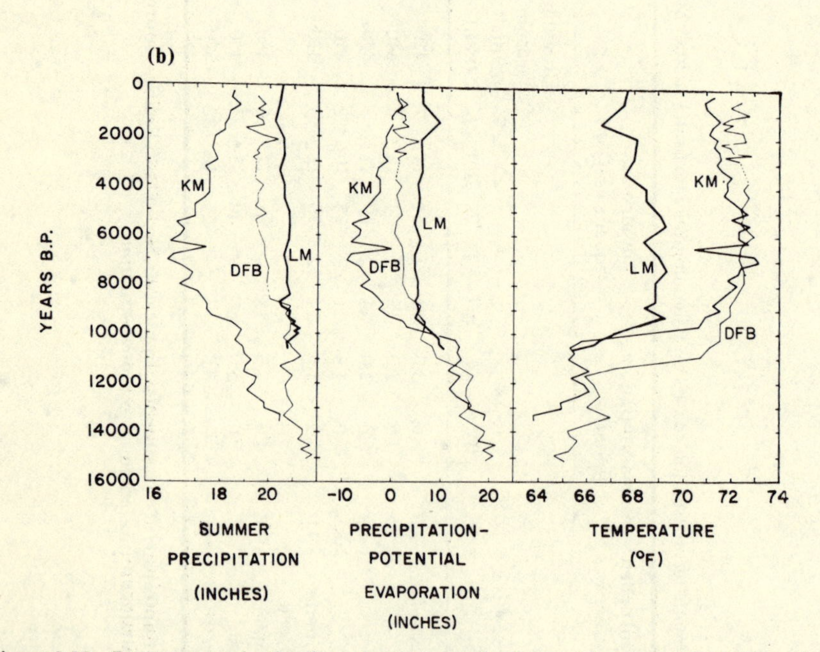

Figure 9.22 Reconstructed paleoclimatic estimates for three midwestern US sites based on transfer functions of pollen data (DF = Disterhaft Farm, south-eastern Wisconsin; KM = Kirchner March, southern Minnesota; LM = Lake Mary, northern Wisconsin). (a) Estimates of the duration of air mass occurrence in southern Minnesota since 14 000 years BP. Air masses are: A, arctic; mT, maritime tropical; R, return polar; Ps, Pacific south; Pn, Pacific north. (b) Estimates of growing season rainfall, precipitation minus potential evapotranspiration, and July mean temperature at all three fossil pollen sites (after Webb & Bryson 1972).

growing season length increased by ~30% and so did the seasonal growing degree day total. At Lake Mary, 200 km to the north, however, severe climatic conditions continued until ~9500 years BP, because of the proximity of the site to the local ice margin at that time. During the ensuing millennia, westerly airflow increased in frequency, reaching a maximum around 7200 years BP when the climate of southern Minnesota (Kirchner Marsh) resembled that of eastern South Dakota today. Since ~4700 years BP climatic conditions have been relatively stable, though somewhat cooler during the last 2000 years. No other study using palynological data has provided such a detailed interpretation of paleoclimatic conditions during the last 15 000 years. However, this does not mean that it is all correct! After all, it is extremely difficult to check if the reconstructions are accurate, because there is such a paucity of other paleoclimatic data, and the entire technique is built on a foundation of debatable assumptions and methodological stumbling blocks (Birks 1981). Nevertheless, Webb and Bryson's work, building on the thesis of Cole (1969), was a major step forward in quantitative paleoclimatic interpretation of pollen data. Subsequent studies have addressed the major assumptions of the method and attempted to define confidence limits for the paleoclimatic estimates (Bryson & Kutzbach 1974, Howe & Webb 1977, Webb & Clark 1977). Perhaps the greatest difficulty involves distinguishing between climatically induced changes in vegetation and changes resulting from varying migration rates of particular taxa from (often distant) refuges (Davis 1978). Vegetation responses to similar climatic fluctuations may also differ from one time period to another, depending on the stage of soil development (Livingstone 1968, Brubaker 1975). These problems are particularly acute when dealing with a recently deglaciated region; considerable skill is required to determine whether the unique pollen assemblages of late-glacial times reflect unique climatic conditions, or whether they simply reflect vegetation communities brought about by differential migration rates of different plant species into areas with only rudimentary soils (Davis 1976, Livingstone 1968). The weight of evidence at present points to the latter conclusion, though no doubt late-glacial climates were quite unusual because of the presence of large ice sheets at that time (Bryson & Wendland 1967).

Recent applications have demonstrated that the transfer function approach may be useful in palynological paleoclimatic reconstructions even for periods when only relatively subtle changes in vegetation communities have occurred. For example, Bernabo (1981) has reconstructed growing season temperature variations at four sites in Michigan spanning the last 1000–3000 years, a period when major vegetation boundaries have not changed, but the relative species composition of each vegetation zone has undergone adjustments in response to climatic fluctuations. Detailed sampling produced paleotemperature estimates at

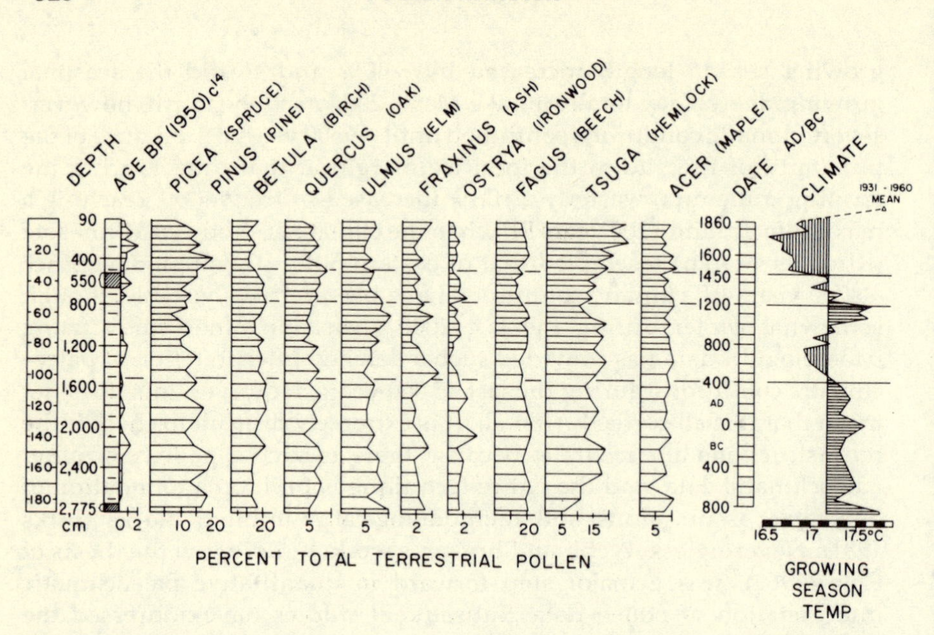

Figure 9.23 Percentage pollen diagram from Marion Lake, Michigan. The record extends back to 2775 years BP. Pollen percentages were used in a transfer function analysis to reconstruct growing season temperatures (at right) (after Bernabo 1981).

5–40 year intervals (Fig. 9.23). Temperatures generally declined from 2800 to ~1100 years BP followed by a warmer interval centered on 850 years BP (AD 1100) when temperatures approached 1931–60 values. A cold interval from AD 1450 to AD 1850 corresponds to the "Little Ice Age", which is well known from European historical data (Lamb 1982). Because of the detailed nature of the reconstruction, direct comparisons can be made between the palynological and other proxy paleotemperature estimates (such as tree rings) as well as with early instrumentally recorded meteorological data. In all cases, the palynologically based paleotemperature estimates gave reasonable values, consistent with data from other areas. This approach of using *independent* paleoclimatic reconstructions, derived from different proxy data sets, as a method of testing results is often the only means of validating the more complex paleoclimatic reconstructions. Alternatively, two independent calibration data sets may be employed and the resulting reconstructions compared. For example, pre-settlement pollen and 19th century climatic data might be used as one data set, and modern pollen and contemporary climatic records as another (cf. Webb 1973). If sufficient data were available, two separate, but spatially overlapping data sets might be used to produce independent paleoclimatic reconstructions which could then be compared (e.g. Blasing & Fritts 1975). Thus various strategies can be adopted to test the paleoclimatic estimates rigorously.

Pollen analysis is undergoing revolutionary methodological changes. Multivariate statistical methods are providing a rigorous means of transforming fossil pollen assemblages into comprehensive paleoclimatic reconstructions. In terms of logic, these methods differ little from the standard qualitative interpretations of the past, but in terms of precision and reproducibility they are far more effective. Application of these techniques is only just beginning (e.g. Heusser & Streeter 1980, Heusser *et al.* 1980, Adam & West 1983) but they would appear to have great potential in reconstructing not only former temperatures and rainfall amounts but in determining the nature of past circulation patterns (e.g. Diaz & Andrews 1982). As Faegri (1950) stated, this "is the primary object of paleoclimatological research . . . once they have been found, many of the problems that baffle us today will find their obvious explanation."

10
Dendroclimatology

10.1 Introduction

Variations in tree-ring widths from one year to the next have long been recognized as an important source of chronological and climatic information. In Europe, studies of tree rings as a potential source of paleoclimatic information go back to the early 18th century when several authors commented on the narrowness of tree rings (some with frost damage) dating from the severe winter of 1708–9. In North America, Twining (1833) first drew attention to the great potential of tree rings as a paleoclimatic index (for a historical review, see Studhalter 1955). However, in the English-speaking world, the "father of tree-ring studies" is generally considered to be A. E. Douglass, an astronomer who was interested in the relationship between sunspot activity and rainfall. To test the idea of a sunspot–climate link, Douglass needed long climatic records and he recognized that ring-width variations in trees of the arid south-western United States might provide a long, proxy record of rainfall variation (Douglass 1914, 1919). His efforts to build long-term records of tree growth were facilitated by the availability of wood from archaeological sites, as well as from modern trees (Robinson 1976). Douglass' early work was crucial for the development of dendrochronology (the use of tree rings for dating) and for dendroclimatology (the use of tree rings as a proxy indicator of climate).

Although much work has been carried out since these early pioneering studies, the greatest strides in dendroclimatology have been made in the last 10–15 years, largely as a result of the work of H. C. Fritts and associates at the Laboratory of Tree Ring Research in the University of Arizona, Tucson; much of this work has been documented at length in the excellent book by Fritts (1976). Latest developments, including discussion of recent dendroclimatic studies of the Southern Hemisphere, are discussed in the volume edited by Hughes *et al.* (1982).

10.2 Fundamentals of dendroclimatology

A cross section of most temperate forest trees will show an alternation of lighter and darker bands, each of which is usually continuous around the tree circumference. These are seasonal growth increments produced by meristematic tissues in the cambium of the tree. When viewed in detail (Fig. 10.1) it is clear that they are made up of sequences of large, thin-walled cells (earlywood) and more densely packed, thick-walled cells (latewood). Collectively, each couplet of earlywood and latewood comprises an annual growth increment, more commonly called a tree ring. The mean width of a ring in any one tree is a function of many variables, including the tree species, tree age, availability of stored food within the tree and of important nutrients in the soil, and a whole complex of climatic factors (sunshine, precipitation, temperature, wind speed, humidity, and their distribution through the year). The problem

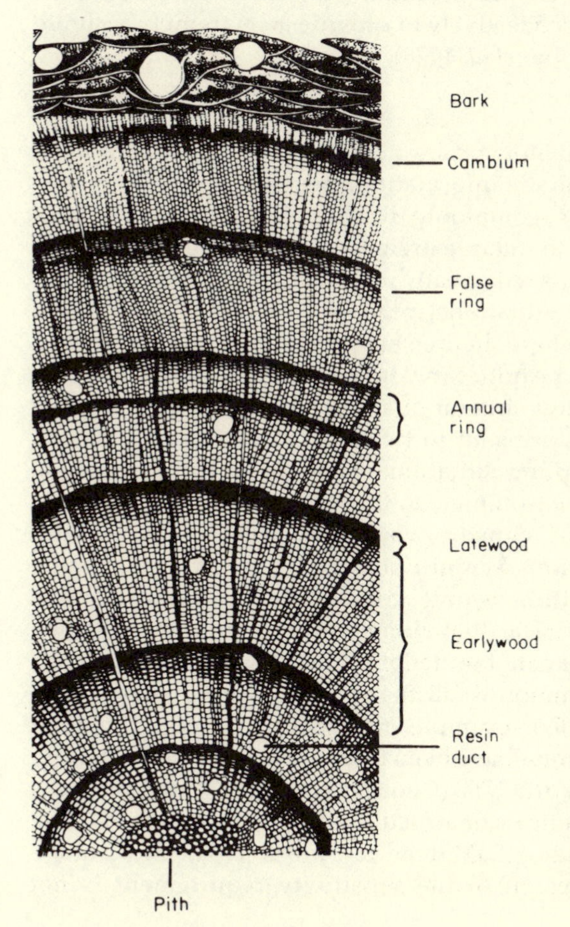

Figure 10.1 Drawing of cell structure along a cross-section of a young stem of a conifer. The earlywood is made up of large and relatively thin-walled cells (tracheids); latewood is made up of small, thick-walled tracheids. Variations in tracheid thickness may produce false rings in either earlywood or latewood (after Fritts 1976).

facing dendroclimatologists is to extract whatever climatic signal is available in the tree-ring data and to distinguish this signal from the background "noise." Furthermore, the dendroclimatologist must know precisely the age of each tree ring if the climatic signal is to be chronologically useful. From the point of view of paleoclimatology, it is perhaps useful to consider the tree as a filter or transducer which, through various physiological processes, converts a given climatic input signal into a certain ring-width output which is stored and can be studied in detail, even thousands of years later (e.g. Yapp & Epstein 1977, Fritts 1976).

Climatic information has most often been gleaned from interannual variations in ring width, but recently there has been a great deal of work on the use of density variations, both inter- and intra-annually (Sec. 10.4). Significant advances have also been made in studying isotopic variations in wood as a proxy of temperature variation through time (Sec. 10.5). These different approaches are complementary and can be used independently to check paleoclimatic reconstructions based on only one of the methods, or collectively to provide an extremely accurate reconstruction (Schweingruber *et al.* 1978).

10.2.1 Sample selection

In conventional dendroclimatological studies, where ring-width variations are the source of climatic information, trees are sampled in sites where they are under stress; commonly, this involves selection of trees which are growing close to their extreme ecological range. In such situations, climatic variations will greatly influence annual growth increments. In more beneficent situations, perhaps nearer the middle of a species range, or in a site where the tree has access to abundant ground water, tree growth may not be noticeably influenced by climate, and this will be reflected in the low interannual variability of ring widths (Fig. 10.2). Such tree rings are said to be complacent. There is thus a spectrum of possible sampling situations, ranging from those where trees are extremely sensitive to climate to those where trees are virtually unaffected by interannual climatic variations. Clearly, for useful dendroclimatic reconstructions, samples close to the sensitive end of the spectrum are favored as these would contain the strongest climatic signal. However, it is now clear that climatic information may also be obtained from trees which are not under obvious climatic stress, providing the climatic signal common to all the samples can be successfully isolated (LaMarche 1982). For example, ring widths of New England deciduous and coniferous trees have been used to reconstruct the history of drought in the area since AD 1700 (Cook & Jacoby 1977) and, recently, reasonably good paleoclimatic reconstructions have been achieved using Tasmanian mesic forest trees (LaMarche & Pittock 1982). For isotope dendroclimatic studies (Sec. 10.6) the sensitivity requirement is not

Figure 10.2 Trees growing on sites where climate seldom limits growth processes produce rings that are uniformly wide (a). Such rings provide little or no record of variations in climate and are termed *complacent*. Trees growing on sites where climatic factors are frequently limiting produce rings that vary in width from year to year depending on how severely limiting climate has been to growth (b). These are termed *sensitive* (from Fritts 1971).

critical and it would, in fact, be preferable to use complacent tree rings for analysis (Gray & Thompson 1978). Sensitivity is also less significant in densitometric studies (Sec. 10.5).

Commonly two types of climatic stress are recognized, moisture stress and temperature stress. Trees growing in semi-arid areas are frequently limited by the availability of water, and ring-width variations primarily reflect this variable. Trees growing near to the latitudinal or altitudinal treeline are mainly under growth limitations imposed by temperature and hence ring-width variations in such trees contain a strong temperature signal. However, other climatic factors may be indirectly involved.

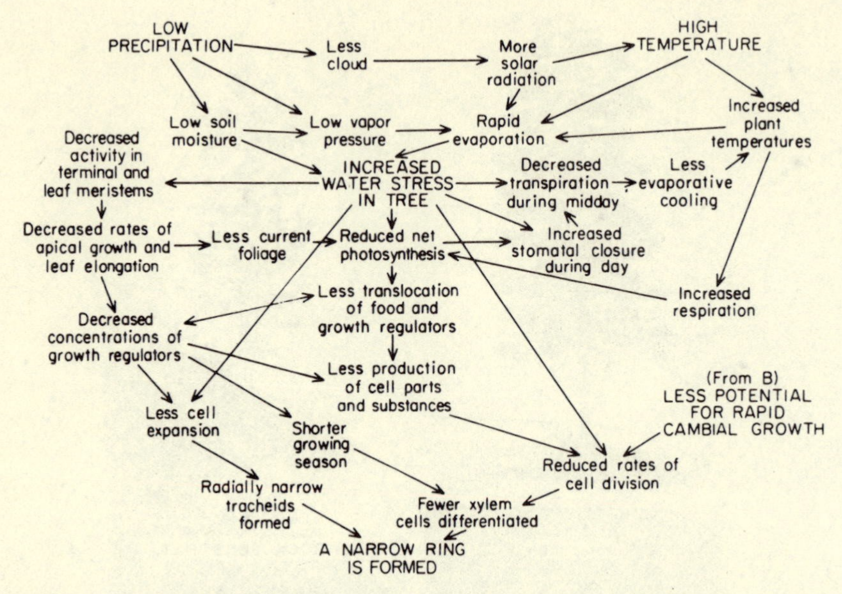

Figure 10.3 A schematic diagram showing how low precipitation and high temperature during the growing season may lead to the formation of a narrow tree ring in arid-site trees. Arrows indicate the net effects and include various processes and their interactions. It is implied that the effects of high precipitation and low temperature are the opposite and may lead to an increase in ring widths (from Fritts 1971).

Biological processes within the tree are extremely complex (Fig. 10.3) and similar growth increments may result from quite different combinations of climatic conditions. Furthermore, climatic conditions *prior* to the growth period may "precondition" physiological processes within the tree and hence strongly influence subsequent growth (Fig. 10.4). For the same reason, tree growth and food production in one year may influence growth in the following year, and lead to a strong serial correlation or autocorrelation in the tree-ring record. Tree growth in marginal environments is thus commonly correlated with a number of different climatic factors in both the growth season (year t_0) and in the preceding months, as well as with the record of prior growth itself (generally in the preceding growth years, t_{-1} and t_{-2}). Indeed in complex dendroclimatic models, tree growth in subsequent years (t_{+1}, t_{+2}, etc.) may also be considered, since they also contain climatic information about year t_0. This will be discussed in more detail in Sections 10.2.4 and 10.3.

Trees are sampled radially using an increment borer which removes a core of wood (generally 4 mm in diameter), leaving the tree unharmed. It is important to realize that dendroclimatic studies are unreliable unless an adequate number of samples are recovered; two or three cores should be taken from each tree and at least 20–30 trees should be sampled at an

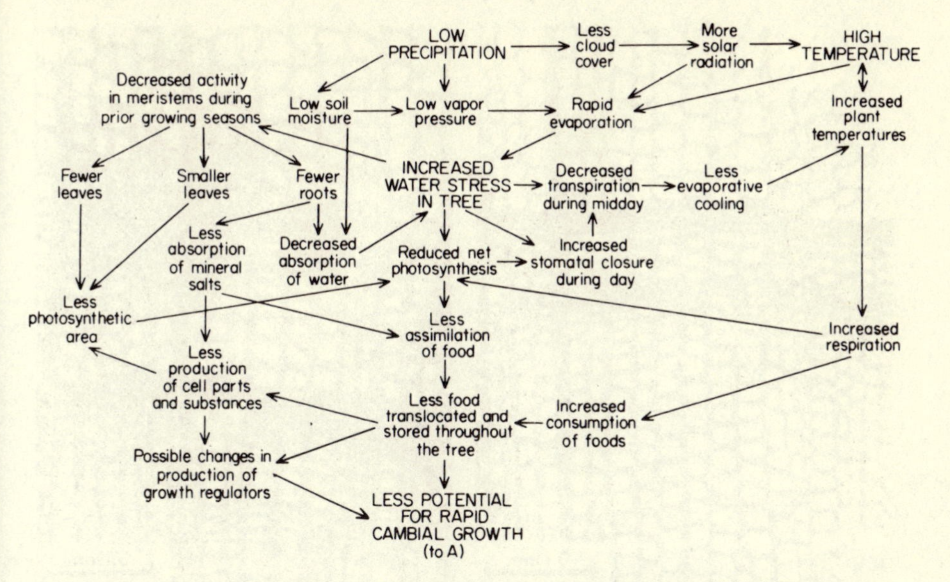

Figure 10.4 A schematic diagram showing how low precipitation and high temperature *before* the growing season may lead to a narrow tree ring in arid-site trees (from Fritts 1971).

individual site, though this is not always possible. Eventually, as discussed below, the cores are used to compile a master chronology of ring-width variation for the site and it is this that is used to derive climatic information.

10.2.2 Cross dating

For tree-ring data to be used for paleoclimatic studies, it is essential that the age of each ring be known precisely. This is necessary in constructing the master chronology from a site where ring widths from modern trees of similar age are being compared, and equally necessary when matching up sequences of overlapping records from modern and archeological specimens to extend the chronology back in time (Stokes & Smiley 1968). Great care is needed because occasionally trees will produce false rings or intra-annual growth bands, which may be confused with the actual earlywood/latewood transition (Fig. 10.5). Furthermore, in extreme years some trees may not produce an annual growth layer at all, or it may be discontinuous around the tree, or so thin as to be indistinguishable from adjacent latewood (i.e. a partial or missing ring). Clearly, such circumstances would create havoc with climatic data correlation and reconstruction, so careful cross dating of tree-ring series is necessary. This involves comparing ring-width sequences from each core so that characteristic patterns of ring-width variation (ring-width "signatures") are correctly matched (Fig. 10.6). If a false ring is present, or if a ring is missing, it will thus be immediately apparent. The same procedure can

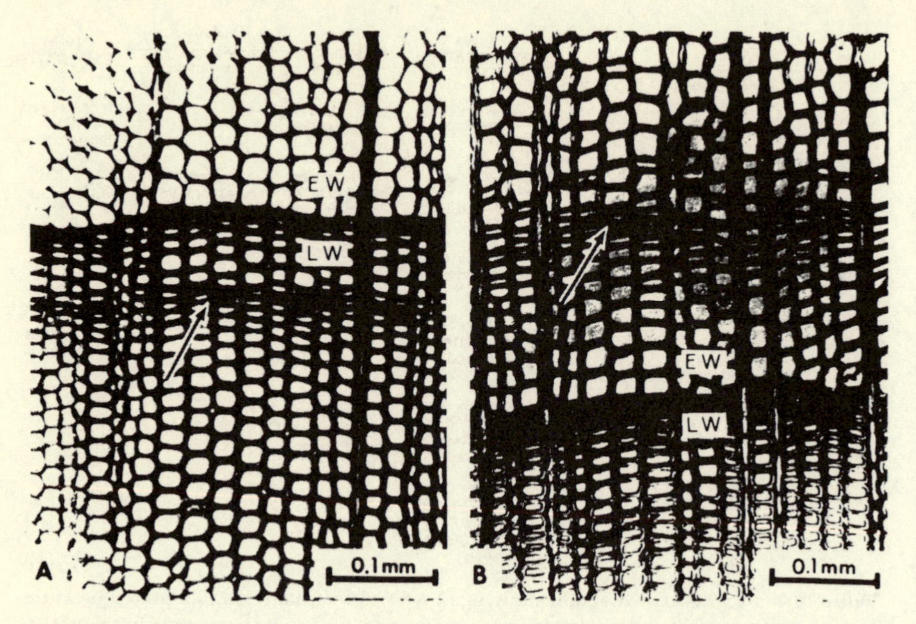

Figure 10.5 Annual growth increments or rings are formed because the wood cells produced early in the growing season (earlywood, EW) are large, thin-walled, and less dense, while the cells formed at the end of the season (latewood, LW) are smaller, thick-walled, and more dense. An abrupt change in cell size between the last-formed cells of one ring (LW) and the first-formed cells of the next (EW) marks the boundary between annual rings. Sometimes growing conditions temporarily become severe before the end of the growing season and may lead to the production of thick-walled cells within an annual growth layer (arrows). This may make it difficult to distinguish where the actual growth increment ends, which could lead to errors in dating. Usually these intra-annual bands or false rings can be identified, but where they can not the problem must be resolved by cross dating (after Fritts 1976).

be used with archeological material; the earliest records from living trees are matched or cross dated with archeological material of the same age and the procedure is repeated many times over to establish a thoroughly reliable chronology. In the south-western USA, the ubiquity of beams or logs of wood used in Indian pueblos has enabled chronologies of up to 2000 years to be constructed. In fact, accomplished dendrochronologists can accurately date wood used in dwellings by comparing their tree-ring widths with master chronologies for the area (Robinson 1976). Similar chronologies are being established in western Europe; Baillie (1977), for example, has used beams of wood from historical and archeological sites to establish an oak chronology for northern Ireland back to AD 1001. In the Netherlands, studies of oak panels used for paintings up to AD 1650, and even wooden sculptures, have provided cross-datable material (Eckstein *et al.* 1975). Tree stumps recovered from Holocene bogs have also been cross dated, forming a "floating chronology," fixed in time by

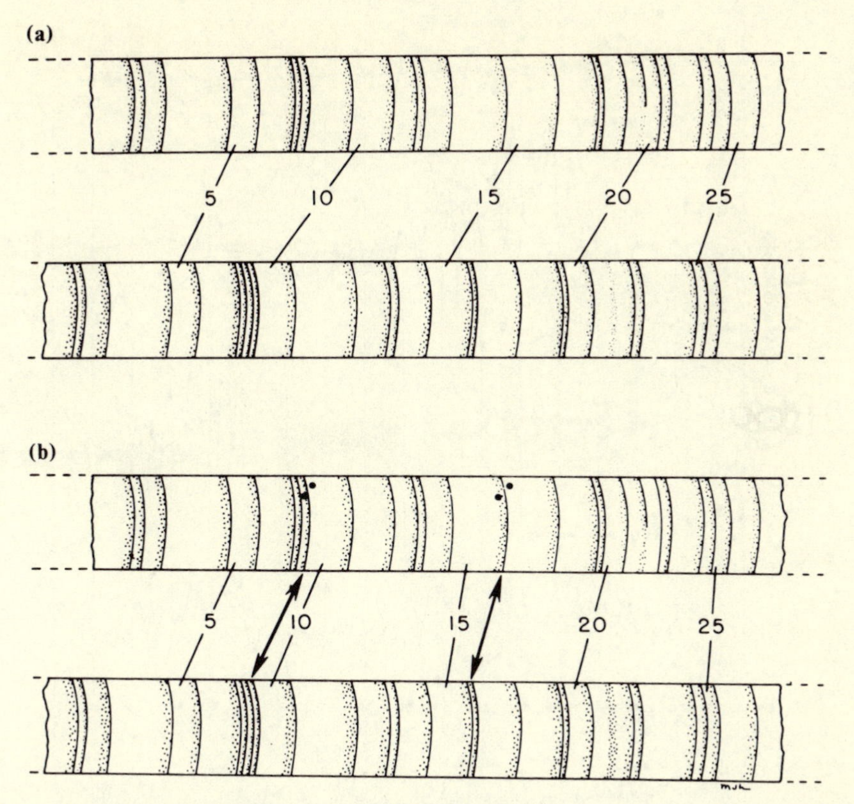

Figure 10.6 Cross dating of tree rings. Comparison of tree-ring widths makes it possible to identify false rings or where rings are locally absent. For example, in (a), strict counting shows a clear lack of synchrony in the patterns. In the lower specimen of (a), rings 9 and 16 can be seen as very narrow, and they do not appear at all in the upper specimen. Also, rings 21 (lower) and 20 (upper) show intra-annual growth bands. In (b), the positions of inferred absence are designated by dots (upper specimen), the intra-annual band in ring 20 is recognized, and the patterns in all ring widths are synchronously matched (after Fritts 1976).

^{14}C dating only, at present (e.g. Pilcher *et al.* 1977). Finally, cross-dating techniques have been most successfully applied to very old living trees (bristlecone pines) and wood fragments from adjacent dead tree stumps. In this way, a chronology extending over 7000 years has been established (Ferguson 1970) which is considered to be so accurate that radiocarbon dates on bristlecone pine samples of known age are used to calibrate the radiocarbon timescale (see Sec. 3.2.1.5).

10.2.3 Standardization of ring-width data

Once the chronology for each core has been established, individual ring widths are measured and plotted to establish the general form of the data (Fig. 10.7a). It is common for time series of ring widths to contain a

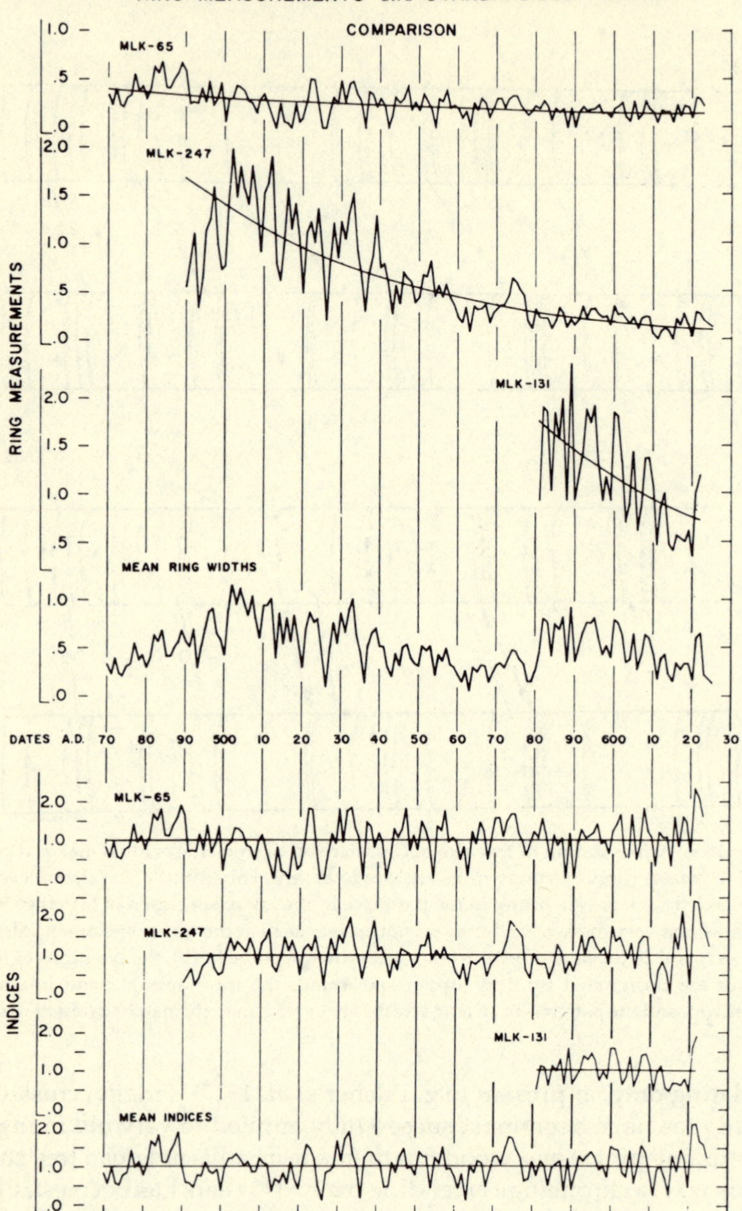

Figure 10.7 Standardization of ring-width measurements is necessary to remove the decrease in size associated with increasing age of the tree. If the ring widths for the three specimens shown in the upper figure are simply averaged by year, without removing the effect of the tree's age, the mean ring-width chronology shown below them exhibits intervals of high and low growth, associated with the varying age of the samples. This age variability is generally removed by fitting a curve to each ring-width series, and dividing each ring width by the corresponding value of the curve. The resulting values, shown in the lower half of the figure, are referred to as indices, and may be averaged among specimens differing in age to produce a mean chronology for a site (lowermost record) (from Fritts 1971).

function resulting entirely from the tree growth itself, with wider rings generally produced during the early life of the tree. In order that ring-width variations from different cores can be compared, it is first necessary to remove the growth function peculiar to that particular tree. Only then can a master chronology be constructed. Growth functions are removed by fitting an exponential or polynomial curve to the data (Fig. 10.7a) and dividing each ring-width value by the "expected" value on the growth curve. This standardization procedure leads to a new time series of ring-width indices, with a mean of one and a variance which is fairly constant through time (Fritts 1971). Ring-width indices are then averaged, year by year, to produce a master chronology for the sample site, independent of growth function and differing sample age (Fig. 10.7b). Averaging the standardized indices also increases the (climatic) signal to noise ratio. This is because climatically related variance, common to all records, is not lost by averaging, whereas non-climatic "noise," which varies from tree to tree, will be partially cancelled in the averaging process. It is thus important that a large enough number of cores be obtained initially to help enhance the climatic signal common to all the samples.

Standardization is an essential prerequisite to the use of ring-width data in dendroclimatic reconstruction but it poses significant methodological problems. Consider, for example, the ring-width chronologies shown in Fig. 10.8. Drought-sensitive conifers from the south-western United States characteristically show ring-width variations like those in Figure 10.8a. For most of the chronology a negative exponential function, of the form $y = ae^{-by} + k$, fits the data well. However, this is not the case for the early section of the record, which must be either discarded or fitted by a separate mathematical function. Obviously, the precise functions selected will have an important influence on the resulting values of the ring width indices. In the case of deciduous trees, the growth curve is often quite variable and unlike the negative exponential values characteristic of arid-site conifers. In such cases (Fig. 10.8b) a polynomial function is fitted to the data and individual ring widths are divided by the local value of this curve to produce a series of ring-width indices. In the case of polynomial functions, given a large enough number of coefficients, it is theoretically possible to describe the raw data quite precisely, which would, of course, remove all the climatic information. It is therefore necessary to restrict the number of coefficients to the minimum; in practice, additional coefficients are not included unless they reduce the variance of the ring-width data by at least a further 5% (Fritts 1976), though this cut-off point is quite arbitrary.

Further problems arise when complex growth functions are observed, such as those in Figure 10.8c. In this case it would be difficult to decide on the use of a polynomial function (dashed line) or a negative exponential

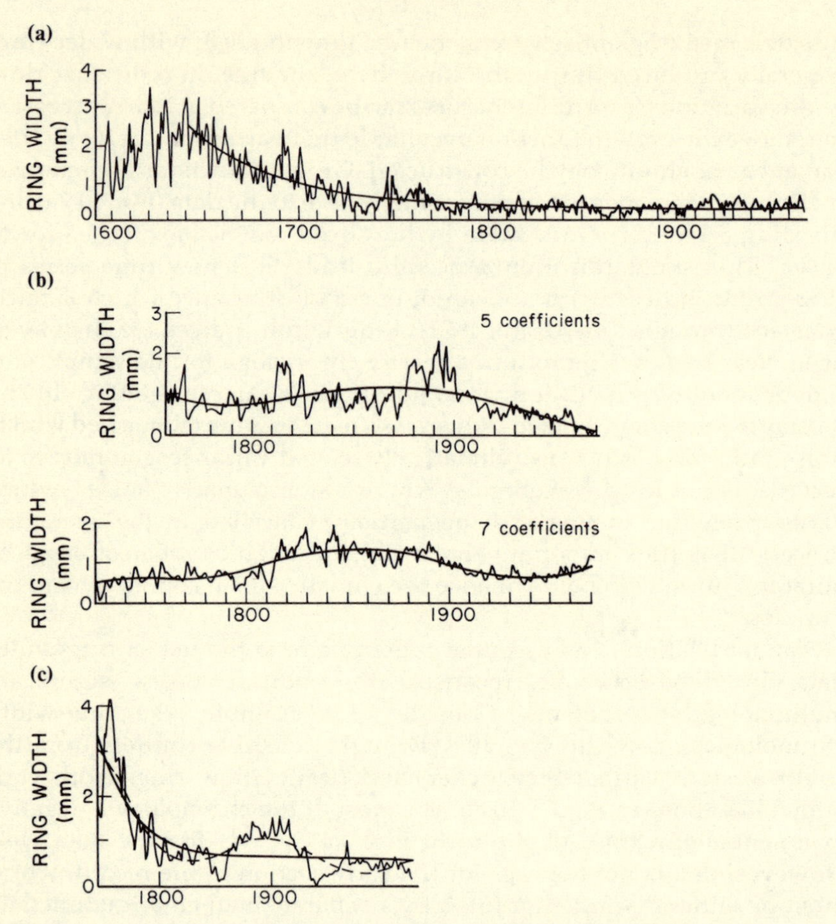

Figure 10.8 Some problems in standardization of ring widths. In (a) most of the tree-ring series can be fitted by the exponential function shown. However, the early part of the record must be discarded. In (b) the two ring-width series required higher-order polynomials to fit the lower frequency variations of each record (the greater the number of coefficients for each equation, the greater the degree of complexity in the shape of the curve). In (c) the series could be standardized using either a polynomial (dashed) or exponential function (solid line). Depending on the function selected and its complexity, low-frequency climatic information may be eliminated. The final ring-width indices depend very much on the standardization procedure employed (examples selected from Fritts 1976).

function (solid line), and in either case the first few observations should perhaps be discarded. It is clear that this standardization procedure is not easy to apply and may actually remove important low-frequency climatic information. It is not possible, *a priori*, to decide if part of the long-term change in ring width is due to a coincident climatic trend. The problem is exacerbated if one is attempting to construct a long-term dendrochronological record, when only tree fragments or historical timbers are available and the corresponding growth function may not be apparent. Such difficulties are less significant in densitometric or isotope dendroclimatic studies because there is little or no growth trend in the density and isotope data (Polge 1970, Schweingruber *et al.* 1979); it is thus likely that these approaches will yield more low-frequency climatic information than is possible in the measurement of ring widths alone.

10.2.4 Calibration of ring-width data

Once a master chronology of standardized ring-width indices has been obtained, the next step is to relate variations in ring-width data to variations in climatic data. This process is known as calibration, whereby a mathematical or statistical procedure is used to convert growth measurement into climatic estimates. If an equation can be developed which accurately describes instrumentally observed climatic variability in terms of tree growth over the same interval, then paleoclimatic reconstructions can be made using only the tree-ring data. A survey of the various methods which have been used to determine the tree growth–climate relationship indicates several different levels of complexity, each level involving more complex statistics (Table 10.1). In this section, a brief summary of each method is given to provide an overview of the various approaches. In Section 10.3, there follows a more detailed discussion of each method, with examples of how they have been applied to dendroclimatic reconstructions. For a more exhaustive treatment of the statistics involved, and more examples of how they have been used, the reader is referred to Fritts (1976 Ch. 7).

At the primary level of calibration (level I in Table 10.1) is the simple linear regression model with only two variables: growth indices and a climatic parameter, perhaps mean summer temperature. This approach

Table 10.1 Methods used to determine relationship between tree growth and climate.

Level	Number of variables of: Tree growth	Climate	Main statistical procedures used
I	1†	1†	Linear regression analysis
II	1†	n†	Multiple stepwise regression analysis
IIIa	1†	n (eigenvector)‡	Principal components and stepwise multiple regression analysis
IIIb	n (eigenvector)‡	1†	Principal components and multiple regression analysis
IV	n (eigenvector)‡	n (eigenvector)‡	Principal components and multiple canonical regression analysis

† Temporal array of data.
‡ Spatial and temporal array of data.

necessitates an assumption, *a priori*, that the climatic variable selected is the main one accounting for most of the variance in the tree-growth record. However, as discussed above, tree growth is generally too complex to be usefully equated with climate using only one variable. A more objective approach (level II in Table 10.1) is the use of multiple regression techniques to select from a variety of climatic variables those which are primarily responsible for variance in the tree-growth record (Ferguson 1977). This empirical approach allows the data to "speak for itself," and involves no *a priori* assumptions other than the selection of variables entered into the regression initially as possible predictors (or independent variables). The analysis results in an equation expressing the response of the tree (the dependent variable) to variations in the most important climatic variables, and this is known as a response function. In practice, response functions are always multivariate, reflecting the complexity of the tree growth – climate relationship.† One of the difficulties in multiple regression is the fact that climatic variables are themselves often highly intercorrelated. For example, July temperature and July precipitation may exhibit a strong negative correlation. In such a case it would be problematical to determine whether incorporating the variable for July temperature in a regression equation would truly reflect the relationship of the tree to temperature in that month or to precipitation amounts or to some combination of both. A way around this is to express the variance of the climatic data in terms of principal components or eigenvectors and to use these as predictors in the regression procedure (level IIIa). Principal components analysis involves statistical transformations of the original (intercorrelated) data set to produce a set of orthogonal (i.e. uncorrelated) eigenvectors (Grimmer 1963, Stidd 1967, Daultrey 1976). Each eigenvector is a variable which expresses part of the variance in the data set (which is usually expressed in terms of "departures from long-term averages" or anomaly patterns). There are as many eigenvectors as original variables, but most of the original variance will be accounted for by only a few of the eigenvectors. The primary eigenvectors can be thought of as preferred modes of distribution of the data set and account for most of its variance (Mitchell *et al.* 1966). Subsequent eigenvectors account for minor amounts of the remaining variance. The value or amplitude of each eigenvector will vary from year to year, being highest in the year when that particular combination of climatic variables which the eigenvector represents, is most apparent. Conversely, it will be

†It is worth noting that level I calibrations may be used successfully *if* response function analysis indicates that the climatic variables influencing ring widths can be conveniently grouped. For example, white oak in Iowa responds to annual precipitation there in all 12 months of the year, according to response function analysis. Thus annual precipitation can be used in a straight-line regression, with growth indices from a white oak chronology as the independent variable (Blasing *et al.* 1981).

lowest in the year when the inverse of this combination is most apparent in the data. By using eigenvector amplitudes as independent (prediction) variables in the stepwise regression procedure, a higher proportion of the dependent data variance can be accounted for by fewer variables than would be possible using the "raw" climatic data themselves.

These methods have all focused on the relationship between tree growth on an individual site (as expressed in terms of the master chronology of ring-width indices) and its response to climate in the area. Similar methodology can be applied to studying the way in which a network of trees responds to a specific climatic, or climatically related, parameter. In this case, variance of the tree-growth data is expressed in the form of eigenvectors, each one thus representing a spatial pattern of growth variation (level IIIb, Table 10.1). Amplitudes of these eigenvectors are then used as independent variables in the multiple regression analysis. The resulting equation is termed a transfer function, whereby spatial patterns of growth records are "transferred" into climatic estimates.

Simple transfer functions express the relationship between *one* climatic variable and *multiple* growth variables. A more complex step (level IV in Table 10.1) is to relate the variance in multiple growth records to that in a multiple array of climatic variables. To do so, each data matrix, made up of data representing variations in both time and space, is converted into its principal components or eigenvectors; these are then related using multiple canonical correlation and regression techniques (Clark 1975). This involves identifying the variance which is common to individual eigenvectors in the two different data sets and defining the relationship between them. The importance of the technique is that it allows spatial arrays (maps) of tree-ring indices to be used to reconstruct maps of climatic variation through time. At present, these are the most complex models used in dendroclimatic reconstruction and result in the most sophisticated year-by-year paleoclimatic reconstructions ever obtained from proxy data (Fritts *et al.* 1979).

Before concluding this section on calibration, it is worth noting that tree-ring indices need not be calibrated directly with climatic data. The ring-width variations contain a climatic signal and this may also be true of other natural phenomena which are dependent in some way on climate. It is thus possible to calibrate such data directly with tree rings and to use the long tree-ring records to reconstruct the other climate-related series. In this way, dendroclimatic analysis has been used to reconstruct runoff records (Stockton 1975, Stockton & Boggess 1980), lake-level variations (Stockton & Fritts 1973), sea-surface temperatures (Douglas 1973), and even albacore tuna populations off the California coast (Clark *et al.* 1975). Some of these applications are discussed in more detail in subsequent sections.

10.3 Dendroclimatic reconstructions

10.3.1 *Models derived by stepwise multiple regression (level II)*

Nearly all modern dendroclimatic studies involve the use of multivariate statistics to define the relationship between climate and tree growth (levels II and III in Table 10.1). In level II models, the basic equation (assuming linear relationships) is of the form

$$y_t = a_1x_{1^t} + a_2x_{2^t} + a_3x_{3^t} \ldots + a_m x_{m^t} + b,$$

where, in the case of response functions (Sec. 10.2.4), y_t is the tree growth index value for year t; b is a constant; $x_1, \ldots x_m$ are climatic variables in year t; and $a_1, \ldots, a_m$ are weights or regression coefficients assigned to each climatic variable in order to obtain the estimate of y_t. In effect, the equation is simply an expansion of the linear equation, $y_t = ax_t + b$, to incorporate a larger number of terms, each additional variable "accounting for" more of the variance in the ring-width data (Ferguson 1977). Theoretically, if all of the factors governing tree growth (including their interactions) could be considered, it would be possible to construct an equation to predict the value of y_t precisely. However, even in this situation the predicted value would only be a point at the center of a zone of probability since each regression coefficient in the equation has an associated standard error. As increasing numbers of variables are added to the multiple regression equation, so the zone of uncertainty in the calculation of y_t increases due to the additive effect of all the coefficient standard errors. There is thus little point in using a large number of independent variables to account for 100% of the variance in the tree-ring record because the confidence limits (or probability range) would be so large as to make the estimate virtually worthless. What is needed is an equation which uses the minimum number of climatic variables to account for the maximum amount of variance in the tree-ring record. Commonly, the procedure of stepwise multiple regression is used to achieve this aim (Fritts 1962, 1965). From a matrix of potentially influential climatic variables, the one which accounts for most of the tree-ring variance is selected; next, the variable which accounts for the largest proportion of the remaining variance is identified and added to the equation, and so on in a stepwise manner. Tests of statistical significance, as each variable is selected, enable the procedure to be terminated when a further increase in the number of variables in the equation results in an insignificant increase in variance explanation. In this way, only the most important variables are selected, objectively, from the large array of potentially important influences on tree growth. This approach was taken by Fritts *et al.* (1965) to interpret tree-ring records from south-western Colorado. A "great drought" at the end of the 13th century was thought to have led to prehistoric people in the area abandoning their settlements and migrating to other regions. Several analyses were performed, using measurements of precipitation, maximum, minimum, and mean temperatures at nearby weather stations, averaged over varying time periods, as the independent (predictor) variables. In each case, the dominant control on tree growth in the area was precipitation, followed by maximum, mean, and minimum temperatures in order of importance. Furthermore, precipitation in the months prior to the growth season was most significant for growth whereas temperatures at the beginning of the growth season, and during the previous season, were significantly

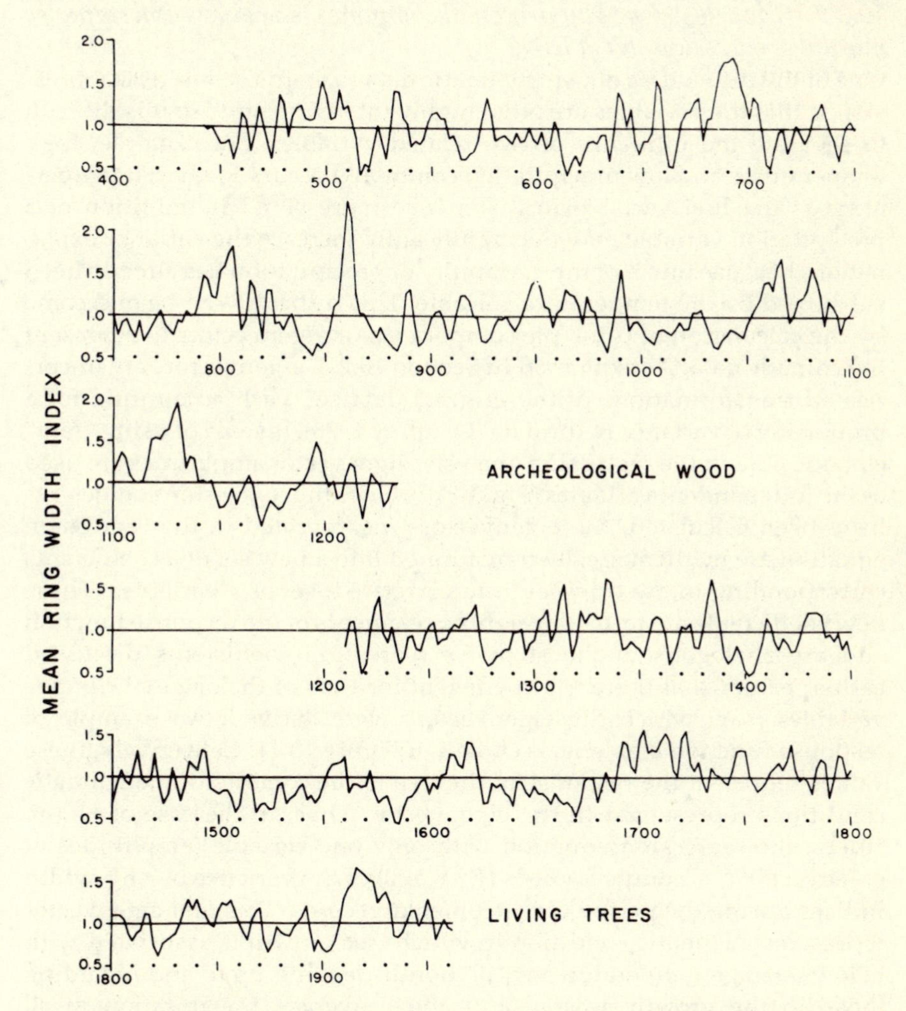

Figure 10.9 Five year running means of ring width indices from *Pseudotsuga menziesii* at Mesa Verde, Colorado, corrected for autocorrelation and plotted on every even year from AD 442 through 1962 (after Fritts *et al.* 1965).

(inversely) related to tree growth. Knowing this, long-term variations of tree growth in the area could be interpreted as being primarily a record of August to May precipitation, with low growth associated with low precipitation amounts and high temperatures. The record of ring-width indices (Fig. 10.9) appears to show that although the drought of the late 13th century was pronounced, it was no more significant than several other similar dry spells in the preceding and subsequent periods. It could thus be concluded the drought was only one of several factors contributing to settlement abandonment in south-western Colorado at this time.

10.3.2 Models derived by principal components analysis and stepwise multiple regression (level III)

One of the difficulties of using climatic data in stepwise multiple regression is that the variables are often highly intercorrelated, so it is difficult to separate the influence of two related variables. For example, high temperatures and low precipitation commonly occur together; if temperature is the first variable in a stepwise regression, the addition of a precipitation variable may not significantly increase the variance explanation, because much of the variability of precipitation has already been subsumed by the temperature variable. This difficulty can be overcome by the calculation of principle components or eigenvectors to represent the climatic data. As explained in Section 10.2.4, eigenvectors are uncorrelated transformations of the original data set, each accounting for a proportion of variance in the data (Daultrey 1976). Instead of using "raw" climatic data in the regression analysis, eigenvector amplitudes are used as the independent variables (Fig. 10.10). Once the regression coefficients have been calculated, the eigenvectors incorporated in the regression equation are mathematically transformed into a new set of n coefficients corresponding to the original (intercorrelated) set of n variables. These new coefficients are termed weights or elements of the response function and are analogous to the stepwise regression coefficients discussed earlier, except that there is a coefficient for each of the original climatic variables from which the eigenvectors were derived. An example of response function elements is shown in Figure 10.11. Collectively, these values represent the response of the tree to the combination of climatic conditions represented in the eigenvector. Thus, in the case of Figure 10.11a, the regression equation with only one variable (amplitudes of eigenvector 1) accounts for 36% ($R^2 \times 100$) of the variance of ring-width indices during the period of instrumental records. This first eigenvector represents a climatic condition in which tree growth is associated with below average temperatures in all months leading up to and including those in the growth season, and above average precipitation in all months. Note that the 95% confidence limits on these weights are small since they are based on only one variable. Figure 10.11b shows the

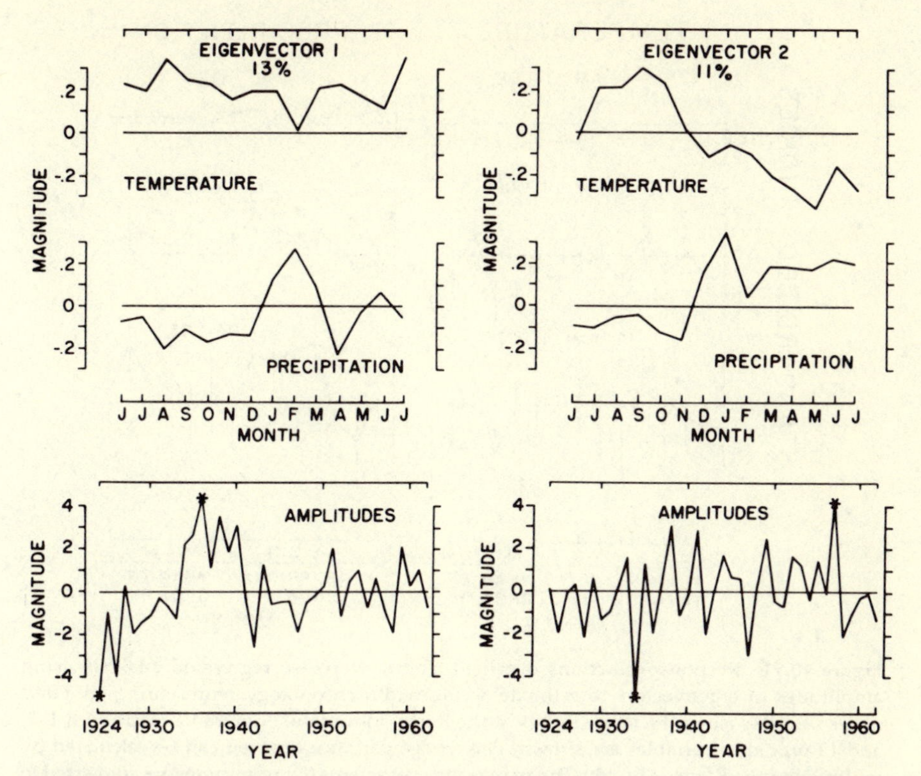

Figure 10.10 Magnitudes of the elements of the first and second eigenvectors of climate at Mesa Verde, southwestern Colorado, and their corresponding amplitude sets. In eigenvector 1 (which reduces 13% of the climatic variance) the eigenvector elements for temperature are all the same sign; the corresponding signs for ten elements for precipitation have the opposite sign. This arises because temperatures throughout the 14 month period are somewhat positively correlated with each other, but they are negatively correlated with precipitation for ten out of 14 months. In eigenvector 2 (which reduces 11% of the climatic variance) the eigenvector expresses a mode of climate in which the departures of temperature for July to November are opposite in sign to those for December to July. All elements for precipitation have signs opposite those for temperature, indicating a generally inverse relationship. The eigenvectors are multiplied with normalized climatic data to obtain the amplitude sets. Asterisks mark those elements with the largest positive and negative values, indicating a climatic regime for the year which most resembles the eigenvector in question (either positively or negatively) (after Fritts 1976).

response function weights resulting from an equation utilizing three eigenvector variables; these account for 67% of the tree-growth variance. Using this equation, ring-width indices are inversely related to temperature in most months, but positively correlated with precipitation. May of the growth year and September of the previous year are the only months for which temperature is positively and significantly related to growth. Note also that the wider 95% confidence limits span zero in many months, making interpretation very difficult. As more eigenvector vari-

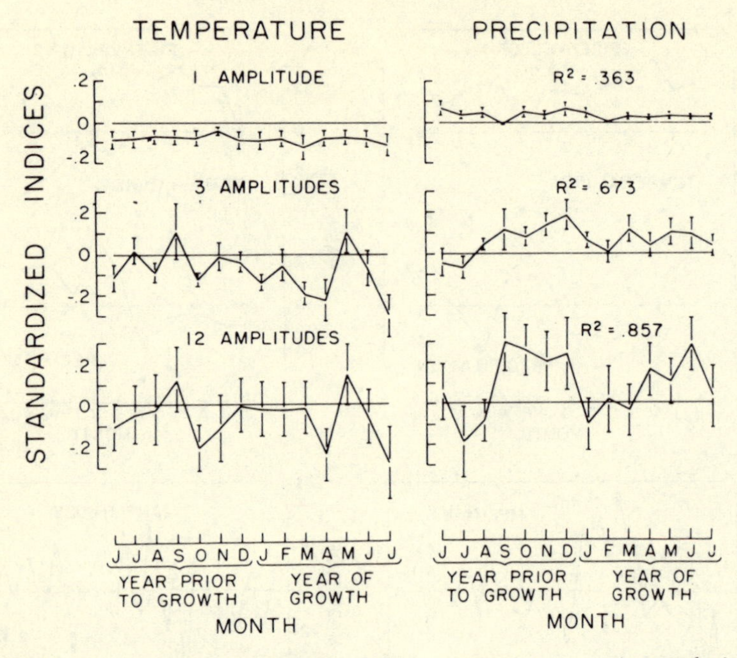

Figure 10.11 Response functions obtained from a stepwise regression analysis using amplitudes of eigenvectors to estimate a ring-width chronology representing six *Pinus ponderosa* sites along the lower slopes of the Rocky Mountains, Colorado. Steps with 1, 3, and 12 predictor variables are shown. Percentage variance reduced can be calculated by multiplying the R^2 value by 100. The regression coefficients for amplitudes are converted to response functions though when response functions are complex, as in this example, a linear combination of many eigenvectors is needed to obtain the best fitting relationship (after Fritts 1976).

ables are added to the equation (Fig. 10.11c) the percentage explanation increases, but the confidence limits increase also and the exact relationship of each response function element to tree growth becomes more uncertain. Ideally then, one would aim to achieve an equation with the minimum number of eigenvectors and the maximum percentage explanation.

As an example of how these complex calculations can be used for paleoclimatic reconstruction, consider the work of LaMarche (1974). LaMarche studied ring-width variations of bristlecone pines in the White Mountains of California. Here, the tree occupies a distinct altitudinal range, on the flanks of the mountains. Ecological studies indicate that trees at the upper and lower forest borders respond differently to climate, so analysis of ring widths in both sites may yield paleoclimatic information unobtainable from either record alone. In order to quantify the tree ring–climate relationship at each site, local climatic data was expressed in terms of eigenvectors and used as independent variables in

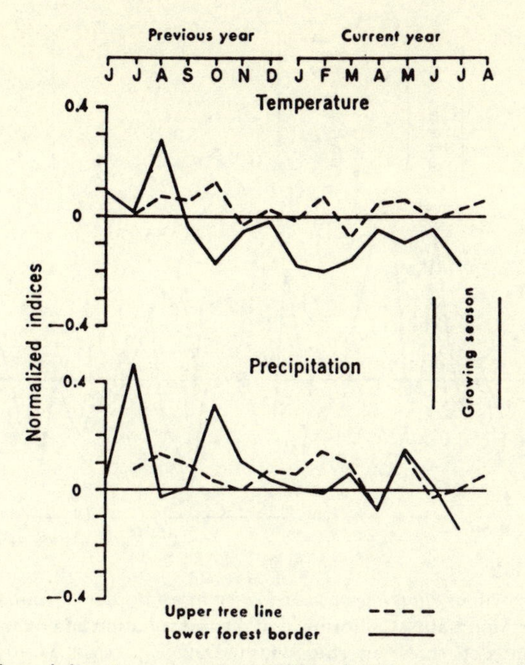

Figure 10.12 Effect of climate on tree-ring width in bristlecone pines (*Pinus longaeva*) of the White Mountains (California) shown by response functions of trees at the upper treeline (dashed line) and lower forest border (solid line). The response functions relate normalized ring-width indices to temperature and precipitation over a 14 month period prior to and including the growing season. The generally positive effect of high temperatures on ring width at the upper treeline contrasts with a predominantly negative effect at the lower forest border. Precipitation is favorable to growth at both sites (after La Marche 1974).

a stepwise multiple regression analysis of ring-width variations. Specifically, monthly and mean temperature and monthly precipitation data from nearby weather stations, for the period from June of the year prior to the growth increment to August of the growth year (30 variables in all), were used to obtain the climate eigenvectors (LaMarche & Stockton 1974). The resulting response functions are shown in Figure 10.12. The width of annual rings in low altitude bristlecone pines is largely dependent on moisture, as shown by the positive effect of precipitation in nearly all months considered, and the negative effect of temperature in most months. High precipitation in the previous summer and autumn and in the current spring favors growth of a wide ring during the short summer growing season. High temperatures lead to depletion of soil moisture and drought stress in the trees, resulting in lower net productivity. By contrast, trees at the upper treeline are less limited by moisture availability and are more directly dependent on monthly temperatures in almost all months (Fig. 10.12). Thus, tree-ring indices from the upper treeline may be interpreted as a record of temperature, whereas tree-ring

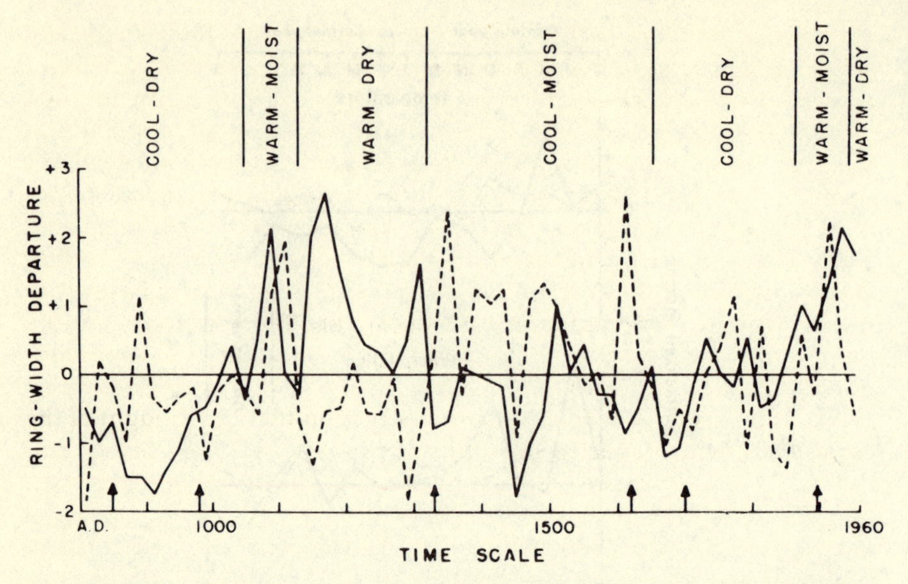

Figure 10.13 Growth of *Pinus longaeva* on lowest forest border (....) and upper treeline (—) sites of the White Mountains, California, and the precipitation and temperature anomalies inferred from the departures in ring width. Data expressed as 20 year averages of standardized normal values. Arrows show dates of glacial moraines in nearby mountains (after LaMarche 1974).

indices from the lower treeline can be considered to be an index of precipitation. Together, then, the ring-width indices enabled combinations of periods of above and below average temperatures as well as above or below average precipitation to be identified (Fig. 10.13). It would appear that conditions similar to those of the last 30–40 years ("warm, dry") were last experienced in the period ~ AD1100–1300, apparently a time of widespread drought (Sec. 10.3.1). This was followed by a period of first "cool, moist," then "cool, dry" climate which collectively spanned a period of ~500 years. In this case, the differing response functions of trees at the upper and lower treelines provided more insight into paleoclimatic conditions than would have been possible using only one set of ring-width data.

In the above examples, ring width was the dependent variable in the multiple regression, with climatic variables (either "raw" or expressed by eigenvectors) used as independent or predictor variables. This may provide a strong mathematical model of ring-width variations but one still has to interpret the early tree-ring record in a qualitative manner. Thus LaMarche (1974) was only able to describe paleoclimates as, for example, "cooler and wetter" but not *how much* cooler or wetter. A more direct calibration of ring-width data is to make climate the dependent variable with ring-width data the predictors. This is usually accomplished by utilizing a number of different ring-width series from a given area and expressing their variance by eigenvectors. These are then used in a stepwise multiple regression to derive an equation which accounts for most of the variance in the climatic variable selected. For example, Cook and Jacoby (1979) selected series of ring-width indices from six different sites in the Hudson Valley, New York, and calculated eigenvectors of their principal characteristics. These were then used as predictors in a multiple regression analysis with Palmer drought severity indices (Palmer 1965)† as the dependent variable. The resulting equation, based on climatic data for the period 1931–70, was then used to reconstruct Palmer indices back to 1694 when the tree-ring records began (Fig. 10.14). It would appear from this reconstruction that the drought of the early 1960s, which affected the entire north-eastern United States, was the most severe the area has experienced in the last three centuries. Further work in this region may provide independent verification of these estimates, using some of the early instrumental records which go back into the 18th century (e.g. Smith *et al.* 1981).

Calibration of the tree-ring records need not be in terms of a single climatic variable or climatic index. Tree rings containing a climatic signal can also be calibrated against other, climatically related time series, and some remarkable work along these lines has been accomplished. Stockton (1975) was interested in reconstructing long-term variations in runoff from the Colorado River Basin, where runoff records only began in 1896. As runoff, like tree growth, is a function of precipitation, temperature, and evapotranspiration, both during the summer and in the preceding months, it was thought that direct calibration of tree-ring widths in terms of runoff might be possible. Using 17 tree-ring chronologies from throughout the watershed, eigenvectors of ring-width variation were computed. Stepwise multiple regression analysis was then used to relate runoff over the period 1896–1960 to eigenvector amplitudes over the same interval. Optimum prediction was obtained using eigenvectors of ring width in the growth year (t_0) and also in years t_{-2}, t_{-1}, and t_{+1}, each of which contained climatic information related to tree growth in year t_0. In this way an equation accounting for 82% of variance in the dependent data set was obtained; the reconstructed and measured runoff values are thus very similar for the calibration period (Fig. 10.15). The equation was then used to reconstruct runoff back to 1564, using the eigenvector amplitudes of ring widths over this period (Fig. 10.16). The reconstruction indicates that the long-term average runoff for 1564–1961 was ~13

†Palmer indices are measures of the relative intensity of precipitation abundance or deficit and take into account soil-moisture storage and evapotranspiration as well as prior precipitation history. Thus they provide, in one variable, an integrated measure of many complex climatic factors.

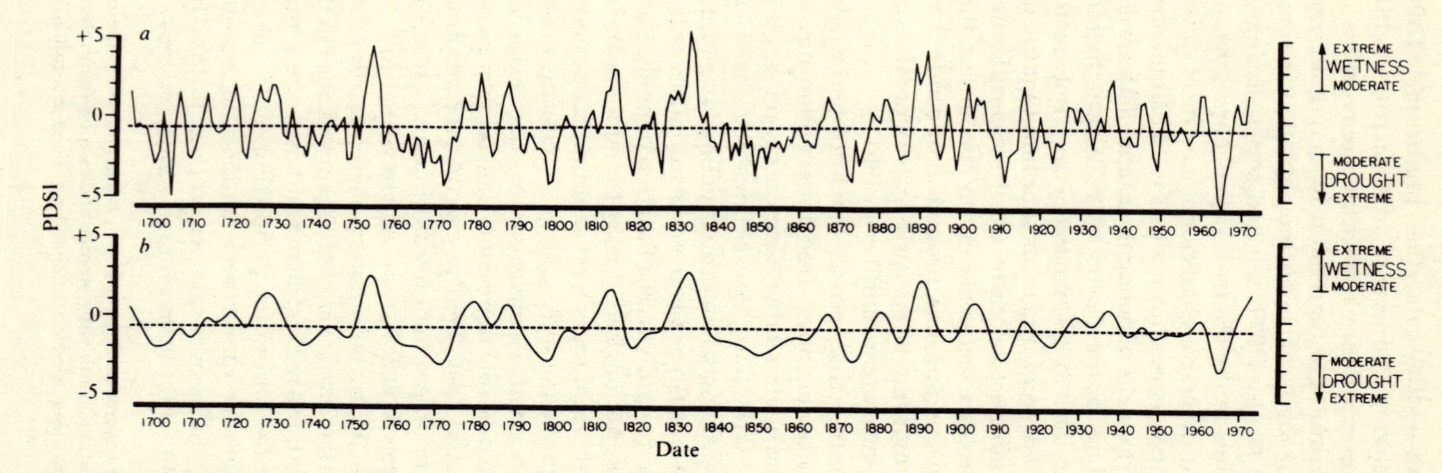

Figure 10.14 July Palmer drought indices for the Hudson Valley, New York, from 1694 to 1972 reconstructed from tree rings. (a) Unsmoothed estimates; (b) a low-pass filtered version of the unsmoothed series that emphasizes periods of ≥10 years (after Cook & Jacoby 1979).

million acre feet (~16 × 10^9 m^3), over 2 million acre feet (2.5 × 10^9 m^3) *less* than during the period of instrumental measurements. Furthermore, it would appear that droughts were more common in this earlier period than during the last century, and the relatively long period of above average runoff from 1905 to 1930 has only one comparable period (1601–21) in the last 400 years. Stockton argues that these estimates, based on a longer time period than the instrumental observations, should be seriously considered in river management plans, particularly in regulating flow through Lake Powell, a large reservoir constructed on the Colorado River. In this case, "dendrohydrological" analysis provided a valuable long-term perspective on the relatively short instrumental record. Similar work has been accomplished by Stockton and Fritts (1973), who used tree-ring eigenvectors calibrated against lake-level data to reconstruct former levels of Lake Athabasca, Alberta, back to 1810 (Fig. 10.17). Their reconstruction indicated that although the long-term average lake level is similar to that recorded over the last 40 years, the long-term variability of lake levels is far greater than could be expected from the short instrumental record. To preserve this pattern of periodic flooding, essential to the ecology of the region, the area is now artificially flooded at intervals which the dendroclimatic analysis suggests have been typical of the last 160 years.

10.3.3 *Models derived by principal components and multiple canonical regression analysis (level IV)*

The models discussed above involve either calculating the response of a single ring-width index series to a variety of climatic indices (i.e. a response function) or transferring ring-width variations from several sites into estimates of a single climatic or climate-related variable (i.e. a

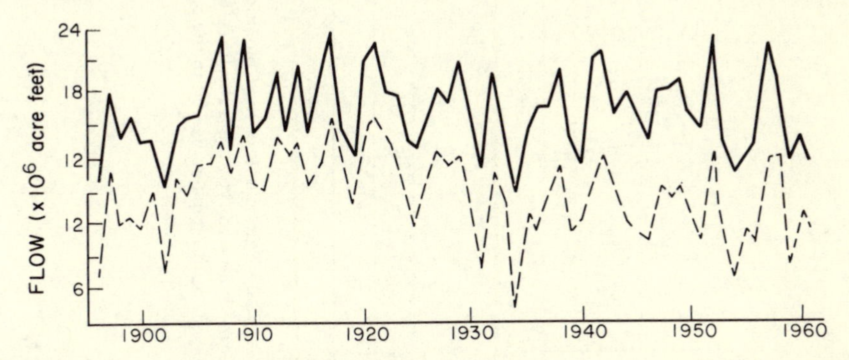

Figure 10.15 Runoff in the Upper Colorado River Basin. Reconstructed values (....) are based on tree-ring width variations in trees on 17 sites in the basin. Actual data, measured at Lee Ferry, Arizona, are shown for comparison (—). Based on this calibration period, an equation relating the two data sets was developed and used to reconstruct the flow of the river back to 1564 (Fig. 10.16) (after Stockton 1975).

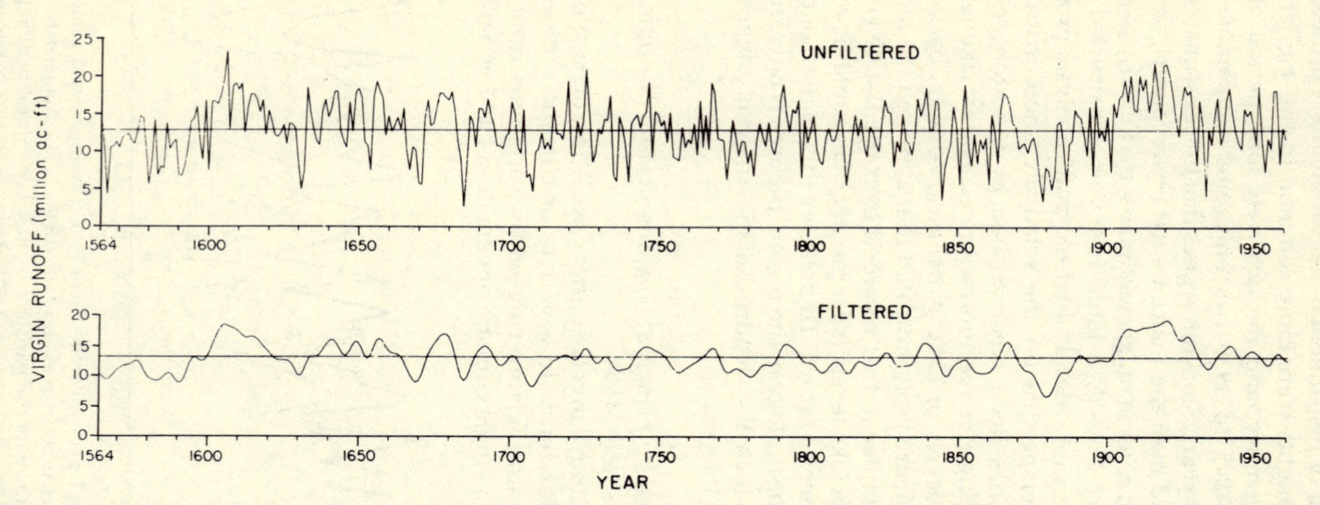

Figure 10.16 Annual virgin runoff of the Colorado River at Lee Ferry, as reconstructed using ring-width index variation, calibrated as shown in Figure 10.15. Growth for each year, and the three following years, was used to estimate river flow statistically. Smooth curve (below) represents essentially a 10 year running mean. Runoff in the period ~1905–25 was exceptional when viewed in the context of the last 400 years (after Stockton 1975).

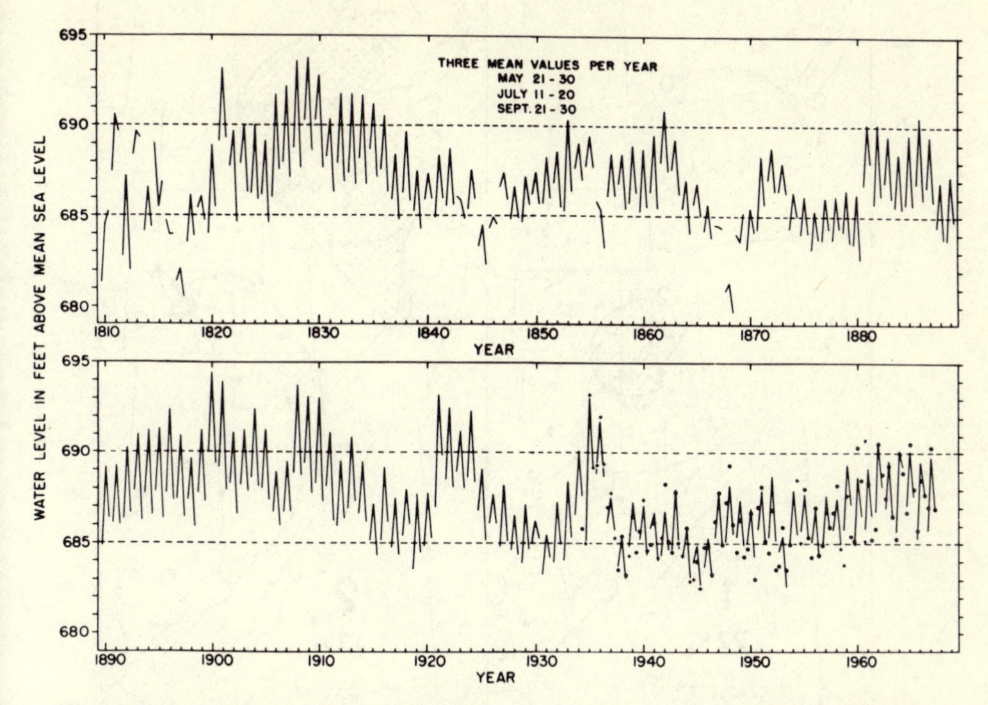

Figure 10.17 Levels of Lake Athabasca, Alberta, Canada, as reconstructed from tree-ring data. Tree rings indicate that prior to 1935 there was greater variability in lake levels during May and July, but there was less variability in lake levels for September than during the recent calibration period. Dots indicate actual lake levels used for calibration. Lines connect the three estimates from tree rings, representing mean lake level for May 21–30, July 11–20, and September 21–30. Points are not connected over the winter season, as calibrations of levels for the frozen lake could not be made (from Stockton & Fritts 1973).

transfer function). At the next level of complexity, a spatial array of tree-ring data is used to reconstruct the spatial distribution of a climatic parameter. This type of analysis has been termed dendroclimatography (Fritts 1976) and was first attempted by Fritts *et al.* (1971) in order to reconstruct variations in sea-level pressure over much of the Northern Hemisphere. Earlier work had shown that, although trees in different parts of western North America responded differently to climate (as indicated by individual response functions), spatial patterns of tree-growth anomaly were quite similar to spatial patterns of precipitation anomaly over the same region and time period (Fig. 10.18; LaMarche & Fritts 1971a). This suggested that anomalous large-scale weather patterns of particular years were associated with tree-growth anomaly patterns for the same years. Large-scale pressure anomaly patterns (as indicators of the anomalous weather patterns) might, therefore, be estimated from corresponding spatial patterns of tree-growth anomaly. To investigate this idea, Fritts *et al.* (1971) used multiple canonical regression analysis to

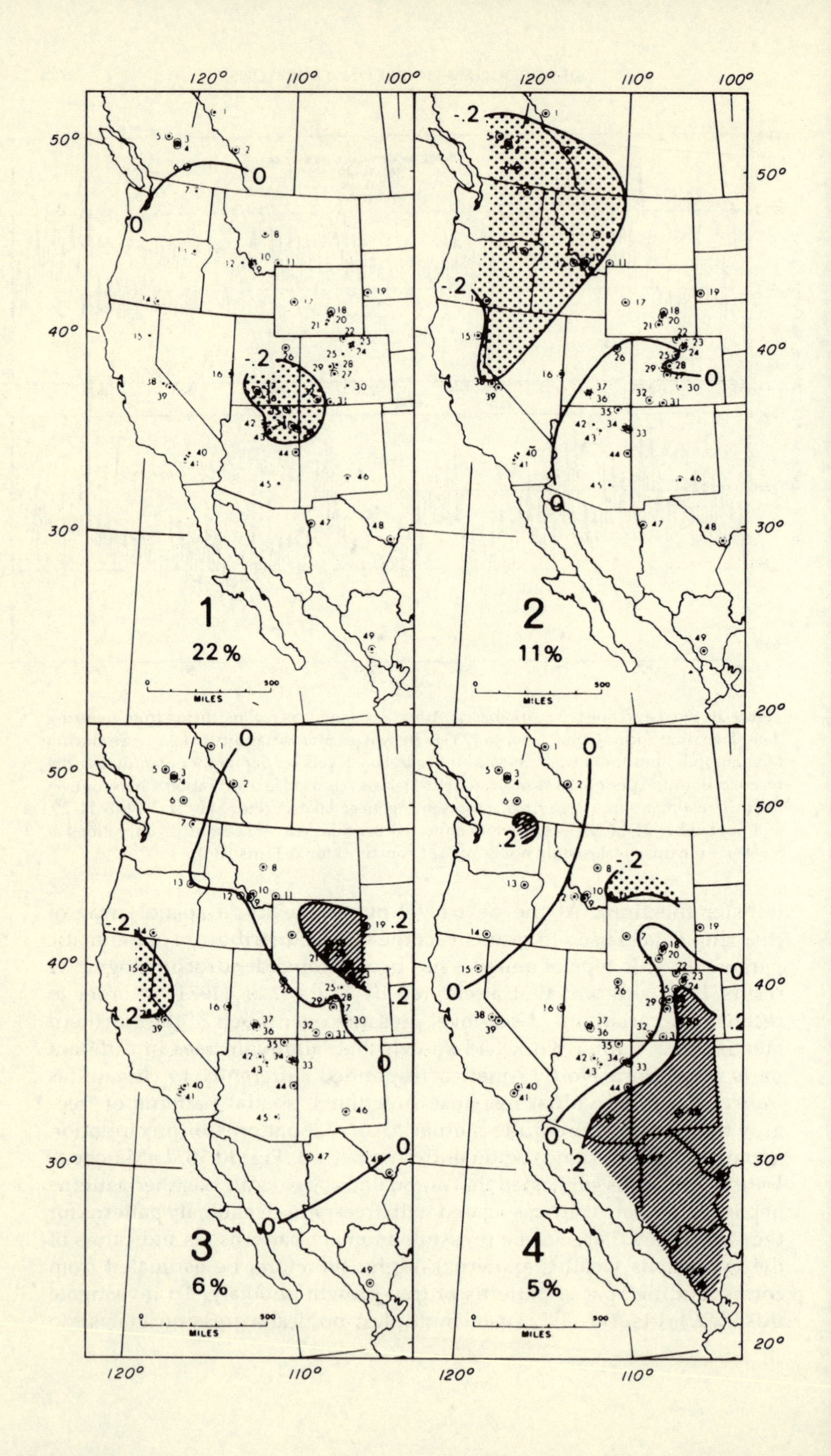
120°
110°
100°
50°
40°
30°
20°
1
22%
2
11%
3
6%
4
5%
MILES
500
0
-.2
.2

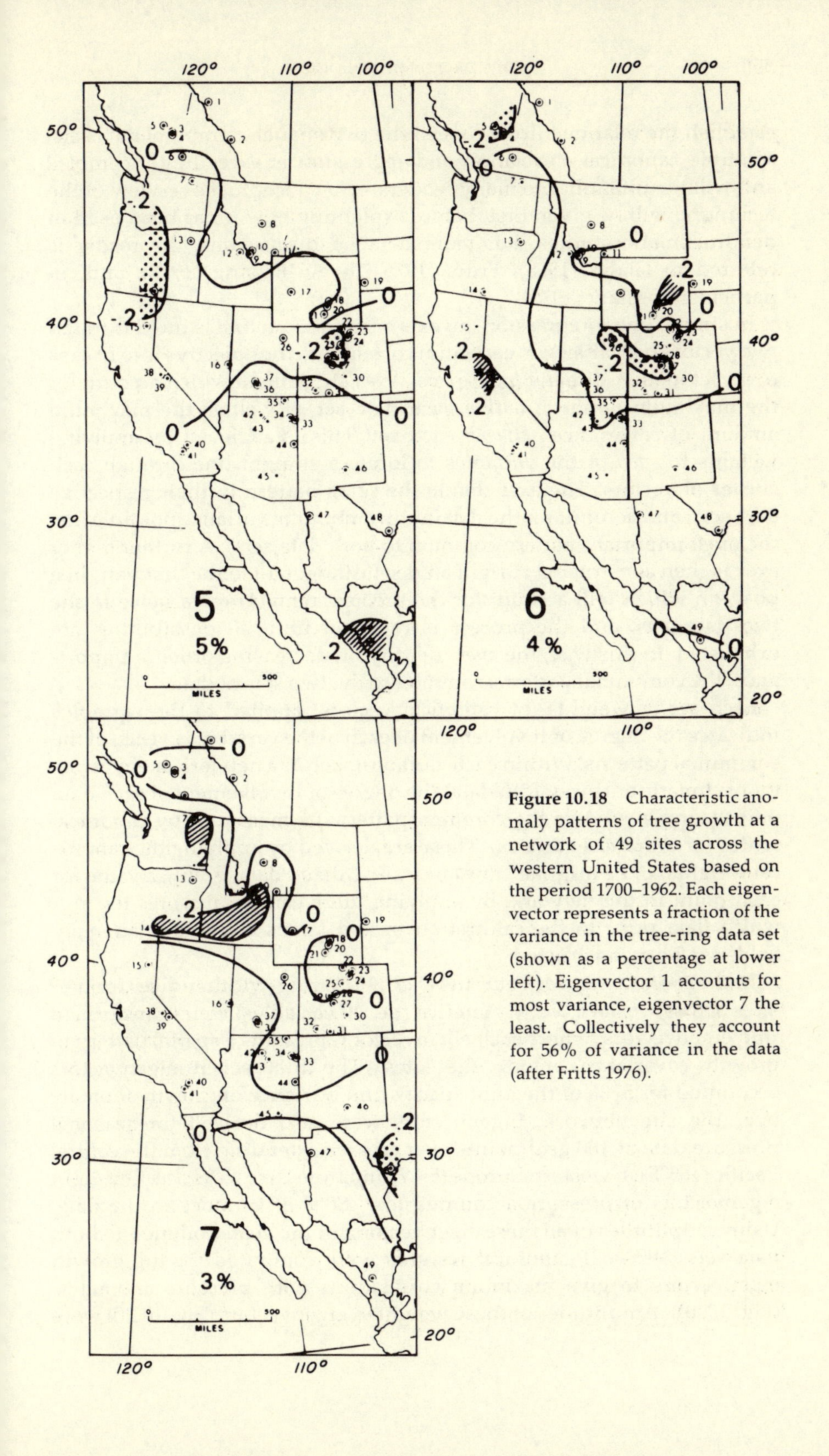

Figure 10.18 Characteristic anomaly patterns of tree growth at a network of 49 sites across the western United States based on the period 1700–1962. Each eigenvector represents a fraction of the variance in the tree-ring data set (shown as a percentage at lower left). Eigenvector 1 accounts for most variance, eigenvector 7 the least. Collectively they account for 56% of variance in the data (after Fritts 1976).

establish the relationships between the two spatial–temporal data sets. Multiple canonical correlation and regression is exceedingly complex and will be unfamiliar to many readers, so a conceptual overview of the technique will be given first, before explaining how it has been used in dendroclimatic studies. For more detailed discussions, the reader is referred to Glahn (1968), Fritts (1976 Ch. 9), Blasing (1978), and, in particular, to Clark (1975).

In simple linear regression, an axis or regression line is fitted through two vectors (variables); in canonical correlation, the objective is to fit axes to n vectors in n-dimensional space. Axes are defined which account for the maximum variance *within* each data set as well as the maximum amount of covariance *between* each set. This is achieved by applying weights to *each* of the variables to locate a straight line through each cluster of vectors. The axes define the same pattern in their respective data sets, and account for the maximum amount of variance possible (i.e. the most important pattern common to both data sets). A further pair of axes is then constructed at right angles (orthogonal) to the first pair, in a position which will account for the greatest remaining variance in the two data sets, and the process is repeated until all possibilities are exhausted. In this way, the axes identify, in decreasing order of importance, the communal patterns common to the two sets of data.

Each set of weights (or canonical vectors) applied to the variables indicates the degree of involvement of each of the variables in each of the communal patterns. Within each variable (such as a network or eigenvector of climatic or ring-width data) the degree of involvement of each data point (sample site) in the common patterns is measured by canonical scores (or canonical variates). These are derived by applying the canonical weights to the original "raw" or standardized data, giving a value for each point in the network. By applying these canonical scores to ring-width data prior to the calibration period, maps of climatic anomaly fields can be reconstructed.

In the pioneering work of Fritts *et al.* (1971) ring-width indices from 49 sites across western North America (i.e. 49 variables) were transformed into eigenvectors, where each eigenvector represents a spatial pattern of growth covariance among the sites. The first seven eigenvectors accounted for 57% of the joint space–time variance of growth anomaly over the site network. Eigenvectors were also derived for seasonal pressure data at 100 grid points over an area extending from the central Pacific (165°E) to western Europe (5°W) and from 25 to 65°N; the first eight eigenvectors of pressure accounted for ~80% of variance in the data. Using amplitudes of all these eigenvectors for the years common to both data sets (1900–62) canonical weights were computed for the growth eigenvectors to give maximum correlations with pressure anomalies (Fig. 10.19). Amplitudes of these weighted eigenvectors (Fig. 10.20) were then used as predictors of normalized pressure departures at each point in the pressure grid network, by applying the canonical weights to the standardized ring-width data. This resulted in estimates of pressure anomaly values at each point in the grid network, for each season, for each year of the ring-width network. Maps of mean pressure anomaly could thus be produced for any interval by simply averaging the individual anomaly values (Fig. 10.21). An estimate of the statistical significance of the departure values is given by identifying those values which exceed two standard errors of the estimate. In the examples given in Figure 10.21, the reconstructions indicate a weak Aleutian low pressure area (above 1900–62 average pressure in the Gulf of Alaska) and a more intense Canadian low pressure area existed in the early 19th century. This was the reverse of conditions in much of the 18th century.

The percentage of pressure variance reconstructed at all grid points in these first attempts ranged from 18.5% in spring to 24% in winter. Comparison of reconstructions made for the calibration period indicated that reconstructions were poor over the North Atlantic, with optimum reconstructions being obtained over North America and areas to the west. Considering that tree growth is affected by synoptic scale systems originating in the Pacific, this was an understandable result. Subsequent work was therefore designed to optimize predictions and various refinements have been made (Fritts 1976). In particular, the tree-ring network has been increased to 65 chronologies and the pressure grid has been "moved" westwards (from 100°E to 80°W) to span more of the "upwind" area. Calibrations now also incorporate eigenvectors of tree growth from the growth year (t_0) as well as years t_{-1}, t_{+1}, and t_{+2}, all of which may contain climatic information relevant to the growth year. In addition, eigenvectors of ring-width series which have been adjusted to remove the effects of the prior year's growth (autocorrelation) have been derived (Fritts *et al.* 1979). These have also been incorporated into the models (designated $M(t_{-1})$ to $M(t_{+2})$), sometimes in addition to the original autocorrelated data. The objective is to use the optimum combination of ring-width eigenvectors, with or without autocorrelation effects, to maximize variance explanation in the calibration data set. In addition to grid-point pressure data, calibrations have now also been made using temperature and precipitation data from the USA and Canada, so reconstructed pressure, temperature, and precipitation anomaly maps can be compared. Figure 10.22 shows the percentage variance accounted for in the three calibrations of winter data, using some of the best models so far derived (Fritts *et al.* 1979). For pressure, the model utilizes as predictors eigenvectors of ring widths for (a) year t_0 combined with year t_{-1}; (b) year t_{-1} combined with the series for year t_0 with autocorrelation removed ($M(t_0)$); (c) a combination of $M(t_0)$ and $M(t_{+1})$; (d) a combination of t_{+1} and $M(t_{+1})$. This very complex analysis yields estimates of winter pressure for

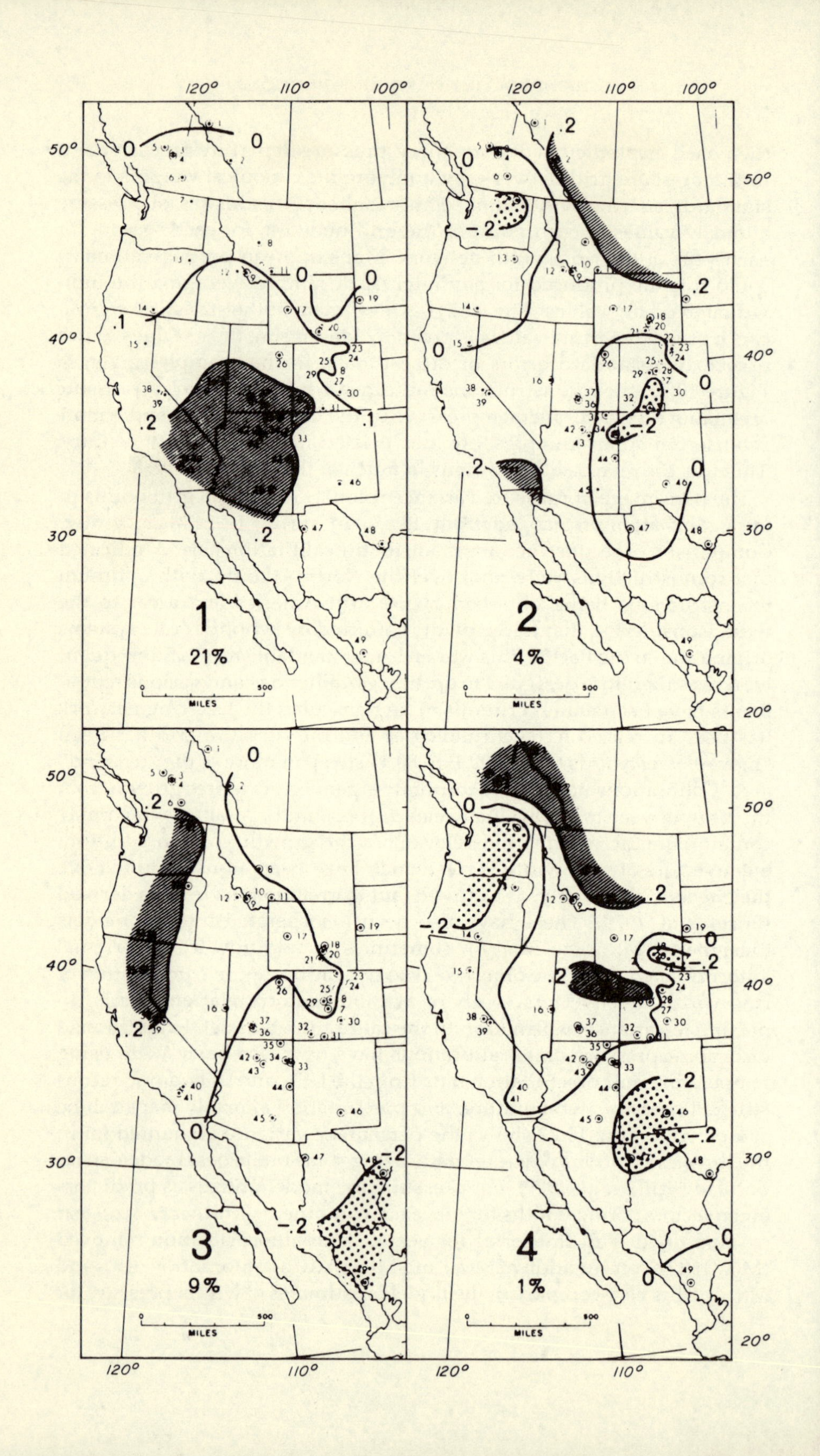

120°
110°
100°
50°
40°
30°
20°
1
21%
2
4%
3
9%
4
1%
MILES
0
500
0
.1
.2
-.2

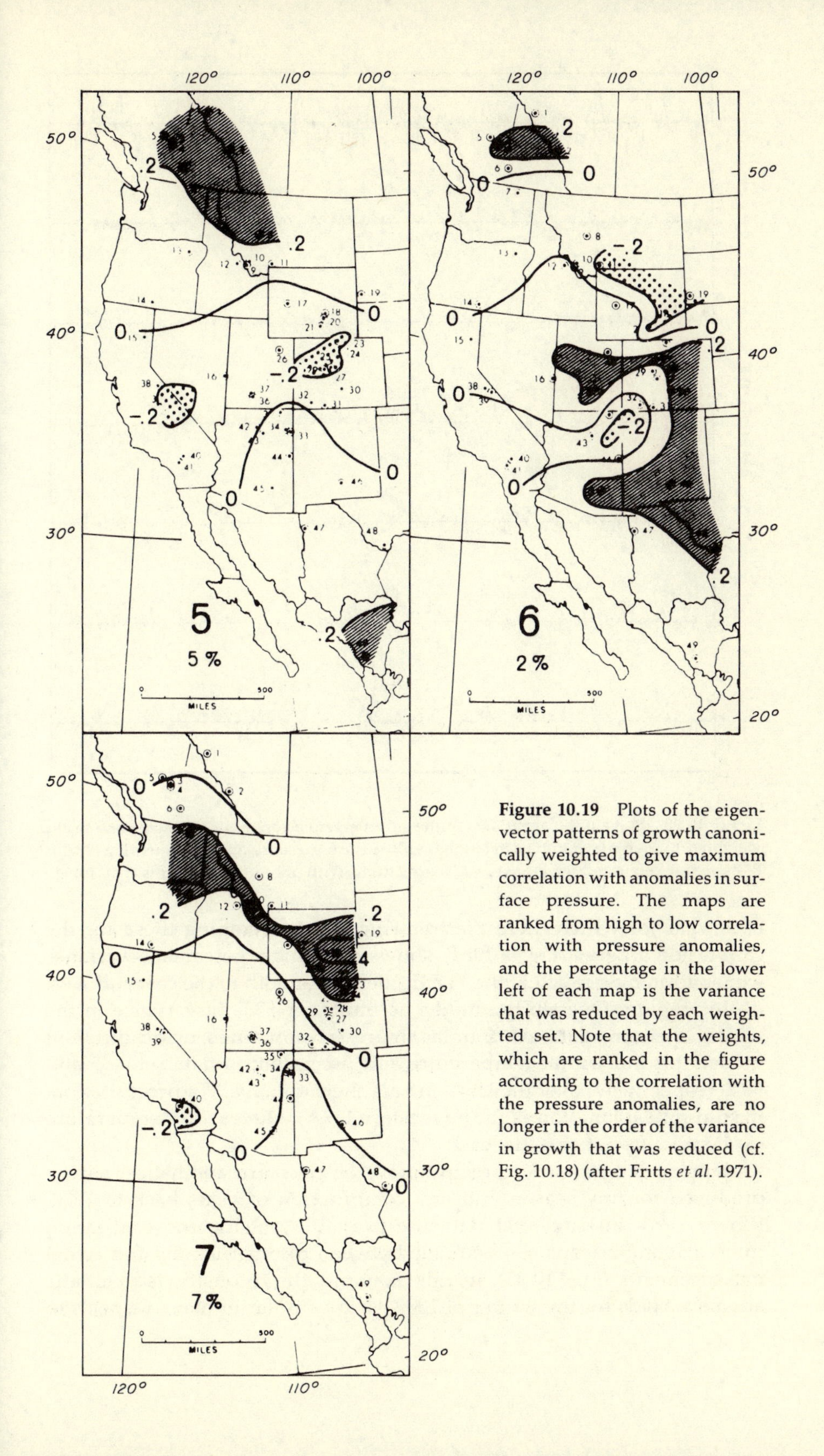

Figure 10.19 Plots of the eigenvector patterns of growth canonically weighted to give maximum correlation with anomalies in surface pressure. The maps are ranked from high to low correlation with pressure anomalies, and the percentage in the lower left of each map is the variance that was reduced by each weighted set. Note that the weights, which are ranked in the figure according to the correlation with the pressure anomalies, are no longer in the order of the variance in growth that was reduced (cf. Fig. 10.18) (after Fritts *et al.* 1971).

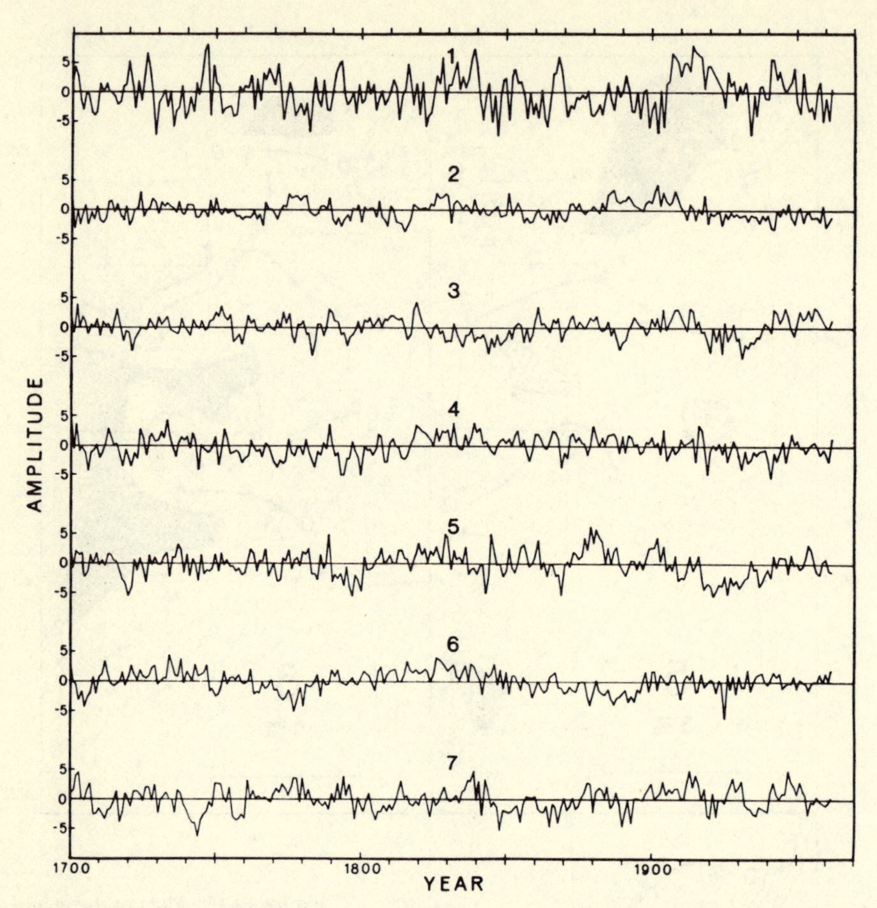

Figure 10.20 Plots of the *amplitudes* of the seven eigenvectors of tree growth canonically weighted to give maximum correlations with surface pressure anomalies (cf. Fig. 10.20). Plots are ranked from high (1) to low (7) correlations with pressure (after Fritts *et al.* 1971).

each year for which there are tree-ring data. Averaging these for the calibration interval (1899–1961) shows very high percentage variance explanation over much of the Pacific, declining both to the east and west of this region. Overall, the model accounts for 53% of variance in the winter pressure data set. Similar figures are obtained using different models for the winter temperature and precipitation data sets (55 and 50% respectively) over the USA and southern Canada. For precipitation, maximum explanation is in the western USA, whereas for temperature explanation is better in the east.

Maps of temperature, precipitation, or pressure anomalies can be produced for any season and any combination of years back to 1600. Where early instrumental data are available outside the calibration interval, comparisons can be made between reconstructions and actual measurements (Fig. 10.23). In this example, the reconstructed climatic anomaly fields for the winter of 1889–90 are similar in many respects to

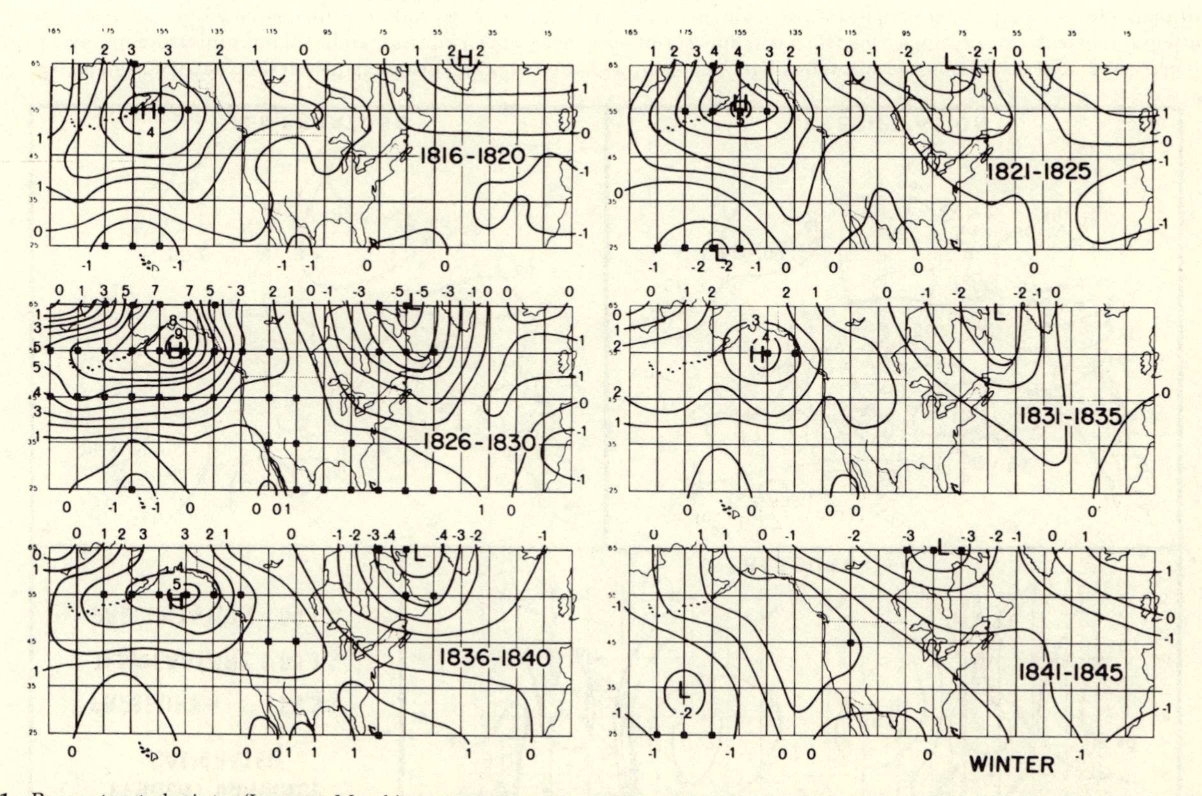

Figure 10.21 Reconstructed winter (January–March) pressure anomaly patterns 1816–45, based on ring width indices from a network of stations in the western United States. Square dots indicate the five year mean anomaly at the grid point was greater than twice the standard error (from Fritts *et al.* 1971).

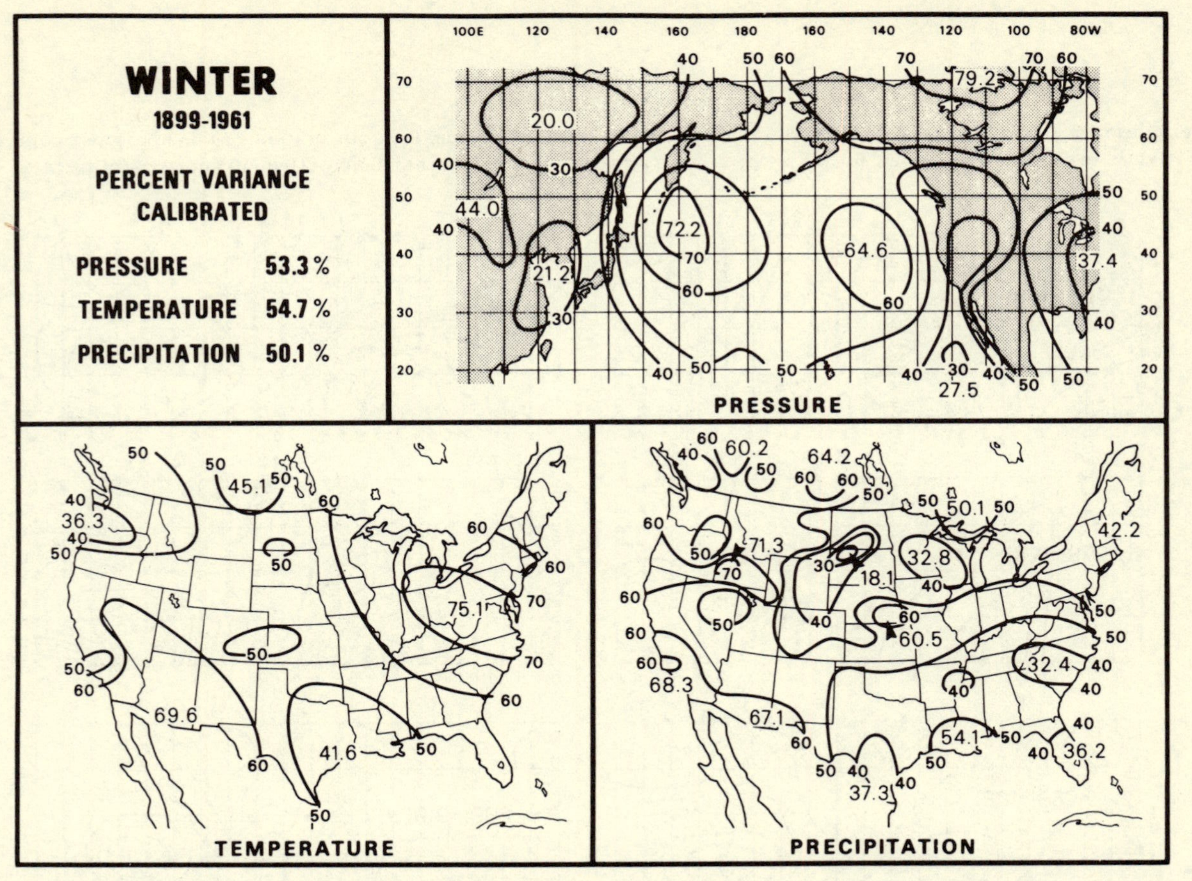

Figure 10.22 Extent to which tree rings model climate in calibration interval 1899–1961. Maps show the percentage variance reduced in climatic data over the United States (temperature and precipitation) and Pacific Basin (pressure) using tree-ring data. Isolines are of equal percentage reduction in variance. Overall variance calibrated shown in upper left panel. Variance in some areas is represented better by tree rings than in others (cf. precipitation in southern California and the upper Mid-West) (from Fritts *et al.* 1979).

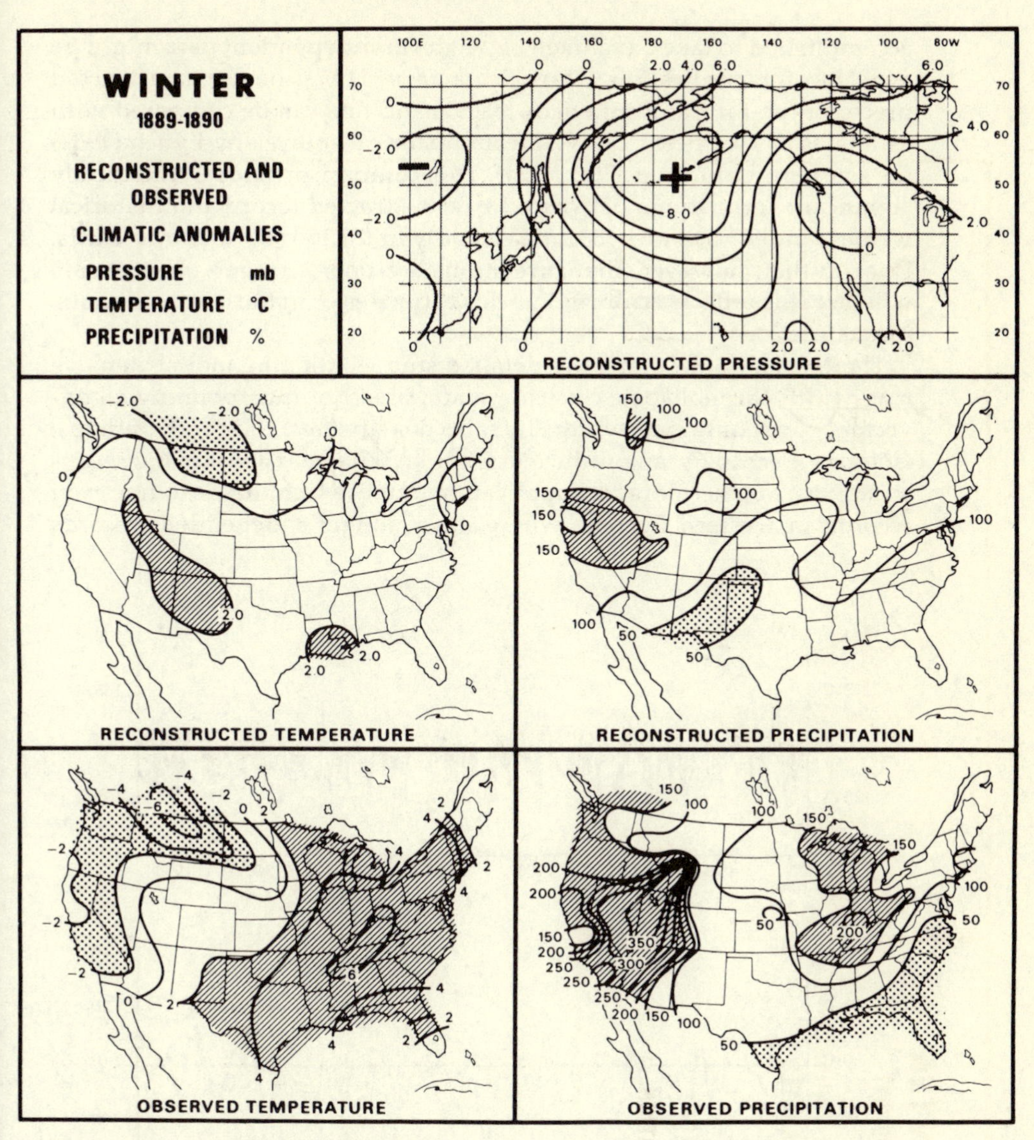

Figure 10.23 Tree-ring reconstructions of pressure, temperature, and precipitation anomalies along with available meteorological observations of temperature and precipitation for winter 1889–90. Anomalies are expressed as departures from the mean period 1901–70 (after Fritts *et al.* 1979).

the (independent) instrumental records. However, early data are quite sparse, particularly in the western USA, and very few records extend from 1889 to the end of the calibration period, so instrumental verification in this area is quite difficult (Bradley 1976a). Reconstructed data can be averaged for any region within the anomaly field and this may be accomplished to take advantage of whatever independent data might be available for verification. Thus Fritts *et al.* (1979) have reconstructed precipitation for California back to 1600 and this can be compared with estimates of "wetness" derived from historical sources by Lynch (1931) for southern California (Fig. 10.24). This comparison shows statistically significant correlations between the reconstructed record and historical wetness index back to ~1831, particularly in the low-frequency records. Prior to that, however, the correlations are poor and one must decide whether this reflects errors in Lynch's estimates or in the dendroclimatic reconstruction.

There is no doubt that more detailed studies utilizing more extensive ring-width chronologies, covering wider areas, or incorporating eigenvectors of not only ring-widths but also density (Sec. 10.5) and, perhaps, isotopes (Sec. 10.6) can produce even more reliable paleoclimatic reconstructions. Furthermore, the application of these techniques to tree-ring records in western Europe, where much longer independent records

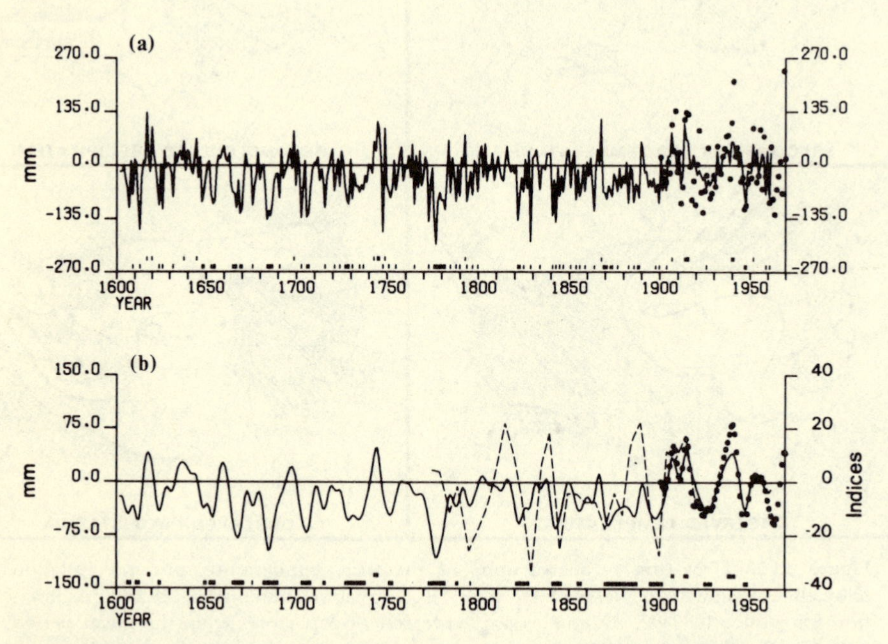

Figure 10.24 Reconstruction of winter precipitation for California, expressed as departures from the 1901–70 observational record. (a) Unfiltered data; (b) data are smoothed by a 13 weight, low-pass digital filter. Dots represent actual rainfall data in the calibration period (1901–70). Vertical marks on the bottom indicate values which equal or exceed the 0.95 confidence level based on the standard deviation of the residuals. Year designations are for January of each winter season. Dashed lines indicate filtered annual rainfall indices derived by Lynch (1931) for southern California, plotted at 5 year intervals from 1775 to 1925 (after Fritts *et al.* 1979).

exist, has tremendous potential. Indeed, early work using simple models, but incorporating both ring-width and densitometric data, has produced reconstructions which are extremely accurate, according to tests with independent data (Schweingruber *et al.* 1978). It is likely that the number and quality of paleoclimatic reconstructions based on dendroclimatic analysis will greatly expand over the next decade.

10.4 Verification of climatic reconstructions

As indicated in the previous section, it is a necessary step in dendroclimatic analysis, to test or verify the reconstructed climatic data in some way. This poses severe problems because, by definition, the paleoclimatic reconstruction being attempted is for a period when no instrumental data are available. Two ways around this problem are generally used. First, when calibrating the tree-ring data, very long instrumental records for the area are sought. Only part of this record is then used in the calibration (often ~60 years of data), leaving the remaining early instrumental data as an independent check on the dendroclimatic reconstruction. If the reconstruction is in the form of a map, several records from different areas may be used to verify the reconstruction, perhaps indicating geographical regions where the reconstructions appear to be most accurate. The main difficulty in this approach has been that the primary areas where tree-ring studies have been carried out (i.e. the western USA and northern treelines) are areas with very few early instrumental records (Bradley 1976b). Recent dendroclimatic studies in New England (Cook & Jacoby 1979) and in western Europe (Serre 1978, Schweingruber *et al.* 1978) can be more exhaustively tested because of the much longer instrumental records in these areas. Indeed, it may be fruitful to conduct two calibrations, with tree-ring and climatic data from different time periods (e.g. 1850–1910 and 1910–70) and to compare the dendroclimatic reconstructions for earlier periods derived from the two data sets. This has not yet been attempted.

A second approach is to use other proxy data as a means of verification. This may involve comparisons with historical records, as mentioned above, glacier advances (LaMarche & Fritts 1971b) or pollen variations in varved lake sediments (Fritts *et al.* 1979), etc. It may even be possible to use an independent tree-ring data set to compare observed growth anomalies with those expected from paleoclimatic reconstructions. Blasing and Fritts (1975), for example, used a network of trees from an area between northern Mexico and southern British Columbia to reconstruct maps of sea-level pressure anomalies over the eastern Pacific and western North America. A separate temperature-sensitive data set from Alaska and the North West Territories of Canada was then used to test the reconstructions. Periods of anomalously low growth in the northern trees were associated with increased northerly airflow as predicted by the pressure reconstructions (Fig. 10.25). Further discussion of verification tests, and many applications, are given in Fritts (1976, Ch. 9) and Fritts *et al.* (1979).

In all verification tests, one is inevitably faced with two questions:

(a) If the verification is poor, does the fault lie with the dendroclimatic reconstruction (and hence the model from which it was derived) or with the proxy or instrumental data used as a test (which may itself

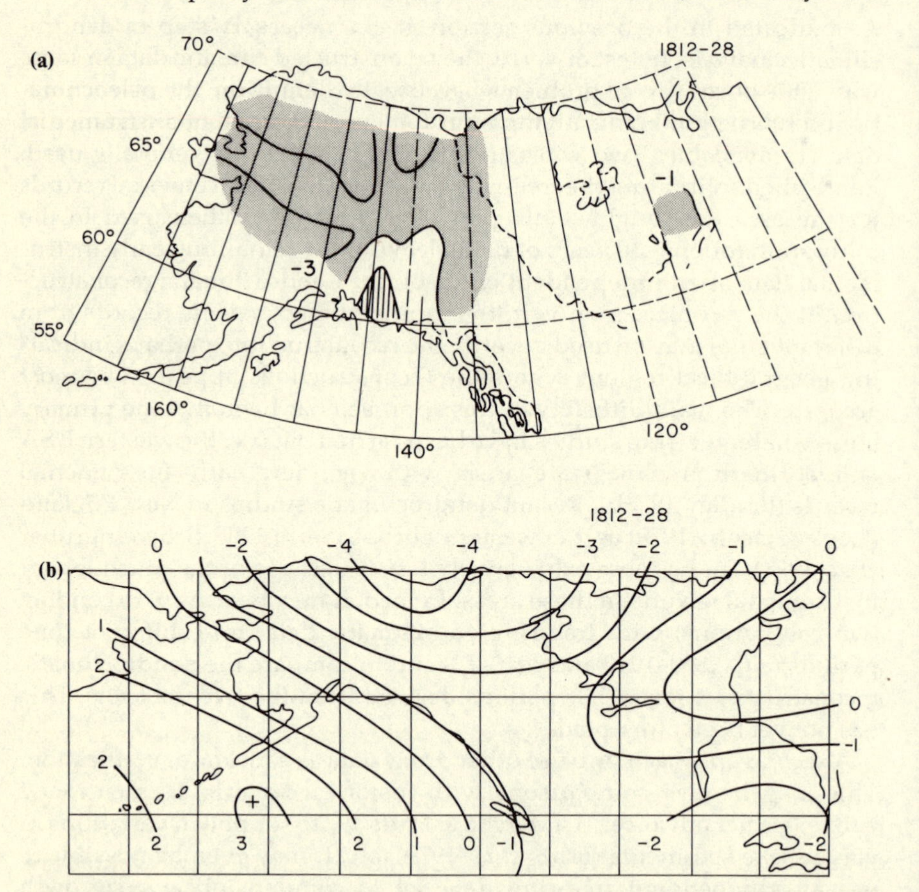

Figure 10.25 One of the major anomaly patterns of Arctic tree growth in the first half of the 19th century (a) and the reconstructed pressure anomalies averaged for the years of that growth type (b) derived from the application of transfer functions to the 49 chronologies of arid North American tree sites. The Arctic tree growth is expressed in standard deviation units as departures from the mean for the period 1800–1939. The years shown are the first and last of those included in the particular growth type (after Blasing & Fritts 1975).

be of poor quality and subject to different interpretations)? In such cases, re-evaluation of the tree-ring data, the model, and the test data must be made before a definitive conclusion can be reached.

(b) Is the dendroclimatic reconstruction for the period when no independent checks are possible as reliable as for the period when verification checks can be made? This might seem an insoluble problem but it is particularly important when one considers the standardizing procedure employed in the derivation of tree-growth indices (Sec. 10.2.3). Errors are most likely to occur in the earliest part of the record, whereas tests using instrumental data are generally made near the end of the tree-growth record, where the slope of the standardization function is lowest, and least likely to involve the incorporation of large error. The optimum solution is for both instrumental and proxy data checks to be made on reconstructions at intervals throughout the record, thereby increasing confidence in the paleoclimatic estimates.

10.5 Densitometric dendroclimatology

Wood density is an integrated measure of several properties including cell wall thickness, lumen diameter, size and density of vessels or ducts, proportion of fibers, etc. (Polge 1970). Tree rings are made up of both earlywood and latewood, which vary markedly in average density and these density variations can be used, like ring-width measurements, to identify annual growth increments and to cross date samples (Fig. 10.26; Parker & Hennoch 1971, Parker 1971). It has also been shown empirically that density variations contain a strong climatic signal and can be used to estimate long-term climatic variations over wide areas (Schweingruber *et al.* 1978, 1979). Density variations are measured on X-ray negatives of prepared core sections (Fig. 10.27) and the optical density of the negatives is inversely proportional to wood density.

Density variations are particularly valuable in dendroclimatology because, like isotopic measurements, they do not change significantly with tree age (i.e. there is no growth function to be removed). Hence they may contain more low-frequency climatic information than can be obtained from standardized ring-width data. Generally, two values are measured in each growth ring: minimum density and maximum density, representing locations within the earlywood and latewood layers, respectively. These values are then calibrated in the same way as with the ring-width data using the statistical procedures described in Section 10.2.4. At present, maximum density appears to be a better climatic indicator than either minimum density or ring widths, at least in the

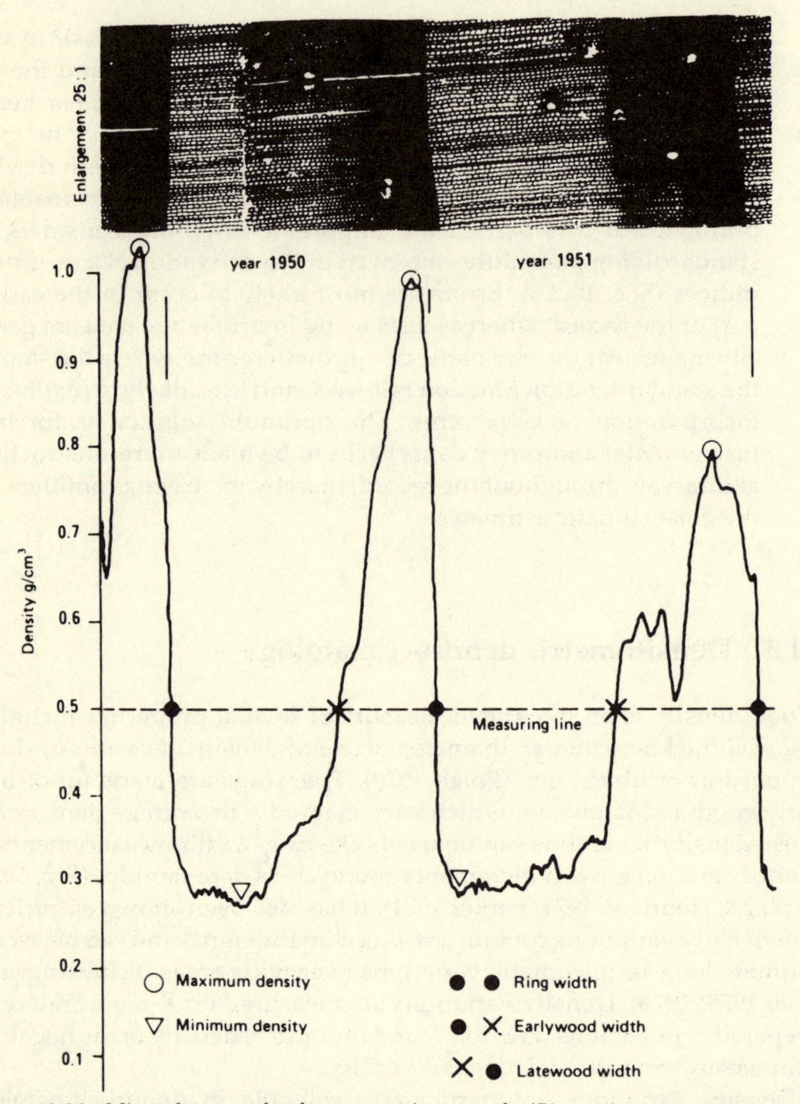

Figure 10.26 Microphotograph of a recent pine sample (*Pinus silvestris*) from northern Germany and the corresponding density profile. The limits of the annual rings, latewood, and earlywood are indicated on the horizontal line (after Schweingruber *et al.* 1979).

relatively complacent samples so far studied in western Europe. However, optimum climatic reconstructions may be achieved by calibrating eigenvectors of both ring widths *and* densitometric data against modern instrumental data, as demonstrated by Schweingruber *et al.* (1978). Using density and ring-width indices, they calculated transfer functions from 60 years of climatic data, then reconstructed temperatures back to the mid-19th century. Comparison with 30 years of independent instrumen-

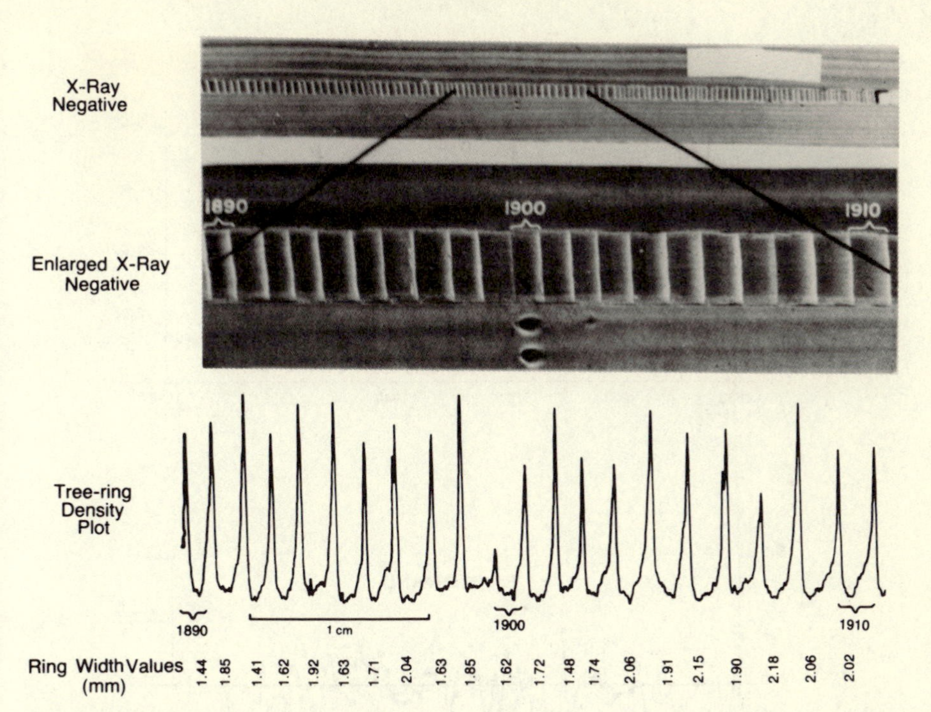

Figure 10.27 Example of a tree-ring density plot based on X-ray negative of a section of wood (above). Minimum and maximum densities in each annual ring are clearly seen (from Jones & Parker 1970).

tal data (Fig. 10.28) shows clearly that the reconstructions are extremely accurate and indeed, "no study has been published with such conclusive verification using independent climatic data" (Schweingruber *et al.* 1978). At present, densitometric dendroclimatology is in its infancy, but the potential value of the field for paleoclimatic reconstructions seems enormous.

10.6 Isotope dendroclimatology

A recent development in dendroclimatology has been the study of isotopic variations in tree rings and their relationships with climate (Jacoby 1980, Gray 1981). The subject is rapidly developing and the results are often equivocal, but the potential at this point seems to be great. The basic premise is succinctly stated by Epstein *et al.* (1976): "since temperature variations over the earth's surface are correlated with D/H and $^{18}O/^{16}O$ variations in meteoric [atmospheric] waters, natural systems which record variations of D/H and $^{18}O/^{16}O$ should preserve a

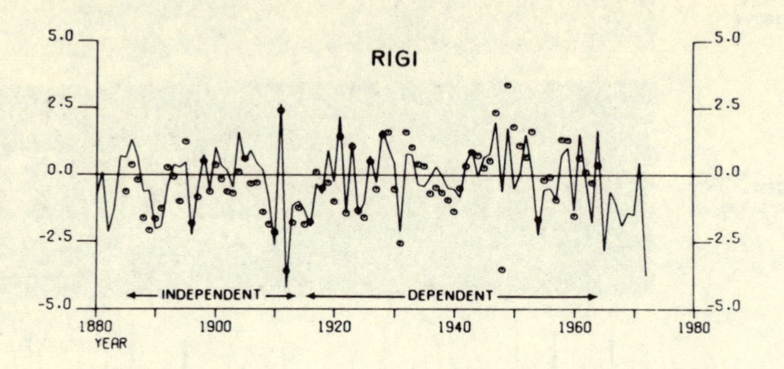

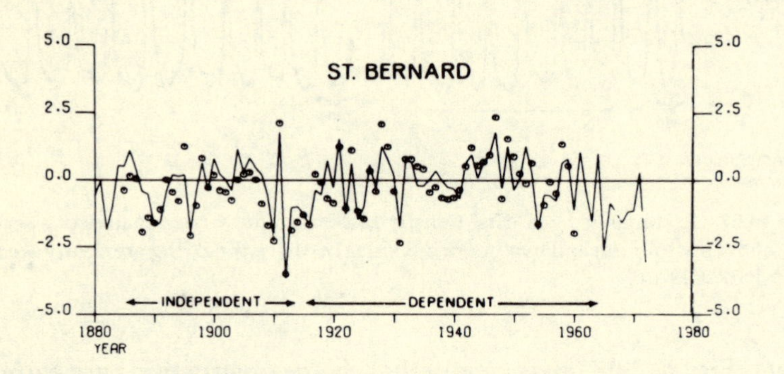

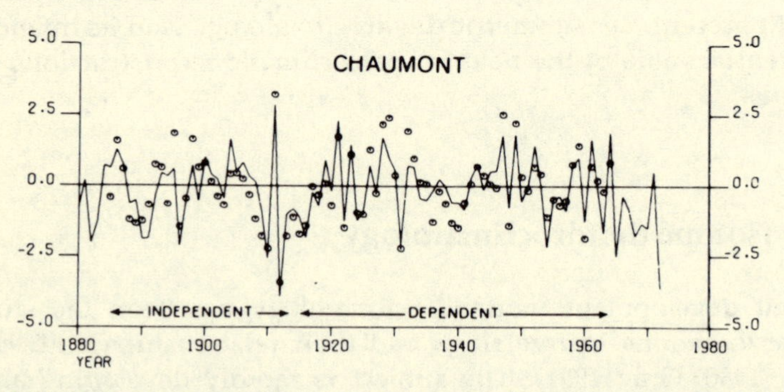

Figure 10.28 Reconstruction of July – August temperatures for locations in Switzerland. Predictors used were measurements of maximum density, minimum density, earlywood width, and latewood width from six tree sites for the year in which the temperature was measured and the year following it (after Schweingruber *et al.* 1978).

record of past temperature variations."† The problem is that additional isotopic fractionation (of hydrogen, oxygen, and also carbon, atoms) occurs within trees during the synthesis of woody material and these fractionations are themselves temperature dependent (Libby 1972). Relative humidity and wind speed (evapotranspiration effects) may also influence oxygen and hydrogen fractionation (Burk & Stuiver 1982). Furthermore, fractionation effects vary from one substance to another and repeated chemical exchanges may radically alter the original isotopic composition of the environmental water taken up by the plant. It has been shown, for example, that lipid materials in plant tissues are 100‰ depleted in deuterium with respect to total plant hydrogen. Analysis of whole wood samples, which might contain varying amounts of lipid material from year to year, could thus show variations in δD which are not climatically dependent.† In fact, repeated work with whole wood has demonstrated large differences in δD within the same species of tree in a homogeneous climatic region (Epstein *et al.* 1976) and long-term variations in whole wood δD show virtually no correlation with climatic data (Epstein & Yapp 1976, Gray and Thompson 1977). Such work is persuasive and makes the significance of earlier studies on climate and isotopic variations in whole wood samples (e.g. Libby & Pandolfi 1974) somewhat questionable.

To avoid the difficulties of chemical heterogeneity in wood samples, a single component, cellulose, is extracted. Cellulose contains both carbon-bound and oxygen-bound (hydroxyl) hydrogen atoms, but the latter exchange readily within the plant. It is therefore necessary to remove all hydroxyl hydrogen atoms to avoid problems of isotopic exchange since the initial period of biosynthesis. When this is done (producing nitrated cellulose) D/H values are highly correlated with D/H ratios of associated environmental waters, with δD values from trees in a wide variety of climatic regions consistently 20–22‰ *lower* in the nitrated cellulose (Fig. 10.29; Epstein *et al.* 1976; Epstein & Yapp 1977). Since δD values of annual precipitation are highly correlated with mean annual temperature (Yapp & Epstein 1982), such a relationship strongly suggests that long-term records of δD from trees can be used as a proxy for temperature variations. Epstein and Yapp (1976) attempted to do this, using two overlapping bristlecone pine records from the White Mountains of California. A

†For a discussion of isotopes, deuterium/hydrogen (D/H), and $^{18}O/^{16}O$ ratios, see Sections 5.2.1 and 5.2.2.

$$\delta D = \left(\frac{(D/H)_{sample}}{(D/H)_{standard}} - 1 \right) \times 1000,$$

where the standard is "standard mean ocean water" (SMOW). For $\delta^{18}O$, substitute $^{18}O/^{16}O$ for D/H in the equation.

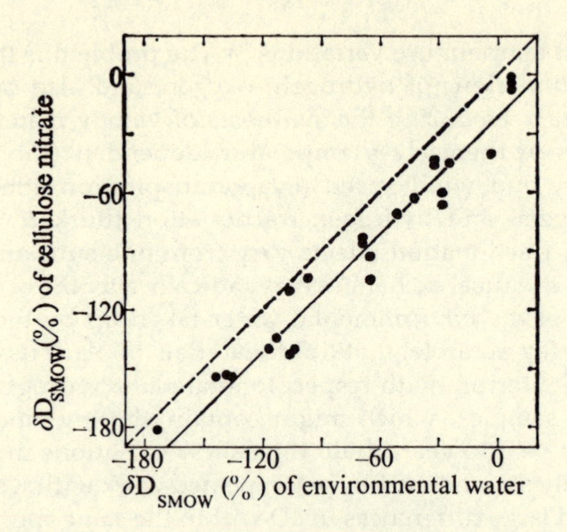

Figure 10.29 Plot of δD of plant-derived cellulose nitrate (containing isotopically non-exchangeable hydrogen) against δD of environmental waters used in plants. Solid line is best-fit linear regression line ($x = 0.97y - 22$; $r = 0.98$). The dashed line is for $y = x$. The samples include varieties of both land and aquatic plants from a wide geographical range (after Epstein & Yapp 1977).

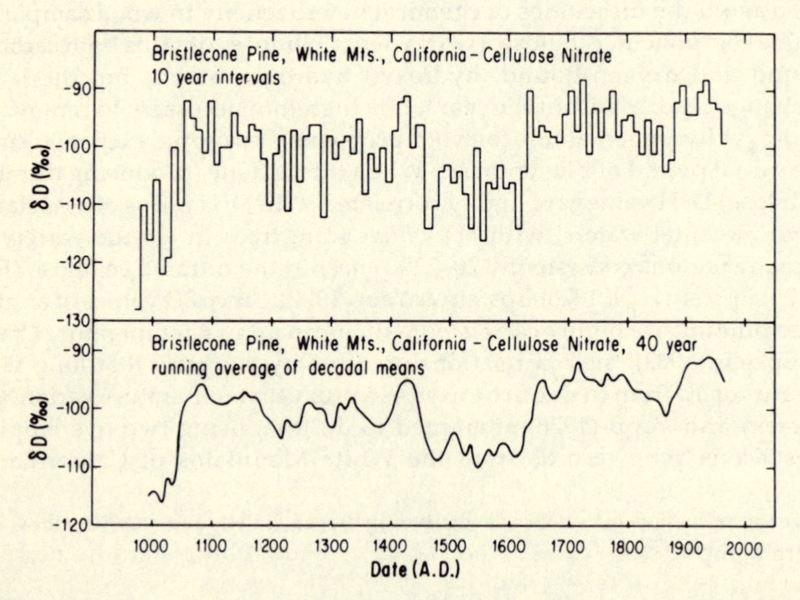

Figure 10.30 The δD record of cellulose nitrate in two bristlecone pines which grew at ~7000 m in the White Mountains of California. Each sample in the upper graph represents a 10 year interval. One tree grew in the interval 970–1650 while the others span 1550–1974. The smoothed curve (below) gives 40 year running means for the 1000 year bristlecone δD record (centered on the third decade in each averaged interval) (after Epstein & Yapp 1976).

record of 10 year mean δD values in cellulose nitrate was obtained (Fig. 10.30). Assuming that higher δD values indicate warmer temperatures, the record shows a period of very cold temperatures prior to AD 1050 and again from AD 1470 to AD 1650. Mean temperatures since then have been slightly above the millennium average, except for a cooler period from ~1810–1890. Unfortunately, the lack of long-term climatic records in the region precludes direct calibration of the record with monthly or seasonal temperature data. Similar analysis of cellulose from Scots pine from Loch Affric, Scotland, shows reasonably good correlation with winter temperatures at Edinburgh (1841–1970), but only when smoothed by 40 year running means. The smoothed Scottish and Californian records also have much in common, but the correlation of high-frequency variations is poor. This led Epstein and Yapp to conclude that both δD records may record long-term hemispheric-scale climatic fluctuations, but that superimposed on these long-term trends there may exist short-term δD variations reflecting temperature changes over a much more restricted area. Of course, an alternative explanation for this high-frequency variability is simply noise in the data, but only further sampling and analysis will settle the issue.

In an analysis of $\delta^{18}O$ variations in cellulose, Gray and Thompson (1976, 1977) also found good correlations between long-term temperature records and isotopic ratios. Using white spruce from central Alberta and climatic data from Edmonton the following relationship was demonstrated (Fig. 10.31):

$$\delta^{18}O = (1.3T \pm 0.1T) + (20.5 \pm 0.2),$$

where T is the mean temperature from the previous September to August of the year of growth. This period correlates much better than summer temperatures, perhaps indicating that soil-stored winter water may play a significant role in cellulose production the following season. Alternatively, the correlation may be higher when winter temperatures are included because of the production of stored foods during winter months (particularly the hydrolysis of starch to glucose) (Perry 1977, Gray & Thompson 1976). It is interesting that similarly high correlations have been found between $\delta^{18}O$ variations in Bavarian fir (*Abies alba*) and German oak (*Quercus petraea*) and winter temperatures from various long-term European stations (Libby *et al.* 1976). However, the $\delta^{18}O$ values were computed for whole wood samples and are thus hard to interpret (Epstein & Yapp 1977). Nevertheless, whole wood analysis of $\delta^{18}O$ in a 1700 year record of Japanese cedar also shows a remarkably good fit with the limited historical evidence available for temperature variations in China over the last two millennia (Libby *et al.* 1976, Chu 1973), as shown in Figure 10.32. Whether one should dismiss such correlations as mere

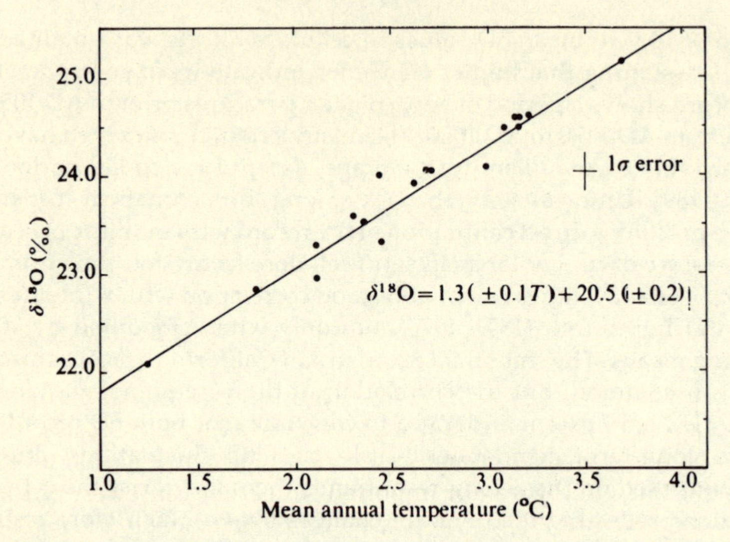

Figure 10.31 Variation of $\delta^{18}O_{SMOW}$ with September–August mean annual temperatures for 5 year periods. The equation of the line was obtained by a least squares fit to the data (after Gray & Thompson 1976).

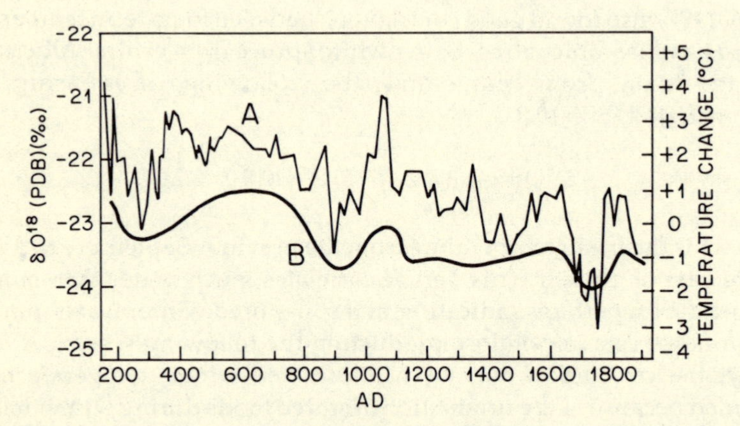

Figure 10.32 $\delta^{18}O$ values in samples of *Cryptomeria japonica* from ~1500 m on Yaku Island, Japan (~30°N), compared to the Chinese paleotemperature record of Chu (1973) (after Libby *et al.* 1976).

coincidence or should accept that there is a strong climatic signal overriding other biochemical factors is a matter of debate; this empirical approach of finding a good correlation, thereby "proving" a causal connection (without understanding it in detail) has many supporters. For example, Grinsted *et al.* (1979) have shown that $\delta^{13}C$ variations in cellulose from the long bristlecone pine series are quite well correlated (when smoothed by a 60 year running mean) with the record of upper treeline ring-width variations, a proxy record considered to reflect summer temperatures (LaMarche 1974). Understandably, Grinsted *et al.* conclude that one of the dominant factors affecting $\delta^{13}C$ variations in *Pinus longaeva* cellulose may be summer temperature, though the mechanism remains unclear. In addition to fractionation effects within the plant, atmospheric ^{13}C variations may result from changes in CO_2 exchange between the atmosphere and the oceans as a result of temperature changes in the (predominantly Southern Hemisphere) oceans (Pearman *et al.* 1976). Temperature changes of the Southern Hemisphere ocean surface may thus be a prime control on $\delta^{13}C$ variations in the atmosphere, and hence in tree rings all over the world. However, $\delta^{13}C$ variations have also been ascribed to the dilution of atmospheric ^{13}C (during the early part of the 20th century) due to the release of biospheric carbon dioxide accompanying increased land cultivation (Farmer & Baxter 1974). Presumably this effect was not important in earlier centuries, but it does illustrate the complexities involved in trying to interpret the temperature–isotope connection. There is a danger of overly simplistic empiricism.

In the foregoing discussion, tree rings have been referred to as though they are isotopically uniform, individually. However, it has been shown in several studies that isotopic concentrations may be quite different in latewood compared to earlywood sections of a tree ring. Epstein and Yapp (1976), for example, found earlywood (winter growth) in a Douglas fir from Arizona to be isotopically heavier than latewood (spring and early summer growth), a difference which is opposite to what one might expect from seasonal differences in meteoric D/H ratios. Again, this may relate to the storage of starch from the previous year and its utilization during earlywood growth, or perhaps reflect variations in ground water supply. If the reason for such differences can be identified there would be the possibility of using the intra-ring isotopic variations to calculate *seasonal* temperature and/or hydrological variations. Wigley *et al.* (1978) consider that such intra-ring isotopic variability may result in the observed interannual isotopic variations being simply a function of ring width. The magnitude of this effect would depend on how different were the isotopic values for earlywood and latewood, and how the relative proportions of earlywood and latewood vary in a sequence of tree rings. This effect could be quite important in trees showing large interannual variations in ring width and for this reason complacent trees rather than those under stress should be selected for study (Gray & Thompson 1978). This is in contrast to conventional dendroclimatology, in which trees under stress are the ones required; thus isotope dendroclimatology may prove a valuable complementary method to conventional dendroclimatology in areas where the latter type of study would find little or no climatic signal.

10.6.1 *Isotopic studies of sub-fossil wood*

It has already been pointed out that, because of fractionation effects, δD values in nitrated cellulose from trees average −20‰ below that of associated meteoric waters (Fig. 10.29). This is supported by mapping δD values from modern plants and comparing them with measured δD values of meteoric waters (Fig. 10.33). It is clear that the plants provide a good proxy measure of spatial variations in δD. Assuming that this relationship has not changed over time, it is possible to reconstruct former δD values of meteoric water by the analysis of radiocarbon-dated subfossil wood samples (Yapp & Epstein 1977). Figure 10.34 shows such a reconstruction, for "glacial age" wood (dated at 22 000–14 000 years BP). It is interesting to find that the ancient δD values at all sites are consistently *higher* than modern values (an average of +19‰). This stands in remarkable contrast to the measured δD values from "glacial age" ice in the ice cores from Greenland and Antarctica (Ch. 5), where δD values average 80‰ *below* those of modern precipitation in the area. As increases in δD values of meteoric waters are generally associated with temperature increases, these results imply that temperatures over the ice-free area of North America were warmer in late Wisconsin times than today. However, there are many other factors which could account for the high δD values observed, notably (a) a reduction in the temperature gradient between the ocean surface and the adjacent precipitation site on land; (b) a change in δD of the ocean waters as a result of ice growth on land (probably corresponding to an increase in oceanic δD of 4–9‰); (c) a change in the ratio of summer to winter precipitation; (d) a positive shift in the average δD value of oceanic water vapor, which at present is not generally evaporating in isotopic equilibrium with the oceans. Whatever the reason it has important implications for the isotopic composition of the Laurentide Ice Sheet, which has generally been assumed to have been composed of ice very depleted in ^{18}O and deuterium. Using δD values from trees growing along the shores of glacial lakes Agassiz and Whittlesey, Yapp and Epstein (1977) calculate that the $\delta^{18}O$ value of the former Laurentide Ice Sheet probably averaged around −12 to −15‰ rather than −30‰ as estimated by Dansgaard and Tauber (1969). This would, in turn, imply that the $\delta^{18}O$ value of the oceans only increased by +0.8‰ compared to modern values. It would be extremely valuable to extend this work to other formerly glaciated areas (particularly Scandinavia) to study "glacial age" δD values in more detail since these results have an important bearing on the interpretation of other paleoclimatic records, particularly ice and ocean cores.

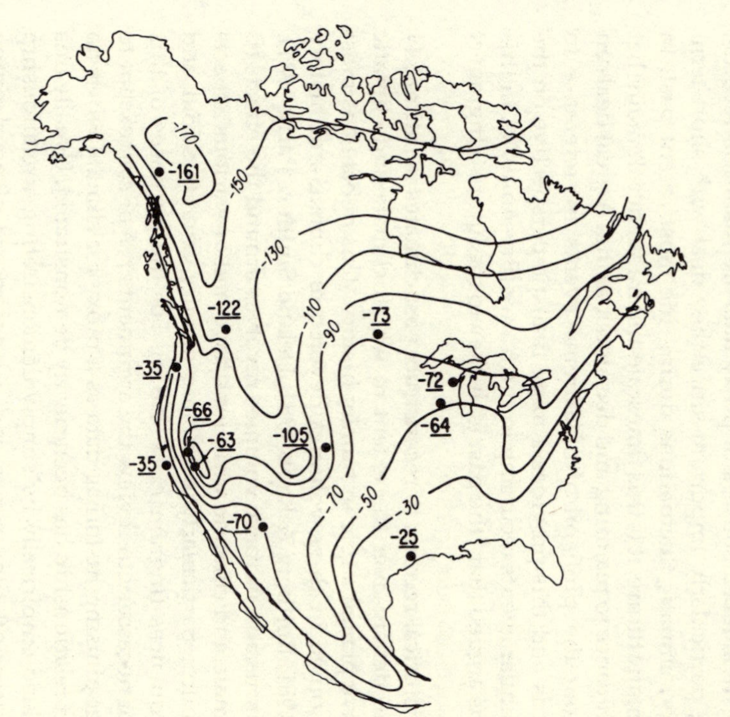

Figure 10.33 Distribution of δD values in modern meteoric waters, as inferred from the cellulose C—H hydrogen (cellulose nitrate) of modern plants. With the exception of only two Sierra Nevada samples, the inferred water δD values (underlined) and corresponding iso-δD contour values differ by an average of only about 4‰ (after Yapp & Epstein 1977).

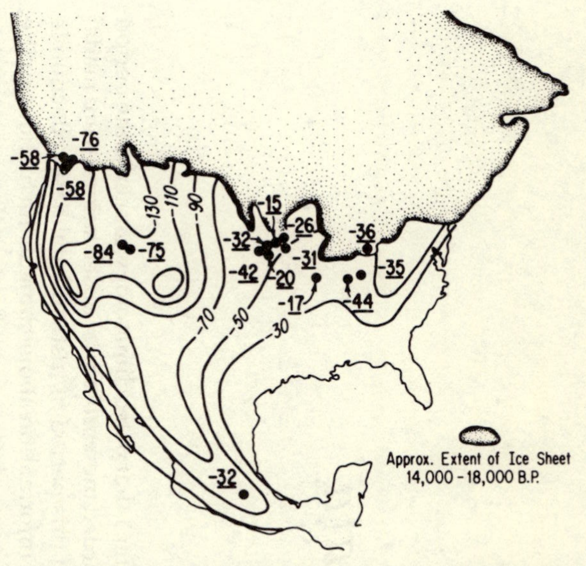

Figure 10.34 Distribution of 16 δD values of meteoric waters from 15 different sites during the late Wisconsin glacial maximum, as inferred (underlined) from δD values of tree cellulose C—H hydrogen. The approximate position of the southern margin of the ice sheet is shown at its maximum extent (hatched line). The "glacial age" meteoric waters of the 15 sites have, on average, δD values which are 19‰ more positive than the corresponding modern meteoric waters at those sites as deduced from the data shown in Figure 10.33. The "glacial age" distribution pattern of δD values is similar to the modern pattern, but is systematically shifted by the positive bias of the ancient waters. The North American coastline shown is that of today and does not take into account the lower sea level at the time of glacial maximum (after Yapp & Epstein 1977).

11

Historical data

11.1 Introduction

It was pointed out in Chapter 1 that paleoclimatology concerns the period before instrumentally recorded meteorological observations were made. In many parts of the world this period of instrumental data collection is very short and virtually all information about climatic fluctuations in the past must be gleaned from proxy data sources. Some of the most diverse, but potentially invaluable, sources of proxy data are historical records. These data are particularly important since they deal with short-term (high-frequency) climatic fluctuations during the most recent past. In terms of the climatic future, it is this timescale and frequency domain that is of most significance to planning and decision making. A great deal can be learned about the probability of extreme events by reference to historical records and this provides a more realistic perspective on the likelihood of similar events recurring in the future (Bryson 1974). In this sense it could be argued that "the past is the key to the future" (Ingram *et al.* 1978).

The use of historical records to reconstruct past climates is intimately linked with the debate over the extent to which climate and climatic fluctuations have played a role in human history. This debate has most recently been joined in three volumes devoted to climate and history (Wigley *et al.* 1981, Rottberg & Rabb 1981, Delano Smith & Parry 1981). Much of the discussion in these volumes revolves around the effects (if any) which climate and climatic variations have had on various aspects of human activity, particularly economic activity and its social and political consequences (Ingram *et al.* 1981a). For the purposes of this chapter it is not necessary to review the arguments *pro* or *con*, except to note the danger of using historical data as a proxy for climate when no cause and effect relationship has been clearly demonstrated. Usually this problem is tackled empirically by simply demonstrating a relationship between the historical time series and a brief record of overlapping climatic data. However, without a model demonstrating the logic of such a relationship, erroneous conclusions may result (de Vries 1981). Models

Table 11.1 Earliest historical records of climate (modified after Ingram *et al.* 1978).

Area	Earliest written evidence (approximate dates)
Egypt	3000 BC
China	1750 BC
Southern Europe	500 BC
Northern Europe	0
Japan	AD 500
Iceland	AD 1000
North America	AD 1500
South America	AD 1550
Australia	AD 1800

should be comprehensive, recognizing not just the role of climate but also the potential significance of other non-climatic factors (Parry 1981). In this way, not only will there be a stronger basis for historical paleoclimatic reconstruction but also a better understanding of the role of climate and climatic fluctuations in history.

Paleoclimatic data from historical sources rely, of course, on written observations. This means that certain parts of the world have been endowed with a much richer heritage of historical paleoclimatic information than other regions (Table 11.1). The longest records come from Egypt, where stone inscriptions relating to the Nile flood levels are available from mid-Holocene times (~5000 years BP), indicating higher rainfall amounts from East African summer monsoons at that time (Bell 1970). In China, the earliest inscriptions on oracle bones (Fig. 11.1) date back to the time of the Shang dynasty (~3700–3100 years BP), when conditions appear to have been slightly warmer than in modern times (Wittfogel 1940, Chu 1973). Such records are, of course, few and far between and for the vast majority of the continental land areas, as well as the oceans, historical observations are generally only available for a few hundred years at the most.

Historical data can be grouped into three major categories. First, there are observations of weather phenomena *per se*, for example, the frequency and timing of frosts on the occurrence of snowfall recorded by early diarists. Secondly, there are records of weather-dependent natural phenomena (sometimes termed parameteorological phenomena), such as droughts, floods, lake or river freeze-up and break-up, etc. Thirdly, there are phenological records, which deal with the timing of recurrent weather-dependent biological phenomena, such as the dates of flowering of shrubs and trees, or the arrival of migrant birds in the spring. We will include in this group of records, observations of the former spatial extent of particular climate-dependent species. Within each of these categories there is a wide range of potential sources and an equally wide range of possible climate-related phenomena. These will be discussed in more detail below,

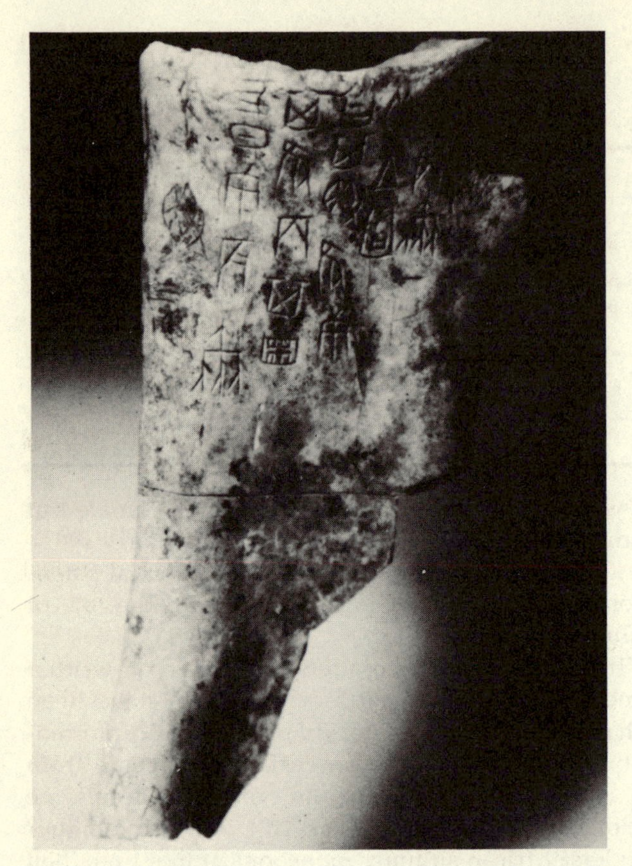

Figure 11.1 Ancient Chinese oracle bone, dating from the Shang dynasty (~1500 BC). These contain some of the earliest written paleoclimatic information.

but first it is worth considering some of the sources of historical data and the difficulties of deriving climatic information from them.

11.2 Historical records and their interpretation

Potential sources of historical paleoclimatic information include: (a) ancient inscriptions, (b) annals, chronicles, etc., (c) governmental records, (d) private estate records, (e) maritime and commercial records, (f) personal papers, such as diaries or correspondence, and (g) scientific or proto-scientific writings, such as (non-instrumental) weather journals (Ingram *et al.* 1978). In all these sources, the historical climatologist is faced with the difficulty of ascertaining exactly what the qualitative description of the past is equivalent to, in terms of modern-day observations. What do the terms "drought," "frost," "frozen over," really mean? How can qualifying terms (e.g. "extreme" frost) be interpreted? Baker (1932), for example, notes that one 17th century diarist recorded three droughts of "unprecedented severity" in the space of only five years! An approach to solving this problem has been to use content analysis (Baron 1982) to assess in quantitative terms, and as rigorously as possible, climatic information in the historical source. Historical sources are examined for the frequency with which key descriptive words were used (e.g. "snow," "frost," "blizzard," etc.) and the use which the writer may have made of modifying language (e.g. "severe frost," "devastating frost," "mild frost," etc.). In this way an assessment can be made of the range of descriptive terms which were used, so they can be ranked in order of increasing severity *as perceived by the original writer*. The ranked terms may then be given numerical values so that statistical analyses can be performed on the data. This may involve simple frequency counts of one variable (e.g. snow) or more complex calculations using combinations of variables. In this way, the original qualitative information may be transformed into more useful quantitative data on the climate of different periods in the past. Non-climatic information is often required to interpret the climatic aspects of the source: where did the event take place (was the event only locally important; was the diarist itinerant or sedentary?) and precisely when did it occur and for how long? This last question may involve difficulties connected with changing calendar conventions as well as trying to define what is meant by terms such as "summer" or "winter," and what time span might be represented by a phrase such as "the coldest winter in living memory." Not all of these problems may be soluble, but content analysis can help to isolate the most pertinent and unequivocal aspects of the historical source (Moody & Catchpole 1975).

Historical sources rarely give a complete picture of former climatic conditions. More commonly, they are discontinuous observations, very much biased towards the recording of extreme events, and even these may pass unrecorded if they fail to impress the observer. Furthermore, long-term trends tend to go unnoticed since they are beyond the temporal perspective of one individual. In a sense, the human observer acts as a high-pass filter, recording short-term fluctuations about an ever-changing norm (Ingram *et al.* 1981b).

Historical records which appear to be rich in parameteorological information were probably not produced with the historical climatologist in mind, and care must be taken to assess the purpose of the writer in making the record. For example, Chinese dynastic histories contain abundant references to floods and droughts, but the purpose of the notation was not to record the vagaries of climate, but to make a record of human suffering and damage caused by changes in rainfall. Such events were considered to convey a supernatural warning to the Emperor of the day (Yao 1944). Because of this, there is a strong seasonal bias towards reporting events in the vitally important growing season, as well as a spatial bias in the data, towards settled areas with large populations. In the Ming dynasty (1368–1643), for example, the number of recorded droughts and floods in the metropolitan province exceeds that of any other province (Chu 1926, Yao 1943). Also, because floods and droughts resulted in tax exemptions in the region afflicted, it is not unlikely that local officials may have been tempted to exaggerate the severity of extreme events. This, of course, is true of many areas throughout history; the Chinese have no monopoly of hyperboles in reporting disasters!

Finally, it is worth noting that not all historical sources are equally reliable. It may be difficult in some cases to determine if the author is writing about events of which he has first-hand experience, or if events have been distorted by rumor, or the passage of time. Ideally, sources should be original documents rather than compilations; many erroneous conclusions about past climate have resulted from climatologists relying on poorly compiled secondary sources which have proved to be quite erroneous when traced back to the original data (Bell & Ogilvie 1978, Wigley 1978, Ingram *et al.* 1981b). Nevertheless, some major compilations such as the voluminous Chinese encyclopedia, the *T'u-shu Chi-cheng*, have proved to be invaluable sources of paleoclimatic information (Chu 1926, 1973, Yao 1942, 1943).

As with all proxy data, historical observations need to be calibrated in some way, in order to make comparisons with recent data possible. This is commonly done by utilizing early instrumental data which may overlap with the proxy record, to develop an equation relating the two data sets. Thus, Bergthorsson (1969) was able to calibrate observations of sea-ice frequency with mean annual temperatures during the 19th century (Fig. 11.2) and then use this equation to assess long-term temperature fluctuations over the last 400 years from sea-ice observations (Fig. 11.3). Similarly, observations of Dutch canal freezing frequencies have been calibrated with winter temperature data by de Vries (1977), as have snow cover data for the Zürich area (Pfister 1977). The resulting

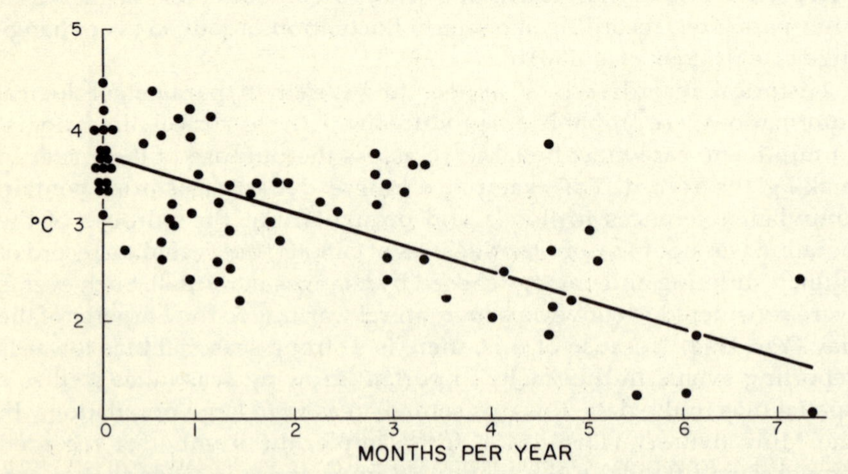

Figure 11.2 Relationship between annual temperature and ice incidence off the coast of Iceland, in months per year. Calibration period 1846–1919 (after Bergthorsson 1969).

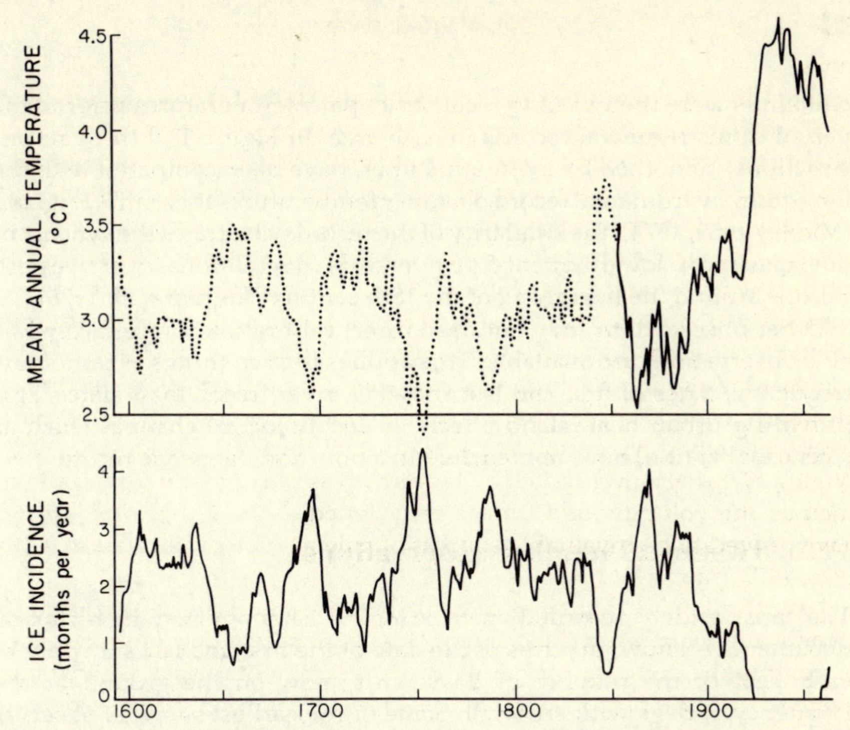

Figure 11.3 Reconstruction of paleotemperature based on calibration shown in Figure 11.2. Decadal running means of mean annual temperature (top) and ice incidence in months (below). Paleotemperatures are shown by dashed line. This record has been extended back to AD 950 but the earlier part of the record is considered to be unreliable (after Bergthorsson 1969).

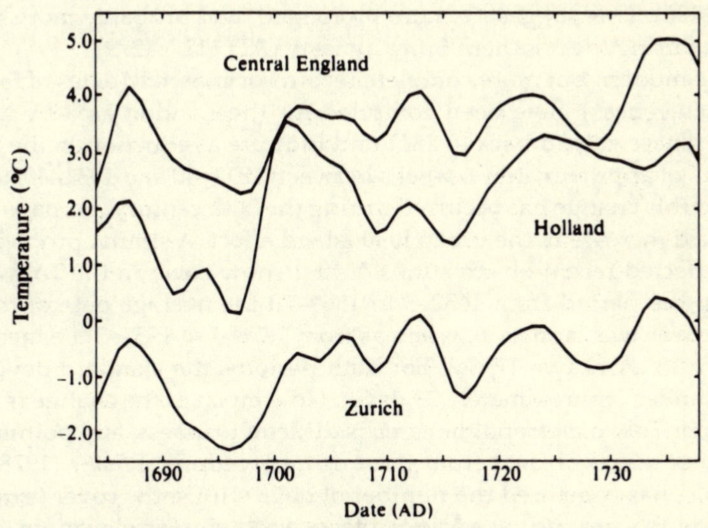

Figure 11.4 Comparison of winter temperatures (December, January, February) from three European sites for the period 1684–1738. Short-term fluctuations have been smoothed by means of a 9 year binomial filter (after Ingram *et al.* 1978).

equations were then used to reconstruct paleotemperatures prior to the period of instrumental records in each area. In Figure 11.4 these reconstructions, smoothed by a binomial filter, have been compared with the long-term instrumental record of winter temperatures for central England (Manley 1959, 1974), the similarity of the records illustrates the synchronous nature of low-frequency temperature fluctuations over western Europe around the beginning of the 18th century (Ingram *et al.* 1978).

Other observations may not need direct calibration if recent comparable observations are available. This applies to such things as rain/snow frequency, dates of first and last snowfall, river freeze/thaw dates, etc., providing urban heat island effects, or technological changes (such as river canalization) have not resulted in a non-homogeneous record.

11.3 Historical weather observations

The most widely recorded meteorological phenomenon in historical documents is snow, in terms of the date of the first and last snowfalls of each year or the number of days with snow on the ground or the frequency of days with snowfall. Some of the earliest snowfall observations are from Hangchow, China, for the period AD 1131–1210, when discontinuous records were made of the dates of the last snowfall in spring. Comparison with Hangchow records for the period 1905–14 indicates that snowfalls commonly occurred 3–4 weeks later in the spring months during the 12th century than was typical of the early 20th century (Chu 1926). This suggests a more prolonged, and probably more severe, winter during the Southern Sung dynasty (AD 1127–1279).

More modern but more complete records of snowfall (dates of first and last occurrences) have been compiled for the London area by Manley (1969). These extend back to 1811 and indicate a reduction in the "snow season" of approximately 6 weeks between 1811–40 and 1931–60, though most of this change has occurred during the 20th century, perhaps due to a marked increase in the urban heat island effect. A similar problem may have affected recent observations of first snow cover in the Tokyo area. During the period from 1632–3 to 1869–70 the average date of the first snow cover was January 6, whereas from 1876–7 to 1954–5 the mean was January 15 (Arakawa 1956a). For both periods, the standard deviations were similar (approximately 20 days). How much of the change is due to a warmer Tokyo metropolitan area is difficult to assess, and points to the particular value of data from more rural locations. Pfister (1978a), for example, has compared the number of days with snow cover (extensive snow on the ground) at various places with similar elevations on the upper and lower plateaux of Switzerland. During the 18th century, snow cover was often far more persistent than in even the harshest winter of

Table 11.2 Number of days with snow cover per year (March to May values in parentheses) for selected localities in Switzerland (after Pfister, 1978a).

Winter of	Lower plateau	Upper plateau
1769–70		126 (45)
1784–5	134 (51)	154 (60)
1788–9	112 (35)	
1962–3†	59 (0)	86 (12)

†Harshest winter of the 20th century. Severe winters were also noted in 1684–5, 1715–16, 1730–1 and 1969–70.

the 20th century, 1962–3 (Table 11.2), and this provides strong support for the record of "days with snow lying" in the Zürich – Winterthur area (Fig. 11.5). These data point to the decade 1691–1700 as having had the highest number of days with snow on the ground (>65 annually) compared to recent averages of only half that number (Pfister 1978b). It is also of significance that this decade was the coldest in central England, according to long-term instrumental records of temperature (Manley 1974). Thus, urban growth effects, while undoubtedly exerting a progressive influence on temperature, can not account for all the marked changes observed.

An interesting method of assessing past winter temperatures has been demonstrated by Flohn (1949), who observed that the ratios of snow days to rain days in winter months correlate well with winter temperatures during the instrumental period. Using 16th century observations of Tycho Brahe in Hven, Denmark (1582–97), and of Haller in Zürich (1546–76), Flohn was able to show that winters after 1564 were increasingly severe, leading to a marked increase in glacial advances in the Alps

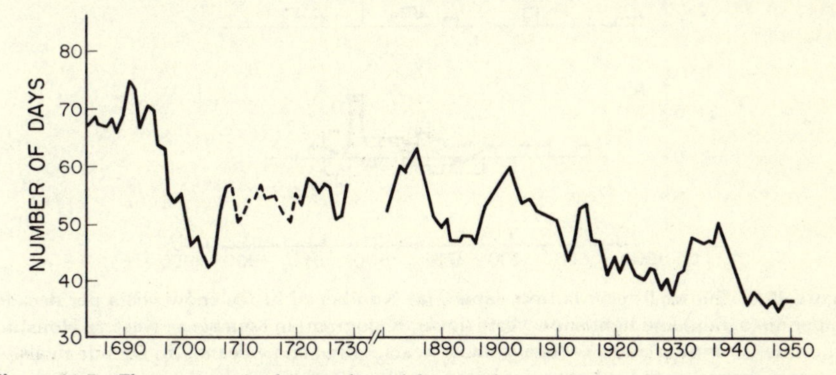

Figure 11.5 Fluctuations in the number of days with snow lying in Zürich (note break in abscissa scale). Prior to 1800, figures have been estimated based on daily non-instrumental observations. Observations for 1721–38 made in Winterthur, 20 km north-east of Zürich. Value plotted as 10 year running means with value at year *n* corresponding to the decade *n* to (n + 9) (after Pfister 1978b).

in the early 17th century. At Hven, winter temperatures from 1582 to 1597 averaged 1.5 °C below those around the early 20th century. It is also interesting that 16th century weather singularities, synoptic events which recur at the same time each year, continue to be observed in the 20th century, indicating an underlying cyclicity in the general circulation which has persisted in spite of fundamental changes in the climate of the region.

In the Kanazawa area of Japan (west central Honshu) historical records of snowfall were maintained by the ruling Maeda family from 1583 until ~1870. Based on detailed analysis of these and subsequent records, Yamamoto (1971) constructed an index of snowfall variation which corresponds reasonably well with instrumentally recorded winter temperatures in recent years. Although no precise calibration has been attempted, the index gives an overall impression of snowier winters, particularly in the first half of the 19th century, and this conclusion seems to be supported by other snowfall indices derived from Japanese historical sources (Fig. 11.6).

Another observation frequently noted in historical records is the

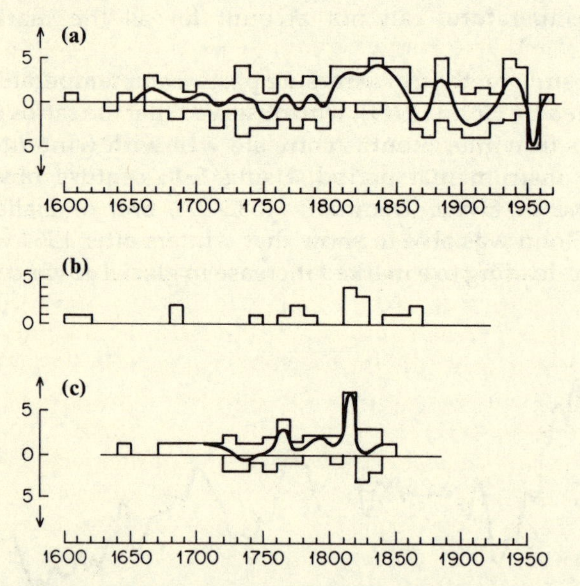

Figure 11.6 Snowfall records from Japan. (a) Number of heavy snow years per decade (upper histogram) and light snow years (lower histogram) in Kanazawa, western Honshu. Solid line is running mean of (heavy snow years − light snow years). (b) Decade totals of heavy snow years in Shiga Prefecture, southern Honshu. (c) Severe winter index for Tottori, south-western Honshu, based on heavy snow years + severe cold winters (upper histogram) and light snow years + extraordinarily warm winters (lower histogram). The solid line is the running mean of upper values − lower values. In all records, the first half of the 19th century appears to have had exceptionally heavy snowfall in Japan (after Yamamoto 1971).

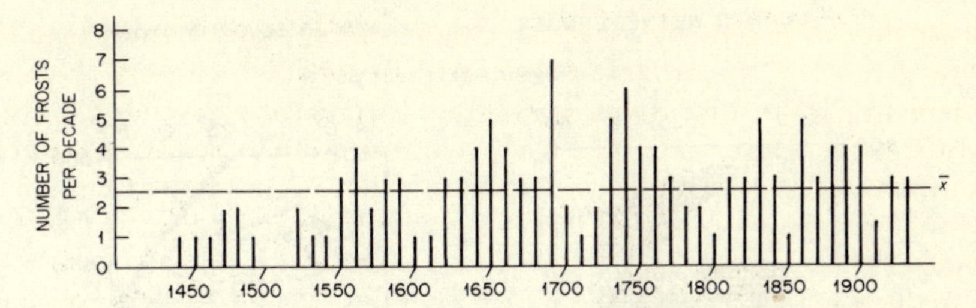

Figure 11.7 Number of frosts per decade in the lower Yellow River region, China from 1440 to 1940 (after Zhang & Gong 1979).

incidence of frost during the growing season, an occurrence of particular significance to agriculturalists. As with snowfall, long records of frost occurrence are available from China, in particular from the farming regions of the mid and lower sections of the Yellow (Huang-ho) River. Although frost occurrence is not necessarily an index of growing season temperature, a reduction in the growing season due to "early" and "late" frosts is a consequence of lower summer temperatures (e.g. Bradley 1980). Thus, frost occurrence may provide a guide to overall summer temperatures. Long-term frost data from China, shown as decadal averages in Figure 11.7, indicate higher frequencies of frost during the periods 1551–1600, 1621–1700, 1731–80, and 1811–1910. Geographically, the historical records indicate that frost-free periods from 1440 to 1900 averaged ~2 months shorter in Inner Mongolia and north-eastern China than during the 20th century. In southern China, the frost-free season in recent decades has been 5–6 weeks longer than the long-term mean from 1440 to 1900 (Zhang & Gong 1979).

European historical archives have provided a variety of non-instrumental records of value to the climatologist. Many of these sources have been evaluated by H. H. Lamb to produce an index of winter severity and summer wetness spanning the last 1000 years (Lamb 1961, 1963, 1977). By carrying out such assessments for different locations near 50°N, the spatial and temporal pattern of winter and summer conditions was mapped (Fig. 11.8). These studies suggest that the period AD 1080–1200 was characterized by dry summers throughout Europe, the like of which has not been seen since (Lamb 1965). Winters at this time were, however, relatively harsh, at least until ~AD 1170 (Alexandre 1977). By contrast, the period 1500–1700 was characterized by both cold winters and wet summers over most of the region. Furthermore, the anomaly patterns show a general westward drift through time from ~1200 to 1500 which was reversed in the next 200 years or so (Fig. 11.8). Lamb (1977) suggests that these changes result from varying wavelengths of the upper westerlies, the mean wavelength shortening during the westward anomaly shift and lengthening as the anomalies shift eastward. If this deduction is correct,

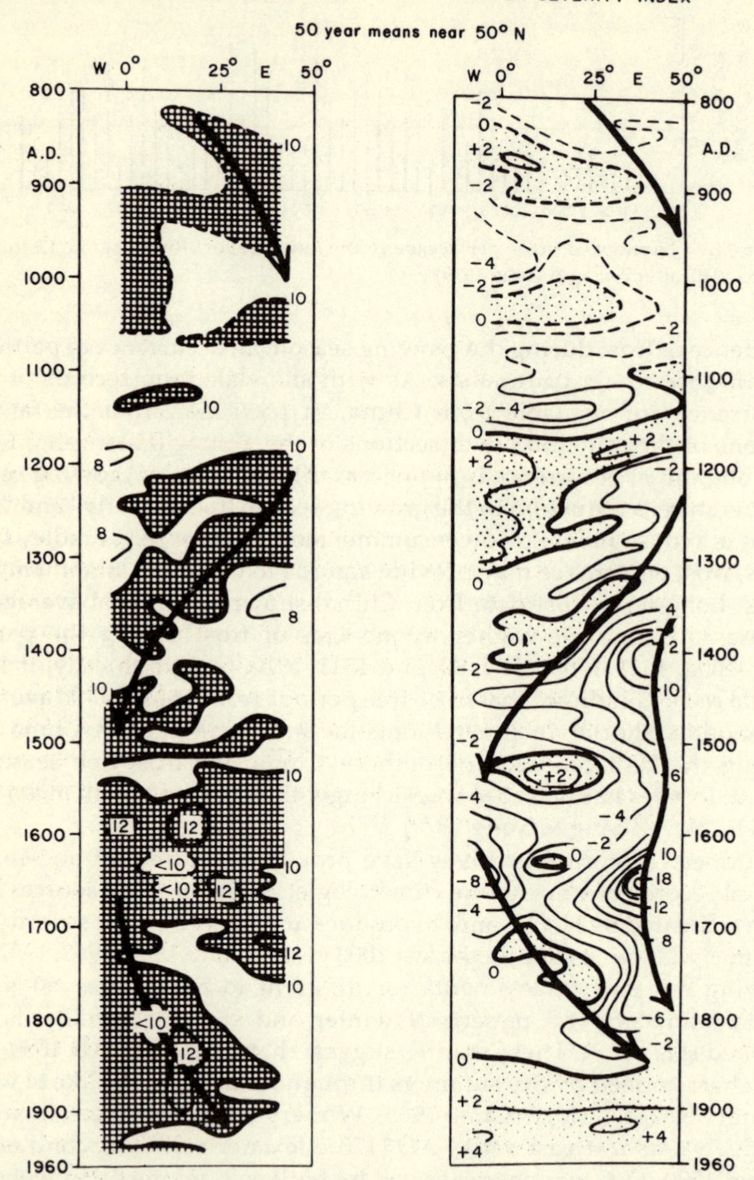

Figure 11.8 Fifty year mean values of summer wetness and winter severity indices at different longitudes in Europe near 50°N, between England and Russia from AD 800 to 1960. Hatched areas indicate excess of wet over dry months in high summer (July and August). Stippled areas indicate more very mild than severe winter months (December, January, February). Arrows show the movement of the greatest anomalies during times of change (after Lamb 1977).

it is possible to use empirically derived equations of the general circulation to make further deductions about upper air flow through time. Thus Lamb (1972) suggests that the main flow of the upper westerlies was displaced southward during the time of colder climate by 3–5° of latitude in the American and Atlantic sectors, and that there was a general weakening of the circulation at this time. This would be associated with a southward displacement of the mean Atlantic depression track entering western Europe, presumably increasing rainfall in southern Europe and the Mediterranean area. Interestingly, the broad patterns of winter temperature anomaly in Europe are also observed in China and Japan, indicating that the changes were probably hemisphere-wide (cf. Figs 11.8 and 11.13; Wang *et al.* n.d.).

One of the most interesting studies of historical material is that of Neuberger (1970), who has examined the changing climate of the Little Ice Age in Western Europe through artists' perceptions of their climatic environment, as depicted in contemporary paintings. Over 12 000 paintings from the period 1400–1967 were examined and wherever possible the intensity of the blue sky, visibility depicted, percentage cloudiness, and cloud type were categorized for each painting. More than half of the paintings contained some sort of meteorological information and the basic characteristics were averaged for different periods within the last 570 years, as shown in Figure 11.9. During the period 1400–1549, paint-

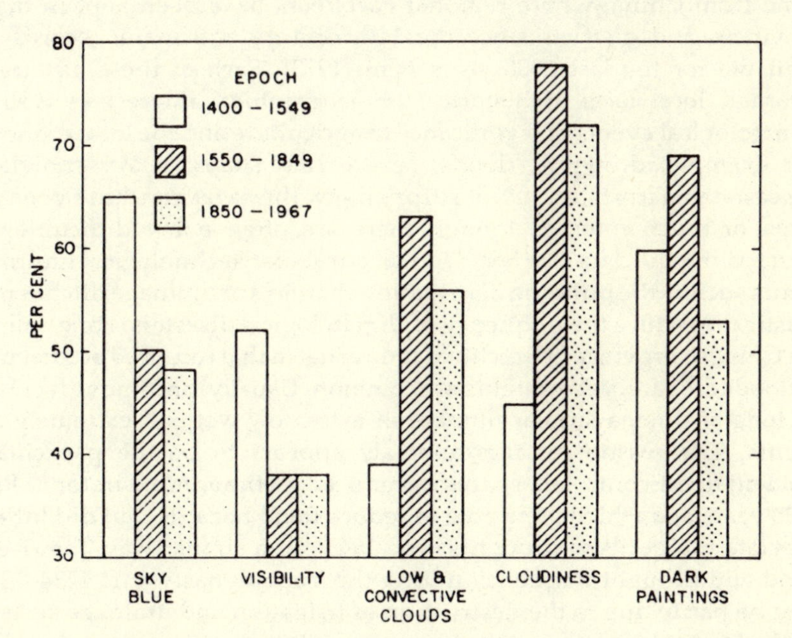

Figure 11.9 Changes in the frequency of certain characteristics of European paintings in the periods 1400–1549, 1550–1849, and 1850–1967 (after Neuberger 1970).

ings have a high percentage of blue sky, good visibility, and little cloud cover. Paintings completed during the next 300 years were generally darker, with less blue sky, showed lower visibilities, and had a much higher percentage of cloud cover and a greater frequency of low and convective type clouds. Over the last 100 years there has been some reduction in cloudiness depicted and a drop in low and convective cloud frequency, though visibilities remain low, perhaps reflecting increasing atmospheric turbidity due to industrial and agricultural activity. It is remarkable that, in spite of many changes in style, the artists have captured in their paintings a significant record of climatic variation through time. Perhaps this perceptual record, more than any cold statistic, indicates the degree to which life during the Little Ice Age was affected by the deteriorating climate.

11.4 Historical records of weather-dependent natural phenomena

Of all the weather-related natural catastrophes, floods and droughts appear to have had the most widespread and persistent impact on human communities, as records of these events are found in historical documents from all over the world. The longest and most detailed records come from China, where regional gazetteers have been kept in many provinces and districts since the 14th century and many records are available for the last 2000 years (Chu 1973). Each of these gazetteers recorded local facts of historical or geographical interest as well as climatological events of significance to agriculture and the local economy (for example, droughts, floods, severe cold snaps, heavy snowfalls, unseasonable frosts, etc.). Not surprisingly, the gazetteers have been the focus of much interest, though there are often many difficulties in interpreting the data (see Sec. 11.2). In particular, technological improvements such as the building of irrigation channels or drainage ditches may drastically reduce the frequency of climatological disasters. For example, the Chinese province of Szechuan is unusual in that recorded occurrences of floods are rare, yet droughts are common. Usually, one finds, over long periods of time, a similar number of extremely wet and extremely dry events. The reason for this anomaly appears to be the particularly efficient flood-control measures introduced by the administrator Li Ping 2100 years ago which were able to reduce flood hazards, but did little to alleviate the perils of droughts (Yao 1943). In a similar way, the rise in flood and drought frequency during the Yuan dynasty (AD 1234–1367) may be partly due to the destruction of irrigation and drainage systems by the Mongol invaders of the time (Chu 1926).

Early work attempted to reduce the bias due to technological changes and increasing observations in recent history by assuming that such changes affected drought and flood observations equally. An index of "raininess" or precipitation anomaly was obtained by expressing the number of droughts (D) and floods (F) as a ratio (D/F), a common approach to reducing data errors in historical climatology. Thus, Yao (1942) was able to characterize each century over the last 2200 years as "wet" or "dry" compared to the mean ratio for different regions of China (Table 11.3).

More recent work by Wang and Zhao (1981) has capitalized on the detailed spatial–temporal data base by classifying each regional record, year by year, into one of five anomaly classes (wet to dry) and subjecting these matrices to principal components analysis. One hundred and eighteen stations from almost the whole of eastern China were used in the analysis of data from 1470 to 1977. The resulting eigenvectors indicated broad-scale patterns of "precipitation anomalies" and these were

Table 11.3 Ratio of droughts to floods in northern, central, and southern China (from Yao 1942). D = number of droughts; F = number of floods; $R = D/F$.

	Northern China				Central China				Southern China			
Century	*D*	*F*	*R*	Remarks	*D*	*F*	*R*	Remarks	*D*	*F*	*R*	Remarks
BC												
2		11	0.0	Wet?		2	0.0	Wet?		2	0.0	Wet?
1		12	0.0	Wet?		1	0.0	Wet?				
AD												
1	5	6	0.83	Dry						1	0.0	Wet?
2	12	17	0.71	Dry	1	3	0.33	Wet		1	0.0	Wet?
3	9	27	0.33	Wet		27	0.0	Wet?	1	4	0.25	Wet
4	4	6	0.67		4	17	0.24	Wet		6	0.0	Wet?
5	7	20	0.35	Wet	8	39	0.21	Wet				
6	13	27	0.48	Wet	4	14	0.29	Wet				
7	26	57	0.46	Wet	11	26	0.42	Wet	1	3	0.33	Wet
8	15	67	0.22	Wet	5	21	0.24	Wet	2	6	0.33	Wet
9	22	56	0.39	Wet	27	60	0.45	Wet	9	12	0.75	Dry
10	74	98	0.76	Dry	20	46	0.44	Wet	8	16	0.50	Wet
11	56	118	0.47	Wet	23	50	0.46	Wet	2	15	0.13	Wet
12	37	34	1.09	Dry	103	131	0.79	Dry	36	48	0.75	Dry
13	60	83	0.72	Dry	55	112	0.49	Wet	25	33	0.76	Dry
14	110	199	0.55	Wet	65	108	0.60	Dry	32	50	0.64	
15	40	75	0.53	Wet	42	115	0.37	Wet	33	46	0.72	Dry
16	52	49	1.06	Dry	60	104	0.58		66	104	0.63	
17	77	107	0.72	Dry	96	145	0.66	Dry	55	81	0.68	Dry
18	116	131	0.89	Dry	107	164	0.65	Dry	42	72	0.58	
19	100	89	1.12	Dry	85	114	0.75	Dry	6	16	0.37	Wet
Totals	835	1289	0.65		716	1299	0.55		318	516	0.62	

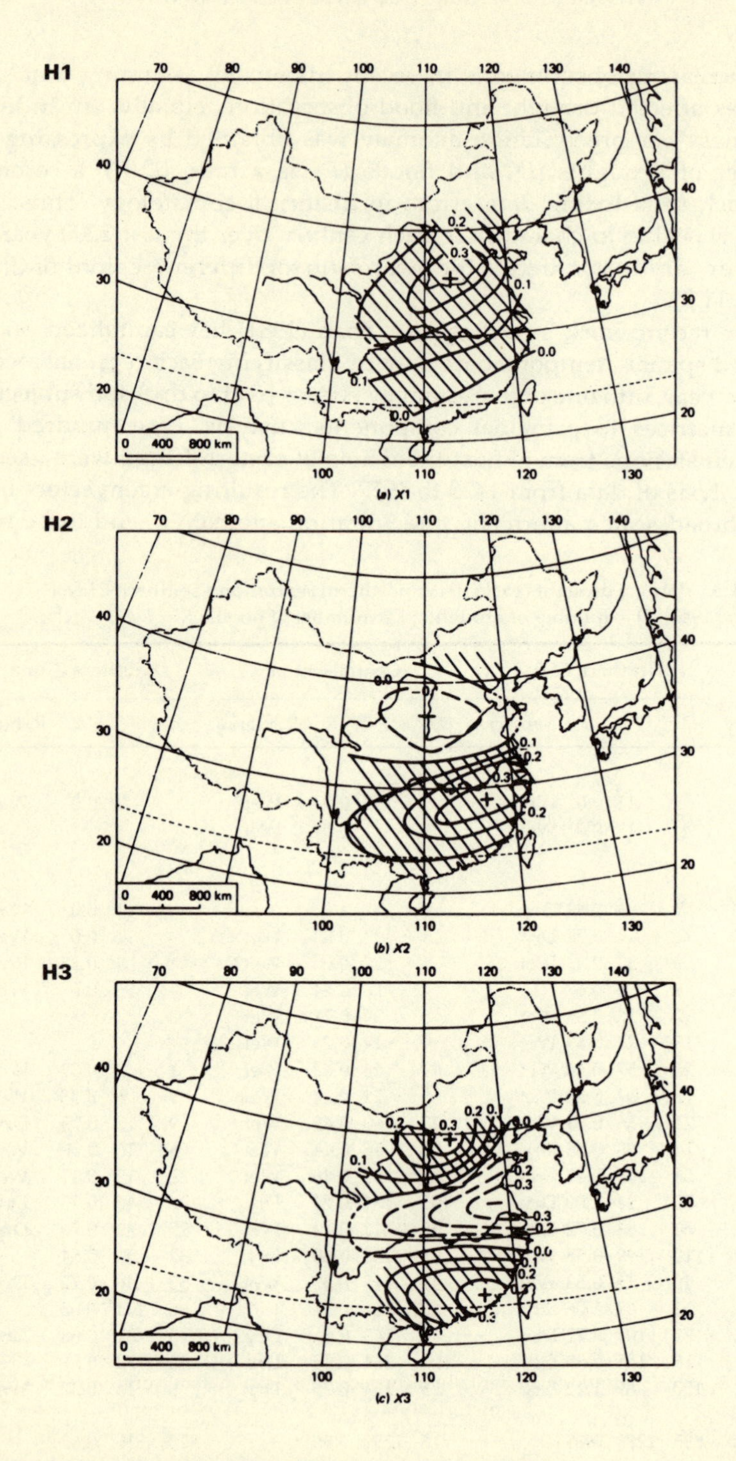
H1
70
80
90
100
110
120
130
140
40
30
20
0.2
0.3
0.1
0.0
0.1
0.0
0 400 800 km
100
110
120
130
H2
70
80
90
100
110
120
130
140
40
30
20
0.0
0.1
0.2
0.3
0.2
0 400 800 km
100
110
120
130
H3
70
80
90
100
110
120
130
140
40
30
20
0.2
0.3
0.2
0.1
0.0
0.1
0.2
0.3
0.3
0.2
0.0
0.1
0.2
0.3
0 400 800 km
100
110
120
130

(a) X1

(b) X2

(c) X3

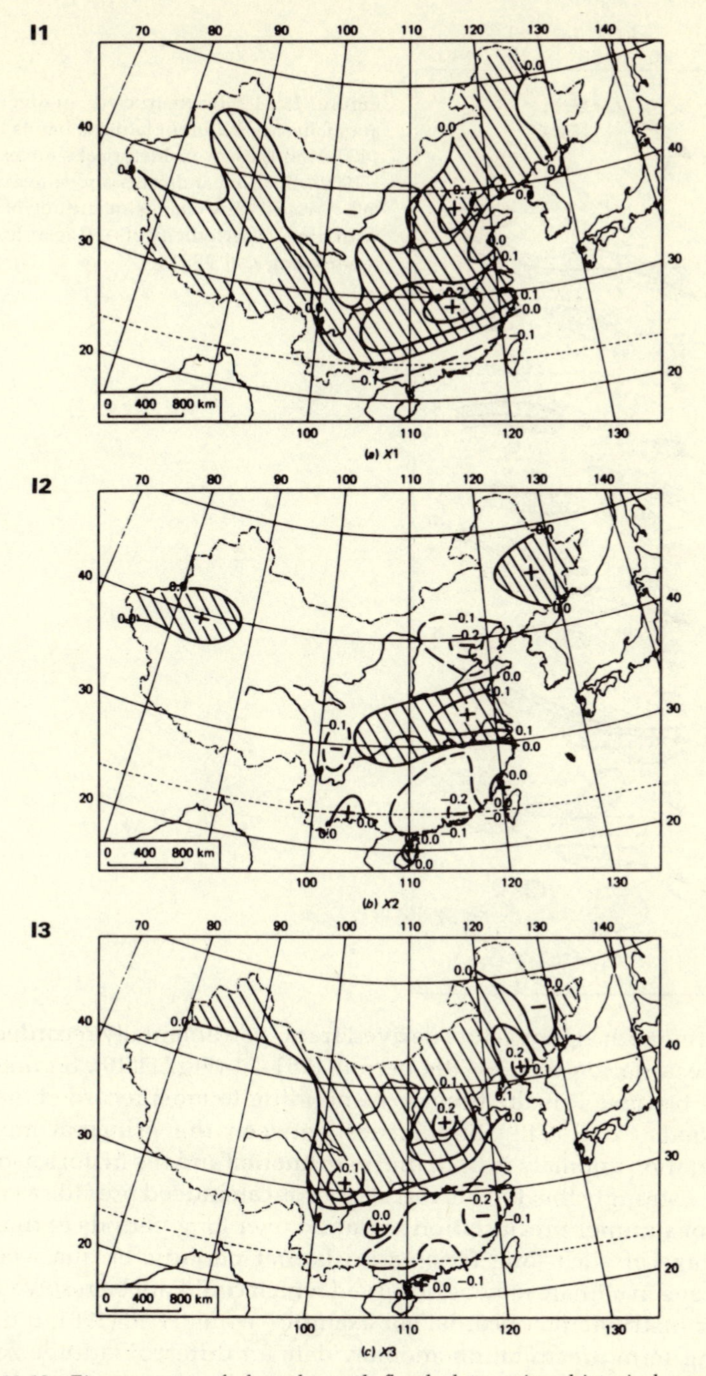

Figure 11.10 Eigenvectors of drought and flood data using historical records for AD 1470–1977 (left) and instrumentally recorded precipitation data from 1951–1974 (right). The first three eigenvectors of historical data (H1 to H3) account for 15, 11, and 7% of variance in the data respectively. The first three eigenvectors of instrumental data (I1 to I3) account for 18, 13, and 11% of variance in the data, respectively. Eigenvectors H1 and I1 are similar; so are H2 and I3 and H3 and I2. This indicates that historical data can be reliable indicators of climatic anomaly patterns (after Wang & Zhao 1981).

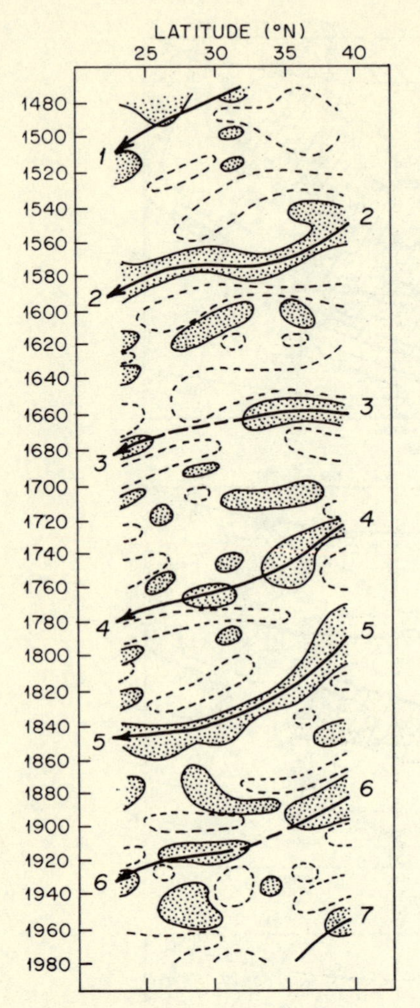

Figure 11.11 Reconstructed summer rainfall anomalies for different latitude bands in China, 1470–1980 (based on historical sources prior to ~1900). Shading indicates above average rainfall. Arrows show change in latitude of anomaly with time. A periodicity of ~80 years is apparent (after Wang *et al.* 1981).

compared with eigenvectors derived from instrumentally recorded summer precipitation data for the period 1951–74 (Fig. 11.10). Summer was chosen because it is the time corresponding to most recorded droughts and floods (Yao 1942). Similarities between the principal modes of precipitation anomaly in both the instrumental and the historical periods indicate strongly that the historical records can indeed provide a valuable proxy of summer precipitation variation over long periods of time. One advantage of such long time series is that periodic or quasi-periodic variations in climate may be observed which could not be resolved in the shorter instrumental records. For example, Wang *et al.* (1981, n.d.) used the long-term precipitation anomaly data for different latitude zones in China to construct a time–space diagram which points to a recurrent pattern of precipitation anomaly beginning in northern China and migrating southward, with a period of ~80 years (Fig. 11.11). More detailed studies of records from different areas have used spectral

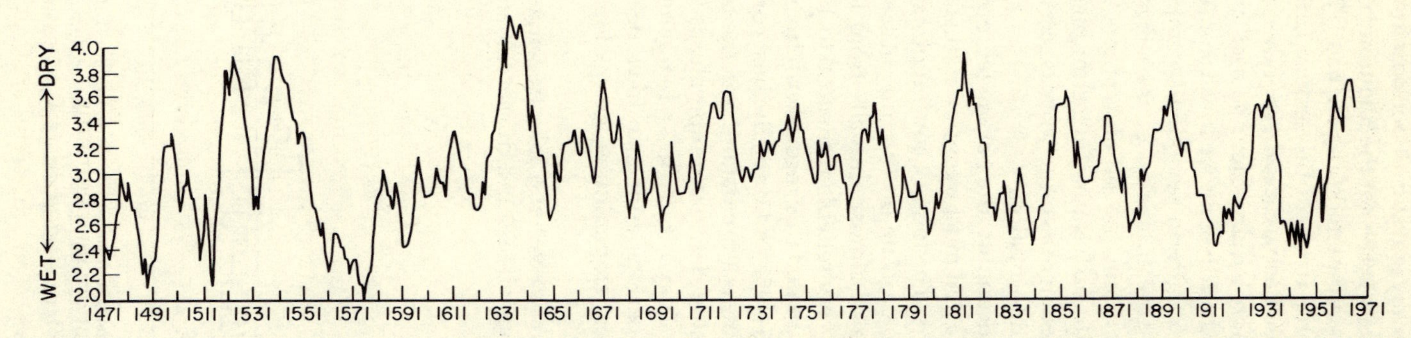

Figure 11.12 Ten year running means of "wetness index" for Shanghai, China, based on historical sources. Spectral analysis of the record reveals a statistically significant periodicity of ~37 years (after Wang & Zhao 1979).

analysis to isolate statistically significant periodicities in the data. Thus, in the precipitation anomaly series for the Shanghai region (Fig. 11.12), a periodicity of 36.7 years is apparent (Wang & Zhao 1979). This periodicity is also seen in other records from south-western China and the eastern part of the Yangtze River Basin. Elsewhere, other periodic variations have occurred, most notably a quasi-biennial oscillation (2–2.5 years) in the Yangtze River area and north of the Yellow River (Wang & Zhao 1981). Both the short-term and long-term periodicities appear to be related to large, synoptic-scale pressure anomalies over eastern Asia and adjacent equatorial regions. At present, no other workers have attempted to analyze historical data with such statistical techniques, but no doubt this approach will be used elsewhere as more work is carried out on historical sources of climatological information.

Apart from floods and droughts, the parameteorological phenomenon which seems to have attracted most attention in historical records is the freezing of lakes and rivers. The longest continuous series is that of Lake Suwa (near Kyoto) in Japan. Data on the time of freezing of this small (~15 km^2) lake are available almost annually from 1444 to the present, though the dates are not very reliable from 1680 to 1740 (Arakawa 1954, 1957). Gray (1974) has calibrated this record with instrumental data from Tokyo for the period 1876–1953 and finds the best correlation with mean December to February temperature data. Using the regression equation relating Lake Suwa freezing dates to temperatures since 1876, Gray is then able to reconstruct Tokyo midwinter temperatures back to 1450 (Fig. 11.13). The coldest periods appear to have been ~1450–1500 and ~1600–1700, when winter temperatures were about 0.5 °C below the mean of the last 100 years.

Long historical records of river and lake freezings are also available from China. Using local gazetteers and diaries, Zhang and Gong (1979)

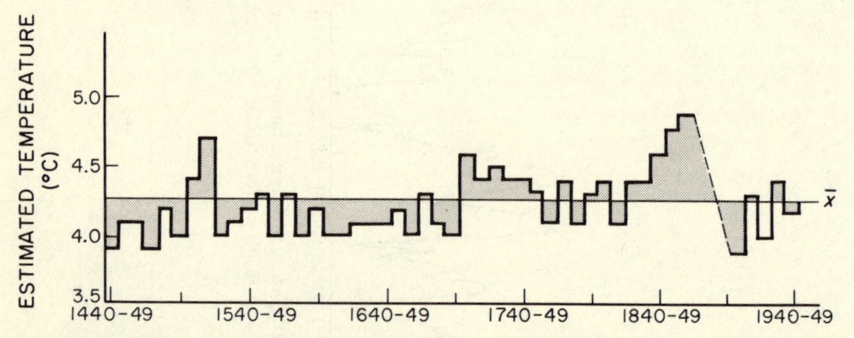

Figure 11.13 December–February temperature estimates for Tokyo reconstructed from Lake Suwa freezing dates, based on a calibration for the period of instrumental records (after Arakawa 1954 and Gray 1974). $\bar{x}$ = mean. Only 7 years of data in the period 1870–99. Other values are decadal averages.

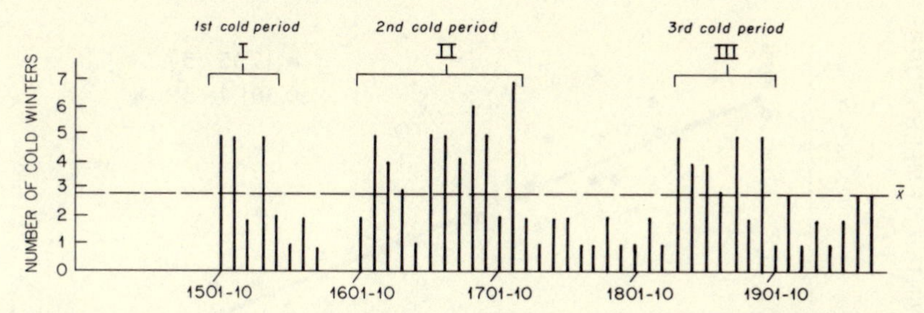

Figure 11.14 Number of cold winters per decade in central and southern China from 1501–10 to 1971–80. Major cold intervals are indicated (after Zhang & Gong 1979).

compiled records of the frequency of freezings of lakes in the mid and lower reaches of the Yangtze River, freezings of rivers and wells in the lower Yellow River Basin, the occurrence of sea ice in the Gulf of Chihli and Kiangsu Province (31–41°N), and snowfall in tropical areas of southern China. Using all this information they were able to calculate the number of exceptionally cold winters per decade from 1500 to 1978 (Fig. 11.14). The highest frequency of cold winters occurred during the periods 1500–50, 1601–1720, and 1830–1900, with the decade 1711–20 being the most severe period in the last 480 years. By mapping the areas most affected by severe winters it was also possible to recognize two main patterns of the anomaly—periods when cold winters predominated east of ~115°E and periods when areas to the west were colder. From modern meteorological studies it appears that such large-scale anomaly patterns result from changes in the position of the upper level trough over East Asia. When the trough is in a more westerly position and fairly deep, cold air sweeps down more frequently over the area west of 115°E. When the trough is weaker and less extensive, westerly flow prevails, leading to milder conditions in the west, with cold air outbreaks more common to the east. Generally speaking, the colder periods shown in Figure 11.14 had more frequent cold outbreaks west of 115°E, indicating that such periods were characterized by a strongly developed upper air trough over eastern Asia. Conversely, the warmer periods were times of stronger westerly flow in winter and weaker upper level trough development (Zhang & Gong 1979).

Historically, one of the most important effects of severe winter temperatures was on water transportation systems, and many records exist regarding the disruptions caused by canals and rivers freezing over for prolonged periods. In the Netherlands, for example, canals were built in the early 17th century to connect major cities, and records of transportation on the canals, including times of freezing over, have been kept since 1633 (de Vries 1977). Using instrumental winter temperature data from De Bilt (Labrijn 1945), the number of days on which the Haarlem – Leiden canal was frozen each winter was calibrated (Fig. 11.15), enabling De Bilt

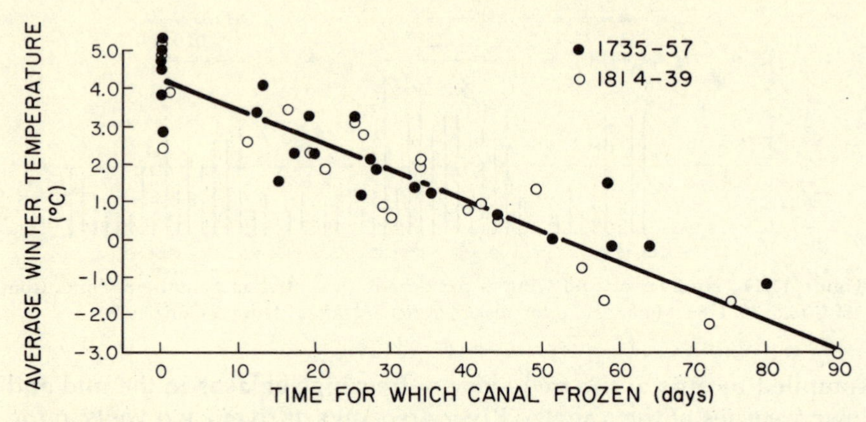

Figure 11.15 Relationship between the number of days with a canal frozen between Haarlem and Leiden (Holland) per winter and average winter temperature. A best-fit linear regression line is shown (after van den Dool *et al.* 1978).

winter temperatures to be reconstructed back to 1657. Further temperature estimates, back to 1634, were possible by calibration of the canal freezing data with barge trip frequency between Haarlem and Amsterdam (1634–82), a service which was commonly suspended due to ice cover on the canal (van den Dool *et al.* 1978). In this way, a complete winter temperature reconstruction for De Bilt has been obtained back to 1634 (Fig. 11.16).

In more northern latitudes, rivers freeze over every year, and in historical time the dates of freeze-up and break-up were both economically and psychologically important. Consequently, diaries and journals from these regions commonly contain frequent reference to the state of icing on nearby rivers and estuaries. A valuable analysis of such data from western Hudson's Bay has been made by Catchpole *et al.* (1976). Using content analysis they analyzed journals kept by Hudson's Bay Company trading post managers from the early 18th to the late 19th

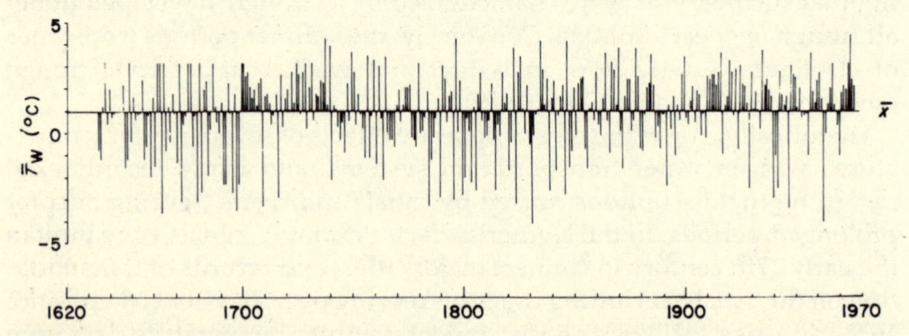

Figure 11.16 Reconstructed winter temperatures at De Bilt, Holland, expressed as departures from the long-term average (+2 °C) based on regression shown in Figure 11.15 and historical records of canal freezing frequencies (after van den Dool *et al.* 1978).

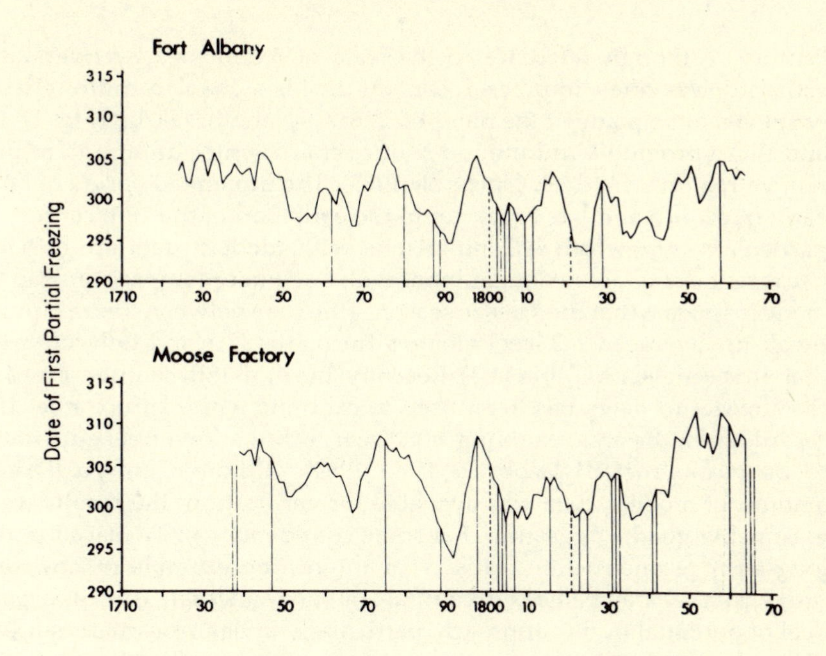

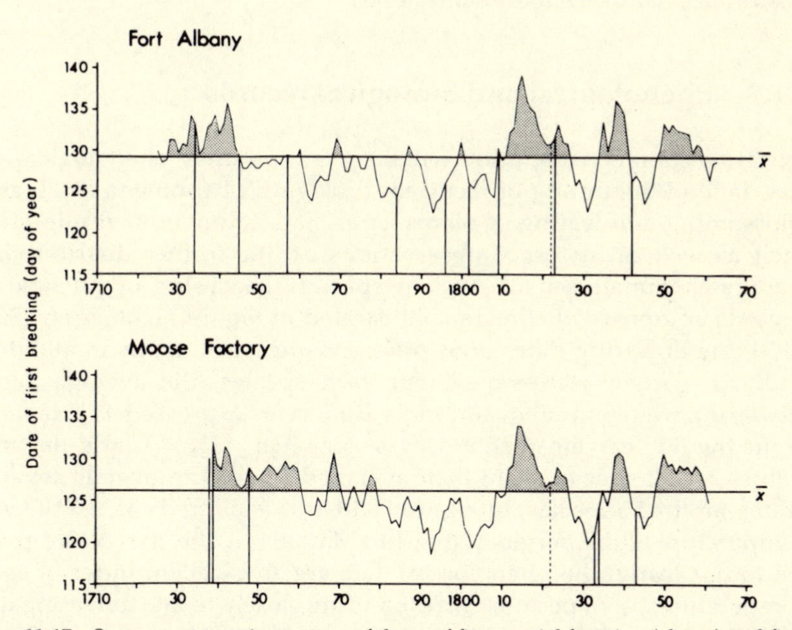

Figure 11.17 Seven year running means of dates of first partial freezing (above) and first breaking (below) of ice in estuaries at the locations indicated (all on west coast of Hudson's Bay, Canada). Data obtained by content analysis of historical sources. See also Table 11.4. Comparable dates for modern conditions shown as dashed line for Moose Factory. Dates given in days after December 31 (after Catchpole *et al.* 1976).

century. Although reference to the state of ice on nearby rivers and estuaries was often imprecise, content analysis enabled quite reliable estimates to be made of the dates of freeze-up and break-up (Fig. 11.17) and these provide a unique index of overall "winter duration" in this remote region (Moody & Catchpole 1975). The prolonged period of both early freeze-up and late break-up in the early part of the 19th century is particularly noteworthy. Comparisons with modern data are difficult because the sites are no longer inhabited, but where comparisons can be made it appears that the "freeze season" (the time between freeze-up and break-up) averaged 2–3 weeks longer during the 18th and 19th centuries than in recent years (Table 11.4). Recently, the mid-18th century record of first freeze-up dates has been used to calibrate white spruce tree-ring records from the area, enabling a 300 year record of first freeze-up dates to be reconstructed (Jacoby & Ulan 1982). Although only a limited amount of modern data was available for verification, the results were reasonably good, suggesting that some confidence can be placed in the long-term reconstruction. This is an interesting example of how one proxy data set may be used to calibrate or verify another. There is a great deal of potential in this approach, particularly in dendroclimatic studies using trees from the eastern United States and western Europe, and historical records from the same areas.

11.5 Phenological and biological records

In this section consideration will be given to purely phenological data, i.e. data on the timing of recurrent biological phenomena (such as the blossoming and leafing of plants, crop maturation, animal migrations, etc.) as well as historical observations on the former distribution of particular climate-sensitive plant species. The value of phenological records as a proxy of climate is illustrated in Figure 11.18. From 1923 to 1953, the flowering dates of 51 different species of plants in a Bluffton, Indiana, garden were noted. For each species, the average date of flowering was computed, and individual years expressed as a departure from the 30 year mean (Londsey & Newman 1956). Yearly departure values for all species were then averaged to give an overall departure index for the 51 species; in Figure 11.18, this is plotted against the mean temperature of the period March 1 to May 16 (i.e. the start of the growth period). Clearly, the phenological data are an excellent index of spring temperatures, cool periods corresponding closely to late flowering dates and vice versa. This example illustrates well the potential paleoclimatic value of phenological observations; if they can be calibrated, they may provide an excellent proxy record of past climatic variation.

One of the longest and best known phenological records comes, like so

Table 11.4 Comparison of historical (H) and modern (M) dates of freeze-up and break-up (in days after December 31) (after Catchpole *et al.* 1976).

Site	First partial freezing (H) or first permanent ice (M)			First complete freezing (H) or complete freezing (M)			First breaking (H) or first deterioration of ice (M)		
	Earliest	Mean	Latest	Earliest	Mean	Latest	Earliest	Mean	Latest
Churchill Factory (H)	272	289	311	293	315	342	142	168	189
Churchill River at Churchill (M)	273	291	318	288	319	336	141	160	169
Fort Prince of Wales	273	292	319	295	321	345	150	168	187
Hudson Bay at Churchill (M)	292	305	319	313		340	124	159	180
Moose Factory (H)	281	304	335	290	319	341	105	126	145
Moose River at Moosonee (M)	304	316	331	217	330	347	103	116	126

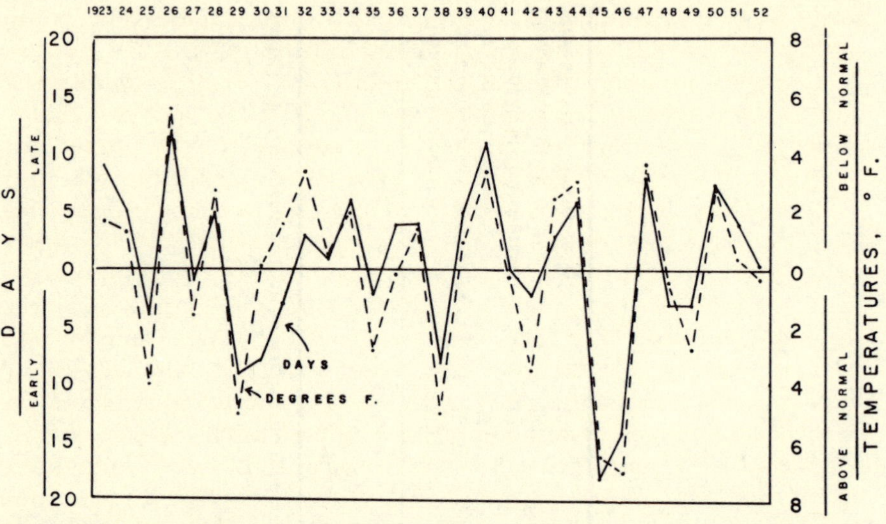

Figure 11.18 Phenological data as a climatic index. Solid line shows the average number of days per year in which the flowering dates of 51 species were earlier (below zero, left ordinate) or later (above zero, left ordinate) than the average. The dashed line shows the average departure from mean daily temperature from March 1 to May 16 for each year (right ordinate scale). Above average temperatures plotted below the zero line, below average plotted above the zero line. Observations from Bluffton, Indiana (after Lindsey & Newman 1956).

many other long historical records, from the Far East. At Kyoto (the capital of Japan until 1869) the Governor or Emperor used to hold a party under the flowering cherry blossoms of his estate, when they were in full bloom (Arakawa, 1956b, 1957). The blooming dates can be considered as an index of spring warmth (February and March), as shown by Sekiguti (1969) using modern phenological records and instrumental data. Higher spring temperatures result in earlier blooming dates. The record is *very* sparse, but nevertheless is of interest as it spans such a long period of time (Table 11.5). It appears from this record that the 11th to 14th centuries were relatively cool, though this impression could be entirely due to the inadequate statistics (only 30 reliable dates in 400 years!). Were it not for similar phenological observations from China, tending to

Table 11.5 Average cherry blossom blooming dates at Kyoto, Japan, by century (from Arakawa 1956b). Mean date, April 14.6; $N = 171$.

Century	9th	10th	11th	12th	13th	14th	15th	16th	17th	18th	19th	20th[†]
Day in April	11	12	18	17	15	17	13	17	12	–	12	14
No. observations	7	14	5	4	8	13	30	39	10	–	5	36

[†] Chu (1973) notes 20th century data (1917–53) for blossoms "in full bloom."

support this idea, particularly of a cooler 12th century in the Far East, one could place little faith in the Kyoto data alone (Chu 1973).

The most important phenological records from Europe concern the date of the wine harvest. Wine harvests are, of course, not only determined by climatic factors but also by economic considerations. For example, an increasing demand for brandy may cause the vineyard owner to delay the harvest in order to obtain a liquor richer in sugar and more desirable from the point of view of spirits production. However, such factors are unlikely to be of general significance, and, providing variation in the dates of wine harvests show regional similarities, it is reasonable to infer that climate is a controlling factor. Thus, Le Roy Ladurie and Baulant (1981) have produced a regionally homogeneous index of wine harvest dates for central and northern France, based on over 100 local harvest-date series, extending back to 1484. For the period of overlap with instrumental records from Paris (1797–1879) the index had a correlation coefficient with mean April to September temperatures of +0.86, indicating that it provides a good proxy of the overall warmth of the growth season (Fig. 11.19; cf. Garnier 1955, Le Roy Ladurie 1971). Indeed, Bray (1982) has shown that the reconstructed summer paleotemperatures also show a strong correlation with the record of alpine glacier advances in western Europe. Periods characterized by temperatures consistently below the median value are generally followed by glacier advances.

Le Roy Ladurie (1971) has used a variety of historical records (economic, phenological, and pictorial) to document climatic fluctuations in western Europe over the last 1000 years. This climatic analysis provides a fascinating backdrop to the social and political events in that region during a period of significant climatic change. In particular, he indicates that the "Little Ice Age" (~AD 1525–1850) was not a period of uniformly severe climate, but rather a period when extreme climate conditions were more prevalent than in the periods before or since. This conclusion was recently reinforced by Pfister (1981). Overall, conditions were most severe from the latter half of the 16th century to the late 18th century, a period which can be considered as the "climatic pessimum" of the last 1000 years. However, even in this period occasional sequences of mild winters and warm summers did occur (e.g. the first few decades of the 18th century) (de Vries 1981).

The deterioration of climate during the Little Ice Age also had important geographical consequences for plants and animals, particularly in marginal environments. In the Lammermuir hills of south-eastern Scotland, for example, oats were cultivated to elevations of over 450 m during the warm interval from AD 1150 to AD 1250. By AD 1300, the uppermost limit had fallen to 400 m and by AD 1600 to only ~265 m, more than doubling the area of uncultivable land (Parry 1975, 1981). These changes

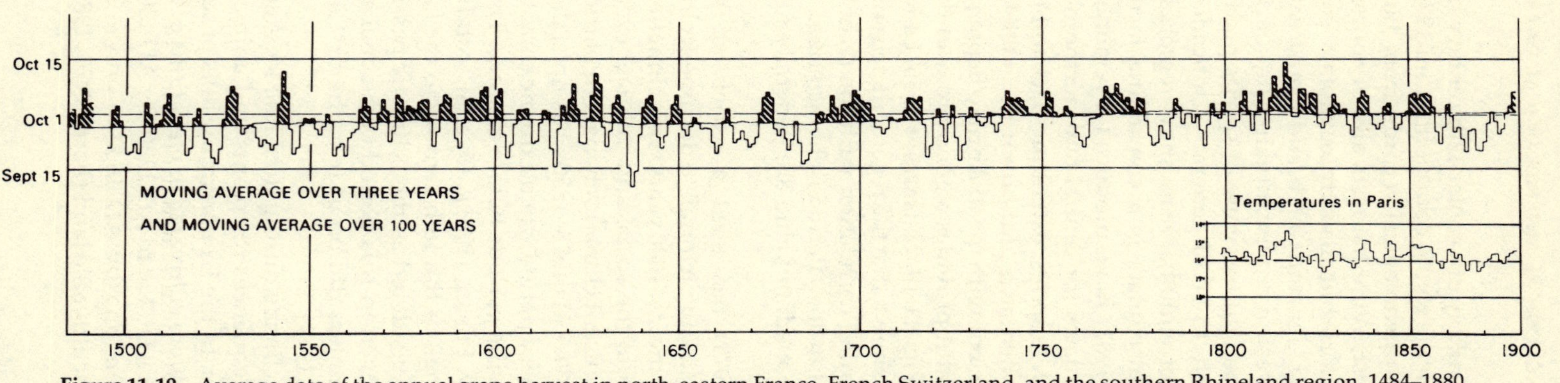

Figure 11.19 Average date of the annual grape harvest in north-eastern France, French Switzerland, and the southern Rhineland region, 1484–1880. At lower right are mean April–September temperatures in Paris during the period of instrumental records. Data smoothed by a 3 year running mean. Harvest data shaded when later than 100 year running mean (continuous slightly rising line) (after Le Roy Ladurie & Baulant 1981).

probably resulted from reduced summer warmth, wetter conditions, and earlier snowfalls in winter months, factors which all combined to increase the probability of crop failure from only one year in 20 in the Middle Ages to one year in two or three during the Little Ice Age. The abandonment of upland field cultivation was also accompanied by the abandonment of upland settlements and resulted in a considerable redistribution of population in the area.

Former plant and animal distributions can provide useful indices of climatic fluctuation, though it is not always possible to quantify the significance of the change in species range. Harper (1961), for example, has documented significant changes in the distribution of flora and fauna in subarctic Canada during the 20th century as a result of the widespread increase in temperature during this period. Similar observations have been made in Finland by Kalela (1952) and the change is not confined to terrestrial species; northward migration of fish in the North Atlantic as a result of increasing water temperatures in the North Atlantic have also been noted (Halme 1952). Trading post records of animal catches around the coasts of Greenland point to the close dependence of animal populations on climatic fluctuations and associated changes in the distribution of sea ice (Vibe 1967). These records rarely extend back beyond the mid-19th century, however, and merely point to the biological significance of climatic fluctuations which instrumental records have documented in considerable detail. Nevertheless, there is great potential in this area of research to use historical records of former plant and animal distributions, the timing of migrations, etc., to document climatic variations during periods for which no instrumental records exist. The possible value of such work is well illustrated by the wide-ranging surveys of former plant and animal populations in China reported by Chu (1973) and Wang *et al.* (n.d.).

12

Quaternary paleoclimatology: current status and future directions

12.1 Introduction

Paleoclimatic research has undergone dramatic changes over the past 20–30 years and our understanding of the paleoclimatic record has improved remarkably. Entirely new methods have emerged, older methods have benefited from new approaches, new multivariate statistical techniques have been developed, and improvements in analytical equipment have made formerly complex measurements routine. There has also been an important shift in public awareness of climatic variability; potential changes of climate in the future are now perceived as being of international significance (Central Intelligence Agency 1972). This has led to more attention being focused on climatic records of the past in order to understand the nature and causes of climatic variability (Committee on Science and Technology 1976). Consequently paleoclimatic research has undergone tremendous growth, as reflected in the methodological developments reviewed in previous chapters. In this final section, the significant developments in paleoclimatology will be briefly reviewed, and probable directions of future research will be assessed.

12.2 Methodological improvements

The most significant advances in paleoclimatic research over the last 20–30 years have resulted from geochemical studies, particularly the temperature-dependent fractionation processes. The analysis of oxygen isotopes in marine sediments has resulted in profound changes in our understanding of Quaternary paleoclimates. Only a decade or so ago the majority view was that four major glaciations characterized the Quaternary period. Today, studies of marine sediments point to more than 20 episodes of continental-scale glaciation during the Pleistocene (van Donk 1976). It is thus apparent that modern climatic conditions are quite atypical in the context of the last 2 million years. The modal climate of the Quaternary has been closer to full glacial than to full interglacial conditions.

Oxygen isotope studies of ice cores have also been of great significance, providing remarkable resolution of climatic variations over at least the Holocene, and perhaps much longer. Detailed analyses of trace substances in ice cores have added a new dimension to paleoclimatology – the reconstruction of past atmospheric composition – and may enable the importance of factors implicated in climatic fluctuations (e.g. volcanic dust and solar activity variations) to be directly evaluated.

Paleoclimatology has benefited greatly from improvements in techniques of multivariate data analysis, using high speed computers. Not only can very large data sets be handled routinely but complex interactions between variables can be assessed. Multivariate analysis has been of major significance in palynological, dendroclimatic, and marine sedimentary research. Of equal importance has been the application of spectral analysis (and other similar analytical techniques) to paleoclimatic time series. This has enabled the spectrum of climate over different timescales to be assessed (e.g. Kutzbach & Bryson 1974, Mitchell 1976). There is now strong evidence that orbital perturbations are a major factor in climatic fluctuations on the scale of 10^4–10^5 years.

Many of the advances made in understanding the paleoclimatic record could not have been achieved without major improvements in dating methods. Radiocarbon dating has been crucial in resolving the paleoclimatic record of the last 20 000–30 000+ years, and the development of a reliable paleomagnetic timescale (based primarily on potassium – argon dates) has been the key to unravelling the long-term paleoclimatic record of marine sediments. Accurate dating has enabled "time-slices" to be studied by mapping paleoclimatic conditions at particular periods in the past (e.g. Peterson *et al.* 1979). This approach has provided a very valuable spatial perspective which greatly increases our understanding of former atmosphere circulation patterns.

One area of paleoclimatic research which has received only brief attention in previous chapters is the use of numerical models to develop a better understanding of the climate system (Schneider & Dickinson 1974). Most models are computer-based and are used to simulate a particular process, test a hypothesis, or examine the interactions between various components of the climate system. Although even the most sophisticated models (multidimensional general circulation models or GCMs) are still quite simplistic representations of the climate system, there have been attempts to use computer models to answer questions of relevance to paleoclimatology (Barry 1975, J. Williams 1978). For example, models have been used to assess the climatic consequences of an open Arctic Ocean (Newson 1973, Warshaw & Rapp 1975) and to evaluate the relative importance of volcanic dust and solar constant variations on hemispheric temperatures (Wetherald & Manabe 1975, Hunt 1977). In this way, factors which may have had an important role in past climatic variations can be evaluated. Models of the general atmospheric circulation have also been used with glacial age boundary conditions (utilizing paleoclimatic data as input to the model) to reconstruct the "equilibrium" general circulation at 18 000 years BP (J. Williams *et al.* 1974, Gates 1976a,b, Manabe & Hahn 1977). Models have recently been used to assess the role of orbital perturbations on climate using modern ("interglacial") boundary conditions, but with changes in solar radiation distribution due to orbital perturbations (e.g. Kutzbach & Otto-Bliesner 1982, Royer *et al.* 1984). In this way, the probable response of the general circulation to such forcing can be assessed. For example, results from models using solar radiation conditions of 125 000 and 115 000 years BP indicate that orbital changes alone may bring about lower temperatures and a significant increase in moisture in those areas of eastern Canada considered to be important centers of ice-sheet growth in the early stages of glaciation (Royer *et al.* 1984). Computer models can thus make significant contributions to paleoclimatic research, both in testing hypotheses and in pointing to new areas of research or controversy which further studies of the paleoclimatic record may be able to resolve. However, it should be noted that many models are still quite crude and the simulations they produce may be of limited value. Poor spatial resolution, preset boundary conditions, and/or inflexibility in the interactions between different components of the climate system (feedback processes) often contribute to uncertainties in the final results. More sophisticated models coupling oceanic, atmospheric, and cryospheric subsystems may eventually give the paleoclimatologist more confidence in numerical model simulations.

12.3 Paleoclimatic reconstruction: future directions

It was pointed out in Chapter 1 that a major goal of paleoclimatic reconstruction is to understand factors which cause climatic fluctuations. There are many hypotheses as to why climate varies but few have been rigorously tested, primarily because the paleoclimatic record is not known in sufficient detail to be able to do so. In order to isolate the causes of climatic variation (bearing in mind that several factors may be operating simultaneously) improvements in several areas are needed: calibration, dating, and geographical coverage.

12.3.1 Calibration

Accurate quantitative paleoclimatic reconstructions are based on a precise calibration of the proxy data set with modern climatic data. Even if a proxy record is well dated it will be of little value unless the climate dependency of the proxy is clearly understood. There is room for considerable improvement in this area. The climate–proxy relationship has to be more than just a statistically significant correlation coefficient. There must be an understanding of the reasons *why* a phenomenon is climate dependent, or more specifically why it is related to a particular climatic parameter at a particular time of year. Climatic parameters are often highly intercorrelated and may exhibit significant (intra-annual) serial correlation; consequently, a strong correlation between a climatic variable and a proxy record does not necessarily indicate that the most meaningful calibration has been obtained. At the present time, many paleoclimatic reconstructions are based on fairly weak models of climate dependency in the proxy phenomenon being used. This is not to suggest that the climate dependency does not exist, but that the precise nature of that dependency may not have been adequately demonstrated or rigorously isolated. Consider, for example, dendroclimatic studies. In conventional dendroclimatology, models of those climatic factors important in determining tree-ring widths have been developed and the biological processes involved have been studied in considerable detail (see Fritts 1976 Ch. 5). Thus, not only can statistically significant correlations between climatic data and tree-ring widths be demonstrated, but the *reasons* for those correlations are generally understood. However, models linking climate and tree-ring *density* are more rudimentary and consequently densitometric dendroclimatic studies have generally been based on a far less satisfactory empirical approach. However, this is a relatively new field and current research is aimed at resolving this matter by placing the climate–proxy relationship on a firm theoretical basis (e.g. Conkey 1982).

A similar problem exists with the paleoclimatic interpretation of oxygen isotopes in polar ice cores. Many of the factors influencing $\delta^{18}O$ in precipitation have been studied, yet relatively little attention has been paid to $\delta^{18}O$ variations in high latitude snowfall *in situ*. As most precipitation in these regions results from relatively few storms, the appropriate level of analysis is to examine daily data, to understand the relationship between $\delta^{18}O$ in precipitation and different synoptic conditions. However, $\delta^{18}O$ values in ice cores are still considered a proxy of mean annual temperature, in spite of the fact that virtually no precipitation occurs on most days of the year. (At Alert, North West Territories, Canada, for example, 80% of annual precipitation occurs on less than 40 days per year.) $\delta^{18}O$ values are interpreted in terms of mean annual temperature simply because there is an empirical correlation between the two variables, not because there is any clear physical dependency. Because the isotopic record of modern snowfall in polar regions has not been well studied on a synoptic basis, it becomes extremely difficult to interpret the down-core isotopic data in paleoclimatic terms.

These examples serve to illustrate both the limitations of many current calibration models and the potential for improvement. Studies of the interactions between proxy phenomena and climate *today* are the key to more meaningful paleoclimatic reconstructions, and significant improvements in this area seem possible.

12.3.2 *Dating*

Considerable attention has been paid to dating methods in this book because of their importance in paleoclimatic reconstruction. Accurate dating remains the nemesis of most paleoclimatic studies dealing with events beyond 30 000 years BP, and many important questions remain unanswered due to inadequate dating control. For example, there is quite a lot of evidence that very significant changes in climate have occurred abruptly at certain times in the past (Flohn 1979). These have been observed in a variety of sedimentary records (e.g. Dansgaard *et al.* 1972, Osborne 1974, Woillard 1979) and, if real, are of major significance. However, without better dating control one can never be certain that such changes are not artifacts of the sedimentary record, brought about, for example, by discontinuities in sedimentation or changes in sedimentation rate. As improvements in dating techniques occur, this question of catastrophic climatic changes should be resolved.

Significant improvement in paleoclimatic research can be expected as new techniques of radiocarbon dating become routinely available. This will enable relatively small samples to be rapidly dated and give fairly precise age assessments on material from the last 70 000 years, at least. However, better dating methods for events in the 100 000–200 000 year time range are badly needed, particularly for land-based deposits. Ice cores, for example, have the potential of providing very long paleoclimatic records, but without better dating methods than those used so far their value is severely limited. There is of course, considerable attention being paid to this problem (Dansgaard 1981) and major progress can be expected over the next few years.

12.3.3 *Geographic coverage*

Paleoclimatic records are by no means uniformly distributed either in time or space. Obviously sources of paleoclimatic information are most abundant for the last few centuries, decreasing in availability and reliability back in time. As noted in Chapter 1, Peterson *et al.* (1979) were only able to make use of approximately one well dated record per 10^6 km^2 in a survey of continental data for 18 000 years BP; for earlier periods the coverage drops off dramatically (and *continuous* well dated time series are quite rare). Improvements in the spatial network of paleoclimatic data would not only provide more confidence in paleoclimatic reconstructions but also enable more rigorous tests to be made of hypotheses of climatic change. In particular, there is a real need for more long-term non-marine records spanning the last interglacial–glacial cycle. With such data, many important questions could be resolved. For example: To what extent do changes on the continents lead those occurring in the oceans and vice versa? What feedback processes are important between components of the climate system, particularly at times of glaciation and deglaciation? To what extent do events in the Northern Hemisphere lag behind events in the Southern Hemisphere (or vice versa)? How do events in the Tropics relate to events at high latitudes? Many questions have arisen from the wealth of long-term paleoclimatic records recovered from the ocean floor (particularly regarding the precise mechanisms involved in ice-sheet growth or decay) yet there are very few long records (>100 000 years) available to examine paleoclimatic events on the continents. A major program of coring non-marine sedimentary deposits is therefore needed to provide many more long and continuous continental paleoclimatic records. The analytical tools and methodological procedures are at hand; more effort needs to be directed at the technical problems of recovering such records.

In the preceding chapters recent developments in Quaternary paleoclimatology have been reviewed. It is obviously a large field, which is becoming more specialized all the time. This growth poses a danger and a challenge. The danger is that the field will become so fragmented and compartmentalized that there will be little interdisciplinary understanding and collaboration. The challenge is obviously to avoid that, by keeping abreast of methodological developments in the various subfields of paleoclimatology. It is hoped that this book will contribute to that goal, and encourage yet more interdisciplinary research on climatic variations of the Quaternary period.

Appendix

A.1 Calculation of radiocarbon age and standardization procedure

Although it is not necessary for a user of radiocarbon dates to know, in detail, how the actual value is arrived at, some understanding of the procedure is enlightening, particularly when considering adjustments for ^{14}C fractionation effects (see Sec. 3.2.1.2). The whole subject is a complex sequence of calibrations, adjustments, and corrections which will soon be further complicated by the use of a new reference standard because the supply of NBS oxalic acid, the original reference material, is exhausted. The following brief explanation is offered for the adventurous.

In order to make dates from different laboratories comparable, a standard material is used by all laboratories for the measurement of "modern" carbon isotope concentrations. This standard is "US National Bureau of Standards oxalic acid" prepared from West Indian sugar cane grown in 1955. Ninety-five percent of the ^{14}C activity of this material is equivalent to the ^{14}C activity of wood grown in 1890, so, by this devious means, all laboratories may standardize their results to a material which has not been contaminated by "atomic bomb" ^{14}C. However, by convention, all dates are given in "years before 1950" (years BP or "before physics"), so dates are adjusted to this temporal standard, rather than the possibly more logical time of 1890. A second adjustment is needed to correct for the fact that oxalic acid undergoes variable ^{13}C and ^{14}C fractionation effects during analysis. To make interlaboratory comparisons of samples possible, standardization is necessary to take these fractionation effects into account. Fortunately, the fractionation effect of ^{14}C is extremely close to twice that of ^{13}C, which is far more abundant and can be measured easily in a mass spectromer.† The necessary standardization is thus achieved by measuring ^{13}C rather than ^{14}C. Following the detailed analysis of oxalic acid samples by Craig (1961a) it was agreed that all standard samples should be adjusted to a $\delta^{13}C$ value of $-19.3‰$, where

$$\delta^{13}C = \frac{(^{13}C/^{12}C)_{ox} - (^{13}C/^{12}C)_{PDB}}{(^{13}C/^{12}C)_{PDB}} \times 10^3; \qquad (A.1)$$

C_{ox} refers to the oxalic acid standard, and C_{PDB} refers to another reference standard, a Cretaceous belemnite (*Belemnitella americana*) from the Peedee Formation of South Carolina (Craig 1957).

†$\delta^{14}C = 2\delta^{13}C + 10^{-3}(\delta^{13}C)^2$.

Standardization of the ^{14}C activity in the oxalic acid reference sample is achieved thus:

$$0.95A_{ox} = 0.95A^{1}_{ox}\left[1 - \frac{2(\delta^{13}C_{ox} + 19)}{1000}\right] \tag{A.2}$$

where A^{1}_{ox} is the ^{14}C activity of the reference oxalic acid and $^{13}C_{ox}$ is based on the ^{13}C activity of the reference oxalic acid (Eq. A.1). The value of $0.95A_{ox}$ then becomes the universal ^{14}C standard activity from which all dates are calculated.

The procedure normally adopted for calculating the radiocarbon date of a sample can be summarized in three equations:

(a) The activity of the sample is expressed as a departure from the reference standard:

$$\delta^{14}C = \frac{A_{sample} - 0.95A_{ox}}{0.95A_{ox}} \times 10^{3},$$

where A_{sample} is the ^{14}C activity of the sample, corrected for background radiation, and A_{ox} is the 1950 activity of NBS oxalic acid, corrected for background and isotope fractionation.

(b) The sample activity is corrected for fractionation by normalizing to $\delta^{13}C = -25‰$ PDB, which is the average for wood (see below for discussion):

$$\Delta = \delta^{14}C - (2[\delta^{13}C + 25])\left(1 + \frac{\delta^{14}C}{1000}\right),$$

where

$$\delta^{13}C = \frac{(^{13}C/^{12}C)_{sample} - (^{13}C/^{12}C)_{PDB}}{(^{13}C/^{12}C)_{PDB}} \times 10^{3}.$$

(c) Age (T) is then calculated using the "Libby" half-life of 5570 years:

$$T = 8033 \log_e (1 + (\Delta/1000))^{-1}.$$

A.2 Fractionation effects

Because isotopic fractionation occurs to differing extents during plant photosynthesis and during shell carbonate deposition, it is necessary to know the magnitude of the effect so that dates on different materials can be compared. Lerman (1972) and Troughton (1972) have studied fractionation effects in modern plants and found a trimodal distribution; the magnitude of fractionation seems to be related to the particular biochemical pathway evolved by different plant species for photosynthesis. Thus, highest ^{13}C depletion corresponds to plants which utilize the Calvin photosynthetic cycle (CAL) and lowest depletion occurs in plants utilizing the Hatch–Slack (HS) cycle. Succulents utilize a third metabolic pathway (crassulacean acid metabolism, CAM) and may fix carbon by either the HS or CAL pathways depending on temperature and photoperiod;

they thus form a third group. The implications for dating are that samples containing, say, a high proportion of HS plants would show $^{14}C/^{12}C$ ratios characteristic of these less depleted plants. An uncorrected date on such material would thus appear younger than a stratigraphically equivalent sample comprising predominantly CAL-type plants. Similar problems presumably exist in dating animal remains which feed predominantly on plants using one or another photosynthetic pathway, though this question has not been studied in detail.

By international agreement, all ^{14}C dates are corrected for fractionation effects by standardization to a $\delta^{13}C$ value of −25‰, the average value for wood. This results in relatively small "corrections" for CAL plants but larger corrections for HS and CAM plants, freshwater and marine shells, and aquatic plants. In particular, corrections of up to 450 years may be necessary in the case of marine shells because they produce carbonate in equilibrium with ocean water which is relatively enriched in ^{13}C. Table A.1 indicates the magnitude of corrections necessary to standardize ^{14}C dates for fractionation effects. Where no such adjustment was made in a dated sample, these values may act as a guide to the appropriate correction needed.

Table A.1 Radiocarbon fractionation errors (after Lerman, 1972).

	Dating material	Photosynthetic pathway[†]	δC 13‰	Years to add to uncorrected date
1.	Wood and wood charcoal	CAL	−25 ± 5	0 ± 80
2.	Tree leaves	CAL	−27 ± 5	−30 ± 80
3.	Peat, humus, soil[‡]	CAL	−27 ± 7	−30 ± 110
4.	Grains[§]	CAL	−23 ± 4	+30 ± 60
5.	Leaves and straw of grasses and sedges (e.g. totora)[§]	CAL	−27 ± 4	−30 ± 60
6.	Bones (European)		−20 ± 4	+80 ± 60
7.	Grains[‖]	HS	−10 ± 3	+240 ± 50
8.	Leaves and straw of grasses and sedges (e.g. papyrus)[‖]	HS	−13 ± 4	+200 ± 60
9.	Succulents[¶]	CAM	−17 ± 8	+130 ± 120
10.	Aquatic plants (freshwater)		−8 to −24	
	Aquatic plants (marine)		−8 to −17	
11.	Fresh and brackish water shells		0 to −12	
12.	Marine shells		+3 to −2	

[†] CAL = Calvin metabolic pathway; HS = Hatch–Slack metabolic pathway; CAM = crassulacean acid metabolism.

[‡] Most nordic and temperate peats, humus and soils are of C-type. Others must be considered as unidentified.

[§] Wheat, oats, barley, rice rye, etc. . . . and related grasses.

[‖] Maize, sorghum, millet, panic, etc. . . . and related grasses.

[¶] Cactus, agave, pineapple, *Tillandsia*, etc. . . .

References

On the following pages the less obvious journal titles are abbreviated thus:

Acta. Bot. Neerl. *Acta Botanica Neerlandica*
Acta Met. Sinica. *Acta Meteorologica Sinica*
Ann. Assoc. Am. Geogr. *Annals of the Association of American Geographers*
Ann. N.Y. Acad. Sci. *Annals of the New York Academy of Science*
Ann. Rev. Earth Planet. Sci. *Annual Review of Earth and Planetary Science*
Ann. Rev. Ecol. Systematics *Annual Reviews of Ecology and Systematics*
Annlr. Met. *Annaler Meteorologisch*
Annls Econ. Soc. Civil. *Annales: Economies, Sociétés, Civilizations*
Annls Géogr. *Annales de Géographie*
Annls Soc. Géol. Belg. *Annales Société Géologie Belgique*
Arch. Met. Geophys. Bioklimatol. *Archiv für Meteorologie, Geophysik und Bioklimatologie*
Aust. J. Bot. *Australian Journal of Botany*

Bonner Met. Abhandl. *Bonner Meteorologische Abhandlungen*
Bot. Rev. *Botanical Review*
Bot. Tidsskr. *Botanisk Tidsskrift*
Bull. Am. Met. Soc. *Bulletin of the American Meteorological Society*
Bull. Int. Acad. Pol. Sci. Lett. *Bulletin International de l'Academie Polonaise des Sciences et des Lettres*
Bull. R. Soc. N.Z. *Bulletin of the Royal Society of New Zealand*

Carnegie Instn Wash. Ybk *Carnegie Institution of Washington Yearbook*
Celestial Mech. *Celestial Mechanics*
Colloques Int. Centre Natn Rech. Scient. *Colloques Internationale du Centre Nationale de la Recherche Scientifique*
Comments Earth Sci. – Geophys. *Comments on Earth Sciences – Geophysics*

Danmarks Geol. Undersogelse *Danmarks Geologiske Undersogelse*

Folia Geogr. Danica *Folia Geographica Danica*

Geochim. Cosmochim. Acta *Geochimica et Cosmochimica Acta*
Geol. Fören. Stockholm Förhandl. *Geologiska Föreningens i Stockholm Förhandlingar*
Geol. Mijnb *Geologie Mijnbouw*
Geogr. Annlr *Geografiska Annaler*
Geogr. Helv. *Geographica Helvetica*
Géogr. Phys. Quat. *Géographie Physique et Quaternaire*
Geogr. Rev. Japan *Geographical Review of Japan*
Geophys. Res. Letters *Geophysical Research Letters*

Geosci. Can. *Geoscience Canada*
Geosci. Man *Geoscience and Man*

Inst. Br. Geogrs. Trans. *Institute of British Geographers Transactions*

J. Appl. Met. *Journal of Applied Meteorology*
J. Interdisc. Hist. *Journal of Interdisciplinary History*
J. Met. *Journal of Meteorology*
J. Met. Soc. Japan *Journal of the Meteorological Society of Japan*

Leidse Geol. Med. *Leidse Geologische Mededelingen*
Limnol. Oceanogr. *Limnology and Oceanography*

Meddr Grønland *Meddelelser øm Grønland*
Meded. Verhandl. *Mededelingen en Verhandlingen*
Mem. Torrey Bot. Club *Memoirs of the Torrey Botanical Club*
Met. Rundschau *Meteorologische Rundschau*
Mich. Acadn *Michigan Academician*

Natn. Speleol. Soc. Bull. *National Speleological Society Bulletin*
New Phytol. *New Phytologist*
Norske Geol. Tidsskr. *Norske Geologisk Tidsskrift*
Nuovo Cim. *Il Nuovo Cimento*
N.Z. J. Bot. *New Zealand Journal of Botany*

Palaeoecol. Afr. *Palaeoecology of Africa*
Phil. Mag. *Philosophical Magazine*
Phil. Trans R. Soc. *Philosophical Transactions of the Royal Society*
Phys. Earth Planet. Interiors *Physics of the Earth and Planetary Interiors*
Phys. Letters *Physics Letters*
Polar Geol. Geogr. *Polar Geology and Geography*
Proc. K. Ned. Akad. Wet. *Proc. Koninkl Nederlandse Akademie Wetensch*
Proc. Natn. Acad. Sci. U.S.A. *Proceedings of the National Academy of Science USA*
Proc. R. Soc. *Proceedings of the Royal Society of London*
Proc. Zool. Soc. Lond. *Proceedings of the Zoological Society of London*
Progr. Oceanogr. *Progress in Oceanography*
Progr. Phys. Geogr. *Progress in Physical Geography*

Q. Jl R. Met. Soc. *Quarterly Journal of the Royal Meteorological Society*
Quat. Sci. Rev. *Quaternary Science Reviews*

Rev. Palaeobot. Palynol. *Review of Palaeobotany and Palynology*
Revue Géogr. Phys. Géol. Dynam. *Revue de Géographie Physique et de Géologie Dynamique*
Revue Géomorph. Dynam. *Revue de Géomorphologie Dynamique*

Scient. Sinica *Scientia Sinica*

Trans. Bot. Soc. Edinb. *Transactions of the Botanical Society of Edinburgh*

Utah Geol. Mineral Survey Bull. *Utah Geological and Mineral Survey Bulletin.*

Vierteljahresheft Naturforsch. Ges. Zürich *Vierteljahresheft der Naturforschenden Gesellschaft in Zürich*

Vistas Astron. *Vistas in Astronomy*

Wiss. Ergebn. *Wissenschaftliche Ergebnisse*

Z. GletscherKde Glazialgeol. *Zeitschrift für Gletscherkunde und Glazialgeologie*

Absy, M. L. 1982. Quaternary palynological studies in the Amazon Basin. In *Biological diversification in the Tropics*, G. T. Prance (ed.), 67–73. New York: Columbia University Press.

Absy, M. L. and T. van der Hammen 1976. Some palaeoecological data from Rondonia, southern part of the Amazon Basin. *Acta Amazonica* **6**, 293–9.

Adam, D. P. 1975. Ice ages and the thermal equilibrium of the Earth. *Quat. Res.* **5**, 161–71.

Adam, D. P. and G. J. West 1983. Temperature and precipitation estimates through the last glacial cycle from Clear Lake, California, pollen data. *Science* **219**, 168–70.

Adam, D. P., J. D. Sims, and C. K. Throckmorton 1981. 130 000 year continuous pollen record from Clear Lake County, California. *Geology* **9**, 373–7.

Adelseck, C. G. and W. H. Berger 1977. On the dissolution of planktonic foraminifera and associated microfossils during settling and on the sea floor. In *Dissolution of deep-sea carbonates*, W. V. Sliter, A. W. H. Bé, and W. H. Berger (eds), 70–81, Cushman Foundation for Foraminiferal Research Special Publication No. 13. Washington, D.C.: Cushman Foundation.

Aitken, M. J. 1974. *Physics and archaeology*, 2nd edn. Oxford: Clarendon Press.

Aitken, M. J. 1978. Archaeological involvements of physics. *Phys. Letters* C 277–351.

Aldaz, L. and S. Deutsch 1967. On a relationship between air temperature and oxygen isotope ratio of snow and firn in the South Pole region. *Earth Planet. Sci. Letters* **3**, 267–74.

Alexandre, P. 1977. Les variations climatiques au Moyen Age (Belgique, Rhénanie, Nord de la France). *Annls Econ. Soc., Civil.* **32**, 183–97.

Allison, I. and P. Kruss 1977. Estimation of recent climatic change in Irian Jaya by numerical modelling of its tropical glaciers. *Arctic Alpine Res.* **9**, 49–60.

Ambach, W., W. Dansgaard, H. Eisner, and J. Møller 1968. The altitude effect on the isotopic composition of precipitation and glacier ice in the Alps. *Tellus* **20**, 595–600.

Andersen, S. Th. 1970. The relative pollen productivity and pollen representation of N. European trees, and correction factors for tree pollen spectra. *Danmarks Geol. Undersogelse*, Series II, No. 96.

Andersen, S. Th. 1973. The differential pollen productivity of trees and its significance for the interpretation of a pollen diagram from a forested region. In *Quaternary plant ecology*, H. J. B. Birks and R. G. West (eds), 109–15, Oxford: Blackwell Scientific Publications.

Andersen, S. Th. 1980. Influence of climatic variation on pollen season severity in wind-pollinated trees and herbs. *Grana* **19**, 47–52.

Andersen, T. F. and J. C. Steinmetz 1981. Isotopic and biostratigraphical records of calcareous nannofossils in a Pleistocene core. *Nature, Lond.* **294**, 741–4.

Andrews, J. T. and R. G. Barry 1978. Glacial inception and disintegration during the last glaciation. *Ann. Rev. Earth Planet. Sci.* **6**, 205–28.

Andrews, J. T., R. G. Barry, R. S. Bradley, G. H. Miller, and L. D. Williams 1972. Past and present glaciological responses to climate in eastern Baffin Island. *Quat. Res.* **2**, 303–14.

Andrews, J. T., S. Funder, C. Hjort, and J. Imbrie 1974. Comparison of the glacial chronology of eastern Baffin Island, east Greenland and the Camp Century accumulation record. *Geology* **2**, 355–8.

Andrews, J. T., P. E. Carrara, F. B. King, and R. Stuckenrath 1975. Holocene environmental changes in the alpine zone, northern San Juan mountains, Colorado: evidence from bog stratigraphy and palynology. *Quat. Res.* **5**, 173–97.

Andrews, J. T., P. T. Davis, and C. Wright 1976. Little Ice Age permanent snowcover in the eastern Canadian Arctic: extent mapped from Landsat-1 satellite imagery. *Geogr. Annlr* **58A**, 71–81.

Antevs, E. 1948. Climatic changes and pre-white man. *Univ. Utah Bull.* **36**, 168–191.

Arakawa, H. 1954. Fujiwhara on five centuries of freezing dates of Lake Suwa in central Japan. *Arch. Met. Geophys. Bioklimatol. B* **6**, 152–66.

Arakawa, H. 1956a. Dates of first or earliest snow covering for Tokyo since 1632. *Q. Jl R. Met. Soc.* **82**, 222–6.

Arakawa, H. 1956b. Climatic change as revealed by the blooming dates of the cherry blossoms at Kyoto. *J. Met.* **13**, 599–600.

Arakawa, H. 1957. Climatic change as revealed by the data from the Far East. *Weather* **12**, 46–51.

Árnason, B. 1969. The exchange of hydrogen isotopes between ice and water in temperate glaciers. *Earth Planet. Sci. Letters* **6**, 423–30.

Arnow, T. 1980. Water budget and water-surface fluctuations of Great Salt Lake. *Utah Geol. Mineral Survey Bull.* **116**, 255–63.

Ashworth, A. C. 1977. A late Wisconsinan Coleopterous assemblage from southern Ontario and its environmental significance. *Can. J. Earth Sci.* **14**, 1625–34.

Ashworth, A. C. 1980. Environmental implications of a beetle assemblage from the Gervais formation (Early Wisconsinan?), Minnesota. *Quat. Res.* **13**, 200–12.

Atkinson, T. C., R. S. Harmon, P. L. Smart, and A. C. Waltham 1978. Palaeoclimatic and geomorphic implications of $^{230}Th/^{234}U$ dates on speleothems from Britain. *Nature, Lond.* **272**, 24–8.

Bada, J. L. 1972. The dating of fossil bones using the racemization of isoleucine. *Earth Planet. Sci. Letters* **15**, 223–31.

Bada, J. L. and P. M. Helfman 1975. Amino acid racemization dating of fossil bones. *World Archaeol.* **7**, 160–73.

Bada, J. L. and R. A. Schroeder 1975. Amino acid racemization reactions and their geochemical implications. *Naturwissenschaften* **62**, 71–9.

Bada, J. L., B. P. Luyendyk, and J. B. Maynard 1970. Marine sediments: dating by the racemization of amino acids. *Science* **170**, 730–2.

Bada, J. L., R. Protsch, and R. A. Schroeder 1973. The racemization reaction of isoleucine used as a palaeotemperature indicator. *Nature, Lond.* **241**, 394–5.

Bada, J. L., R. A. Schroeder, R. Protsch, and R. Berger 1974a. Concordance of collagen-based radiocarbon and aspartic-acid racemization ages. *Proc. Natn. Acad. Sci. U.S.A.* **71**, 914–17.

Bada, J. L., R. A. Schroeder, and G. F. Carter 1974b. New evidence for the antiquity of Man in North America deduced from aspartic acid racemization. *Science* **184**, 791–3.

Baillie, M. G. L. 1977. The Belfast oak chronology to AD 1001. *Tree Ring Bull.* **37**, 1–12.

Baker, J. N. L. 1932. The climate of England in the 17th century. *Q. Jl R. Met. Soc.* **58**, 421–36.

Bandy, O. L. 1960. The geologic significance of coiling ratios in the foraminifer *Globigerina pachyderma* (Ehrenberg). *J. Paleont.* **34**, 671–81.

Barbetti, M. 1980. Geomagnetic strength over the last 50 000 years and changes in atmospheric ^{14}C concentration: emerging trends. *Radiocarbon* **22**, 192–9.

Barkov, N. I., Ye. S. Korotkevich, F. G. Gordiyenko, and V. M. Kotlyakov 1977. The isotope analysis of ice cores from Vostok station (Antarctica) to the depth of 950 m. In *Isotopes and impurities in snow and ice*, 382–387, International Association of Scientific Hydrology Publication No. 118, Washington, D.C.: IASH.

Baron, W. R. 1982. The reconstruction of eighteenth century temperature records through the use of content analysis. *Climatic Change* **4**, 385–98.

Barry, R. G. 1975. Climate models in paleoclimatic reconstruction. *Palaeogeogr. Palaeoclimatol. Palaeoecol.* **17**, 123–37.

Barry, R. G. 1982. Approaches to reconstructing the climate of the steppe-tundra biome. In *Paleoecology of Beringia*, D. M. Hopkins, J. V. Matthews, C. E. Schweger and S. B. Young (eds), 195–204. New York: Academic Press.

Barry, R. G., J. T. Andrews, and M. A. Mahaffy 1975. Continental ice sheets: conditions for growth. *Science* **190**, 979–81.

Barton, C. E., R. T. Merrill, and M. Barbetti 1979. Intensity of the Earth's magnetic field over the last 10 000 years. *Phys. Earth Planet. Interiors* **20**, 96–110.

Bassett, I. J. and J. Terasmae 1962. Ragweeds, *Ambrosia* species in Canada and their history in postglacial time. *Can. J. Bot.* **40**, 141–50.

Baumgartner, A. 1979. Climatic variability and forestry. In *Proceedings of the World Climate Conference*, 581–607, WMO Publication No. 537. Geneva: World Meteorological Organization.

Baumgartner, A., H. Mayer, and W. Metz 1976. Globale Verteilung der oberflächen Albedo. *Met. Rundschau* **29**, 38–43.

Bé, A. W. H. 1977. An ecological, zoogeographic and taxonomic review of recent planktonic foraminifera. In *Oceanic micropalaeontology*, A. T. S. Ramsay (ed.), 1–88. London: Academic Press.

Bé, A. W. H. and J. C. Duplessy 1976. Subtropical convergence fluctuations and Quaternary climates in the middle latitudes of the Indian Ocean. *Science* **194**, 419–22.

Bé, A. W. H., S. M. Harrison, and L. Lott 1973. *Orbulina universa* d'Orbigny in the Indian Ocean. *Micropaleontology* **19**, 150–92.

Bé, A. W. H., J. E. Damuth, L. Lott, and R. Free 1976. Late Quaternary climatic record in western equatorial atlantic sediment. In *Investigation of late Quaternary paleooceanography and paleoclimatology*, R. M. CLine and J. D. Hays (eds), 165–200, Geological Society of America Memoir no. 145. Boulder, Colo. Geological Society of America.

Beadle, L. C. 1974. *The inland waters of tropical Africa*. London: Longmans.

Becker, B. 1980. Tree-ring dating and radiocarbon calibration in south-central Europe. *Radiocarbon* **22**, 219–26.

Bell, B. 1970. The oldest records of the Nile floods. *Geogr. J.* **136**, 569–73.

Bell, W. T. and A. E. J. Ogilvie 1978. Weather compilations as a source of data for the reconstruction of European climate during the Medieval Period. *Climatic Change* **1**, 331–48.

Bender, M. L. 1974. Reliability of amino acid racemization dating and palaeotemperature analysis on bones. *Nature* **252**, 378–9.

Benedict, J. B. 1967. Recent glacial history of an alpine area in the Colorado Front Range, USA. I. Establishing a lichen growth curve. *J. Glaciol.* **6**, 817–32.

Benedict, J. B. 1973. Chronology of cirque glaciation, Colorado Front Range. *Quat. Res.* **3**, 584–99.

Benson, C. S. 1961. Stratigraphic studies in the snow and firn of the Greenland Ice Sheet. *Folia Geogr. Danica* **9**, 13–37.

Benson, L. V. 1981. Paleoclimatic significance of lake-level fluctuations in the Lahontan Basin. *Quat. Res.* **16**, 390–403.

Benson, W. W. 1982. Alternative models for infrageneric diversification in the humid Tropics: tests with Passion Vine Butterflies. In *Biological diversification in the Tropics*, G. T. Prance (ed.), 608–40. New York: Columbia University Press.

Berger, A. 1977a. Long-term variations of the Earth's orbital elements. *Celestial Mech.* **15**, 53–74.

Berger, A. 1977b. Support for the astronomical theory of climatic change. *Nature* **269**, 44–5.

Berger, A. 1977c. Power and limitation of an energy-balance climate model as applied to the astronomical theory of paleoclimates. *Palaeogeogr. Palaeoclimatol. Palaeoecol.* **21**, 227–35.

Berger, A. 1978. Long-term variations of caloric insolation resulting from the Earth's orbital elements. *Quat. Res.* **9**, 139–67.

Berger, A. 1979. Insolation signatures of Quaternary climatic changes. *Nuovo Cim.* **2**(c), 63–87.

Berger, A. 1980. The Milankovitch astronomical theory of paleoclimates. A modern review. *Vistas Astron.* **24**, 103–122.

Berger, R., A. G. Horney, and W. F. Libby 1964. Radiocarbon dating of bone and shell from their organic components. *Science* **144**, 999–1001.

Berger, W. H. 1968. Planktonic foraminifera: selective solution and paleoclimatic interpretation. *Deep-sea Res.* **15**, 31–43.

Berger, W. H. 1970. Planktonic foraminifera: selective solution and lysocline. *Marine Geol.* **8**, 111–38.

Berger, W. H. 1971. Sedimentation of planktonic foraminifera. *Marine Geol.* **11**, 325–58.

Berger, W. H. 1973a. Deep-sea carbonates: evidence for a coccolith lysocline. *Deep-sea Res.* **20**, 917–921.

Berger, W. H. 1973b. Deep-sea carbonates: Pleistocene dissolution cycles. *J. Foram. Res.* **3**, 187–95.

Berger, W. H. 1975. Deep-sea carbonates: dissolution profiles from foraminiferal preservation. In *Dissolution of deep-sea carbonates,* W. V. Sliter, A. W. H. Bé, and W. H. Berger (eds), 82–6, Cushman Foundation for Foraminiferal Research Special Publication No. 13. Washington, D. C.: Cushman Foundation.

Berger, W. H. 1977. Deep-sea carbonate and the deglaciation preservation spike in pteropods and foraminifera. *Nature, Lond.* **269**, 301–4.

Berger, W. H. and J. V. Gardner 1975. On the determination of Pleistocene temperatures from planktonic foraminifera. *J. Foram.* **5**, 102–13.

Berger, W. H. and J. S. Killingley 1977. Glacial – Holocene transition in deep-sea carbonates: selective dissolution and the stable isotope signal. *Science* **197**, 563–6.

Berger, W. H. and E. L. Winterer 1974. Plate stratigraphy and the fluctuating carbonate line. In *Pelagic sediments on land and in the ocean,* K. J. Hsü and H. Jenkins (eds), 11–48, International Association of Sedimentologists, Special Publication No. 1. Oxford: Blackwell Scientific, on behalf of International Association of Sedimentologists.

Berger, W. H., R. F. Johnson,. and J. S. Killingley 1977. "Unmixing" of the deep-sea record and the deglacial meltwater spike. *Nature, Lond.* **269**, 661–3.

Berggren, W. A., L. H. Burckle, M. B. Cita, H. B. S. Cooke, B. M. Funnell, S. Gartner, J. D. Hays, J. P. Kennett, N. D. Opdyke, L. Pastouret, N. J. Shackleton, and I. Y. Takayanagi 1980. Towards a Quaternary time scale. *Quat. Res.* **13**, 277–302.

Bergthorsson, P. 1969. An estimate of drift ice and temperature in Iceland in 1000 years. *Jökull* **19**, 94–101.

Bernabo, J. C. 1981. Quantitative estimates of temperature changes over the last 2700 years in Michigan based on pollen data. *Quat. Res.* **15**, 143–59.

Bernabo, J. C. and T. Webb, III 1977. Changing patterns in the Holocene pollen record of northeastern North America: a mapped summary. *Quat. Res.* **8**, 69–96.

Berner, W., H. Oeschger, and B. Stauffer 1980. Information on the CO_2 cycle from ice core studies. *Radiocarbon* **22**, 227–34.

Berner, R. A. 1977. Sedimentation and dissolution of pteropods in the ocean. In *The fate of fossil fuel CO_2 in the ocean*, N. R. Anderson and A. Malahoff (eds), 243–60. New York: Plenum Press.

Beschel, R. E. 1950. Flechten als Altersmasstab rezenter Moränen. *Z. GletscherKde Glazialgeol.* **1**, 152–161 (translation in *Arctic Alpine Res.* **5**, 303–9).

Beschel, R. 1961. Dating rock surfaces by lichen growth and its application in glaciology and physiography (lichenometry). In *Geology of the Arctic*, Vol. II, G. O. Raasch (ed.), 1044–62. Toronto: University of Toronto Press.

Bezinge, A. and R. Vivian 1976. Sites sous-glaciares et climat de la période holocène en Europe. *La Houille Blanche* **226**, 441–59.

Bien, G. S., N. W. Rakestraw, and H. E. Suess 1963. Radiocarbon dating of deep water of the Pacific and Indian oceans. In *Radioactive dating*, 159–73. Vienna: International Atomic Energy Authority.

Birkeland, P. W. 1974. *Pedology, weathering and geomorphological Research*. New York: Oxford University Press.

Birks, H. J. B. 1973. Modern pollen rain studies in some arctic and alpine environments. In *Quaternary plant ecology*, H. J. B Birks and R. G. West (eds), 143–68. Oxford: Blackwell Scientific Publications.

Birks, H. J. B. 1978. Numerical methods for the zonation and correlation of biostratigraphical data. In *Palaeohydrological changes in the temperate zone in the last 15 000 years*, B. E. Berglund (ed.), 99–119, Vol. 1, International Geological Correlation Program, Project 158B, Subproject B. University of Lund, Sweden.

Birks, H. J. B. 1981. The use of pollen analysis in the reconstruction of past climates: a review. In *Climate and history*, T. M. L. Wigley, M. J. Ingram, and G. Farmer (eds), 111–38. Cambridge: Cambridge University Press.

Birks, H. J. B. and B. E. Berglund 1979. Holocene pollen stratigraphy of southern Sweden: a reappraisal using numerical methods. *Boreas* **8**, 257–79.

Birks, H. J. B. and H. H. Birks 1981. *Quaternary palaeoecology*. London: Edward Arnold.

Birks, H. J. B. and M. Saarnisto 1975. Isopollen maps and principal component analysis of Finnish pollen data for 4000, 6000 and 8000 years ago. *Boreas* **4**, 77–96.

Birks, H. J. B., T. Webb, III, and A. A. Berti 1975a. Numerical analysis of surface pollen samples from central Canada: a comparison of methods. *Rev. Palaeobot. Palynol.* **20**, 133–69.

Birks, H. J. B., J. Deacon, and S. Peglar 1975b. Pollen maps for the British Isles 5000 years ago. *Proc. R. Soc. B* **189**, 87–105.

Blake, W. 1977. Glacial sculpture along the east-central coast of Ellesmere Island, Arctic archipelago. In *Geological Survey of Canada Paper 77-1C*, 107–15. Ottawa: Geological Survey of Canada.

Blake, W. 1979. Age determinations on marine and terrestrial materials of Holocene age, southern Ellesmere Island, Arctic archipelago. In *Geological Survey of Canada Paper 79-1C*, 105–9. Ottawa: Geological Survey of Canada.

Blasing, T. J. 1978. Time series of multivariate analysis in paleoclimatology. In *Times series and ecological processes*, H. H. Shugart (ed.), 211–26, SIAM-SIMS Conference Series No. 5. Philadelphia: Society for Industrial and Applied Mathematics.

Blasing, T. J. and H. C. Fritts 1975. Past climate of Alaska and northwestern Canada as reconstructed from tree-rings. In *Climate of the Arctic*, G. Weller and S. A. Bowling (eds), 48–58. Fairbanks: University of Alaska Press.

Blasing, T. J., D. N. Duvick, and D. C. West 1981. Dendroclimatic calibration and verification using regionally averaged and single station precipitation data. *Tree Ring Bull.* **41**, 37–43.

Blinman, E., P. J. Mehringer, and J. C. Sheppard 1979. Pollen influx and the deposition of Mazama and Glacier Peak tephras. In *Volcanic activity and human ecology*, P. D. Sheets and D. K. Grayson (eds), 393–425. New York: Academic Press.

Bloom, A. L., W. S. Broecker, J. M. A. Chappell, R. K. Matthews, and K. J. Mesolella 1974. Quaternary sea-level fluctuations on a tectonic coast: new $^{230}Th/^{234}U$ dates from the Huon Peninsula, New Guinea. *Quat. Res.* **4**, 185–205.

Boersma, A. 1978. Foraminifera. In *Introduction to marine micropaleontology*, B. U. Haq and A. Boersma (eds), 19–77. New York: Elsevier/North Holland.

Bolin, B. 1981. The carbon cycle. In *Climatic variations and variability: facts and theories*, A. Berger (ed.), 623–39. Dordrecht: D. Riedel.

Bonhommet, N. and J. Babkine 1969. Palaeomagnetism and potassium – argon age determinations of the Laschamp geomagnetic polarity event. *Earth Planet. Sci. Letters* **6**, 43–46.

Bonny, A. P. 1972. A method for determining absolute pollen frequencies in lake sediments. *New Phytol.* **71**, 391–403.

Borchardt, G. A., M. E. Harwood, and R. A. Schmitt 1970. Correlation of volcanic ash deposits by activation analyses of glass separates. *Quat. Res.* **1**, 247–60.

Boulton, G. S. 1979. A model of Weichselian glacier variation in the North Atlantic region. *Boreas* **8**, 373–95.

Boulton, G. S., C. T. Baldwin, J. D. Peacock, A. M. McCabe, G. H. Miller, J. Jarvis, B. Horsefield, P. Worsley, N. Eyles, P. N. Chroston, T. E. Day, P. Gibbard, P. E. Hare, and V. Von Brunn 1982. A glacio-isostatic facies model and amino acid stratigraphy for late Quaternary events in Spitsbergen and the Arctic. *Nature, Lond.* **298**, 437–41.

Bowler, J. M. 1976. Aridity in Australia: age origins and expression in aeolian landforms and sediments. *Earth. Sci. Rev.* **12**, 279–310.

Bowles, F. A. 1975. Paleoclimatic significance of quartz/illite variations in cores from the eastern equatorial North Atlantic. *Quat. Res.* **5**, 225–35.

Bradbury, J. P., B. Leyden, M. Salgado-Labouriau, W. M. Lewis, C. Schubert, M. W. Binford, D. G. Frey, D. R. Whitehead, and F. H. Weibezahn 1981. Late Quaternary environmental history of Lake Valencia, Venezuela. *Science* **214**, 1299–1305.

Bradley, R. S. 1976a. *Precipitation history of the Rocky Mountain states.* Boulder, Colo.: Westview Press.

Bradley, R. S. 1976b. Seasonal precipitation fluctuations in the western United States during the late nineteenth century. *Monthly Weather Rev.* **104**, 501–12.

Bradley, R. S. 1980. Secular fluctuations of temperature in the Rocky Mountain states and a comparison with precipitation fluctuations. *Monthly Weather Rev.* **108**, 873–85.

Bradley, R. S. and G. H. Miller 1972. Recent climatic change and increased glacierization in the eastern Canadian Arctic. *Nature, Lond.* **237**, 385–7.

Bradley, R. S. and J. England 1978a. Recent climatic fluctuations of the Canadian High Arctic and their significance for glaciology. *Arctic Alpine Res.* **10**, 715–31.

Bradley, R. S. and J. England 1978b. Influence of volcanic dust on glacier mass balance at high latitudes. *Nature* **271**, 736–8.

Brakenridge, G. R. 1978. Evidence for a cold, dry full-glacial climate in the American Southwest. *Quat. Res.* **9**, 22–40.

Bray, J. R. 1974. Glacial advance relative to volcanic activity since AD 1500. *Nature, Lond.* **248**, 42–3.

Bray, J. R. 1976. Volcanic triggering of glaciation. *Nature, Lond.* **260**, 414–15.

Bray, J. R. 1977. Pleistocene volcanism and glacial initiation. *Science* **197**, 251–4.

Bray, J. R. 1979. Surface albedo increase following massive Pleistocene explosive eruptions in western North America. *Quat. Res.* **12**, 204–11.

Bray, J. R. 1982. Alpine glacier advance in relation to a proxy summer temperature index based mainly on wine harvest dates, AD 1453–1973. *Boreas* **11**, 1–10.

Broecker, W. 1971. Calcite accumulation rates and glacial to interglacial changes in oceanic mixing. In *The late Cenozoic glacial ages,* K. K. Turekian (ed.), 239–65. New Haven, Conn.: Yale University Press.

Broecker, W. 1975. Climatic change: are we on the brink of a pronounced global warming? *Science* **189**, 460–3.

Broecker, W. S. and M. L. Bender 1972. Age determinations on marine strandlines. In *Calibration of hominoid evolution*, W. W. Bishop and J. A. Miller (eds), 19–38. Edinburgh: Scottish Academic Press.

Broecker, W. S. and S. Broecker 1974. Carbonate dissolution on the eastern flank of the East Pacific Rise. In *Studies in paleo-oceanography* W. W. Hay (ed.), 44–58, Society of Economic Paleontologists and Mineralogists Special Publication No. 20. Tulsa, Okl.: Society of Economic Paleontologists and Mineralogists.

Broecker, W. S. and J. van Donk 1970. Insolation changes, ice volumes and the ^{18}O record in deep-sea cores. *Rev. Geophys. Space Phys.* **8**, 169–98.

Broecker, W. S., T. Takahashi, H. J. Simpson, and T. H. Peng 1979. The fate of fossil fuel CO_2 and the global carbon budget. *Science* **206**, 409–18.

Broecker, W. S., D. L. Thurber, J. Goddard, T.-L. Ku, R. K. Matthews, and K. J. Mesolella 1968. Milankovitch hypothesis supported by precise dating of coral reefs and deep-sea sediments. *Science* **159**, 297–300.

Brooks, C. E. P. 1926. *Climate through the ages.* New Haven, Conn.: Yale University Press.

Brown, K. S. 1982. Paleoecology and regional patterns of evolution in Neotropical forest butterflies. In *Biological diversification in the Tropics*, G. T. Prance (ed.), 255–308. New York: Columbia University Press.

Brown, K. S. and A. N. Ab'Saber 1979. Ice-age forest refuges and evolution in the Neotropics: correlation of paleoclimatological, geomorphological and pedological data with modern biological endemism. *Paleoclimas* **5**, 1–30.

Brown, K. S., P. M. Sheppard, and J. R. G. Turner 1974. Quaternary refugia in tropical America: evidence from race formation in Heliconius butterflies. *Proc. R. Soc. B* **187**, 369–78.

Brubaker, L. B. 1975. Postglacial forest patterns associated with till and outwash in north central Upper Michigan. *Quat. Res.* **5**, 499–528.

Bryson, R. A. 1966. Air masses, streamlines, and the boreal forest. *Geogr. Bull.* **8**, 228–69.

Bryson, R. A. 1974. A perspective on climatic change. *Science* **184**, 753–60.

Bryson, R. A. and D. A. Baerreis 1967. Possibilities of major climatic modification and their implications: northwest India, a case for study. *Bull. Am. Met. Soc.* **48**, 136–42.

Bryson, R. A. and J. E. Kutzbach 1974. On the analysis of pollen – climate canonical transfer functions. *Quat. Res.* **4**, 162–74.

Bryson, R. A. and T. J. Murray 1977. *Climates of hunger.* Madison, Wisc: University of Wisconsin Press.

Bryson, R. A. and W. M. Wendland 1967. Tentative climatic patterns for some late-glacial and post-glacial episodes in central North America. In *Life, land and water*, W. J. Mayer-Oakes (ed.), 271–98. Winnipeg: University of Manitoba Press.

Bryson, R. A., W. N. Irving, and J. A. Larsen 1965. Radiocarbon and soil evidence of former forest in the southern Canadian forest. *Science* **147**, 46–8.

Bryson, R. A., D. A. Baerreis, and W. M. Wendland 1970. The character of late-glacial and post-glacial climatic changes. In *Pleistocene and Recent environments of the central Great Plains*, W. Dort, Jr and J. K. Jones (eds), 53–74. Lawrence, Kansas: University of Kansas Press.

Bucha, V. 1969. Changes of the Earth's magnetic moment and radiocarbon dating. *Nature* **224**, 681–2.

Bucha, V. 1970. Influence of the Earth's magnetic field on radiocarbon dating. In *Radiocarbon variations and absolute chronology*, I. U. Olsson (ed.), 501–11. New York: Wiley.

Budd, W. F. and I. N. Smith 1981. The growth and retreat of ice sheets in response to orbital radiation changes. In *Sea level, ice and climatic change*, 369–409. International Association of Scientific Hydrology Publication No. 131. Washington, D.C.: IASH.

Budyko, M. I. 1978. The heat balance of the earth. In *Climatic Change*, J. Gribbin (ed.), 85–113, Cambridge: Cambridge University Press.

Burbank, D. W. 1981. A chronology of Late Holocene glacier fluctuations on Mount Rainier, Washington. *Arctic Alpine Res.* **13**, 369–81.

Burk, R. L. and M. Stuiver 1981. Oxygen isotope ratios in trees reflect mean annual temperature and humidity. *Science* **211**, 1417–19.

Burrows, C. J. and V. L. Burrows 1976. *Procedures for the study of snow avalanche chronology using growth layers of woody plants.* University of Colorado INSTAAR Occasional Paper No. 23.

Butzer, K. W., G. L. Isaac, J. L. Richardson, and C. Washbourn-Kamau 1972. Radiocarbon dating of East African lake levels. *Science* **175**, 1069–76.

Calkin, P. E. and J. M. Ellis 1980. A lichenometric dating curve and its application to Holocene glacier studies in the Central Brooks Range, Alaska. *Arctic Alpine Res.* **12**, 245–64.

Carney, T. F. 1972. *Content analysis: a technique for systematic inference from communications.* Winnipeg: University of Manitoba Press.

Carrara, P. E. 1979. The determination of snow avalanche frequency through tree-ring analysis and historical records at Ophir, Colorado. *Geol. Soc. Am. Bull. 1*, **90**, 773–80.

Carrara, P. E. and J. T. Andrews 1975. Holocene glacial/periglacial record northern San Juan Mountains, southwestern Colorado. *Z. GletscherKde Glazialgeol.* **11**, 155–74.

Catchpole, A. J. W., D. W. Moodie, and D. Milton 1976. Freeze-up and break-up of estuaries on Hudson Bay in the 18th and 19th centuries. *Can. Geogr.* **20**, 279–97.

Central Intelligence Agency 1974. *Potential implication of trend in world population, food production and climate.* Washington, D.C.: Directorate of Intelligence, Office of Political Research, CIA.

Chappell, J. M. A. and H. A. Polach 1972. Some effects of partial recrystallisation on ^{14}C dating of late Pleistocene corals and molluscs. *Quat. Res.* **2**, 244–52.

Charney, J. G., P. H. Stone, and W. J. Quirk 1975. Drought in the Sahara: a biogeophysical feedback mechanism. *Science* **187**, 434–5.

Chen, C. 1968. Pleistocene pteropods in pelagic sediments. *Nature, Lond.* **219**, 1145–7.

Chester, R. and L. R. Johnson 1971. Atmospheric dusts collected off the Atlantic coasts of North Africa and the Iberian Peninsula. *Marine Geol.* **11**, 251–60.

Chinn, T. J. H. 1981. Use of rock weathering-rind thickness for Holocene absolute age-dating in New Zealand. *Arctic Alpine Res.* **13**, 33–45.

Christodoulides, C. and J. H. Fremlin 1971. Thermoluminescence of biological materials. *Nature, Lond.* **232**, 257–8.

Chu, K'o-chen 1926. Climate pulsations during historical times in China. *Geogr. Rev.* **16**, 274–82.

Chu, K'o-chen 1973. A preliminary study on the climatic fluctuations during the last 5000 years in China. *Scient. Sinica* **16**, 226–256.

Churchill, D. M. 1968. The distribution and prehistory of *Eucalyptus diversicolor* F. Muell, *E. marginata* Donn *ex* Sm., and *E. calophylla* R. Br in relation to rainfall. *Aust. J. Bot.* **16**, 125–51.

Clark, D. 1975. *Understanding canonical correlation analysis,* Concepts and Techniques in Modern Geography No. 3. Norwich: University of East Anglia.

Clark, N., T. J. Blasing, and H. C. Fritts 1975. Influence of interannual climatic fluctuations on biological systems. *Nature, Lond.* **256**, 302–4.

CLIMAP Project Members 1976. The surface of the ice-age Earth. *Science* **191**, 1131–44.

Cole, H. 1969. *Objective reconstruction of the paleoclimatic record through application of eigenvectors of present-day pollen spectra and climate to the late-Quaternary pollen stratigraphy*, PhD thesis, Univ. of Wisconsin, Madison. Ann Arbor, Mich.: University Microfilms.

Colinvaux, P. A. 1972. Climate and the Galapagos Islands. *Nature, Lond.* **240**, 17–20.

Colinvaux, P. A. 1978. The Ice-age Amazon. *Nature, Lond.* **278**, 399–400.

Colman, S. M. and K. L. Pierce 1981. *Weathering rinds on andesitic and basaltic stones as a Quaternary age indicator, western United States.* US Geological Survey Professional Paper 1210. Washington, D.C.: US Government Printing Office.

Committee on Science and Technology 1976. *The National Climate Program Act: Hearings before the Subcommittee on the Environment and the Atmosphere of the Committee on Science and Technology*, US House of Representatives, 94th Congress, 2nd Session (no 78). Washington, D.C.: US Government Printing Office.

Conkey, L. E. 1982. *Eastern US tree-ring widths and densities as indicators of past climate*, PhD thesis, University of Arizona. Ann Arbor, Mich.: University Microfilms.

Cook, E. R. and G. C. Jacoby 1977. Tree-ring – drought relationships in the Hudson Valley, New York. *Science* **198**, 399–401.

Cook, E. R. and G. C. Jacoby 1979. Evidence for the quasi-periodical July drought in the Hudson Valley, New York. *Nature, Lond.* **282**, 390–2.

Cooke, R. U. and A. Warren 1977. *Geomorphology in deserts*. Berkeley and Los Angeles: University of California Press.

Coope, G. R. 1959. A late Pleistocene insect fauna from Chelford, Cheshire. *Proc. R. Soc. B* **151**, 70–86.

Coope, G. R. 1967. The value of Quaternary insect faunas in the interpretation of ancient ecology and climate. In *Quaternary paleoecology*, E. J. Cushing and H. E. Wright (eds), 359–80. New Haven, Conn.: Yale University Press.

Coope, G. R. 1974. Interglacial coleoptera from Bobbitshole, Ipswich. *J. Geol. Soc. Lond.* **130**, 333–40.

Coope, G. R. 1975a. Climatic fluctuations in north-west Europe since the last Interglacial, indicated by fossil assemblages of coleoptera. In *Ice Ages ancient and modern*, A. E. Wright and F. Moseley (eds), 153–68, Geological Journal Special Issue No. 6. Liverpool: Liverpool University Press.

Coope, G. R. 1975b. Mid-Weichselian climatic changes in western Europe, reinterpreted from coleopteran assemblages. *Bull. R. Soc. N.Z.* **13**, 101–10.

Coope, G. R. 1977a. Quaternary coleoptera as aids in the interpretation of environmental history. In *British Quaternary studies*, F. W. Shotton (ed.), 55–68. Oxford: Clarendon Press.

Coope, G. R. 1977b. Fossil coleopteran assemblages as sensitive indicators of climatic changes during the Devensian (Last) cold stage. *Proc. R. Soc. B* **280**, 313–37.

Coope, G. R. and J. A. Brophy 1972. Late glacial environmental changes indicated by a coleopteran succession from North Wales. *Boreas* **1**, 97–142.

Coope, G. R. and W. Pennington 1977. The Windermere interstadial of the late Devensian. *Phil. Trans R. Soc. B* **208**, 337–9.

Cox, A. 1969. Geomagnetic reversals. *Science* **163**, 237–45.

Cragin, J. H., M. H. Herron, C. C. Langway, Jr, and G. Klouda 1977. Interhemispheric comparison of changes in the composition of atmospheric precipitation during the late Cenozoic era. In *Polar oceans*, M. J. Dunbar (ed.), 617–30, Calgary: Arctic Institute of North America.

Craig, H. 1953. The geochemistry of the stable carbon isotopes. *Geochim. Cosmochim. Acta* **3**, 53–92.

Craig, H. 1957. Isotopic standards for carbon and oxygen and correction factors for mass spectrometric analysis of CO_2. *Geochim. Cosmochim. Acta* **12**, 133–49.

Craig, H. 1961a. Mass spectrometer analysis of radiocarbon standards. *Radiocarbon* **3**, 1–3.

Craig, H. 1961b. Standard for reporting concentrations of deuterium and oxygen-18 in natural waters. *Science* **133**, 1833–4.

Craig, H. 1965. The measurement of oxygen isotope paleotemperature. In *Proceedings of the Spoleto Conference on Stable Isotopes in Oceanographic Studies and Paleotemperatures*, 3–24. Pisa: Consiglio Nazionale delle Ricerche Laboratoriodi Geologia Nucleare.

Crane, A. J. 1981. Comments on recent doubts about the CO_2 greenhouse effect. *J. Appl. Met.* **20**, 1547–9.

Creer, K. M., T. W. Anderson, and C. F. M. Lewis 1976. Late Quaternary geomagnetic stratigraphy recorded in Lake Erie sediments. *Earth Planet. Sci. Lett.* **31**, 37–47.

Croll, J. 1867a. On the eccentricity of the Earth's orbit, and its physical relations to the glacial epoch. *Phil. Mag.* **33**, 119–31.

Croll, J. 1867b. On the change in the obliquity of the ecliptic, its influence on the climate of the polar regions and on the level of the sea. *Phil. Mag.* **33**, 426–45.

Croll, J. 1875. *Climate and time*. New York: Appleton and Co.

Crozaz, G. and C. C. Langway, Jr 1966. Dating Greenland firn-ice cores with ^{210}Pb. *Earth Planet. Sci. Letters* **1**, 194–6.

Crozaz, G., E. Picciotto, and W. De Breuck 1964. Antarctic snow chronology with Pb-210. *J. Geophys. Res.* **69**, 2597–604.

Crozaz, G., C. C. Langway, Jr, and E. Picciotto 1966. Artificial radioactivity reference horizons in Greenland firn. *Earth Planet. Sci. Letters* **1**, 42–8.

Curry, R. R. 1969. Holocene climatic and glacial history of the Sierra Nevada, California. *Geol. Soc. Am. Spec. Paper* no. 123, 1–47.

Curtis, G. H. 1975. Improvements in potassium–argon dating, 1962–1975. *World Archaeol.* **7**, 198–207.

Cushing, E. J. 1967. Evidence for differential pollen preservation in late Quaternary sediments in Minnesota. *Rev. Paleobot. Palynol.* **4**, 87–101.

Dale, M. B. and D. Walker 1970. Information analysis of pollen diagrams I. *Pollen et Spores* **12**, 21–37.

Dalrymple, G. B. 1972. Potassium argon dating of geomagnetic reversals and north American glaciations. In *Calibration of hominoid evolution*, W. W. Bishop and J. A. Miller (eds), 107–34. Edinburgh: Scottish Academic Press.

Dalrymple, G. B. and M. A. Lanphere 1969. *Potassium–argon dating: principles, techniques and applications to geochronology*. San Francisco: W. H. Freeman.

Damon, P. E. 1970a. Radiocarbon as an example of the unity of science. In *Radiocarbon variations and absolute chronology*, I. U. Olsson (ed.), 641–4. New York: Wiley.

Damon, P. E. 1970b. Climatic versus magnetic perturbation of the atmospheric ^{14}C reservoir. In *Radiocarbon variations and absolute chronology*, I. U. Olsson (ed.), 571–93. New York: Wiley.

Damon, P. E. and W. M. Kunen 1976. Global cooling? *Science* **193**, 447–53.

Damon, P. E., A. Long, and D. C. Grey 1966. Fluctuation of atmospheric ^{14}C during the last six millennia. *J. Geophys. Res.* **71**, 1055–63.

Damon, P. E., J. C. Lerman, and A. Long 1978. Temporal fluctuations of atmospheric ^{14}C: causal factors and implications. *Ann. Rev. Earth Planet. Sci.* **6**, 457–94.

Damuth, J. E. and R. W. Fairbridge 1970. Equatorial Atlantic deep-sea arkosic sands and ice-age aridity in tropical South America. *Geol. Soc. Am. Bull.* **81**, 189–206.

Dansgaard, W. 1953. The abundance of O^{18} in atmospheric water and water vapour. *Tellus* **5**, 461–9.

Dansgaard, W. 1961. The isotopic composition of natural waters with special reference to the Greenland Ice Cap. *Meddr Grønland* **165**, 1–120.

Dansgaard, W. 1964. Stable isotopes in precipitation. *Tellus* **16**, 436–68.

Dansgaard, W. 1981. Ice core studies: dating the past to find the future. *Nature, Lond.* **290**, 360–1.

Dansgaard, W. and S. J. Johnsen 1969. A flow model and a time scale for the ice core from Camp Century. *J. Glaciol.* **8**, 215–23.

Dansgaard, W. and H. Tauber 1969. Glacier oxygen-18 content and Pleistocene ocean temperatures. *Science* **166**, 499–502.

Dansgaard, W., H. B. Clausen, and A. Aarkrog 1966. A Si^{32} fallout in Scandinavia, a new method for ice dating. *Tellus* **18**, 187–91.

Dansgaard, W., S. J. Johnsen, J. Møller, and C. C. Langway, Jr 1969. One thousand centuries of climatic record from Camp Century on the Greenland ice sheet. *Science* **166**, 377–81.

Dansgaard, W., S. J. Johnsen, H. B. Clausen, and C. C. Langway 1971. Climatic record revealed by the Camp Century ice core. In *The late Cenozoic Ice Ages*, K. K. Turekian (ed.), 37–56. New Haven, Conn.: Yale University Press.

Dansgaard, W., S. J. Johnsen, H. B. Clausen, and C. C. Langway, Jr 1972. Speculations about the next glaciation. *Quat. Res.* **2**, 396–8.

Dansgaard, W., S. J. Johnsen, H. B. Clausen, and N. Gundestrup 1973. Stable isotope glaciology. *Meddr Grønland* **197**, 1–53.

Dansgaard, W., S. J. Johnsen, N. Reeh, N. Gundestrup, H. B. Clausen, and C. U. Hammer 1975. Climatic changes, Norsemen and modern man. *Nature, Lond.* **255**, 24–8.

Dansgaard, W., H. B. Clausen, N. Gundestrup, C. U. Hammer, S. F. Johnsen, P. M. Kristinsdottir, and N. Reeh 1982. A new Greenland deep ice core. *Science* **218**, 1273–7.

Dansgaard, W., S. J. Johnsen, H. B. Clausen, D. Dahl-Jensen, N. Gundestrup, C. U. Hammer, and H. Oeschger 1984. North Atlantic climatic oscillations revealed by deep Greenland ice cores. In *Climate processes and climate sensitivity*, J. E. Hansen and T. Takahashi (eds), 288–98. Washington, D.C.: American Geophysical Union.

Daultrey, S. 1976. *Principal components analysis*, Concepts and Techniques in Modern Geography No. 8. Norwich: University of East Anglia.

Davis, M. B. 1963. On the theory of pollen analysis. *Am. J. Sci.* **261**, 899–912.

Davis, M. B. 1967a. Late-glacial climate in northern United States: a comparison of New England and the Great Lakes Region. In *Quaternary paleoecology*, E. J. Cushing and H. E. Wright, Jr (eds), 11–43. New Haven, Conn.: Yale University Press.

Davis, M. B. 1967b. Pollen accumulation rates at Rogers Lake, Conn. during late and postglacial time. *Rev. Palaeobot. Palynol.* **2**, 219–230.

Davis, M. B. 1969. Palynology and environmental history during the Quaternary period. *Am. Scient.* **57**, 317–32.

Davis, M. B. 1973. Redeposition of pollen grains in lake sediment. *Limnol. Oceanogr.* **18**, 44–52.

Davis, M. B. 1976. Pleistocene biogeography of temperate deciduous forests. *Geosci. Man* **13**, 13–26.

Davis, M. B. 1978. Climatic interpretation of pollen in Quaternary sediments. In *Biology and Quaternary environments*, D. Walker and J. C. Guppy (eds), 35–51. Canberra: Australian Academy of Science.

Davis, M. B. and L. B. Brubaker 1973. Differential sedimentation of pollen grains in lakes. *Limnol. Oceanogr.* **18**, 635–46.

Davis, M. B. and E. S. Deevey, Jr 1964. Pollen accumulation rates: estimates from late glacial sediment of Rogers Lake. *Science* **145**, 1293–5.

Davis, M. B., L. B. Brubaker, and T. Webb, III 1973. Calibration of absolute pollen influx. In *Quaternary plant ecology*, H. J. B. Birks and R. G. West (eds), 9–25. Oxford: Blackwell Scientific Publications.

Davis, R. B. 1974. Stratigraphic effects of tubificids in profundal lake sediments. *Limnol. Oceanogr.* **19**, 466–88.

Davis, R. B. and T. Webb, III 1975. The contemporary distribution of pollen in eastern North America: a comparison with the vegetation. *Quat. Res.* **5**, 395–434.

Deevey, E. S. and R. F. Flint 1957. Postglacial Hypsithermal interval. *Science* **125**, 182–4.

Defant, A. 1961. *Physical oceanography*, Vol. 1. New York: Pergamon/Macmillan.

de Jong, A. F. M., W. G. Mook, and B. Becker 1980. Confirmation of the Suess wiggles, 3200–3700 BP. *Nature, Lond.* **280**, 48–9.

Delano Smith, C. and M. Parry (eds) 1981. *Consequences of climatic change*. Nottingham: Department of Geography, University of Nottingham.

Delmas, R. J., J.-M. Ascencio, and M. Legrand 1980. Polar ice evidence that atmospheric CO_2 20 000 years BP was 50% of present. *Nature* **284**, 155–7.

de Martonne, E. and L. Aufrère 1928. L'extension des régions privées d'écoulement vers l'ocean. *Annls Géogr.* **38**, 1–24.

Denton, G. H. and W. Karlén 1973a. Holocene climatic variations – their pattern and possible cause. *Quat. Res.* **3**, 155–205.

Denton, G. H. and W. Karlén 1973b. Lichenometry: its application to Holocene moraine studies in southern Alaska and Swedish Lapland. *Arctic Alpine Res.* **5**, 347–72.

Denton, G. H. and W. Karlén 1977. Holocene glacial and tree-line variations in the White River Valley and Skolai Pass, Alaska and Yukon Territory. *Quat. Res.* **7**, 63–111.

de Vries, H. L. 1958. Variation in concentration of radiocarbon with time and location on earth. *Proc. K. Ned. Akad. Wet. B* **61**, 94–102.

de Vries, J. 1977. Histoire du climat et économie: des faits nouveaux, une interpretation différente. *Annls Econ. Soc. Civil.* **32**, 198–228.

de Vries, J. 1981. Measuring the impact of climate on history: the search for appropriate methodologies. In *Climate and history: studies in interdisciplinary history*, R. I. Rotberg and T. K. Rabb (eds), 19–50. Princeton: Princeton University Press.

Diaz, H. F. and J. T. Andrews 1982. Analysis of the spatial pattern of July temperature departures (1943–1972) over Canada and estimates of the 700 mb midsummer circulation during Middle and Late Holocene. *J. Climatol.* **2**, 251–65.

Diester-Haas, L. 1976. Late Quaternary climatic variations in northwest Africa deduced from east Atlantic sediment cores. *Quat. Res.* **6**, 299–314.

Donn, W. L. and M. Ewing 1968. The theory of an ice-free Arctic Ocean. *Met. Monogr.* **8**, 100–5.

Donn, W. L. and D. Shaw 1975. The evolution of climate. In *Proceedings of the WMO/IAMAP Symposium on Long-term Climatic Fluctuations*, 53–62, WMO Publication No. 421. Geneva: World Meteorological Organization.

Douglas, A. V. 1973. *Past air–sea interactions off southern California as revealed by coastal tree-ring chronologies*. MS thesis, University of Arizona, Tucson.

Douglass, A. E. 1914. A method of estimating rainfall by the growth of trees. In *The climatic factor*, E. Huntingdon (ed.), 101–22. Carnegie Institution of Washington Publication No. 192. Washington, D.C.: Carnegie Institution.

Douglass, A. E. 1919. *Climatic cycles and tree growth*, Vol. 1, Carnegie Institution of Washington Publication No. 289. Washington, D.C.: Carnegie Institution.

Dreimanis, A., G. Hütt, A. Raukas, and P. W. Whippey 1978. Dating methods of Pleistocene deposits and their problems: I. Thermoluminescence dating. *Geosci. Can.* **5**, 55–60.

Dudley, W. C. and D. E. Goodney 1979. Stable isotope analysis of calcareous nanno-plankton: a paleo-oceanographic indicator of surface water conditions. In *Evolution des atmosphères planétaires et climatologie de la terre*, 133–48. Toulouse: Centre National d'Etudes Spatiales.

Duplessy, J.-C. 1978. Isotope studies. In *Climatic change*, J. Gribbin (ed.), 46–67. Cambridge: Cambridge University Press.

Duplessy, J.-C., C. Lalou, and A. C. Vinot. 1970a. Differential isotopic fractionation in benthic foraminifera and paleotemperatures re-assessed. *Science* **168**, 250–1.

Duplessy, J.-C., J. Labeyrie, C. Lalou, and H. V. Nguyen 1970b. Continental climatic variations between 130 000 and 90 000 years BP. *Nature, Lond.* **226**, 631–2.

Duplessy, J.-C., J. Labeyrie, C. Lalou, and H. V. Nguyen 1971. La mesure des variations climatique continentales: application à la période comprise entre 130 000 et 90 000 ans BP. *Quat. Res.* **1**, 162–74.

Duplessy, J.-C., P.-L. Blanc, and A. W. H. Bé 1981. Oxygen-18 enrichment of planktonic foraminifera due to gametogenic calcification below the euphotic zone. *Science* **213**, 1247–50.

Duval, P. and C. Lorius 1980. Crystal size and climatic record down to the last ice age from Antarctic ice. *Earth Planet. Sci. Letters* **48**, 59–64.

Dyakowska, J. 1936. Researches on the rapidity of the falling down of pollen of some trees. *Bull. Int. Acad. Pol. Sci. Lett. B1* 155–68.

Dylik, J. 1975. The glacial complex in the notion of the late Cenozoic cold ages. *Biuletyn Peryglacjalny* **24**, 219–31.

Eckstein, D., J. A. Brongers, and J. Bauch 1975. Tree-ring research in the Netherlands. *Tree Ring Bull.* **35**, 1–14.

Eddy, J. A. 1976. The Maunder minimum. *Science* **192**, 1189–202.

Eddy, J. A. 1977. Climate and the changing sun. *Climatic Change* **1**, 173–90.

Elliott, D. L. 1979. The current regenerative capacity of the northern Canadian trees, Keewatin, NWT, Canada: some preliminary observations. *Arctic Alpine Res.* **11**, 243–51.

Emiliani, C. 1954. Depth habitats of some species of pelagic foraminifera as indicated by oxygen isotope ratios. *Am. J. Sci.* **252**, 149–58.

Emiliani, C. 1955. Pleistocene temperatures. *J. Geol.* **63**, 538–78.

Emiliani, C. 1966. Paleotemperature analysis of Caribbean cores, P6304-8 and P6304-9 and a generalized temperature curve for the past 425 000 years. *J. Geol.* **74**, 109–26.

Emiliani, C. 1969. A new paleontology. *Micropaleontology* **15**, 265–300.

Emiliani, C. 1971a. The amplitude of Pleistocene climatic cycles at low latitudes and the isotopic composition of glacial ice. In *The late Cenozoic glacial ages*, K. K. Turekian (ed.), 183–97. New Haven, Conn.: Yale University Press.

Emiliani, C. 1971b. Depth habitats of growth stages of pelagic foraminifera. *Science* **173**, 1122–4.

Emiliani, C. 1972. Quaternary paleotemperatures and the duration of the high temperature intervals. *Science* **178**, 398–401.

Emiliani, C. 1977. Oxygen isotopic analysis of the size fraction between 62 and 250 micrometers in Caribbean cores P6304-8 and P6304-9 *Science* **198**, 1255–6.

Emiliani, C. 1978. The cause of the Ice Ages. *Earth Planet. Sci. Letters* **37**, 349–52.

Emiliani, C. and N. J. Shackleton 1974. The Brunhes epoch: isotopic paleotemperatures and geochronology. *Science* **183**, 511–4.

Emiliani, C., S. Gartner, B. Lidz, K. Eldridge, D. K. Elvey, T. C. Huang, J. J. Stipp, and M. F. Swanson 1975. Paleoclimatological analysis of late Quaternary cores from the northeastern Gulf of Mexico. *Science* **189**, 1083–8.

Endler, J. A. 1977. *Geographic variation, speciation and clines*. Princeton, N.J.: Princeton University Press.

Endler, J. A. 1982. Pleistocene forest refuges: fact or fancy. In *Biological diversification in the Tropics*, G. T. Prance (ed.), 641–57. New York: Columbia University Press.

England, J. and R. S. Bradley 1978. Past glacial activity in the Canadian High Arctic. *Science* **200**, 265–70.

Epstein, S. 1956. Variations of the O^{18}/O^{16} ratios in freshwater and ice. In *Nuclear processes in geologic settings*, 20–28, National Academy of Science/National Research Council Publication 400: IV. Washington, D.C.: National Academy of Science.

Epstein, S. and T. Mayeda 1953. Variation of O^{18} content of waters from natural sources. *Geochim. Cosmochim. Acta* **4**, 213–24.

Epstein, S. and R. P. Sharp 1959. Oxygen isotope variations in the Malaspina and Saskatchewan glaciers. *J. Geol.* **67**, 88–102.

Epstein, S. and C. J. Yapp 1976. Climatic implications of the D/H ratio of hydrogen in C/H groups in tree cellulose. *Earth Planet. Sci. Letters* **30**, 252–66.

Epstein, S. and C. J. Yapp 1977. Isotope tree thermometers (comment). *Nature, Lond.* **266**, 477–8.

Epstein, S., R. Buchsbaum, H. A. Lowenstam, and H. C. Urey 1953. Revised carbonate–water isotopic temperature scale. *Geol. Soc. Am. Bull.* **64**, 1315–26.

Epstein, S., R. P. Sharp, and A. J. Gow 1970. Antarctic ice sheet: stable isotope analyses of Byrd station cores and interhemispheric climatic implications. *Science* **168**, 1570–2.

Epstein, S., C. J. Yapp, and J. H. Hall 1976. The determination of the D/H ratio of non-exchangeable hydrogen in cellulose extracted from aquatic and land plants. *Earth Planet. Sci. Letters* **30**, 241–51.

Erdtman, E. 1969. *Handbook of palynology*. Copenhagen: Munksgaard.

Ericson, D. B. 1959. Coiling direction of *Globigerina pachyderma* as a climatic index. *Science* **130**, 219–20.

Ericson, D. B. and G. Wollin 1968. Pleistocene climates and chronology in deep-sea sediments. *Science* **162**, 1227–34.

Ericson, D. B., G. Wollin, and J. Wollin 1954. Coiling direction of *Globoratalia truncatulinoides* in deep sea cores. *Deep-sea Res.* **2**, 152–8.

Ericson, D. B., M. Ewing, and G. Wollin 1963. Pliocene–Pleistocene boundary in deep sea sediments. *Science* **139**, 727–37.

Ericson, D. B., M. Ewing, and G. Wollin 1964. The Pleistocene epoch in deep-sea sediments. *Science* **146**, 723–32.

Faegri, K. 1950. On the value of palaeoclimatological evidence. In *Centenary Proceedings of the Canadian Royal Meteorological Society*, 188–95. Toronto: Canadian Royal Meteorological Society.

Faegri, K. and J. Iversen 1975. *Textbook of pollen analysis*, 3rd ed. New York: Hafner Press.

Fagerlind, F. 1952. The real signification of pollen diagrams. *Botaniska notiser*, 185–224.

Farmer, J. G. and M. J. Baxter 1974. Atmospheric carbon dioxide levels as indicated by the stable isotope record in wood. *Nature, Lond.* **247**, 273–5.

Faul, H. and G. A. Wagner 1971. Fission track dating. In *Dating techniques for the archaeologist*, H. N. Michael and E. K. Ralph (eds), 152–6. Cambridge, Mass.: MIT Press.

Ferguson, C. W. 1970. Dendrochronology of bristlecone pine, *Pinus aristata*. Establishment of a 7484 year chronology in the White Mountains of eastern-central California, USA. In *Radiocarbon variations and absolute chronology*, I. U. Olsson (ed.), 237–59. New York: Wiley.

Ferguson, R. 1977. *Linear regression in geography*, Concepts and techniques in modern geography No. 15. Norwich: University of East Anglia.

Field, W. O. (ed.) 1975. *Mountain glaciers of the Northern Hemisphere*. Vols 1 and 2. Hanover, Ontario: Cold Regions Research and Engineering Laboratory.

Fisher, D. A. 1979. Comparison of 10^5 years of oxygen isotope and insoluble impurity profiles from the Devon Island and Camp Century ice cores. *Quat. Res.* **11**, 299–305.

Fisher, D. A. and R. M. Koerner 1981. Some aspects of climatic change in the High Arctic during the Holocene as deduced from ice cores. In *Quaternary paleoclimate*, W. C. Mahaney (ed.), 249–71. Norwich: University of East Anglia.

Fisher, D. A., R. M. Koerner, W. S. B. Paterson, W. Dansgaard, N. Gundestrup, and N. Reeh 1983. Effect of wind-scouring on climatic records from ice-core oxygen-isotope profiles. *Nature, Lond.* **301**, 205–9.

Fitch, J. R. 1972. Selection of suitable material for dating and the assessment of geological error in potassium – argon age determination. In *Calibration of hominoid evolution*, W. W. Bishop and J. A. Miller (eds), 77–91. Edinburgh: Scottish Academic Press.

Fleischer, R. L. 1975. Advances in fission track dating. *World Archaeol.* **7**, 136–50.

Fleischer, R. L. and H. R. Hart 1972. Fission track dating techniques and problems. In *Calibration of hominoid evolution*, W. W. Bishop and J. A. Miller (eds), 135–70. Edinburgh: Scottish Academic Press.

Fleming, S. 1976. *Dating in Archaeology: a guide to scientific techniques*. London: J. M. Dent.

Fleming, S. 1979. *Thermoluminescence techniques in archaeology*. Oxford: Clarendon Press.

Flenley, J. R. 1978. *The equatorial rainforest: a geological history*. London: Butterworth.

Flenley, J. R. 1979. The late Quaternary vegetational history of the equatorial mountains. *Progr. Phys. Geogr.* **3**, 488–509.

Flint, R. F. 1943. Growth of North American Ice Sheet during the Wisconsin Age. *Geol. Soc. Am. Bull.* **54**, 325–62.

Flint, R. F. 1971. *Glacial and Quaternary geology*. New York: Wiley.

Flint, R. F. 1976. Physical evidence of Quaternary climatic change. *Quat. Res.* **6**, 519–28.

Flock, J. W. 1978. Lichen – bryophyte distribution along a snow-cover/soil-moisture gradient, Niwot Ridge, Colorado. *Arctic Alpine Res.* **10**, 31–47.

Flohn, H. 1978. Comparison of Antarctic and Arctic climate and its relevance to climate evolution. In *Antarctic glacial history and world palaeoenvironments*, E. M. Van Zinderen Bakker (ed.), 3–13, Rotterdam: A. A. Belkema.

Flohn, H. 1949. Klima und Witterungsablauf in Zürich im 16 Jahrhundert *Vierteljahresheft Naturforsch. Ges. Zürich* **94**, 28–41.

Flohn, H. 1975. Tropische Zirkulationsformen im Lichte der Satellitenaufnahmen *Bonner Met. Abhandl.* **21**.

Flohn, H. 1979. On time scales and causes of abrupt paleoclimatic events. *Quat. Res.* **12**, 135–49.

Frakes, L. A. 1979. *Climates through geologic time*. Amsterdam: Elsevier.

Fredskild, B. and P. Wagner 1974. Pollen and fragments of plant tissue in core samples from the Greenland Ice Cap. *Boreas* **3**, 105–8.

Fritts, H. C. 1962. An approach to dendroclimatology screening by means of multiple regression techniques. *J. Geophys. Res.* **67**, 1413–20.

Fritts, H. C. 1965. Tree-ring evidence for climatic changes in western North America. *Monthly Weather Rev.* **93**, 421–43.

Fritts, H. C. 1971. Dendroclimatology and dendroecology. *Quat. Res.* **1**, 419–49.

Fritts, H. C. 1976. *Tree rings and climate*. London: Academic Press.

Fritts, H. C., D. G. Smith, and M. A. Stokes 1965. The biological model for paleoclimatic interpretation of Mesa Verde tree-ring series. *Am. Antiquity* **31**, no. 2 (part 2), 101–21.

Fritts, H. C., T. J. Blasing, B. P. Hayden, and J. E. Kutzbach 1971. Multivariate techniques for specifying tree-growth and climate relationships and for reconstructing anomalies in paleoclimate. *J. Appl. Met.* **10**, 845–64.

Fritts, H. C., G. R. Lofgren, and G. A. Gordon 1979. Variations in climate since 1602 as reconstructed from tree rings. *Quat. Res.* **12**, 18–46.

Fromm, E. 1970. An estimation of errors in the Swedish varve chronology. In *Radiocarbon variations and absolute chronology*, I. U. Olsson (ed.), 163–72. New York: Wiley.

Galloway, R. W. 1970. The full glacial climate in the southwestern United States. *Ann. Assoc. Am. Geogr.* **60**, 245–56.

Gardner, J. V. 1975. Late Pleistocene carbonate dissolution cycles in the eastern Equatorial Atlantic. In *Dissolution of deep-sea carbonates*, W. V. Sliter, A. W. H. Bé, and W. H. Berger (eds), 129–41, Cushman Foundation for Foraminiferal Research Special Publication No. 13. Washington, D.C.: Cushman Foundation.

Gardner, J. V. and J. D. Hays 1976. Responses of sea-surface temperature and circulation to global climatic change during the past 200 000 years in the eastern Equatorial Atlantic Ocean. In *Investigation of late Quaternary paleooceanography and paleoclimatology*, R. M. Cline and J. D. Hays (eds), 221–46, Geological Society of America Memoir 145. Boulder, Colo.: Geological Society of America.

Garnier, M. 1955. Contribution de la phênologie à l'étude des variations climatiques. *La Météorologie* **40**, 291–300.

Gates, W. L. 1976a. Modelling the ice age climate. *Science* **191**, 1138–44.

Gates, W. L. 1976b. The numerical simulation of ice-age climate with a global general circulation model. *J. Atmos. Sci.* **33**, 1844–73.

Geitzenauer, K. R., M. B. Roche, and A. McIntyre 1976. Modern Pacific coccolith assemblages: derivation and application to late Pleistocene paleotemperature analysis. In *Investigation of late Quaternary paleooceanography and paleoclimatology*, R. M. Cline and J. D. Hay (eds), 423–48, Geological Society of America Memoir 145. Boulder, Colo.: Geological Society of America.

Gillot, P. Y., J. Labeyrie, C. Laj, G. Valladas, G. Guérin, G. Poupeau, and G. Delibrias 1979. Age of the Laschamp paleomagnetic excursion revisited. *Earth Planet. Sci. Letters* **42**, 444–50.

Glahn, H. R. 1968. Canonical correlation and its relationship to discriminant analysis and multiple regression. *J. Atmos. Sci.* **25**, 23–31.

Godwin, H. 1956. *The history of the British flora*. Cambridge: Cambridge University Press.

Godwin, H. 1962. Half-life of radiocarbon. *Nature, Lond.* **195**, 984.

Gonfiantini, R., V. Togliatti, E. Tongiorgi, W. de Breuck, and E. Picciotto 1963. Snow stratigraphy and oxygen isotope variations in the glaciological pit of King Baudoin station, Queen Maud Land, Antarctica. *J. Geophys. Res.* **68**, 3791–8.

Gordiyenko, F. G. and N. I. Barkov 1973. Variations of ^{18}O content in the present precipitation of Antarctica. *Soviet Antarctic Expedition Inf. Bull.* **8**, 495–6.

Gordiyenko, F. G., V. M. Kotlyakov, Ya.-K. M. Punning, and R. Vairmae 1981. Study of a 200 m core from the Lomonosov Ice Plateau on Spitzbergen and the paleoclimatic implications. *Polar Geol. Geogr.* **5**, 242–51.

Gordiyenko, F. G., V. M. Kotlyakov, Ye. S. Korotkevich, N. I. Barkov, and D. Nikolaev 1983. New results of oxygen isotope investigations on ice cores from the borehole at Vostok to a depth of 1412 m (in Russian). *Materialy Glyatsiologicheskikh Issledovaniy. Khronika. Obsuzhdeniya* **46**, 168–74.

Gordon, A. D. and H. J. B. Birks 1972. Numerical methods in Quaternary paleoecology I. Zonation of pollen diagrams. *New Phytol.* **71**, 961–79.

Gordon, A. D. and H. J. B. Birks 1974. Numerical methods in Quaternary paleoecology II. Comparison of pollen diagrams. *New Phytol.* **73**, 221–49.

Gow, A. J. and T. Williamson 1971. Volcanic ash in the Antarctic ice sheet and its possible climatic implications. *Earth Planet. Sci. Letters* **13**, 210–18.

Gow, A. J., S. Epstein, and W. Sheehy 1979. On the origin of stratified debris in ice cores from the bottom of the Antarctic Ice Sheet. *J. Glaciol.* **23**, 185–92.

Grant-Taylor, T. L. 1972. Conditions for the use of calcium carbonate as a dating material. In *Proceedings of the 8th International Conference on Radiocarbon Dating,* Vol. 2, 592–6. Wellington: Royal Society of New Zealand.

Gray, B. M. 1974. Early Japanese winter temperatures. *Weather* **29**, 103–7.

Gray, J. 1981. The use of stable isotope data in climate reconstruction. In *Climate and history,* T. M. L. Wigley, M. J. Ingram, and G. Farmer (eds), 53–81. Cambridge: CUP.

Gray, J. and P. Thompson 1976. Climatic information from $^{18}O/^{16}O$ ratios of cellulose in tree rings. *Nature, Lond.* **262**, 481–2.

Gray, J. and P. Thompson 1977. Climatic information from $^{18}O/^{16}O$ analysis of cellulose, lignin and whole wood from tree rings. *Nature, Lond.* **270**, 708–9.

Gray, J. and P. Thompson 1978. Climatic interpretation of $\delta^{18}O$ and δD in tree rings. *Nature, Lond.* **271**, 93–4.

Grichuk, M. P. 1967. The study of pollen spectra from recent and ancient alluvium. *Rev. Paleobot. Palynol.* **4**, 107–12.

Grichuk, V. P. 1969. An attempt to reconstruct certain elements of the climate of the northern hemisphere in the Atlantic period of the Holocene. In *Golotsen,* M. I. Neishtadt (ed.) 41–57, 8th INQUA Congress. Moscow: Izd-vo Nauka (in Russian). Translated by G. M. Peterson, Center for Climatic Research, University of Wisconsin, Madison.

Griffey, N. J. 1976. Stratigraphical evidence for an early Neoglacial glacier maximum at Steikvasbreen, Okstiden, north Norway, *Norske Geol. Tidsskr.* **56**, 187–94.

Griffey, N. J. and J. A. Matthews 1978. Major neoglacial glacier expansion episodes in southern Norway: evidence from moraine ridge stratigraphy with ^{14}C dates on buried palaeosols and moss layers. *Geogr. Annlr* **60A**, 73–90.

Griffin, J. J., H. Windom, and E. D. Goldberg 1968. The distribution of clay minerals in the world ocean. *Deep-sea Res.* **15**, 433–59.

Griggs, R. F. 1938. Timberlines in the northern Rocky Mountains. *Ecology* **19**, 548–64.

Grimmer, M. 1963. The space-filtering of monthly surface temperature data in terms of pattern, using empirical orthogonal functions. *Q. Jl R. Met. Soc.* **39**, 395–408.

Grinsted, M. J., A. T. Wilson, and C. W. Ferguson 1979. $^{13}C/^{12}C$ ratio variations in *Pinus longaeva* (Bristlecone Pine) cellulose during the last millennium. *Earth Planet. Sci. Letters* **42**, 251–3.

Grove, A. T. and A. Warren 1968. Quaternary landforms and climate on the south side of the Sahara. *Geogr. J.* **134**, 194–208.

Grove, J. M. 1966. The little ice age in the massif of Mont Blanc. *Inst. Br. Geogrs Trans.* No. 40, 129–46.

Grove, J. M. 1979. The glacial history of the Holocene. *Progr. Phys. Geogr.* **3**, 1–54.

Grubb, P. 1982. Refuges and dispersal in the speciation of African forest mammals. In *Biological diversification in the Tropics,* G. T. Prance (ed.), 537–53. New York: Columbia University Press.

Haeserts, P. 1974. Sequence paléoclimatique du Pléistocène Supérieur du Bassin de la Haine (Belgique). *Annls Soc. Géol. Belg.* **97**, 105–37.

Haffer, J. 1969. Speciation in Amazonian forest birds. *Science* **165**, 131–7.

Haffer, J. 1974. *Avian speciation in tropical South America*, Publication of the Nuttall Ornithological Club No. 14. Cambridge, Mass.: Nuttall Ornithological Club.

Haffer, J. 1982. General aspects of the refuge theory. In *Biological diversification in the Tropics*, G. T. Prance (ed.), 6–24. New York: Columbia University Press.

Hage, K. D., J. Gray, and J. C. Linton 1975. Isotopes in precipitation in western North America. *Monthly Weather Rev.* **103**, 958–66.

Hall, C. M. and D. York 1978. K–Ar and ^{40}Ar–^{39}Ar ages of the Laschamp geomagnetic polarity reversal. *Nature, Lond.* **274**, 462–464,

Halme, E. 1952. On the influence of climatic variation on fish and fishery. *Fennia* **75**, 89–96.

Hamilton, A. C. 1976. The significance of patterns of distribution shown by forest plants and animals in tropical Africa for the reconstruction of upper Pleistocene palaeoenvironments: a review. *Palaeoecol. Afr.* **9**, 63–97.

Hamilton, A. C. and R. C. Perrott 1979. Aspects of the glaciation of Mt Elgon, East Africa. *Palaeoecol. Afr.* **11**, 153–62.

Hamilton, A. C. and R. A. Perrott 1980. Modern pollen deposition on a tropical African mountain. *Pollen et Spores* **22**, 437–68.

Hamilton, W. L. and C. C. Langway, Jr 1967. A correlation of microparticle concentrations with oxygen isotope ratios in 700 year old Greenland ice. *Earth Planet. Sci. Letters* **3**, 363–6.

Hammen, T. van der 1963. A palynological study on the Quaternary of British Guiana. *Leidse Geol. Med.* **29**, 125–80.

Hammen, T. van der 1974. The Pleistocene changes of vegetation and climate in tropical South America. *J. Biogeogr.* **1**, 3–26.

Hammen, T. van der, G. C. Maarleveld, J. C. Vogel, and W. H. Zagwijn 1967. Stratigraphy, climatic succession and radiocarbon dating of the last glacial in the Netherlands. *Geol. Mijnb.* **46**, 79–95.

Hammen, T. van der, T. A. Wijmstra, and W. H. Zagwijn 1971. The floral record of the late Cenozoic of Europe. In *The late Cenozoic glacial ages*, K. K. Turekian (ed.), 391–424. New Haven, Conn.: Yale University Press.

Hammer, C. U. 1977a. Dating Greenland ice cores by microparticle concentration analysis. In *Symposium on Isotopes and Impurities in Snow and Ice*, 297–301, International Association of Scientific Hydrology, Publication No. 118. Washington, D.C.: IASH.

Hammer, C. U. 1977b. Past volcanism revealed by Greenland Ice Sheet impurities. *Nature, Lond.* **270**, 482–6.

Hammer, C. U. 1980. Acidity of polar ice cores in relation to absolute dating, past volcanism and radio echoes. *J. Glaciol.* **25**, 359–72.

Hammer, C. U., H. B. Clausen, W. Dansgaard, N. Gundestrup, S. J. Johnsen, and N. Reeh 1978. Dating of Greenland ice cores by flow models, isotopes, volcanic debris and continental dust. *J. Glaciol.* **20**, 3–26.

Hammer, C. U., H. B. Clausen, and W. Dansgaard 1980. Greenland ice sheet evidence of post-glacial volcanism and its climatic impact. *Nature, Lond.* **288**, 230–5.

Haq, B. U. 1978. Calcareous nannoplankton. In *Introduction to marine micropaleontology*, B. U. Haq and A. Boersma (eds), 79–107. New York: Elsevier/North Holland.

Haq, B. U. and A. Boersma (eds) 1978. *Introduction to marine micropaleontology*. New York: Elsevier/North Holland.

Hare, F. K. 1979. Climatic variation and variability: empirical evidence from meteorological and other sources. In *Proceedings of the World Climate Conference*, 51–87, World Meteorological Organization Publication No. 537. Geneva: WMO.

Hare, P. E. and R. M. Mitterer 1968. Laboratory simulation of amino acid diagenesis in fossils. *Carnegie Instn Wash. Ybk* **67**, 205–8.

Hare, P. E., T. C. Hoering, and K. King, Jr 1980. *Biogeochemistry of amino acids*. New York: Wiley.

Harmon, R. S. 1976. Late Pleistocene glacial chronology of the South Nahanni River Region, Northwest Territories, Canada. *Mich. Acadn* **9**, 147–56.

Harmon, R. S., T. Thompson, H. P. Schwarcz, and D. C. Ford 1975. Uranium-series dating of speleothems. *Nat. Speleol. Soc. Bull.* **37**, 21–33.

Harmon, R. S., D. C. Ford, and H. P. Schwarcz 1977. Interglacial chronology of the Rocky and Mackenzie mountains based on ^{230}Th and ^{234}U dating of calcite speleothems. *Can. J. Earth Sci.* **14**, 2543–52.

Harmon, R. S., P. Thompson, H. P. Schwarcz, and D. C. Ford 1978a. Late Pleistocene paleoclimates of North America as inferred from stable isotope studies of speleothems. *Quat. Res.* **9**, 54–70.

Harmon, R. S., H. P. Schwarcz, and D. C. Ford 1978b. Late Pleistocene sea level history of Bermuda. *Quat. Res.* **9**, 205–18.

Harmon, R. S., H. P. Schwarcz, and J. R. O'Neil 1979, D/H ratios in speleothem fluid inclusions: a guide to variations in the isotopic composition of meteoric precipitation? *Earth Planet. Sci. Letters* **42**, 254–66.

Harper, F. 1961. Changes in climate, faunal distribution and life zones in the Ungava Peninsula. *Polar Notes, Dartmouth College, N.H.* No. III, 20–41.

Harrison, C. G. A. 1964. Evolutionary processes and reversals of the earth's magnetic field. *Nature, Lond.* **217**, 46–7.

Harrison, C. G. A. 1974. The paleomagnetic record from deep-sea cores. *Earth Sci. Rev.* **10**, 1–36.

Harrison, C. G. A., I. McDougall, and N. D. Watkins 1979. A geomagnetic field reversal time-scale back to 13.0 million years before present. *Earth Planet. Sci. Letters* **42**, 143–52.

Hastenrath, S. 1967. Observations on the snow line in the Peruvian Andes. *J. Glaciol.* **6**, 541–50.

Hastenrath, S. 1971. On the Pleistocene snow line depression in the arid regions of the South American Andes. *J. Glaciol* **10**, 255–67.

Hay, W. W. 1974. Introduction. In *Studies in paleo-oceanography*, W. W. Hay (ed.), 1–5, Society of Economic Paleontologists and Mineralogists Special Publication No. 20. Tulsa, Okl.: Society of Economic Paleontologists and Mineralogists.

Hays, J. D. 1971. Faunal extinctions and reversals of the earth's magnetic field. *Geol. Soc. Am. Bull.* **82**; 2433–47.

Hays, J. D. 1978. A review of the late Quaternary history of Antarctic Seas. In *Antarctic glacial history and world palaeoenvironments*, E. M. Van Zinderen Bakker (ed.), 57–71. Rotterdam: A. A. Balkema.

Hays, J. D. and A. Perruzza 1972. The significance of calcium carbonate oscillations in eastern equatorial Atlantic deep-sea sediments for the end of the Holocene warm interval. *Quat. Res* **2**, 355–362.

Hays, J. D. and N. J. Shackleton 1976. Globally synchronous extinction of the radiolarian *Stylatractus universus*. *Geology* **4**, 649–52.

Hays, J. D., J. Imbrie, and N. J. Shackleton 1976. Variations in the earth's orbit: pacemaker of the ice ages. *Science* **194**, 1121–32 (see also *Science* **198**, 528–30).

Healy-Williams, N. and D. F. Williams 1981. Fourier analysis of test shape of planktonic foraminifera. *Nature, Lond.* **289**, 485–7.

Hecht, A. 1973. A model for determining Pleistocene paleotemperatures from planktonic foraminiferal assemblages. *Micropalaeontology* **19**, 68–77.

Hecht, A. 1976. The oxygen isotope record of foraminifera in deep sea sediments. In *Foraminifera*, Vol. 2, R. H. Hedley and C. G. Adams (eds), 1–43. New York: Academic Press.

Hecht, A. D. and S. M. Savin 1970. Oxygen-18 studies of recent planktonic foraminifera: comparisons of phenotypes and of test parts. *Science* **170**, 69–71 (see also **173**, 167–9).

Hecht, A. D. and S. M. Savin 1972. Phenotypic variation and oxygen isotope ratios in recent planktonic foraminifera. *J. Foram. Res.* **2**, 55–67.

Hecht, A. D., A. W. H. Bé, and L. Lott 1976. Ecologic and paleoclimatic implications of morphologic variation of *Orbulina universa* in the Indian Ocean. *Science* **194**, 422–4.

Hecht, A., R. G. Barry, H. C. Fritts, J. Imbrie, J. Kutzbach, J. M. Mitchell, Jr, and S. M. Savin 1979. Paleoclimatic research: status and opportunities. *Quat. Res.* **12**, 6–17.

Hedges, R. E. M. 1979. Radioisotope clocks in archaeology. *Nature, Lond.* **281**, 19–23.

Henderson-Sellers, A. 1982. Climatic sensitivity to variations in vegetated land surface albedoes. In *Proceedings of the Sixth Annual Climate Diagnostics Workshop*, 135–44. Washington: D.C.: US Department of Commerce.

Hendy, C. H. 1970. The use of C-14 in the study of cave processes. In *Radiocarbon variations and absolute chronology*, I. U. Olsson (ed.), 419–42. New York: Wiley.

Hendy, C. H. and A. T. Wilson 1968. Paleoclimatic data from speleothems. *Nature, Lond.* **219**, 48–51.

Herron, M. M. and C. C. Langway 1979. Dating of Ross Ice Shelf cores by chemical analysis. *J. Glaciol.* **24**, 345–57.

Herron, M. M. and C. C. Langway 1980. Firn densification: an empirical model. *J. Glaciol.* **25**, 373–86.

Herron, M. M., S. L. Herron, and C. C. Langway 1981. Climatic signal of ice melt features in southern Greenland. *Nature, Lond.* **293**, 389–91.

Herron, S. and C. C. Langway, Jr 1979. The debris-laden ice at the bottom of the Greenland Ice Sheet. *J. Glaciol.* **23**, 193–207.

Heuberger, H. 1974. Alpine Quaternary glaciation. In *Arctic and alpine environments*, J. D. Ives and R. G. Barry (eds), 319–38. London: Methuen.

Heusser, L. and N. Shackleton 1979. Direct marine-continental correlation: 150 000 year oxygen-isotope pollen record from the North Pacific. *Science* **204**, 837–9.

Heusser, C. J. and S. S. Streeter 1980. A temperature and precipitation record of the past 16 000 years in southern Chile. *Science* **210**, 1345–7.

Heusser, C. J., L. E. Heusser, and S. S. Streeter 1980. Quaternary temperatures and precipitation for the north-west coast of North America. *Nature, Lond.* **286**, 702–4.

Hibler, W. D., III, and C. C. Langway, Jr 1977. Ice core stratigraphy as a climatic indicator. In *Polar Oceans*, M. J. Dunbar (ed.), 589–601. Calgary: Arctic Institute of North America.

Hjort, C. 1973. A sea correction for East Greenland. *Geol. Fören. Stockholm Förhandl.* **95**, 132–4.

Hoinkes, H. C. 1968. Glacier variation and weather. *J. Glaciol.* **7**, 3–19.

Hollin, J. T. 1980. Climate and sea-level in isotope-stage 5: an East Antarctic ice surge at 75 000 BP? *Nature, Lond.* **283**, 629–34.

Hollin, J. T. and D. H. Schilling 1981. Late Wisconsin–Weichselian mountain glaciers and small ice caps. In *The last great ice sheets*, G. H. Denton and T. J. Hughes (eds), 179–206. New York: Wiley.

Hope, G. S. and J. A. Peterson 1975. Glaciation and vegetation in the high New Guinea mountains. *Bull. R. Soc. N.Z.* **13**, 155–62.

Howe, S. and T. Webb 1977. Testing the statistical assumptions of paleoclimatic calibration functions. In *Fifth Conference on Probability and Statistics* (preprint volume), 152–7. Boston: American Meteorological Society.

Hughes, T. J., G. H. Denton, B. G. Anderson, D. H. Schilling, J. L. Fastook, and C. S. Lingle 1981. The last great ice sheets: a global view. In *The last great ice sheets*, G. H. Denton and T. J. Hughes (eds), 275–318. New York: Wiley.

Hughes, M. K., P. M. Kelley, J. R. Pilcher, and V. C. LaMarche 1982. *Climate from tree rings*. Cambridge: Cambridge University Press.

Hummel, J. and R. Reck 1979. A global surface albedo model. *J. Appl. Met.* **18**, 239–53.

Hunt, B. G. 1977. A simulation of the possible consequences of a volcanic eruption on the general circulation of the atmosphere. *Monthly Weather Rev.* **105**, 247–60.

Hurford, A. J. and P. F. Green 1982. A user's guide to fission track dating calibration. *Earth Planet. Sci. Letters* **59**, 343–54.

Hutson, W. H. 1977. Transfer functions under no-analog conditions: experiments with Indian Ocean planktonic foraminifera. *Quat. Res.* **3**, 355–67.

Hutson, W. H. 1978. Application of transfer functions to Indian Ocean planktonic foraminifera. *Quat. Res.* **3**, 87–112.

Huxtable, J., M. J. Aitken, and N. Bonhommet 1978. Thermoluminescence dating of sediment baked by lava flows of the Chaine des Puys. *Nature, Lond.* **275**, 207–9.

Idso, S. B. 1980. The climatological significance of a doubling of the Earth's atmospheric carbon dioxide concentration. *Science* **207**, 1462–3.

Idso, S. B. 1982. An empirical evaluation of Earth's surface air temperature response to an increase in atmospheric carbon dioxide concentration. In *Interpretation of climate and photochemical models, ozone and temperature measurements*, R. A. Reck and J. R. Hummel (eds), 119–34. New York: American Institute of Physics.

Imbrie, J. and K. P. Imbrie 1979. *Ice ages: solving the mystery*. London: Macmillan.

Imbrie, J. and J. Z. Imbrie 1980. Modeling the climatic response to orbital variations. *Science* **207**, 943–53.

Imbrie, J. and N. G. Kipp 1971. A new micropalaeontological method for quantitative paleoclimatology: application to late Pleistocene Caribbean core V28-238. In *The late Cenozoic glacial ages*, K. K. Turekian (ed.), 77–181. New Haven, Conn.: Yale University Press.

Imbrie, J., J. van Donk, and N. G. Kipp 1973. Paleoclimatic investigation of a late Pleistocene Caribbean deep-sea core: comparison of isotopic and faunal methods. *Quat. Res.* **3**, 10–38.

Ingle, J. C. 1973. Neogene foraminifera from the northeastern Pacific Ocean, leg 18, Deep-Sea Drilling Project. In *Initial reports of the Deep-sea Drilling Project*, Vol. 18, 517–68. Washington, D.C.: US Government Printing Office.

Ingram, M. J., D. J. Underhill, and T. M. L. Wigley 1978. Historical climatology. *Nature, Lond.* **276**, 329–34.

Ingram, M. J., G. Farmer, and T. M. L. Wigley 1981a. Past climates and their impact on Man: a review. In *Climate and history*, T. M. L. Wigley, M. J. Ingram, and G. Farmer (eds), 3–50. Cambridge: Cambridge University Press.

Ingram, M. J., D. J. Underhill, and G. Farmer 1981b. The use of documentary sources for the study of past climates. In *Climate and history*, T. M. L. Wigley, M. J. Ingram, and G. Farmer (eds), 180–213. Cambridge: Cambridge University Press.

Innes, J. L. 1982. Lichenometric use of an aggregated *Rhizocarpon* "species." *Boreas* **11**, 53–8.

Iversen, J. 1944. *Viscum, Hedera* and *Ilex* as climate indicators. A contribution to the study of past-glacial temperature climate. *Geol. Fören. Stockholm Förhandl.* **66**, 463–83.

Iversen, J. 1949. The influence of prehistoric man on vegetation. *Danmarks Geol. Undersogelse*, Series IV **3** (6).

Ives, J. D. 1974. Permfrost. In *Arctic and alpine environments*, J. D. Ives and R. G. Barry (eds), 159–94. London: Methuen.

Ives, J. D. 1978. Remarks on the stability of timberline. In *Geoecological relations between the Southern Temperate Zone and the tropical high mountains*, C. Trooo and W. Lauer (eds), 313–7. Wiesbaden: Franz Steiner Verlag.

Ives, J. D., J. T. Andrews, and R. G. Barry 1975. Growth and decay of the Laurentide Ice Sheet and comparisons with Fenno-Scandinavia. *Naturwissenschaften* **62**, 118–25.

Jacobson, G. and R. Bradshaw 1981. The selection of sites for paleovegetational studies. *Quat. Res.* **16**, 80–96.

Jacoby, G. C. (ed.) 1980. *Proceedings of the International Meeting on Stable Isotopes in Tree-Ring Research*, Carbon Dioxide Effects Research and Assessment Program. Washington, D.C.: US Department of Energy.

Jacoby, G. C. and L. D. Ulan 1982. Reconstruction of past ice conditions in a Hudson Bay estuary using tree rings. *Nature, Lond.* **248**, 637–9.

Jaenicke, R. 1981. Atmospheric aerosols and climate. In *Climatic variations and variability: facts and theories*, A. Berger (ed.), 577–9. Dordrecht: D. Riedel.

Jagannathan, P., R. Arléry, H. Ten Kate, and M. V. Zavarina 1967. *A note on climatological normals.* Technical Note 84, WMO No. 208, TP 108. Geneva: World Meteorological Organization.

Janssen, C. R. 1966. Recent pollen spectra from the deciduous and coniferous–deciduous forests and northeastern Minnesota: a study in pollen dispersal. *Ecology* **47**, 804–25.

Jochimsen, M. 1973. Does the size of lichen thalli really constitute a valid measure for dating glacial deposits? *Arctic Alpine Res.* **5**, 417–24.

Johnsen, S. J. 1977. Stable isotope homogenization of polar firn and ice. In *Symposium on Isotopes and Impurities in Snow and Ice*, 210–19, International Association of Scientific Hydrology Publication No. 118. Washington, D.C.: IASH.

Johnsen, S. J., W. Dansgaard, H. B. Clausen, and C. C. Langway, Jr 1970. Climatic oscillations 1200–2000 AD. *Nature, Lond.* **227**, 482–3.

Johnsen, S. J., W. Dansgaard, H. B. Clausen, and C. C. Langway, Jr 1972. Oxygen isotope profiles through the Antarctic and Greenland ice sheets. *Nature, Lond.* **235**, 429–34 (see also *Nature, Lond.* **236**, 249).

Johnsen, S. J., C. U. Hammer, N. Reeh, and W. Dansgaard 1976. Microparticles in "Byrd" Station ice core: comments on the paper by Thompson *et al.* (1975). *J. Glaciol.* **17**, 361 (see also reply and further comments in *J. Glaciol.* **18**, 161–4).

Jones, G. A. and W. F. Ruddiman 1982. Assessing the global meltwater spike. *Quat. Res.* **17**, 148–72.

Junge, C. 1972. The cycle of atmospheric gases – natural and man-made. *Q. Jl R. Met. Soc.* **98**, 711–29.

Jungerius, P. D. 1969. Soil evidence of post-glacial tree line fluctuations in the Cypress Hills area, Alberta, Canada. *Arctic Alpine Res.* **1**, 235–45.

Kalela, O. 1952. Changes in the geographic distribution of Finnish birds and mammals in relation to recent changes in climate. *Fennia* **75**, 38–51.

Karlén, W. 1976. Lacustrine sediments and tree-limit variations as indicators of Holocene climatic fluctuations in Lappland, northern Sweden. *Geogr. Annlr* **58A**, 1–34.

Karlén, W. 1979. Glacier variations in the Svartisen area, northern Norway. *Geogr. Annlr* **61A**, 11–28.

Karlén, W. 1980. Reconstruction of past climatic conditions from studies of glacier-front variations. *World Met. Org. Bull.* **29**, 100–4.

Karlén, W. 1982. Holocene glacier fluctuations in Scandinavia. *Striae* **18**, 26–34.

Karrow, P. F. and T. W. Anderson 1975. Palynological studies of lake sediment profiles from SW New Brunswick: discussion. *Can. J. Earth Sci.* **12**, 1808–12.

Karte, J. and H. Liedtke 1981. The theoretical and practical definition of the term "periglacial" in its geographical and geological meaning. *Biuletyn Peryglacjalny* **28**, 123–35.

Kasahara, A., T. Sasamori, and W. M. Washington 1979. General circulation experiments with a six-layer NCAR model, including orography, cloudiness and surface temperature calculations. *J. Atmos. Sci.* **28**, 657–701.

Kato, K. 1978. Factors controlling oxygen isotope composition of fallen snow in Antarctica. *Nature, Lond.* **272**, 46–8.

Kaufman, A., W. S. Broecker, T. L. Ku, and D. L. Thurber 1971. The status of U-series methods of mollusc dating. *Geochim. Cosmochim. Acta* **35**, 1155–83.

Kellogg, T. B. 1975. Late Quaternary climatic changes in the Norwegian and Greenland Seas. In *Climate of the Arctic*, G. Weller and S. A. Bowling (eds), 3–36. Fairbanks: University of Alaska Press.

Kellogg, T. B. 1980. Paleoclimatology and paleo-oceanography of the Norwegian and Greenland seas: glacial – interglacial contrasts. *Boreas* **9**, 115–37.

Kellogg, W. W. 1975. Climatic feedback mechanisms involving the polar region. In *Climate of the Arctic* G. Weller and S. A. Bowling (ed), 111–16. Fairbanks: Univ. of Alaska Press.

Kellogg, W. W. 1977. *Effects of human activities on global climate*. WMO Technical Note 156 (WMO No. 486). Geneva: World Meteorological Organization.

Kellogg, W. W. and S. H. Schneider 1974. Climate stabilization: for better or worse? *Science* **186**, 1163–72.

Kennett, J. P. 1968. *Globerotalia truncatulinoides* as a paleooceanographic index. *Science* **159**, 1461–3.

Kennett, J. P. 1970. Pleistocene paleoclimates and foraminiferal biostratigraphy in subantarctic deep-sea cores. *Deep-sea Res.* **17**, 125–40.

Kennett, J. P. 1976. Phenotypic variation in some recent and late Cenozoic planktonic foraminifera. In *Foraminifera*, Vol. 2, R. H. Hedley and C. G. Adams (eds), 111–70. New York: Academic Press.

Kennett, J. P. and P. Huddlestun 1972. Late Pleistocene paleoclimatology foraminiferal biostratigraphy and tephrochronology, western Gulf of Mexico. *Quat. Res.* **2**, 38–69.

Kennett, J. P. and N. J. Shackleton 1975. Laurentide ice sheet meltwater recorded in Gulf of Mexico deep sea cores. *Science* **188**, 147–50.

Kennett, J. P. and N. D. Watkins 1970. Geomagnetic polarity change, volcanic maxima and faunal extinction in the South Pacific. *Nature, Lond.* **227**, 930–4.

Kent, D. V. 1982. Apparent correlation of paleomagnetic intensity and climatic records in deep-sea sediments. *Nature, Lond.* **299**, 538–9.

Kershaw, A. P. 1978. Record of last interglacial – glacial cycle from northeastern Queensland. *Nature* **272**, 159–61.

King, J. E. and T. R. Van Devender 1977. Pollen analysis of fossil packrat middens from the Sonoran Desert. *Quat. Res.* **8**, 191–204.

King, J. W. 1982. Use of paleomagnetic secular variation records for correlation and dating of rapid environmental changes. In *Abstracts, American Quaternary Association, 7th Biennial Conference*. Seattle: American Quaternary Association.

King, K. and C. Neville 1977. Isoleucine epimerization for dating marine sediments: the importance of analyzing monospecific samples. *Science* **195**, 1333–5.

King, L. and R. Lehmann 1973. Beobachtung zur oekologie und morphologie von *Rhizocarpon Geographicum* (L) D. C. und *Rhizocarpon Alpicola* (Hepp.) Rabenh. in gletschervorfeld des steingletschers. *Berichte der Schweizerischen Botanischen Gesellschaft* **83**, 139–46.

Kipp, N. G. 1976. New transfer function for estimating past sea-surface conditions from sea-bed distribution of planktonic foraminiferal assemblages in the North Atlantic. In *Investigation of late Quaternary paleooceanography and paleoclimatology*, R. M. Cline and J. D. Hays (eds), 3–42, Geological Society of American Memoir 145. Boulder, Colo.: Geological Society of America.

Kittleman, L. R. 1979. Geologic methods in studies of Quaternary tephra. In *Volcanic activity and human ecology*, D. D. Sheets and D. K. Grayson (eds), 49–82. New York: Academic Press.

Klein, J., J. C. Lerman, P. E. Damon, and E. K. Ralph 1982. Calibration of radiocarbon dates. *Radiocarbon* **24**, 103–50.

Koerner, R. M. 1968. Fabric analysis of a core from the Meighen Ice Cap, NWT, Canada. *J. Glaciol.* **7**, 421–30.

Koerner, R. M. 1977a. Distribution of microparticles in a 299 m core through the Devon Island ice cap, North West Territories, Canada. In *Symposium on Isotopes and Impurities in Snow and Ice*, 371–6, International Association of Scientific Hydrology Publication No. 118. Washington, D.C.: IASH.

Koerner, R. M. 1977b. Devon Island Ice Cap: core stratigraphy and paleoclimate. *Science* **196**, 15–18.

Koerner, R. M. 1979. Accumulation, ablation and oxygen isotope variations on the Queen Elizabeth Islands Ice Caps, Canada. *J. Glaciol.* **22**, 25–41.

Koerner, R. M. 1980. The problem of lichen-free zones in Arctic Canada. *Arctic Alpine Res.* **12**, 87–94.

Koerner, R. M. and D. A. Fisher 1979. Discontinuous flow, ice texture and dirt content in the basal layers of the Devon Island Ice Cap. *J. Glaciol.* **23**, 209–22.

Koerner, R. M. and D. A. Fisher 1981. Studying climatic change from Canadian High Arctic ice cores. In *Climatic Change in Canada – 2.* C. R. Harington (ed.), 195–217, Syllogeus No. 33. Ottawa: National Museum of Natural Science.

Koerner, R. M. and W. S. B. Paterson 1974. Analysis of a core through the Meighen Ice Cap, Arctic Canada, and its paleoclimatic implications. *Quat. Res.* **4**, 253–63.

Koerner, R. M. and R. P. Russell 1979. $\delta^{18}O$ variations in snow on the Devon Island Ice Cap, North West Territories, Canada. *Can. J. Earth Sci.* **16**, 1419–27.

Koerner, R. M. and H. Taniguchi 1976. Artificial radioactivity layers in the Devon Island Ice Cap, North West Territories. *Can. J. Earth Sci.* **13**, 1251–5.

Kohler, M. A., T. J. Nordenson, and D. R. Baker 1966. *Evaporation maps for the United States*, US Weather Bureau Technical Paper No. 37. Washington, D.C.: US Department of Commerce.

Koide, M., K. W. Bruland, and E. D. Goldberg 1973. $^{238}Th/^{232}Th$ amd ^{210}Pb geochronologies in marine and lake sediments. *Geochim. Cosmochim. Acta* **37**, 1171–87.

Kolla, V., P. E. Biscaye, and A. F. Hanley 1979. Distribution of quartz in late Quaternary Atlantic sediments in relation to climate. *Quat. Res.* **11**, 261–77.

Kominz, M. A., G. R. Heath, T.-L. Ku, and N. G. Pisias 1979. Brunhes time scales and the interpretation of climatic change. *Earth Planet. Sci. Letters* **45**, 394–410.

Korff, H. Cl. and H. Flohn 1969. Zusammenhang zwischen dem Temperaturgefälle Äquator-Pol und den planetarischen Luftdruckgürteln. *Annlr Met.* **4**, 163–4.

Krebs, J. S. and R. G. Barry 1970. The arctic front and the tundra–taiga boundary in Eurasia. *Geogr. Rev.* **60**, 548–54.

Krinsley, D. H. 1978. The present state and future prospects of environmental discrimination by scanning electron microscopy. In *Scanning electron microscopy in the study of sediments*, W. B. Whalley (ed.), 169–79. Norwich: University of East Anglia.

Krinsley, D. H. and J. Donahue 1968. Environmental interpretation of sand grain surface textures by electron microscopy. *Geol. Soc. Am. Bull.* **78**, 743–8.

Krinsley, D. H. and S. Margolis 1969. A study of quartz sand grain surface textures with the scanning electron microscope. *Trans. N.Y. Acad. Sci.* **31**, 457–77.

Krinsley, D. H. and W. Newman 1965. Pleistocene glaciation: a criterion for recognition of its onset. *Science* **149**, 442–3.

Krog, H. and H. Tauber 1973. ^{14}C chronology of late and post-glacial marine deposits in North Jutland. *Danmarks Geol. Undersogelse*, 93–105.

Kruse, H. H., T. W. Linick, and H. E. Suess 1980. Computer-matched radiocarbon dates of floating tree-ring series. *Radiocarbon* **22**, 260–6.

Ku, T.-L. 1976. The uranium series method of age determination. *Ann. Rev. Earth Planet. Sci.* **4**, 347–80.

Ku, T.-L. and T. Oba 1978. A method for quantitative evaluation of carbonate dissolution in deep-sea sediments and its application to paleooceanographic reconstruction. *Quat. Res.* **10**, 112–29.

Kukla, G. J. 1975. Missing link between Milankovitch and climate. *Nature, Lond.* **253**, 600–3.

Kukla, G. J. 1977. Pleistocene land–sea correlations. I. Europe. *Earth Sci. Rev.* **13**, 307–74.

Kukla, G. J. 1978. Recent changes in snow and ice. In *Climatic change* J. Gribbin (ed.), 114–30. Cambridge: Cambridge University Press.

Kukla, G. 1979. Climatic role of snow covers. In *Sea level, ice and climatic change*, 79–107, International Association of Scientific Hydrology Publication No. 131, Washington D.C.: IASH.

Kukla, G. J. 1981. Surface albedo. In *Climatic variations and variability*, A. Berger (ed.), 85–109. Dordrecht: D. Reidel.

Kukla, G. J. and J. Gavin 1979. Snow and sea-ice in 1979. In *Proceedings of the Fourth Annual Climate Diagnostics Workshop*, 60–71. Washington, D. C.: US Department of Commerce (NOAA).

Kukla, G. J. and H. J. Kukla 1974. Increased surface albedo in the northern hemisphere. *Science* **183**, 709–14.

Kukla, G. and D. Robinson 1980. Annual cycle of surface albedo. *Monthly Weather Rev.* **108**, 56–68.

Kukla, G. J., J. K. Angell, J. Korshover, H. Dronia, M. Hoshiai, J. Namias, M. Rodewald, R. Yamamoto, and T. Iwashima 1977. New data on climatic trends. *Nature* **270**, 573–80.

Kutzbach, J. E. 1974. Fluctuations of climate – monitoring and modelling. *World Met. Org. Bull.* **23**, 155–63.

Kutzbach, J. E. 1975. Diagnostic studies of past climate. In *The physical basis of climate and climate modelling*, 119–26, GARP Publication No. 16. Geneva: World Meteorological Organization.

Kutzbach, J. E. 1976. The nature of climate and climatic variations. *Quat. Res.* **6**, 471–80.

Kutzbach, J. E. 1980. Estimates of past climate at Paleolake Chad, North Africa, based on a hydrological and energy balance model. *Quat. Res.* **14**, 210–23.

Kutzbach, J. E. and R. A. Bryson 1974. Variance spectrum of Holocene climatic fluctuations in the North Atlantic sector. *J. Atmos. Sci.* **31**, 1958–63.

Kutzbach, J. E. and P. J. Guetter 1980. On the design of paleoenvironmental data networks for estimating large-scale patterns of climate. *Quat. Res.* **14**, 169–87.

Kutzbach, J. E. and B. L. Otto-Bliesner 1982. The sensitivity of the African–Asian monsoonal climate to orbital parameter changes for 9000 years BP in a low-resolution general circulation model. *J. Atmos. Sci.* **39**, 1177–88.

Labeyrie, L. 1974. New approach to surface seawater palaeotemperatures using $^{18}O/^{16}O$ ratios in silica of diatom frustules. *Nature, Lond.* **248**, 40–1.

Labrijn, A. 1945. Het Klimaat van Nederland gedurende de laatste twee en een halve eeuw. *Meded. Verhandl.* **49**, 11–105 (Koninklijk Nederlands Met. Inst. No. 102).

Lajoux, J.-D. 1963. *The rock paintings of the Tassili*. London: Thames and Hudson.

LaMarche, V. C. 1973. Holocene climatic variations inferred from treeline fluctuations in the White Mountains, California. *Quat. Res.* **3**, 632–60.

LaMarche, V. C., Jr 1974. Paleoclimatic inferences from long tree-ring records. *Science* **183**, 1043–8.

LaMarche, V. C. 1982. Sampling strategies. In *Climate from tree rings* M. K. Hughes, P. M. Kelly, J. R. Pilcher, and V. C. LaMarche (eds), 2–6. Cambridge: Cambridge University Press.

LaMarche, V. C., Jr and H. C. Fritts 1971a. Anomaly patterns of climate over the Western United States, 1700–1930, derived from principal component analysis of tree ring data. *Monthly Weather Rev.* **99**, 138–42.

LaMarche, V. C., Jr and H. C. Fritts 1971b. Tree rings, glacial advance and climate in the Alps. *Z. GletscherKde Glazialgeol.* **7**, 125–31.

LaMarche, V. C. and H. A. Mooney 1967. Altithermal timberline advance in western United States. *Nature, Lond.* **213**, 980–2.

LaMarche, V. C. and H. A. Mooney 1972. Recent climatic change and development of the bristlecone pine (*P. longaeva* Bailey) Krummholz zone, Mt Washington, Nevada. *Arctic Alpine Res.* **4**, 61–72.

LaMarche, V. C. and A. B. Pittock 1982. Preiiminary temperature reconstructions for Tasmania. In *Climate from tree rings*, M. K. Hughes, P. M. Kelly, J. R. Pilcher, and V. C. LaMarche (eds), 177–85. Cambridge: Cambridge University Press.

LaMarche, V. C. and C. W. Stockton 1974. Chronologies from temperature-sensitive

Bristlecone Pines at upper treeline in western United States. *Tree Ring Bull.* **34**, 21–45.

Lamb, H. H. 1959. The southern westerlies: a preliminary survey, main characteristics and apparent associations. *Q. J. R. Met. Soc.* **85**, 1–23.

Lamb, H. H. 1961. Climatic change within historical time as seen in circulation maps and diagrams. *Ann. N.Y. Acad. Sci.* **95**, 124–61.

Lamb, H. H. 1963. On the nature of certain climatic epochs which differed from the modern (1900–1939) normal. In *Changes of climate* (Proceedings of the WMO–UNESCO Rome 1961 Symposium on Changes of Climate), 125–50, UNESCO Arid Zone Research Series XX. Paris: UNESCO.

Lamb, H. H. 1964. Trees and climate in Scotland. *Q. Jl R. Met. Soc.* **90**, 382–94.

Lamb, H. H. 1965. The early Medieval warm epoch and its sequel. *Palaeogeogr. Palaeoclimatol. Palaeoecol.* **1**, 13–37.

Lamb, H. H. 1970. Volcanic dust in the atmosphere; with a chronology and assessment of its meteorological significance. *Phil. Trans R. Soc. A* **266**, 425–533.

Lamb, H. H. 1972. *Climate, present, past and future*, Vol. 1. London: Methuen.

Lamb, H. H. 1977. *Climate, present, past and future*, Vol. 2. London: Methuen.

Lamb, H. H. 1982. *Climate history and the modern world*. London: Methuen.

Landsberg, H. E. 1980. Variable solar emissions, the "Maunder Minimum" and climatic temperature fluctuations. *Arch. Met. Geophys. Bioklimatol. B* **28**, 181–91.

Langbein, W. B., *et al.* 1949. *Annual runoff in the United States*, US Geological Survey Circular 52. Washington, D. C.: US Geological Survey.

Langbein, W. B. 1961. *Salinity and hydrology of enclosed lakes*, US Geological Survey Professional Paper 412, Washington, D.C.: US Geological Survey.

Langway, C. C., Jr 1970. *Stratigraphic analysis of a deep ice core from Greenland*, Geological Society of America Special Paper 125. Boulder, Colo.: Geological Society of America.

Langway, C. C., G. A. Klouda, M. A. Herron, and J. H. Craig 1977. Seasonal variations of chemical constituents in annual layers of Greenland deep ice deposits. In *Symposium on isotopes and impurities in snow and ice*, 302–6, International Association of Scientific Hydrology Publication No. 118. Washington, D.C.: IASH.

Larsen, J. A. 1974. Ecology of the northern continental forest border. In *Arctic and alpine environments*, J. D. Ives and R. G. Barry (eds), 341–69. London: Methuen.

Lawrence, D. B. 1950. Estimating dates of recent glacial advances and recession rates by studying tree growth layers. *Trans. Am. Geophys. Union* **31**, 243–8.

Leopold, L. B. 1951. Pleistocene climate in New Mexico. *Am. J. Sci.* **249**, 152–68.

Lerbemko, J. F., J. A. Westgate, D. G. W. Smith, and G. H. Denton 1975. New data on the character and history of the White River volcanic eruption, Alaska. In *Quaternary studies*, R. P. Suggate and M. M. Cresswell (eds), 203–9, Royal Society of New Zealand Bulletin No. 13. Wellington: Royal Society of New Zealand.

Lerman, J. C. 1972. Carbon-14 dating: origin and correction of isotope fractionation errors in terrestrial living matter. In *Proceedings of the 8th International Conference on Radiocarbon Dating*, Vol. 2, 613–24. Wellington: Royal Society of New Zealand.

Le Roy Ladurie, E. 1971. *Times of feast, times of famine*. New York: Doubleday.

Le Roy Ladurie, E. and M. Baulant 1981. Grape harvests from the fifteenth through the nineteenth centuries. In *Climate and history: studies in interdisciplinary history*, R. I. Rotberg and T. K. Rabb (eds), 259–69. Princeton: Princeton University Press.

Lettau, H. 1969. Evapotranspiration climatonomy, I. A new approach to numerical prediction of monthly evapotranspiration, runoff and soil moisture storage. *Monthly Weather Rev.* **97**, 691–9.

Lhote, H. 1959. *The search for the Tassili frescoes*. London: Hutchinson.

Libby, L. M. 1972. Multiple thermometry in paleoclimate and historic climate. *J. Geophys. Res.* **77**, 4310–17.

Libby, L. M. and L. J. Pandolfi 1974. Temperature dependence of isotope ratios in tree rings. *Proc. Natn Acad. Sci. U.S.A.* **71**, 2482–6.

Libby, L. M., L. J. Pandolfi, P. H. Payton, J. Marshall, III, B. Becker, and V. Giertz-Siebenlist 1976. Isotopic tree thermometers. *Nature, Lond.* **261**. 284–8.

Libby, W. F. 1955. *Radiocarbon dating*. Chicago: University of Chicago Press.

Libby, W. F. 1970. Radiocarbon dating. *Phil. Trans R. Soc. A* **269**, 1–10.

Lichti-Federovich, S. 1975. Pollen analysis of ice core samples from the Devon Island Ice Cap. *Geol. Survey Can. Paper 75-1*, Part A, 441–4.

Lichti-Federovich, S. and J. C. Ritchie 1968. Recent pollen assemblages from the western interior of Canada. *Rev. Paleobot. Palynol.* **7**, 297–344.

Lidz, L. 1966. Deep sea biostratigraphy. *Science* **154**, 1448–52.

Lieth, H. 1975. Primary productivity in ecosystems: comparative analysis of global patterns. In *Unifying concepts in ecology: report of plenary sessions, 1st International Congress on Ecology*, W. H. van Dobben and R. H. Lowe-McConnell (eds), 67–88. The Hague: Dr W. Junk BV, Publishers.

Lindsey, A. A. and J. E. Newman 1956. Use of official data in spring time temperature analysis of Indiana phenological record. *Ecology* **37**, 812–23.

Livingstone, D. A. 1968. Some interstadial and postglacial pollen diagrams from eastern Canada. *Ecol. Monogr.* **38**, 87–125.

Livingstone, D. A. 1975. Late Quaternary climatic change in Africa. *A. Rev. Ecol. Systematics* **6**, 249–80.

Livingstone, D. A. 1982. Quaternary geography and Africa and the refuge theory. In *Biological diversification in the Tropics*, G. T. Prance (ed.), 523–36. New York: Columbia University Press.

Locke, C. W. and W. W. Locke 1977. Little ice age snow-cover extent and paleoglaciation thresholds: north-central Baffin Island, NWT, Canada. *J. Arctic Alpine Res.* **9**, 291–300.

Locke, W. W., J. T. Andrews, and P. J. Webber 1979. *A manual for lichenometry*, British Geomorphological Research Group, Technical Bulletin No. 26. Norwich: University of East Anglia.

Lockwood, J. G. 1978. "Unmixing" of the deep-sea record and the deglacial meltwater spike (comment). *Nature, Lond.* **272**, 188.

Löffler, E. 1976. Potassium–argon dates and pre-Würm glaciations of Mount Giluwe volcano, Papua, New Guinea. *Z. GletscherKde Glazialgeol.* **12**, 55–62.

Lorenz, E. N. 1968. Climatic determinism. *Met. Monogr.* **8**, 1–3.

Lorenz, E. N. 1970. Climatic change as a mathematical problem. *J. Appl. Met.* **9**, 235–9.

Lorenz, E. N. 1976. Non-deterministic theories of climatic change. *Quat. Res.* **6**, 495–507.

Lorius, C., L. Merlivat, J. Jouzel, and M. Pourchet 1979. A 30 000 year isotope climatic record from Antarctic ice. *Nature, Lond.* **280**, 644–8.

Lozano, J. A. and J. D. Hays 1976. Relationship of radiolarian assemblages to sediment types and physical oceanography in the Atlantic and western Indian Ocean sectors of the Antarctic Ocean. In *Investigation of late Quaternary paleooceanography and paleoclimatology*, R. M. CLine and J. D. Hays (eds), 303–36. Geological Society of America Memoir 145. Boulder, Colo.: Geological Society of America.

Lund, S. P. and S. K. Banderjee 1979. Paleosecular geomagnetic variations from lake sediments. *Rev. Geophys. Space Phys.* **17**, 244–9.

Lundquist, G. 1959. C-14 daterade tallstubbar frän fjällen (^{14}C dated pine stumps from the high mountains in western Sweden) *Sveriges Geol. Undersökning* **53** (3).

Luz, B. 1977. Paleoclimates of the South Pacific based on statistical analysis of planktonic foraminifers. *Palaeogeogr. Palaeoclimatol. Palaeoecol.* **22**, 61–78.

Luz, B. and N. J. Shackleton 1975. $CaCO_3$ solution in the tropical East Pacific during the past 130 000 years. In *Dissolution of deep-sea carbonates*, W. V. Sliter, A. W. H. Bé, and W. H. Berger (eds), 142–50, Cushman Foundation for Foraminiferal Research Special Publication No. 13. Washington, D.C.: Cushman Foundation.

Lynch, H. B. 1931. *Rainfall and stream run-off in southern California since 1769*. Los Angeles: Metropolitan Water District of Southern California.

Maarleveld, G. C. 1976. Periglacial phenomena and the mean annual temperature during the last glacial time in the Netherlands. *Biuletyn Peryglacjalny* **26**, 57–78.

Mackereth, F. H. 1971. On the variation in direction of the horizontal component of remanent magnetization in lake sediments. *Earth Planet. Sci. Letters* **12**, 332–8.

MacKinnon, P. K. (ed.) 1980. *Ice cores*, Glaciological Data Report GD-8. Boulder, Colo.: World Data Center 'A' for Glaciology, University of Colorado.

Maher, L. J. 1963. Pollen analyses of surface materials from the southern San Juan Mountains, Colorado. *Geol. Soc. Am. Bull.* **74**, 1485–504.

Malmgren, B. A. and J. P. Kennett 1976. Biometric analysis of phenotypic variation in recent *Globigerina bulloides* d'Orbigny in the southern Indian Ocean. *Marine Micropaleont.* **1**, 3–25.

Malmgren, B. A. and J. P. Kennett 1978a. Test size variation in *Globigerina bulloides* in response to Quaternary palaeo-oceanographic changes. *Nature, Lond.* **275**, 123–4.

Malmgren, B. A. and J. P. Kennett 1978b. Late Quaternary paleoclimatic applications of mean size variations in *Globigerina bulloides* d'Orbigny in the southern Indian Ocean. *J. Paleont.* **52**, 1195–207.

Manabe, S. and D. G. Hahn 1977. Simulation of the tropical climate of an ice age. *J. Geophys. Res.* **82**, 3889–911.

Manabe, S. and R. T. Wetherald 1975. The effects of doubling the CO_2 concentration on the climate of a general circulation model. *J. Atmos. Sci.* **32**, 3–15.

Manabe, S., R. T. Wetherald, and R. J. Stouffer 1981. Summer dryness due to an increase of atmospheric CO_2 concentration. *Climatic Change* **3**, 347–86.

Mangerud, J. 1972. Radiocarbon dating of marine shells including a discussion of apparent age of recent shells from Norway. *Boreas* **1**, 143–72.

Mangerud, J. and S. Gulliksen 1975. Apparent radiocarbon ages of recent marine shells from Norway, Spitsbergen and Arctic Canada. *Quat. Res.* **5**, 263–74.

Mankinen, E. A. and G. B. Dalrymple 1979. Revised geomagnetic polarity time scale for the interval 0–5 m.y. BP. *J. Geophys. Res.* **84**, 615–26.

Manley, G. 1953. Mean temperature of Central England 1698–1952. *Q. Jl R. Met. Soc.* **79**, 242–61.

Manley, G. 1959. Temperature trends in England. *Arch. Met. Geophys. Bioklimatol. B* **9**, (3/4).

Manley, G. 1969. Snowfall in Britain over the past 300 years. *Weather, Lond.* **24**, 428–37.

Manley, G. 1974. Central England temperatures: monthly means 1659 to 1973. *Q. Jl R. Met. Soc.* **100**, 389–405.

Margolis, S. V., P. M. Kroopnick, D. E. Goodney, W. C. Dudley, and M. E. Mahoney 1975. Oxygen and carbon isotopes from calcareous nannofossils as paleo-oceanographic indicators. *Science* **189**, 555–7.

Markgraf, V. 1980. Pollen dispersal in a mountain area. *Grana* **19**, 127–46.

Martin, P. S., B. E. Sabel, and D. Shutler, Jr 1961. Rampart cave coprolite and ecology of the Shasta ground sloth. *Am. J. Sci.* **259**, 102–27.

Mason, B. J. 1976. Towards the understanding and prediction of climatic variations. *Q. Jl R. Met. Soc.* **102**, 473–98.

Matthes, F. E. 1940. Report of the Committee on glaciers. *Trans Am. Geophys. Union* **21**, 396–406.

Matthes, F. E. 1942. Glaciers. In *Hydrology*, O. E. Meinzer (ed.), 149–219, New York: Dover/McGraw-Hill.

Matthews, J. A. 1980. Some problems and implications of ^{14}C dates from a podzol buried beneath an end moraine at Haugabreen, southern Norway. *Geogr. Annlr* **62A**, 185–208.

Matthews, J. V. 1968. A paleoenvironmental analysis of three late Pleistocene coleopterous assemblages from Fairbanks, Alaska. *Quaestiones Entomologicae* **4**, 202–24.

Matthews, J. V. 1974. Quaternary environments at Cape Deceit (Seward Peninsula, Alaska): evolution of a tundra ecosystem. *Geol. Soc. Am. Bull.* **85**, 1353–84.

Matthews, J. V. 1975. Insects and plant macrofossils from two Quaternary exposures in the

Old Crow–Porcupine region, Yukon Territory, Canada. *Arctic Alpine Res.* **7**, 249–59.

McAndrews, J. H. 1966. Postglacial history of prairie, savanna, and forest in northwestern Minnesota. *Mem. Torrey Bot. Club* **22**, 1–72.

McCoy, W. D. 1981. *Quaternary aminostratigraphy of the Bonneville and Lahontan Basins, western US with paleoclimatic implications.* PhD thesis, Department of Geography, University of Colorado.

McCoy, W. D. 1982. Isoleucine epimerization in fossil gastropods shells and the late Quaternary paleoclimates of the northern Bonneville Basin. In *Abstracts of the American Quaternary Association, 7th Biennial Conference,* 135. Seattle, Wash.: American Quaternary Association.

McCulloch, D. and D. Hopkins 1966. Evidence for an early recent warm interval in northwestern Alaska. *Geol. Soc. Am. Bull.* **77**, 1089–108.

McDougall, I. 1979. The present status of the geomagnetic polarity time scale. In *The Earth: its origin, structure and evolution,* M. W. McElhinny (ed.), 543–66. London: Academic Press.

McIntyre, A., W. F. Ruddiman, and R. Jantzen 1972. Southward penetrations of the North-Atlantic Polar Front: faunal and floral evidence of large scale surface water mass movements over the last 225 000 years. *Deep-sea Res.* **19**, 61–77.

McIntyre, A., A. W. H. Bé, J. D. Hays, J. V. Gardner, J. A. Lozano, B. Molfino, W. Prell, H. R. Thierstein, T. Crowley, J. Imbrie, T. Kellogg, N. Kipp, and W. F. Ruddiman 1975. Thermal and oceanic structures of the Atlantic through a glacial–interglacial cycle. In *Proceedings of the WMO Symposium on Long-term Climatic Fluctuations,* 75–80, WMO No. 421. Geneva: World Meteorological Organization.

McIntyre, A., N. G. Kipp, A. W. H. Bé, T. Crowley, T. Kellogg, J. V. Gardner, W. Prell, and W. F. Ruddiman 1976. Glacial North Atlantic 18 000 years ago: a CLIMAP reconstruction. In *Investigation of late Quaternary paleooceanography and paleoclimatology,* R. M. Cline and J. D. Hays (eds), 43–76, Geological Society of America Memoir 145. Boulder, Colo.: Geological Society of America.

McManus, D. A. 1970. Criteria of climatic change in the inorganic components of marine sediments. *Quat. Res.* **1**, 72–102.

Meggers, B. J. 1982. Archeological and ethnographic evidence compatible with the model of forest fragmentation. In *Biological diversification in the Tropics,* G. T. Prance (ed.), 483–96. New York: Columbia University Press.

Mercer, J. H. 1976. Glacial history of southernmost South America. *Quat. Res.* **6**, 125–66.

Mesolella, K. J., R. K. Matthews, W. S. Broecker, and D. L. Thurber 1969. The astronomical theory of climatic change: Barbados data. *J. Geol.* **77**, 250–74.

Messerli, B., P. Messerli, C. Pfister, and H. J. Zumbuhl 1978. Fluctuations of climate and glaciers in the Bernese Oberland, Switzerland, and their geological significance, 1600–1975. *Arctic Alpine Res.* **10**, 247–60.

Michel, P. 1973. *Les bassins des fleuves Sénégal et Gambie: étude géomorphologie,* Memoires no. 63. Paris: Office de la Recherche Scientifique et Technique d'Outre-mer.

Michels, J. W. and C. A. Bebrich 1971. Obsidian hydration dating. In *Dating techniques for the archaeologist,* H. N. Michael and E. K. Ralph (eds), 164–221. Cambridge: MIT Press.

Mifflin, M. D. and M. M. Wheat 1979. *Pluvial lakes and estimated pluvial climates of Nevada,* Bulletin 94, Nevada Bureau of Mines and Geology. Reno: University of Nevada.

Migliazza, E. C. 1982. Linguistic prehistory and the refuge model in Amazonia. In *Biological diversification in the Tropics,* G. T. Prance (ed.), 497–519. New York: Columbia University Press.

Milankovitch, M. M. 1941. *Canon of insolation and the ice-age problem.* Beograd: Königlich Serbische Akademie. English translation by the Israel Program for Scientific Translations, published for the US Department of Commerce, and the National Science Foundation, Washington, D.C. (1969).

Millar, D. H. M. 1981. Radio-echo layering in polar ice sheets and past volcanic activity. *Nature, Lond.* **292**, 441–3.

Miller, C. D. 1973. Chronology of Neoglacial deposits in the northern Sowatch Range, Colorado. *Arctic Alpine Res.* **5**, 385–400.

Miller, G. H. 1973. Late Quaternary glacial and climatic history of northern Cumberland Peninsula, Baffin Island, NWT, Canada. *Quat. Res.* **3**, 561–83.

Miller, G. H. 1973. Variations in lichen growth from direct measurements: preliminary curves for *Alectoria minuscula* from eastern Baffin Island, NWT, Canada. *Arctic Alpine Res.* **5**, 333–9.

Miller, G. H. and J. T. Andrews 1973. Quaternary history of northern Cumberland Peninsula, east Baffin Island, NWT, Canada, Part VI. Preliminary lichen growth curve for *Rhizocarpon geographicum*. *Geol. Soc. Am. Bull.* **83**, 1133–8.

Miller, G. H. and P. E. Hare 1975. Use of amino acid reactions in some arctic marine fossils as stratigraphic and geochronologic indicators. *Carnegie Instn Wash. Ybk* **74**, 612–17.

Miller, G. H. and P. E. Hare 1980. Amino acid geochronology: integrity of the carbonate matrix and potential of molluscan fossils. In *Biogeochemistry of amino acids*, P. E. Hare, T. C. Hoering, and K. King, Jr (eds), 415–43. New York: Wiley.

Miller, G. H., R. S. Bradley, and J. T. Andrews 1975. Glaciation level and lowest equilibrium line altitude in the High Canadian Arctic: maps and climatic interpretation. *Arctic Alpine Res.* **7**, 155–68.

Miller, G. H., J. T. Andrews, and S. K. Short 1977. The last interglacial–glacial cycle, Clyde foreland, Baffin Island, NWT: stratigraphy, biostratigraphy and chronology. *Can. J. Earth Sci.* **14**, 2824–57.

Miller, G. H., J. T. Hollin, and J. T. Andrews 1979. Aminostratigraphy of UK Pleistocene deposits. *Nature, Lond.* **281**, 539–43.

Miller, J. A. 1972. Dating Pliocene and Pleistocene strata using the potassium argon and argon-40/argon-39 methods. In *Calibration of hominoid evolution*, W. W. Bishop and J. A. Miller (eds), 63–73. Edinburgh: Scottish Academic Press.

Miroshnikov, L. D. 1958. Ostatki drevney lesnoy rastitel'nosti na Taymyrskom poluostrove. *Priroda, Mosk.* **2**, 106–7.

Mitchell, J. M. 1965. Theoretical paleoclimatology. In *Quaternary of the United States*, H. Wright and D. Frey (eds), 881–901. Princeton, N.J.: Princeton University Press.

Mitchell, J. M. 1976. An overview of climatic variability and its causal mechanisms. *Quat. Res.* **6**, 481–93.

Mitchell, J. M., B. Dzerdzeevski, H. Flohn, W. L. Hofmeyr, H. H. Lamb, K. N. Rao, and C. C. Wallén 1966. *Climate change*, WMO Technical Note No. 79. Geneva: World Meteorological Organization.

Mitchell, J. M., C. W. Stockton, and D. M. Meko 1979. Evidence of a 22 year rhythm of drought in the western United States related to the Hale Solar Cycle since the 17th century. In *Solar–terrestrial influences on weather and climate*, B. M. McCormac and T. A. Seliga (eds), 125–44. Dordrecht: D. Reidel.

Mitterer, R. M. 1974. Pleistocene stratigraphy in southern Florida based on amino acid diagenesis in fossil *Mercenaria*. *Geology* **2**, 425–9.

Molfino, B., N. G. Kipp, and J. J. Morley 1982. Comparison of foraminiferal, coccolithophorid and Radiolarian paleotemperature equations: assemblage coherency and estimate concordancy. *Quat. Res.* **17**, 279–313.

Monod, Th. 1963. The late Tertiary and Pleistocene in the Sahara. In *African ecology and human evolution*, F. C. Howell and F. Boulière (eds), 117–229, Viking Publications in Anthropology No. 36. New York: Wenner-Gren Foundation.

Moody, D. W. and A. J. W. Catchpole 1975. *Environmental data from historical documents by content analysis: freeze-up and break-up of estuaries on Hudson Bay, 1714–1871*. Manitoba Geographical Studies No. 5. Winnipeg: University of Winnipeg.

Moore, P. D. and J. A. Webb 1978. *An illustrated guide to pollen analysis*. London: Hodder and Stoughton.

Moore, T. C. 1978. The distribution of radiolarian assemblages in the modern and ice-age Pacific. *Marine Micropaleont.* **4**, 229–66.

Moore, T. C., L. H. Burckle, K. Geitzenauer, B. Luz, A. Molina-Cruz, J. H. Robertson, H. Sachs, C. Sancetta, J. Thiede, P. Thompson, and C. Wenkam 1980. The reconstruction of sea surface temperatures in the Pacific Ocean of 18 000 BP. *Marine Micropaleont.* **5**, 215–47.

Moreau, R. E. 1963. Vicissitudes of the African biomes in the late Pleistocene. *Proc. Zool. Soc. Lond.* **141**, 395–421.

Morgan, Anne 1973. Late Pleistocene environmental changes indicated by fossil insect faunas of the English Midland. *Boreas* **2**, 173–212.

Morgan, A. V. and Anne Morgan 1979. The fossil coleoptera of the Two Creeks forest bed, Wisconsin. *Quat. Res.* **12**, 226–40.

Morgan, A. V. and A. Morgan 1981. Paleoentomological methods of reconstructing paleoclimate with reference to interglacial and interstadial insect faunas of southern Ontario. In *Quaternary paleoclimate*, W. C. Mahaney (ed.) 173–92. Norwich: University of East Anglia.

Morgan, V. I. 1982. Antarctic Ice Sheet surface oxygen isotope values. *J. Glaciol.* **28**, 315–23.

Morley, J. J. and J. D. Hays 1979. Comparison of glacial and interglacial oceanographic conditions in the South Atlantic from variations in calcium carbonate and radiolarian distributions. *Quat. Res* **12**, 396–408.

Morley, J. J. and J. D. Hays 1981. Towards a high-resolution, global, deep-sea chronology for the last 750 000 years. *Earth Planet. Sci. Letters* **53**, 279–95.

Morley, J. J. and N. J. Shackleton 1978. Extension of the radiolarian *Stylatractus universus*, as a biostratigraphic datum to the Atlantic Ocean. *Geology* **6**, 309–11.

Morrison, R. 1965. Quaternary geology of the Great Basin. In *The Quaternary of the United States*, H. E. Wright and D. G. Frey (eds), 265–86. Princeton, NJ: Princeton University Press.

Morrison, R. B. 1969. Quaternary geology of the Great Basin. In *The Quaternary of the United States*, H. E. Wright and D. G. Frey (eds), 265–85. Princeton, N.J.: Princeton University Press.

Morton, F. I. 1967. Evaporation from large, deep lakes. *Water Resour. Res.* **3**, 181–200.

Mosley-Thompson, E. and L. G. Thompson 1982. Nine centuries of microparticle deposition at the South Pole. *Quat. Res.* **17**, 1–13.

Mott, R. J. 1975. Palynological studies of lake sediment profiles from southwestern New Brunswick. *Can. J. Earth Sci.* **12**, 273–88.

Müller, F. 1958. Eight months of glacier and soil research in the Everest region. In *The mountain world*, M. Barnes (ed.), 191–208. New York: Harper and Bros.

Mullineaux, D. R. 1974. *Pumice and other pyroclastic deposits in Mount Rainier National Park, Washington*, US Geological Survey Bulletin 1326. Washington, D.C.: US Geological Survey.

Munn, R. E. and L. Machta 1979. Human activities that affect climate In *Proceedings of the World Climate Conference*, 170–209, WMO No. 537. Geneva: World Meteorological Organization.

Naeser, C. W., N. D. Briggs, J. D. Obradovich, and G. A. Izett 1981. Geochronology of tephra deposits. In *Tephra studies*, S. Self and R. J. S. Sparks (eds), 13–47. Dordrecht: D. Reidel.

Neftel, A., H. Oeschger, J. Schwander, B. Stauffer, and R. Zumbrunn 1982. Ice core sample measurements give atmospheric CO_2 content during the past 40 000 years. *Nature, Lond.* **295**, 220–3.

Neuberger, H. 1970. Climate in art. *Weather* **25**, 46–56.

Newell, R. E. 1974. Changes in the poleward energy flux by the atmosphere and ocean as a possible cause for ice ages. *Quat. Res.* **4**, 117–27.

Newell, R. E. and L. S. Chiu. 1981. Climatic changes and variations: a geophysical problem. In *Climate variations and variability: facts and theories*, A. Berger (ed.), 21–61. Dordrecht: D. Reidel.

Newell, R. E., G. F. Herman, S. Gould-Stewart, and M. Tanaka 1975. Decreased global rainfall during the past Ice Age. *Nature, Lond.* **253**, 33–4.

Newson, R. 1973. Response of a general circulation model of the atmosphere to removal of the Arctic ice-cap. *Nature, Lond.* **241**, 39–40.

Nichols, H. 1967. The postglacial history of vegetation and climate at Ennadai Lake, Keewatin and Lynn Lake, Manitoba. *Eiszeitalter und Gegenwart* **18**, 176–97.

Nichols, H. 1975. *Palynological and paleoclimatic study of the late Quaternary displacement of the Boreal forest-tundra ecotone in Keewatin and Mackenzie, NWT, Canada*, INSTAAR Occasional Paper No. 15. Boulder, Colo.: University of Colorado.

Nichols, H. 1976. Historical aspects of the northern Canadian treeline. *Arctic* **29**, 38–47.

Nicholson, S. and H. Flohn 1980. African environmental and climatic changes and the general circulation in late Pleistocene and Holocene. *Climatic Change* **2**, 313–48.

Ninkovich, D. and N. J. Shackleton 1975. Distribution, stratigraphic position and age of ash layer "L" in the Panama Basin region. *Earth Planet. Sci. Letters* **27**, 20–34.

Nix, H. A. and J. D. Kalma 1972. Climate as a dominant control in the biogeography of northern Australia and New Guinea. In *Bridge and barrier: the natural and cultural history of Torres Strait*, D. Walker (ed.), 61–91, Department of Biogeography and Geomorphology, Publication No. BG3. Canberra: Australian National University.

Nye, J. F. 1959. The motion of ice sheets and glaciers. *J. Glaciol.* **3**, 493–507.

Nye, J. F. 1965. A numerical method of inferring the budget history of a glacier from its advance and retreat. *J. Glaciol.* **5**, 589–607.

Oerlemans, J. 1981. Modeling of Pleistocene European Ice Sheets: some experiments with simple mass balance parameterizations. *Quat. Res.* **15**, 77–85.

Oeschger, H., B. Alder, H. Loosli, C. C. Langway, Jr, and A. Renaud 1966. Radiocarbon dating of ice. *Earth Planet. Sci. Letters* **1**, 49–54.

Oeschger, H., B. Stauffer, P. Bucher, and H. H. Loosli 1977. Extraction of gases and dissolved and particulate matter from ice in deep boreholes. In *Isotopes and impurities in snow and ice*, 307–11, International Association of Scientific Hydrology Publication No. 118. Washington, D.C.: IASH.

Olausson, E. 1965. Evidence of climatic changes in deep sea cores with remarks on isotopic palaeotemperature analysis. *Progr. Oceanogr.* **3**, 221–52.

Olausson, E. 1967. Climatological, geoeconomical and paleooceanographical aspects of carbonate deposition. *Progr. Oceanogr.* **4**, 245–65.

Olsson, I. U. 1968. Modern aspects of radiocarbon dating. *Earth Sci. Rev.* **4**, 203–18.

Olsson, I. U. 1974. Some problems in connection with the evaluation of ^{14}C dates. *Geol. Fören. Stockholm Förhandl.* **96**, 311–20.

Olsson, I. U. and F. A. N. Osadebe 1974. Carbon isotope variations and fractionation corrections in ^{14}C dating. *Boreas* **3**, 139–46.

Olsson, I. U., M. F. A. F. El-Daoushy, A. I. Abd-El-Mageed, and M. Klasson 1974. A comparison of different methods for pretreatment of bones I. *Geol. Fören. Stockholm Förhandl.* **96**, 171–81.

Opdyke, N. D. 1972. Paleomagnetism of deep-sea cores. *Rev. Geophys. Space Phys.* **101**, 213–49.

Opdyke, N. D., B. Glass, J. D. Hays, and J. Foster 1966. Paleomagnetic study of Antarctic deep-sea cores. *Science* **154**, 349–57.

Osborne, P. J. 1974. An insect assemblage of early Flandrian age from Lea Marston, Warwickshire, and its bearing on the contemporary climate and ecology. *Quat. Res.* **4**, 471–86.

Osborne, P. J. 1980. The late Devensian-Flandrian transition depicted by serial insect faunas from West Bromwich, Staffordshire, England. *Boreas* **9**, 139–47.

Osmaston, H. A. 1975. Models for the estimation of firnlines of present and Pleistocene glaciers. In *Processes in physical and human geography: Bristol essays,* R. F. Peel, M. Chisholm, and P. Haggett (eds), 218–45. London: Heinemann.

Osmond, J. K. 1979. Accumulation models of ^{230}Th and ^{231}Pa in deep sea sediments. *Earth Sci. Rev.* **15**, 95–150.

Østrem, G. 1974. Present alpine ice cover. In *Arctic and alpine environments,* J. D. Ives and R. G. Barry (eds), 225–50. London: Methuen.

Palmer, W. C. 1965. *Meteorological drought,* US Weather Bureau Research Paper No. 45. Washington, D. C.: US Weather Bureau.

Parker, B. C. and E. J. Zeller 1980. Nitrogenous chemical composition of antarctic ice and snow. *Antarctic J. U.S.* **15**, 79–81.

Parker, B. C., E. J. Zeller, and A. J. Gow 1981. Nitrogenous chemical composition of antarctic ice and snow. *Antarctic J. U.S.* **16**, 79–81.

Parker, F. L. and W. H. Berger 1971. Faunal and solution patterns of planktonic foraminifera in surface sediments of the South Pacific. *Deep-sea Res.* **18**, 73–107.

Parker, M. L. 1971. *Dendrochronological techniques used by the Geological Survey of Canada,* Geological Survey of Canada Paper 71-25. Ottawa: Geological Survey of Canada.

Parker, M. L. and W. E. S. Hennoch 1971. The use of Engelmann spruce latewood density for dendrochronological purposes. *Can. J. Forest Res.* **1**, 90–8.

Parkin, D. W. 1974. Trade winds during glacial cycles. *Proc. R. Soc. A* **337**, 73–100.

Parkin, D. W. and N. J. Shackleton 1973. Trade wind and temperature correlations down a deep-sea core off the Saharan coast. *Nature, Lond.* **245**, 455–7.

Parmenter, C. and D. W. Folger 1974. Eolian biogenic detritus in deep sea sediments: a possible index of Equatorial Ice Age aridity. *Science* **185**, 695–8.

Parry, M. L. 1975. Secular climatic change and marginal agriculture. *Trans Inst. Br. Geogrs* No. 64, 1–13.

Parry, M. L. 1981. Climatic change and the agricultural frontier: a research strategy. In *Climate and history,* T. M. L. Wigley, M. J. Ingram, and G. Farmer (eds), 319–36. Cambridge: Cambridge University Press.

Paterson, W. S. B. 1977. Extent of the late-Wisconsin glaciation in northwest Greenland and northern Ellesmere Island: a review of the glaciological and geological evidence. *Quat. Res.* **8**, 180–90.

Paterson, W. S. B. 1981. *The physics of ice,* 2nd edn. Oxford: Pergamon.

Paterson, W. S. B., R. M. Koerner, D. Fisher, S. J. Johnsen, H. B. Clausen, W. Dansgaard, P. Bucher, and H. Oeschger 1977. An oxygen isotope climatic record from the Devon Island Ice Cap, Arctic Canada. *Nature, Lond.* **266**, 508–11.

Patzelt, G. 1973. Die neuzeitlichen Gletscherschwankungen in der Vendigeruppe (Hohe Tauern, Ostalpen). *Z. GletscherKde Glazialgeol.* **9**, 5–57.

Patzelt, G. 1974. Holocene variations of glaciers in the Alps. *Colloques Int. Centre Natn Rech. Scient.* **219**, 51–9.

Pearman, G. I., R. J. Francey, and P. J. B. Fraser 1976. Climatic implications of stable carbon isotopes in tree rings. *Nature, Lond.* **260**, 771–3.

Pears, N. V. 1969. Post-glacial tree-lines of the Cairngorm Mountains, Scotland: some modifications based on radiocarbon dating. *Trans Bot. Soc. Edinb.* **40**, 536–44.

Pearson, G. W., J. R. Pilcher, M. G. L. Baillie, and J. Hillam 1977. Absolute radiocarbon dating using a low altitude European tree-ring calibration. *Nature, Lond.* **270**, 25–8.

Pennington, W. 1973. Absolute pollen frequencies in the sediments of lakes of different morphometry. In *Quaternary plant ecology,* H. J. B. Birks and R. West (eds), 79–104. Oxford: Blackwell Scientific Publications.

Perry, D. A. 1977. Oxygen isotope ratios in spruce cellulose. *Nature, Lond.* **266**, 476–7.

Petersen, G. M., T. Webb, J. E. Kutzbach, T. van der Hammen, T. A. Wijmstra, and F. A. Street 1979. The continental record of environmental conditions at 18 000 yr BP: an initial evaluation. *Quat. Res.* **12**, 47–82.

Petit, J.-R, M. Briat, and R. A. Royer 1981. Ice age aerosol content from East Antarctic ice core samples and past wind strength. *Nature, Lond.* **293**, 391–4.

Péwé, T. L. and R. D. Reger, 1972. Modern and Wisconsinan snowlines in Alaska. In *Proceedings No. 24, Section 12, Quaternary Geology*, 187–97. Montreal: International Geological Congress.

Pfister, C. 1977. Zum klima des Raumes Zürich im späten 17 und frühen 18 Jahrhundert. *Vierteljahresheft Naturforsch. Ges. Zürich* **122**, 447–71.

Pfister, C. 1978a. Climate and economy in eighteenth century Switzerland. *J. Interdisc. Hist.* **9**, 223–43.

Pfister, C. 1978b. Fluctuations in the duration of snow-cover in Switzerland since the late seventeenth century. In *Proceedings of the Nordic Symposium on Climatic Changes and Related Problems*, K. Frydendahl (ed.), 1–6, Danish Meteorological Institute Climatological Papers No. 4. Copenhagen: Danish Meteorological Institute.

Pfister, C. 1981. An analysis of the Little Ice Age climate in Switzerland and its consequences for agricultural production. In *Climate and history*, T. M. L. Wigley, M. J. Ingram, and G. Farmer (eds), 214–48. Cambridge: Cambridge University Press.

Picciotto, E., X. De Maere, and I. Friedman 1960. Isotopic composition and temperature of formation of Antarctic snows. *Nature, Lond.* **187**, 857–9.

Picciotto, E., G. Crozaz, and W. De Breuck 1971. Accumulation on the South Pole – Queen Maud Land Traverse 1964–1968. In *Antarctic snow and ice studies II*, A. P. Crary (ed.), 257–316, Antarctic Research Series. Washington, D.C.: American Geophysical Union.

Pierce, K. L., J. D. Obradovich, and I. Friedman 1976. Obsidian hydration dating and correlation of Bull Lake and Pinedale Glaciations near west Yellowstone, Montana. *Geol. Soc. Am. Bull.* **87**, 703–10.

Pilcher, J. T., J. Hillam, M. G. L. Baillie, and G. W. Pearson 1977. A long sub-fossil oak tree-ring chronology from the North of Ireland. *New Phytol.* **79**, 713–29.

Pillow, M. Y. 1931. Compression wood records hurricane. *J. Forestry* **29**, 575–8.

Pisias, N. G. and T. C. Moore 1981. The evolution of Pleistocene climate: a time series approach. *Earth Planet. Sci. Letters* **52**, 450–8.

Pittock, A. B. 1978. Long-term sun – weather relations. *Rev. Geophys. Space Phys.* **16**, 400–20.

Polge, H. 1970. The use of X-ray densitometric methods in dendrochronology *Tree Ring Bull.* **30**, 1–10.

Pollard, D. 1982. A simple ice sheet model yields realistic 100 000 year glacial cycles. *Nature, Lond.* **296**, 334–8.

Porter, S. C. 1977. Present and past glaciation thresholds in the Cascade Range, Washington, USA: topographic and climatic controls and paleoclimatic implications. *J. Glaciol.* **18**, 101–16.

Porter, S. C. 1978. Glacier Peak tephra in the North Cascade Range, Washington: stratigraphy, distribution and relationship to late-glacial events. *Quat. Res.* **10**, 30–41.

Porter, S. C. 1979. Hawaiian glacial ages. *Quat. Res.* **12**, 161–87.

Porter, S. C. 1981a. Glaciological evidence of Holocene climatic change. In *Climate and history*, T. M. L. Wigley, M. J. Ingram, and G. Farmer (eds), 82–110. Cambridge: Cambridge University Press.

Porter, S. C. 1981b. Recent glacier variations and volcanic eruptions. *Nature* **291**, 139–42.

Porter, S. C. 1981c. Tephrochronology in the Quaternary geology of the United States. In *Tephra studies*, S. Self and R. J. S. Sparks (eds), 135–60. Dordrecht: D. Reidel.

Porter, S. C. 1981d. Lichenometric studies in the Cascade Range of Washington: establishment of *Rhizocarpon geographicum* growth curves at Mount Rainier. *Arctic Alpine Res.* **13**, 11–23.

Porter, S. C. and G. H. Denton 1967. Chronology of neoglaciation in the North American Cordillera. *Am. J. Sci.* **265**, 177–210.

Potter, G. L., H. W. Elsaesser, M. MacCracken, and F. M. Ruther 1975. Possible impact of tropical deforestation. *Nature, Lond.* **258**, 697–8.

Potter, N. 1969. Tree-ring dating of snow avalanche tracks and the geomorphic activity of avalanches, northern Absaroka Mountains, Wyoming. In *US Contributions to Quaternary Research*, S. A. Schumm and W. C. Bradley (eds), 141–65, Geographical Society of America Special Paper 123. Boulder, Colo.: Geological Society of America.

Prance, G. T. 1974. Phytogeographic support for the theory of Pleistocene forest refuges in the Amazon Basin, based on evidence from distribution patterns in Caryocaraceae, Chrysobalanaceae, Dichapetalaceae and Lecythidaceae. *Acta Amazonica* **3**, 5–26.

Prance, G. T. 1982. Forest refuges: evidence from woody angiosperms. In *Biological diversification in the Tropics*, G. T. Prance (ed.), 137–58. New York: Columbia University Press.

Prell, W. L. and W. H. Huston 1979. Zonal temperature-anomaly maps of Indian Ocean surface waters: modern and ice age patterns. *Science* **206**, 454–6.

Prell, W. L., J. V. Gardner, A. W. H. Bé, and J. D. Hays 1976. Equatorial Atlantic and Caribbean foraminiferal assemblages, temperatures and circulation: interglacial and glacial comparisons. In *Investigation of late Quaternary paleooceanography and paleoclimatology*, R. M. Cline and J. H. Hays (eds), 247–66, Geological Society of America Memoir 145. Boulder, Colo.: Geological Society of America.

Prell, W. L., W. H. Hutson, D. F. Williams, A. W. H. Bé, K. Keitzenauer, and B. Molfino 1980. Surface circulation of the Indian Ocean during the last glacial maximum approximately 18 000 yr BP. *Quat. Res.* **14**, 309–36.

Prentice, I. C. 1978. Modern pollen spectra from lake sediments in Finland and Finnmark, North Norway. *Boreas* **7**, 131–53.

Ramanathan, V. 1981. The role of ocean – atmosphere interactions in the CO_2 climate problem. *J. Atmos. Sci.* **38**, 918–30.

Rampton, V. 1971. Late Quaternary vegetational and climatic history of the Snag – Klutlan area, southwestern Yukon Territory, Canada. *Geol. Soc. Am. Bull.* **82**, 959–78.

Rateev, M. A., Z. N. Gorbunova, A. P. Lisitzyn, and G. L. Nosov 1969. The distribution of clay minerals in the oceans. *Sedimentology* **13**, 21–43.

Raynaud, D. and B. Lebel 1979. Total gas content and surface elevation of polar ice sheets. *Nature, Lond.* **281**, 289–91.

Raynaud, D. and C. Lorius 1973. Climatic implications of total gas content in ice at Camp Century. *Nature, Lond.* **243**, 283–4.

Raynaud, D. and I. M. Whillans 1979. Total gas content of ice and past changes of the northwest Greenland Ice Sheet. In *Sea level, ice and climatic change*, 235–6, International Association of Scientific Hydrology Publication No. 131. Washington, D.C.: IASH.

Reeh, N., H. B. Clausen, W. Dansgaard, N. Gundestrup, C. U. Hammer, and S. J. Johnsen 1978. Secular trends of accumulation rates at three Greenland stations. *J. Glaciol.* **20**, 27–30.

Reeves, C. C. 1965. Pleistocene climate of the Llano Estacado. *J. Geol.* **73**, 181–9.

Reid, G. C., I. S. A. Isaksen, T. E. Holzer, and P. J. Crutzen 1976. Influence of ancient solar-proton events on the evolution of life. *Nature, Lond.* **259**, 177–9.

Richmond, G. M. 1965. Glaciation of the Rocky Mountains. In *The Quaternary of the United States*, H. E. Wright and D. G. Frey (eds), 217–30. Princeton, N.J.: Princeton University Press.

Ritchie, J. C. 1976. The late-Quaternary vegetational history of the western interior of Canada. *Can. J. Bot.* **54**, 1793–818.

Ritchie, J. C. 1977. The modern and late Quaternary vegetation of the Campbell–Dolomite Uplands near Inuvik, NWT, Canada. *Ecol. Monogr.* **47**, 410–23.

Ritchie, J. C. and F. K. Hare 1971. Late Quaternary vegetation and climate near the Arctic treeline of northwestern North America. *Quat. Res.* **1**, 331–42.

Ritchie, J. C. and S. Lichti-Federovich 1967. Pollen dispersal phenomena in arctic–subarctic Canada. *Rev. Paleobot. Palynol.* **3**, 255–66.

Robin, G. de Q. 1977. Ice cores and climatic change. *Phil. Trans. R. Soc. B* **280**, 143–68.

Robinson, W. J. 1976. Tree-ring dating and archaeology in the American Southwest. *Tree Ring Bull.* **36**, 9–20.

Robock, A. 1978. Internal and externally caused climate change. *J. Atmos. Sci.* **35**, 1111–22.

Rognon, P. 1976. Essai d'interpretation des variations climatiques au Sahara depuis 40 000 ans. *Revue Géogr. Phys. Géol. Dynam.* **18**, 251–82.

Rognon, P. and M. A. J. Williams, 1977. Late Quaternary climatic changes in Australia and North Africa: a preliminary interpretation. *Palaeogeogr. Palaeoclimatol. Palaeoecol.* **21**, 285–327.

Rotberg, R. I. and T. K. Rabb (eds) 1981. *Climate and history: studies in interdisciplinary history*. Princeton, N.J.: Princeton University Press.

Röthlisberger, F. 1976. Gletscher- und Klimaschwankungen im Raun Zermatt, Ferpècle und Arolla. *Die Alpen* **52**, 59–132.

Röthlisberger, F., P. Haas, H. Holzhauser, W. Keller, W. Bircher, and F. Renner 1980. Holocene glacier fluctuations – radiocarbon dating of fossil soils (FAL) and woods from moraines and glaciers in the Alps. *Geogr. Helv.* **35**, 21–52.

Royer, J. F., M. Deque, and P. Pestiaux 1984. A sensitivity experiment to astronomical forcing with a spectral GCM: simulation of the annual cycle at 125 000 BP and 115 000 BP. In *Milankovitch and climate*, A. L. Berger, J. Imbrie, J. Hays, G. Kukla and B. Saltzman (eds). Dordrecht: D. Reidel.

Ruddiman, W. F. 1971. Pleistocene sedimentation in the equatorial Atlantic: stratigraphy and faunal paleoclimatology. *Geol. Soc. Am. Bull.* **82**, 283–302.

Ruddiman, W. F. 1977a. Investigations of Quaternary climate based on planktonic foraminifera. In *Oceanic micropaleontology*, A. T. S. Ramsay (ed.), 101–61. New York: Academic Press.

Ruddiman, W. F. 1977b. Late Quaternary deposition of ice-rafted sand in the sub-polar North Atlantic (lat. 40–65°N). *Geol. Soc. Am. Bull.* **88**, 1813–27.

Ruddiman, W. F. 1977c. North Atlantic ice-rafting: a major change at 75 000 years before the present. *Science* **196**, 1208–11.

Ruddiman, W. F. and B. C. Heezen 1967. Differential solution of planktonic foraminifera. *Deep-sea Res.* **14**, 801–8.

Ruddiman, W. F. and A. McIntyre 1976. Northeast Atlantic paleoclimatic changes over the past 600 000 years. In *Investigation of late Quaternary paleooceanography and paleoclimatology*, R. M. Cline and J. D. Hays (eds), 111–46, Geological Society of America Memoirs 145. Boulder, Colo.: Geological Society of America.

Ruddiman, W. F. and A. McIntyre 1979. Warmth of the subpolar North Atlantic Ocean during Northern Hemisphere ice-sheet growth. *Science* **204**, 173–5.

Ruddiman, W. F. and A. McIntyre 1981a. Oceanic mechanisms for amplification of the 23 000 year ice-volume cycle. *Science* **212**, 617–27.

Ruddiman, W. F. and A. McIntyre 1981b. The mode and mechanism of the last deglaciation: oceanic evidence. *Quat. Res.* **16**, 125–34.

Ruddiman, W. F. and A. McIntyre 1981c. The North Atlantic Ocean during the last glaciation. *Palaeogeogr. Palaeoclimatol. Palaeoecol.* **35**, 145–214.

Ruddiman, W. F., A. McIntyre, V. Niebler-Hunt, and J. T. Durazzi 1980a. Oceanic evidence for the mechanism of rapid northern hemisphere glaciation. *Quat. Res.* **13**, 33–64.

Ruddiman, W. F., B. Molfino, A. Esmay, and E. Pokras 1980b. Evidence bearing on the mechanism of rapid deglaciation. *Climatic Change* **3**, 65–87.

Rutter, N. W., R. J. Crawford, and R. D. Hamilton 1979. Dating methods of Pleistocene deposits and their problems: IV. Amino acid racemization dating. *Geosci. Can.* **6**, 122–8.

Sachs, M. H., T. Webb, and D. R. Clark 1977. Paleoecological transfer functions. *Ann. Rev. Earth Planet. Sci.* **5**, 159–78.

Salgado-Labouriau, M. L. 1979. Modern pollen deposition in the Venezuelan Andes. *Grana* **18**, 53–68.

Salgado-Labouriau, M. L., C. Schubert, and S. Valastro Jr. 1978. Paleoecologic analysis of a Late-Quaternary terrace from Mucubaji, Venezuelan Andes. *J. Biogeogr.* **4**, 313–25.

Sancetta, C. 1979. Oceanography of the North Pacific during the last 18 000 years: evidence from fossil diatoms. *Marine Micropalaeont.* **4**, 103–23.

Sancetta, C. 1983. Fossil diatoms and the oceanography of the Bering Sea during the last glacial event. In *Siliceous deposits in the Pacific region*, A. Iijima, J. R. Hein and R. Fiever (eds.), 333–46. Amsterdam: Elsevier.

Sancetta, C. and S. W. Robinson 1983. Diatom evidence on Wisconsin and Holocene events in the Bering Sea. *Quat. Res.* **20**, 232–45.

Sancetta, C., J. Imbrie, and N. G. Kipp 1973a. Climatic record of the past 130 000 years in the North Atlantic deep-sea core V23-83: correlation with the terrestrial record. *Quat. Res.* **3**, 110–16.

Sancetta, C., J. Imbrie, and N. G. Kipp 1973b. The climatic record of the past 14 000 years in North Atlantic deep-sea core V23-82: correlation with the terrestrial record. In *Mapping the atmospheric and oceanic circulations and other climatic parameters at the time of the last glacial maximum about 17 000 years ago*, Climatic Research Unit Publication No. 2, 62–5. Norwich: University of East Anglia.

Sarntheim, M. 1978. Sand deserts during glacial maximum and climatic optimum. *Nature, Lond.* **272**, 43–6.

Savin, S. M. and F. G. Stehli 1974. Interpretation of oxygen isotope paleotemperature measurements: effect of the $^{18}O/^{16}O$ ratio of sea water depth stratification of foraminifera, and selective solution. In *Les méthodes quantitatives d'étude des variations du climat au cours du pleistocène*, 183–191, Colloques Internationaux du Centre National de la Recherche Scientifique No. 219. Paris: CNRS.

Schmidt, F. H. 1967. Palynology and meteorology. *Rev. Paleobot. Palynol.* **3**, 27–45.

Schneebeli, W. 1976. Untersuchungen von Gletscherschwankungen in Val de Bagnes. *Die Alpen* **52**, 5–58.

Schneider, S. 1975. On the carbon dioxide–climate confusion. *J. Atmos. Sci.* **32**, 2060–6.

Schneider, S. H. and R. E. Dickinson 1974. Climate modeling. *Rev. Geophys. Space Phys.* **12**, 447–93.

Schneider, S. H. and L. E. Mesirow 1976. *The genesis strategy: climate and global survival.* New York: Plenum Press.

Schneider, S. H. and R. L. Temkin 1978. Climatic changes and human affairs. In *Climatic change*, J. Gribbin (ed.), 228–46. Cambridge: Cambridge University Press.

Schneider, S. H., W. W. Kellogg, and V. Ramanathan 1980. Carbon dioxide and climate. *Science* **210**, 6–7.

Scholander, P. F., H. Devries, W. Dansgaard, L. K. Coachman, D. C. Nutt, and E. Hemmingsen 1962. Radiocarbon age and oxygen-18 content of Greenland icebergs. *Meddr Grønland* **165** (1).

Schott, W. 1935. Die Foraminiferen in dem aequatorialen Teil des Atlantischen Ozeans. Deutsche Atlantische Expedition "Meteor" 1925–1927. *Wiss. Ergebn.* **3** (3), 43–134.

Schroeder, R. A. and J. L. Bada 1973. Glacial–postglacial temperature difference deduced from aspartic acid racemization in fossil bones. *Science* **182**, 479–82.

Schroeder, R. A. and J. L. Bada 1976. A review of the geochemical applications of the amino acid racemization reaction. *Earth Sci. Rev.* **12**, 347–91.

Schwarcz, H. P., R. S. Harmon, P. Thompson, and D. C. Ford 1976. Stable isotope studies of fluid inclusions in speleothems and their paleoclimatic significance. *Geochim. Cosmochim. Acta* **40**, 657–65.

Schweingruber, F. H., H. C. Fritts, O. U. Bräker, L. G. Drew, and E. Schär 1978. The X-ray technique as applied to dendroclimatology. *Tree Ring Bull.* **38**, 61–91.

Schweingruber, F. H., O. U. Bräker, and E. Schär 1979. Dendroclimatic studies on conifers from central Europe and Great Britain. *Boreas* **8**, 427–52.

Sears, P. B. 1964. The goals of paleoecological reconstruction. In *The reconstruction of past environments*, J. H. Hester and J. Schoenwetter (ed), 4–5, Fort Burgwin Research Center Publication No. 3. Fort Burgwin, New Mexico: Fort Burgwin Research Center.

Sekiguti, T. 1969. The historical dates of Japanese cherry festivals since the 8th century and her climatic changes. *Geogr. Rev. Japan* **35**, 67–76.

Self, S. and R. J. S. Sparks (eds) 1981. *Tephra studies* (Proceedings of the NATO Advanced Studies Institute, "Tephra Studies as a Tool in Quaternary Research"). Dordrecht: D. Reidel.

Serre, F. 1978. The dendroclimatological value of the European Larch (*Larix decidua* Mill.) in the French Maritime Alps. *Tree Ring Bull.* **38**, 25–34.

Seurat, L. G. 1934. *Etudes zoologiques sur le Sahara Central.* Memoires de la Société d'Histoire Naturelle de l'Afrique du Nord, No. 4, Mission du Hoggar III.

Shackleton, N. J. 1967. Oxygen isotope analyses and Pleistocene temperatures re-assessed. *Nature* **215**, 15–17.

Shackleton, N. J. 1969. The last interglacial in the marine and terrestrial records. *Proc. R. Soc. B* **174**, 135–54.

Shackleton, N. J. 1974. Attainment of isotopic equilibrium between ocean water and the benthonic foraminifera Genus *Uvigerina*: isotopic changes in the ocean during the last glacial. In *Les méthodes quantitatives d'étude des variations du climat au cours du Pleistocène*, 4–5, Colloques Internationaux de Centre National de la Recherce Scientifique No. 219. Paris: CNRS.

Shackleton, N. J. 1977a. The oxygen isotope stratigraphic record of the late Pleistocene. *Phil. Trans R. Soc. B* **280**, 169–79.

Shackleton, N. J. 1977b. Carbon-13 in *Uvigerina:* tropical rainforest history and the equatorial Pacific carbonate dissolution cycles. In *The fate of fossil fuel CO_2 in the oceans*, N. R. Anderson and A. Malahoff (eds), 401–28. New York: Plenum Press.

Shackleton, N. J. and R. K. Matthews 1977. Oxygen isotope stratigraphy of late Pleistocene coral terraces in Barbados. *Nature, Lond.* **268**, 618–20.

Shackleton, N. J. and N. D. Opdyke 1973. Oxygen isotope and paleomagnetic stratigraphy of equatorial Pacific core V28-238: oxygen isotope temperatures and ice volumes on a 10^5 year and 10^6 year scale. *Quat. Res.* **3**, 39–55.

Shackleton, N. J. and N. D. Opdyke 1976. Oxygen-isotope and paleomagnetic stratigraphy of Pacific core V28-239. Late Pliocene to latest Pleistocene. In *Investigation of late Quaternary paleooceanography and paleoclimatology*, R. M. Cline and J. D. Hays (eds), 449-64, Geological Society of America Memoir 145. Boulder, Colo.: Geological Society of America.

Shackleton, N. J., J. D. H. Wiseman, and H. A. Buckley 1973. Non-equilibrium isotopic fractionation between sea-water and planktonic foraminiferal tests. *Nature* **242**, 177–9.

Sheets, P. D. and D. K. Grayson (eds) 1979. *Volcanic activity and human ecology*. New York: Academic Press.

Shotton, F. W. 1972. An example of hard water error in radiocarbon dating of vegetable matter. *Nature, Lond.* **240**, 460–1.

Shroder, J. F. 1980. Dendrogeomorphology: review of new techniques of tree-ring dating. *Progr. Phys. Geogr.* **4**, 161–88.

Sigafoos, R. S. and E. L. Hendricks 1961. *Botanical evidence of the modern history of Nisqually Glacier, Washington*, US Geological Survey Professional Paper 387-A. Washington, D.C.: US Geological Survey.

Simpson, I. M. and R. G. West 1958. On the stratigraphy and palaeobotany of a late-Pleistocene organic deposit at Chelford, Cheshire. *New Phytol.* **57**, 239–50.

Smagornisky, J. 1981. CO_2 and climate – a continuing study. In *Climatic variations and variability*. A. Berger (ed.), 661–87. Dordrecht: D. Reidel.

Smith, D. C., H. W. Borns, W. R. Baron, and A. E. Bridges 1981. Climatic stress and Maine agriculture 1785–1885. In *Climate and history*, T. M. L. Wigley, M. J. Ingram, and G. Farmer (eds), 450–64. Cambridge: Cambridge University Press.

Snyder, C. T. and W. B. Langbein 1962. The Pleistocene lake in Spring Valley, Nevada, and its climatic implications. *J. Geophys. Res.* **67**, 2385–94.

Solomon, A. M., T. J. Blasing, and J. A. Solomon 1982. Interpretation of floodplain pollen in alluvial sediments from an arid region. *Quat. Res.* **18**, 52–71.

Sorenson, C. J. 1977. Holocene bioclimates. *Ann. Assoc. Am. Geogr.* **67**, 214–22.

Sorenson, C. J. and J. C. Knox 1974. Paleosols and paleoclimate related to late Holocene forest-tundra border migrations: Mackenzie and Keewatin, NWT. In *International Conference on Prehistory and Paleoecology of Western North American Arctic and Subarctic*, S. Raymond and P. Schledermann (eds), 187–203. Calgary: Archaeological Association, University of Calgary.

Stidd, C. K. 1967. The use of eigenvectors for climatic estimates. *J. Appl. Met.* **6**, 255–64.

Stockmarr, J. 1971. Tablets with spores used in pollen analysis. *Pollen et Spores* **13**, 615–21.

Stockton, C. W. 1975. *Long term streamflow records reconstructed from tree rings*, Laboratory for Tree Ring Research, Paper 5. Tucson, Ariz.: University of Arizona Press.

Stockton, C. W. and W. R. Boggess 1980. Augmentation of hydrologic records using tree rings. In *Improved hydrologic forecasting*, 239–265. New York: American Society of Civil Engineers.

Stockton, C. W. and H. C. Fritts 1973. Long-term reconstruction of water level changes for Lake Athabasca by analysis of tree rings. *Water Resour. Bull.* **9**, 1006–27.

Stokes, M. A. and T. L. Smiley 1968. *An introduction to tree-ring dating*. Chicago: University of Chicago Press.

Stothers, R. 1980. Giant solar flares in antarctic ice. *Nature, Lond.* **287**, 365.

Street, F. A. 1981. Tropical paleoenvironments. *Progr. Phys. Geogr.* **5**, 157–85.

Street, F. A. and A. T. Grove 1976. Environmental and climatic implications of late Quaternary lake-level fluctuations in Africa. *Nature* **261**, 285–390.

Street, F. A. and A. T. Grove 1979. Global maps of lake-level fluctuations since 30 000 yr BP. *Quat. Res.* **12**, 83–118.

Studhalter, R. A. 1955. Tree growth: I. Some historical chapters. *Bot. Rev.* **21**, 1–72.

Stuiver, M. 1965. Carbon-14 content of 18th and 19th century wood: variations correlated with sunspot activity. *Science* **149**, 533–5.

Stuiver, M. 1970. Long-term ^{14}C variations. In *Radiocarbon variations and absolute chronology*, I. U. Olsson (ed.), 197–213. New York: Wiley.

Stuiver, M. 1976. The ^{14}C distribution in West Atlantic abyssal waters. *Earth Planet. Sci. Letters* **32**, 322–30.

Stuiver, M. 1978a. Carbon-14 dating: a comparison of beta and ion counting. *Science* **202**, 881–3.

Stuiver, M. 1978b. Radiocarbon timescale tested against magnetic and other dating methods. *Nature, Lond.* **273**, 271–4.

Stuiver, M. 1980. Solar variability and climatic change during the current millennium. *Nature, Lond.* **286**, 868–87.

Stuiver, M. 1982. A high-precision calibration of the AD radiocarbon time-scale. *Radiocarbon* **24**, 1–26.

Stuiver, M. and P. D. Quay 1980. Changes in atmospheric carbon-14 attributed to a variable sun. *Science* **207**, 11–19.

Stuiver, M., C. J. Heusser, and I. C. Yang 1978. North American glacial history extended to 75 000 years ago. *Science* **200**, 16–21.

Stuiver, M., P. D. Quay, and H. G. Ostlund 1983. Abyssal water carbon-14 distribution and the age of the world oceans. *Science* **219**, 849–51.

Suess, H. E. 1965. Secular variations of the cosmic-ray produced carbon-14 in the atmosphere and their interpretations. *J. Geophys. Res.* **70**, 5937–52.

Suess, H. E. 1970. Bristlecone pine calibration of the radiocarbon time-scale 5200 BC to the present. In *Radiocarbon variation and absolute chronology*, I. U. Olsson (ed.), 595–605. New York: Wiley.

Suess, H. E. 1980. The radiocarbon record in tree rings of the last 8000 years. *Radiocarbon* **22**, 200–9.

Swain, A. M. 1973. A history of fire and vegetation in northeastern Minnesota as recorded in lake sediments. *Quat. Res.* **3**, 383–96.

Swain, A. M. 1978. Environmental changes during the last 2000 years in north-central Wisconsin: analysis of pollen, charcoal and seeds from varved lake sediments. *Quat. Res.* **10**, 55–68.

Szabo, B. J. 1979a. Uranium-series age of coral reef growth on Rottnest Island, western Australia. *Marine Geol.* **29**, M11–M15.

Szabo, B. J. 1979b. ^{230}Th, ^{231}Pa and open system dating of fossil corals and shells. *J. Geophys. Res.* **84**, 4927–30.

Szabo, B. J. and D. Collins 1975. Age of fossil bones from British interglacial sites. *Nature, Lond.* **254**, 680–2.

Szabo, B. J. and J. N. Rosholt 1969. Uranium-series dating of Pleistocene molluscan shells from southern California – an open system model. *J. Geophys. Res.* **74**, 3253–60.

Szabo, B. J., G. H. Miller, J. T. Andrews, and M. Stuiver 1981. Comparison of uranium series, radiocarbon and amino acid data from marine molluscs, Baffin Island, Arctic Canada. *Geology* **9**, 451–7.

Szafer, W. 1935. The significance of isopollen lines for the investigation of geographical distribution of trees in the post-glacial period. *Bull. Int. Acad. Polon. Sci. Lett. B***1**, 235–9.

Tarling, D. H. 1975. Archeomagnetism: the dating of archaeological materials by their magnetic properties. *World Archaeol.* **7**, 185–97.

Tarling, D. H. 1978. The geological–geophysical framework of ice ages. In *Climatic change*, J. Gribbin (ed.), 3–24. Cambridge: Cambridge University Press.

Tarr, R. S. 1897. Difference in the climate of the Greenland and American side of Davis' and Baffin's Bay. *Am. J. Sci.* **3**, 315–20.

Tauber, H. 1965. Differential pollen dispersion and the interpretation of pollen diagrams. *Danmarks Geol. Undersogelse*, Series II **89**.

Tauber, H. 1967. Differential pollen dispersion and filtration. In *Quaternary paleoecology*, E. J. Cushing and H. E. Wright (eds), 131–41. New Haven, Conn.: Yale University Press.

Tauber, H. 1970. The Scandinavian varve chronology and ^{14}C dating. In *Radiocarbon variation and absolute chronology*, I. U. Olsson (ed.), 173–96. New York: Wiley.

Ten Brink, N. W. 1973. Lichen growth rates in west Greenland. *Arctic Alpine Res.* **5**, 323–31.

Thierstein, H. R., K. R. Geitzenauer, B. Molfino, and N. J. Shackleton 1977. Global synchroneity of late Quaternary coccolith datum levels: validation by oxygen isotopes. *Geology* **5**, 400–4.

Thompson, L. G. and E. Mosley-Thompson 1981. Microparticle concentration variations linked with climatic change: evidence from polar ice cores. *Science* **212**, 812–15.

Thompson, L. G., W. L. Hamilton, and C. Bull 1975. Climatological implications of microparticle concentrations in the ice core from Byrd Station, western Antarctica. *J. Glaciol.* **14**, 433–44.

Thompson, P. R. and T. Saito 1974. Pacific Pleistocene sediments: planktonic foraminifera dissolution cycles and geochronology. *Geology* **2**, 333–5.

Thompson, P., H. P. Schwarcz, and D. C. Ford 1976. Stable isotope geochemistry, geothermometry and geochronology of speleothems from West Virginia. *Geol. Soc. Am. Bull.* **87**. 1730–8.

Thompson, R. 1977. Stratigraphic consequences of palaeomagnetic studies of Pleistocene and Recent sediments. *J. Geol. Soc. Lond.* **133**, 51–9.

Thompson, R. and T. Wain-Hobson 1979. Paleomagnetic and stratigraphic study of the Loch Shiel marine regression and overlying gyttja. *J. Geol. Soc. Lond.* **136**, 383–8.

Thorarinsson, S. 1981. The application of tephrochronology in Iceland. In *Tephra studies*, S. Self and R. J. S. Sparks (eds), 109–34. Dordrecht: D. Reidel.

Tikhomirov, B. A. 1961. The changes in biogeographical boundaries in the north of USSR as related with climatic fluctuations and activity of man. *Bot. Tidsskr.* **56**, 285–92.

Tricart, J. 1974. Existence des périods sèches au Quaternaire en Amazonie et dans les régions voisines. *Revue Géomorph. Dynam.* **23**, 145–58.

Troll, C. 1973. The upper timberlines in different climatic zones. *Arctic Alpine Res.* **5**, A3–A18.

Troughton, J. H. 1972. Carbon isotope fractionation by plants. In *Proceedings of the 8th International Conference on Radiocarbon Dating*, Vol. 2, T. A. Rafter and T. Grant-Taylor (eds), 421–38. Wellington: Royal Society of New Zealand.

Tsukada, M. 1966. Late Pleistocene vegetation and climate in Taiwan (Formosa). *Proc. Nat. Acad. Sci. U.S.A.* **55**, 543–8.

Twining, A. C. 1833. On the growth of timber. *Am. J. Sci. Arts* **24**, 391–3.

US Committee for GARP (Global Atmospheric Research Program) 1975. *Understanding climatic change.* Washington, D.C.: National Academy of Sciences.

Urey, H. C. 1947. The thermodynamic properties of isotopic substances. *J. Chem. Soc.* **152**, 190–219.

Urey, H. C. 1948. Oxygen isotopes in nature and in the laboratory. *Science* **108**, 489–96.

van den Dool, H. M., H. J. Krijnen, and C. J. E. Schuurmans 1978. Average winter temperatures at De Bilt (The Netherlands), 1634–1977. *Climatic Change* **1**, 319–30.

Van Devender, T. R. 1977. Holocene woodland in the southwestern deserts. *Science* **198**, 189–92.

Van Devender, T. R. and W. G. Spaulding 1979. Development of vegetation and climate in the southwestern United States. *Science* **204**, 701–10.

van Donk, J. 1976. An ^{18}O record of the Atlantic Ocean for the entire Pleistocene. In *Investigation of late Quaternary paleooceanography and paleoclimatology*, R. M. Cline and J. D. Hays (eds), 145–63. Geological Society of America Memoir 145. Boulder, Colo.: Geological Society of America.

Van Loon, H. and J. C. Rogers 1978. The seesaw in winter temperatures between Greenland and northern Europe. Part 1. General Description. *Monthly Weather Rev.* **106**, 296–310.

Van Loon, H. and J. Williams 1976. The connection between trends of mean temperature and circulation at the surface. Part 1 – Winter. *Monthly Weather Rev.* **104**, 365–80.

Van Loon, H., J. J. Taljaard, T. Sasamori, J. London, D. V. Hoyt, K. Labitze, and C. W. Newton 1972. Meteorology of the Southern Hemisphere. *Met. Monogr.* **13** (35).

van Woerkom, A. J. J. 1953. Astronomical theory of climatic change. In *Climatic change*, H. Shapley (ed.), 145–57. Cambridge, Mass.: Harvard University Press.

van Zeist, W. 1967. Archaeology and palynology in the Netherlands. *Rev. Paleobot. Palynol.* **4**, 45–65.

van Zinderen Bakker, E. M. and J. A. Coetzee 1972. A re-appraisal of late Quaternary climatic evidence from Tropical Africa. *Palaeoecol. Afr.* **7**, 151–81.

Vanzolini, P. E. 1973. Paleoclimates, relief, and species multiplication in Equatorial forests. In *Tropical forest ecosystems in Africa and South America: a comparative review.* B. J. Meggers, E. S. Ayensu, and W. D. Duckworth (eds), 255–8. Washington, D.C.: Smithsonian Institution.

Vanzolini, P. E. and E. E. Williams 1970. South American anoles: the geographic differentiation and evolution of the *Anolis chrysolepsis* species group (Sauria, Iguanidae). *Arquivos de Zoologia* **19**, 1–298.

Vernekar, A. D. 1972. Long-period global variations of incoming solar radiation. *Met. Monog.* **12** (34).

Veeh, H. H. and J. M. A. Chappell 1970. Astronomical theory of climatic change: support from New Guinea. *Science* **167**, 862–5.

Verosub, K. L. 1975. Paleomagnetic excursions as magnetostratigraphic horizons: a cautionary note. *Science* **190**, 48–50.

Verosub, K. L. 1977. Depositional and postdepositional processes in the magnetization of sediments. *Rev. Geophys. Space Phys.* **15**, 129–43.

Verosub, K. L. and S. K. Banerjee 1977. Geomagnetic excursions and their paleomagnetic record. *Rev. Geophys. Space Phys.* **15**, 145–55.

Vibe, C. 1967. Arctic animals in relation to climatic fluctuations. *Meddr Grønland* **170** (5).

Vinot-Bertouille, A. C. and J. Duplessy 1973. Individual isotopic fractionation of carbon and oxygen in benthic foraminifera. *Earth Planet. Sci. Letters* **18**, 247–52.

Vuilleumier, B. S. 1971. Pleistocene changes in the flora and fauna of South America. *Science* **173**, 771–80.

Wahl, E. W. and R. A. Bryson 1975. Recent changes in Atlantic surface temperatures. *Nature, Lond.* **254**, 45–6.

Wang, Shao-wu and Zong-ci Zhao 1979. The 36 year wetness oscillation in China and its mechanism. *Acta Met. Sinica* **37**, 64–73.

Wang, Shao-wu and Zong-ci Zhao 1981. Droughts and floods in China, 1470–1979. In *Climate and history*, T. M. L. Wigley, M. J. Ingram, and G. Farmer (eds), 271–88. Cambridge: Cambridge University Press.

Wang, Shao-wu, Zong-ci Zhao, and Zhen-hua Chen 1981. Reconstruction of the summer rainfall regime for the last 500 years in China. *Geojournal* **5**, 117–22.

Wang, Shao-wu, Pei-yuan Zhang, and De-er Zhang n.d. Further studies on the climatic change during historical times in China. Unpublished MS.

Warburton, J. A. and L. G. Young 1981. Estimating ratios of snow accumulation in Antarctica by chemical methods. *J. Glaciol.* **27**, 347–58.

Wardle, P. 1973. Variation of the glaciers of Westland National Park and the Hooker Range, New Zealand. *N. Z. J. Bot.* **11**, 349–88.

Wardle, P. 1974. Alpine timberlines. In *Arctic and alpine environments*, J. D. Ives and R. G. Barry (eds), 371–402. London: Methuen.

Warshaw, M. and R. R. Rapp 1973. An experiment on the sensitivity of a global circulation model. *J. Appl. Met.* **12**, 43–9.

Washburn, A. L. 1979a. Permafrost features as evidence of climatic change. *Earth Sci. Rev.* **15**, 327–402.

Washburn, A. L. 1979b. *Geocryology*. London: Arnold.

Watanabe, O., K. Kato, K. Satow, and F. Okuhira 1978. Stratigraphic analyses of firn and ice at Mizuho Station. In *Ice-coring project at Mizuho Station, east Antarctica*, 25–47, Memoirs of the National Institute of Polar Research (Tokyo), Special Issue No. 10. Tokyo: National Institute of Polar Research.

Watkins, N. D. 1971. Geomagnetic polarity events and the problem of "The Reinforcement Syndrome". *Comments Earth Sci. – Geophys.* **2**, 36–43.

Watkins, N. D. 1972. Review of the development of the geomagnetic polarity time scale and discussion of prospects for its finer definition. *Geol. Soc. Am. Bull.* **83**, 551–74.

Watkins, N. D. and H. G. Goodell 1967. Geomagnetic polarity change and faunal extinction in the southern ocean. *Science* **156**, 1083–7.

Watson, E. 1977. The periglacial environment of Great Britain during the Devensian. *Phil. Trans R. Soc. B* **280**, 183–98.

Webb, T., III 1973. A comparison of modern and pre-settlement pollen from S. Michigan (USA). *Rev. Paleobot. Palynol.* **16**, 137–56.

Webb, T., III 1974. Corresponding patterns of pollen and vegetation in lower Michigan: a comparison of quantitative data. *Ecology* **55**, 17–28.

Webb, T. and R. A. Bryson 1972. Late and post-glacial climatic change in the Northern Mid-West USA: quantitative estimates derived from fossil pollen spectra by multivariate statistical analysis. *Quat. Res.* **2**, 70–115.

Webb, T., III and D. R. Clark 1977. Calibrating micropaleontological data in climatic terms: a critical review. *Ann. N.Y. Acad. Sci.* **288**, 93–118.

Webb, T., III and J. H. McAndrews 1976. Corresponding patterns of contemporary pollen and vegetation in central North America. In *Investigation of late Quaternary paleooceanography and paleoclimatology*, R. M. Cline and J. D. Hays (eds), 267–99, Geological Society of America Memoir 145. Boulder, Colo.: Geological Society of America.

Webb, T., G. Y. Yeracaris, and P. Richard 1978. Mapped patterns in sediment samples of modern pollen from southeastern Canada and northeastern United States. *Géogr. Phys. Quat.* **32**, 163–76.

Webber, P. J. and J. T. Andrews 1973. Lichenometry: a commentary. *Arctic Alpine Res.* **1**, 181–94.

Wehmiller, J. F. and P. E. Hare 1971. Racemization of amino acids in marine sediments. *Science* **173**, 907–11.

Wells, P. V. 1966. Late Pleistocene vegetation and degree of Pluvial climatic change in the Chihuahuan desert. *Science* **153**, 970–5.

Wells, P. V. 1976. Macrofossil analysis of wood rat (*Neotoma*) middens as a key to the Quaternary vegetational history of arid America. *Quat. Res.* **6**, 223–48.

Wells, P. V. 1979. An equable glaciopluvial in the West: pleniglacial evidence of increased precipitation on a gradient from the Great Basin to the Sonoran and Chihuahuan Deserts. *Quat. Res.* **12**, 311–25.

Wells, P. V. and R. Berger 1967. Late Pleistocene history of coniferous woodland in the Mohave Desert. *Science* **155**, 1640–7.

Wells, P. V. and C. D. Jorgensen 1964. Pleistocene wood rat middens and climatic change in the Mohave desert: a record of Juniper woodlands. *Science* **143**, 1171–3.

Wendland, W. M. and R. A. Bryson 1974. Dating climatic episodes of the Holocene. *Quat. Res.* **4**, 9–24.

Westgate, J. A. and M. P. Gorton 1981. Correlation techniques in tephra studies. In *Tephra studies*, S. Self and R. J. S. Sparks (eds), 73–94. Dordrecht: D. Reidel.

Wetherald, R. T. and S. Manabe 1975. The effects of changing the solar constant on the climate of a general circulation model. *J. Atmos. Sci.* **32**, 2044–59.

Weyl, P. K. 1968. The role of the oceans in climatic change: a theory of the ice ages. *Met. Monogr.* **8**, 37–62.

Wigley, T. M. L. 1976. Spectral analysis and the astronomical theory of climatic change. *Nature, Lond.* **264**, 629–31.

Wigley, T. M. L. 1978. Climatic change since 1000 AD. In *Evolution des atmosphères planétaires et climatologie de la terre*, 313–24. Toulouse: Centre National D'Etudes Spatiales.

Wigley, T. M. L., B. M. Gray, and P. M. Kelly 1978. Climatic interpretation of δ^{18}O and δD in tree rings. *Nature, Lond.* **271**, 92–3.

Wigley, T. M. L., M. J. Ingram, and G. Farmer (eds) 1981. *Climate and history: studies in past climates and their impact on man.* Cambridge: Cambridge University Press.

Wijmstra, T. A. 1969. Palynology of the first 30 metres of a 120 m deep section in northern Greece. *Acta Bot. Neerl.* **18**, 511–27.

Wijmstra, T. A. 1978. Palaeobotany and climatic change. In *Climatic change*, J. Gribbin (ed.). Cambridge: Cambridge University Press.

Wijmstra, T. A. and T. van der Hammen 1966. Palynological data on the history of tropical savannas in northern South America. *Leidse Geol. Meded.* **38**, 71–90.

Wiles, W. W. 1967. Pleistocene changes in the pore concentration of a planktonic foraminiferal species from the Pacific Ocean. *Progr. Oceanogr.* **4**, 153–60.

Williams, D. F. and W. C. Johnson 1975. Diversity of recent planktonic foraminifera in the southern Indian Ocean and late Pleistocene paleotemperatures. *Quat. Res.* **5**, 237–50.

Williams, J. 1978. The use of numerical models in studying climatic change. In *Climatic change*, J. Gribbin (ed.), 178–90. Cambridge: Cambridge University Press.

Williams, J., R. G. Barry, and W. M. Washington 1974. Simulation of the atmospheric circulation using the NCAR global circulation model with ice age boundary conditions. *J. Appl. Met.* **13**, 305–17.

Williams, K. M. and G. G. Smith 1977. A critical evaluation of the application of amino acid racemization to geochronology and geothermometry. *Origins of Life* **8**, 91–144.

Williams, L. D. 1975. The variation of corrie elevation and equilibrium line altitude with aspect in eastern Baffin Island, NWT, Canada. *Arctic Alpine Res.* **7**, 169–81.

Williams, L. D. 1979. An energy balance model of potential glacierization of northern Canada. *Arctic Alpine Res.* **11**, 443–56.

Williams, L. D. and T. M. L. Wigley 1983. A comparison of evidence for late Holocene summer temperature variations in the Northern Hemisphere. *Quat. Res.* (in press).

Williams, L. D., T. M. L. Wigley, and P. M. Kelly 1981. Climatic trends at high northern latitudes during the last 4000 years compared with ^{14}C fluctuations. In *Sun and climate*. Toulouse: Centre National D'Etudes Spatiales.

Williams, M. A. J. 1975. Late Pleistocene tropical aridity synchronous in both hemispheres? *Nature* **253**, 617–18.

Williams, R. B. G. 1975. The British climate during the last glaciation; an interpretation based on periglacial phenomena. In *Ice ages: ancient and modern*, A. E. Wright and F. Moseley (eds), 95–120, Geological Journal Special Issue No. 6. Liverpool: Liverpool University Press.

Wilson, A. T. and C. H. Hendy 1971. Past wind strength from isotope studies. *Nature, Lond.* **243**, 344–6.

Wilson, A. T. and C. H. Hendy 1981. The chemical stratigraphy of polar ice sheets – a method of dating ice cores. *J. Glaciol.* **27**, 3–9.

Wilson, A. T. and D. A. House 1965. Fixation of nitrogen by aurora and its contribution to the nitrogen balance of the Earth. *Nature, Lond.* **205**, 793–4.

Wilson, A. T., C. H. Hendy, and C. P. Reynolds 1979. Short-term climatic change and New Zealand temperatures during the last millennium. *Nature, Lond.* **279**, 315–17.

Wilson, C. L. and W. H. Matthews (eds) 1971. *Inadvertent climate modification report of the study of man's impact on climate (SMIC)*. Cambridge, Mass.: MIT Press.

Windom, H. L. 1975. Eolian contributions to marine sediments. *J. Sediment. Petrol.* **45**, 520–9.

Wintle, A. G. 1973. Anomalous fading of thermoluminescence in mineral samples *Nature, Lond.* **244**, 143–4.

Wintle, A. G. 1978. A thermoluminescence dating study of some Quaternary calcite: potential and problems. *Can. J. Earth Sci.* **15**, 1977–86.

Wintle, A. G. 1981. Thermoluminescence dating of late Devensian loesses in southern England. *Nature* **289**, 479–81.

Wintle, A. G. and M. J. Aitken 1977. Thermoluminescence dating of burnt flint: application to a lower palaeolithic site, Terra Amata. *Archaeometry* **19**, 111–30.

Wintle, A. G. and D. J. Huntley 1979. Thermoluminescence dating of a deep-sea ocean core. *Nature, Lond.* **279**, 710–12.

Wintle, A. G. and D. J. Huntley 1980. Thermoluminescence dating of ocean sediments. *Can. J. Earth Sci.* **17**, 348–60.

Wintle, A. G. and D. J. Huntley 1982. Thermoluminescence dating of sediments. *Quat. Sci. Rev.* **1**, 31–54.

Wittfogel, M. A. 1940. Meteorological records from the divination inscriptions of Shang. *Geogr. Rev.* **30**, 110–33.

Woillard, G. M. 1978. Grande Pile Peat Bog: a continuous pollen record for the past 140 000 years. *Quat. Res.* **9**, 1–21.

Woillard, G. M. 1979. Abrupt end of the last interglacial SS in north-east France. *Nature, Lond.* **281**, 558–62.

Woillard, G. M. and W. G. Mook 1982. Carbon-14 dates at Grande Pile: correlation of land and sea chronologies. *Science* **215**, 159–61.

Wollin, G., D. B. Ericson, W. B. F. Ryan, and J. H. Foster 1971. Magnetism of the Earth and climatic changes. *Earth Planet. Sci. Letters* **12**, 175–83.

Wollin, G., W. B. Ryan, and D. B. Ericson 1977. Paleoclimate, paleomagnetism and the eccentricity of the Earth's orbit. *Geophys. Res. Letters* **4**, 267–70.

Wollin, G., W. B. F. Ryan, and D. B. Ericson 1978. Climatic changes, magnetic intensity variations and fluctuations of the eccentricity of the Earth's orbit during the past 2 000 000 years and a mechanism which may be responsible for the relationship. *Earth Planet. Sci. Letters* **41**, 395–7.

Woodwell, G. M., R. H. Whittaker, W. A. Reiners, G. E. Likens, C. C. Delwiche, and D. B. Botkin 1978. The biota and the world carbon budget. *Science* **199**, 141–6.

Worthington, L. V. 1968. Genesis and evolution of water masses. *Met. Monogr.* **8**, 63–7.

Wright, H. E. 1964. Aspects of the early post-glacial forest succession in the Great Lakes region. *Ecology* **45**, 439–48.

Wright, H. E. and H. L. Patten 1963. The pollen sum. *Pollen et Spores* **5**, 445–50.

Wyjmstra, T. A. 1969. Palynology of the first 30 metres of a 120 m deep section in Northern Greece. *Acta Bot. Neerl.* **18**, 511–26.

Yamamoto, R. and T. Iwashima 1975. Change of the surface air temperature averaged over the Northern Hemisphere and large volcanic eruptions during the years 1951–1972. *J. Met. Soc. Japan* **53**, 482–6.

Yamamoto, T. 1971. On the nature of the climatic change in Japan since the "Little Ice Age" around 1800 AD. *J. Met. Soc. Japan* **49** (special issue), 798–812.

Yao, Shan-yu 1942. The chronological and seasonal distribution of floods and droughts in Chinese history 206 BC–1911 AD. *Harvard J. Asiatic Studies* **6**, 273–312.

Yao, Shan-yu 1943. The geographical distribution of floods and droughts in Chinese history 206 BC–AD 1911. *Far East Qtly* **2**, 357–78.

Yao, Shan-yu 1944. Flood and drought data in the *T'u Shu Chi Ch'eng* and the *Ch'ing Shi Kao. Harvard J. Asiatic Studies* **8**, 214–26.

Yapp, C. J. and S. Epstein 1977. Climatic implications of D/H ratios of meteoric water over North America (9500–22 000 BP) as inferred from ancient wood cellulose C–H hydrology. *Earth Planet Sci. Letters* **34**, 333–50.

Yapp, C. J. and S. Epstein 1982. Climatic significance of the hydrogen isotope ratios in tree cellulose. *Nature, Lond.* **297**, 636–9.

Yarranton, G. A. and J. C. Ritchie 1972. Sequential correlation as an aid in placing pollen zone boundaries. *Pollen et Spores* **14**, 213–23.

Yoshino, M. M. 1981. Orographically-induced atmospheric circulations. *Progr. Phys. Geogr.* **5**, 76–98.

Young, M. and R. S. Bradley 1984. Insolation gradients and the paleoclimatic record. In *Milankovitch and climate*, Part 2, A. L. Berger, J. Imbrie, J. Hays, G. Kukla and B. Saltzman (eds), 707–13. Dordrecht: D. Reidel.

Zagwijn, W. H. and R. Paepe 1968. Die stratigraphie der weichselzeitiger Ablagerungen der Niederlande und Belgiens. *Eiszeitalter und Gegenwart* **19**, 129–46.

Zeller, E. J. and B. C. Parker 1981. Nitrate ion in Antarctic firn as a marker for solar activity. *Geophys. Res. Letters* **8**, 895–8.

Zhang, Pei-yuan and Gao-fa Gong 1979. Some characteristics of climatic fluctuations in China since the 16th century. *Acta Met. Sinica* **34**, 238–47 (in Chinese with English summary: translation available from E. J. Bradley, c/o author).

Acknowledgements

In the preparation of a book of this kind, it is necessary to draw together illustrative material from many sources, and I would like to thank those individuals and organizations who have given permission for the reproduction of their copyright material. Every effort has been made to contact copyright holders: I apologize in advance for any inadvertent omissions. In the list that follows, numbers in parentheses refer to text figures:

Figure 1.1 reproduced from *Climates through geological time* (L. A. Frakes) by permission of the author; World Meteorological Organisation (2.1); Figures 2.3 and 2.8 reproduced with permission from *Understanding climate change*, National Academy Press, Washington, DC, 1975; E. M. Van Zinderen Bakker (2.5, 8.9); G. Kukla (2.6); Figures 2.6, 2.11–13 reproduced from *Climatic change* (J. Gribbin, ed.) by permission of Cambridge University Press; Figure 2.9 reproduced by G. Kukla & D. Robinson, *Monthly Weather Rev.* **108**, 56–68 by permission of G. Kukla and the American Meteorological Society; Figure 2.10 reproduced with permission from G. Kukla, IASH Publication No. 131; J. E. Kutzbach (2.14 & 15); Figures 2.16–17, 6.14–16, 6.31, 7.15–18, 9.5, 9.14, and 9.23 by permission of the Editor, *Quaternary Research*; A. Berger (2.17, 2.20–21); Figures 2.20–21 also by permission of the Editor, *Il Nuovo Cimento*; J. Mangerud and the Editor, *Boreas* (3.2 & 7); Figure 3.3 reproduced from I. U. Olsson, *Earth Sci. Rev.* **4**, 203–18 by permission of the author and Elsevier Scientific; I. U. Olsson and the Editor, *Geologiska Foereningens I Stockholm Foertlandlinger* (3.4–5); D. Reidel Publishing (3.7, 3.10, 11.15–16); C. W. Ferguson (3.8); Figure 3.9 by permission of the Editor, *Radiocarbon*; Figure 3.11 reproduced from M. Stuiver & P. D. Quay, *Science* **207**, 11–19 by kind permission of the publisher and M. Stuiver, © 1980 by the American Association for the Advancement of Science; Figures 3.12, 4.12 and 11.10 reproduced from *Climate and history* (T. M. L. Wigley *et al.*, eds) by permission of Cambridge University Press; Figure 3.15 reproduced from *Calibration of hominoid evolution* (Bishop & Miller) by kind permission of Scottish Academic Press; Figure 3.16 reproduced by permission of the American Geophysical Union; S. Fleming (3.20 & 21); Figure 4.2 reproduced from K. L. Verosub, *Science* **190**, 48–50 by permission of the publisher and author, © 1975 by the American Association for the Advancement of Science; Figure 4.4 reproduced from N. D. Opdyke, *Rev. Geophys. Space Phys.* **10**, 213–49, © 1972 by the American Geophysical Union; Figure 4.5 reproduced from J. D. Hays, *Geol Soc. Am. Bull.* **82**, 2433–47 by permission of the author and the Geological Society of America; Figure 4.6 reproduced from G. Wollin *et al.*, *Earth Planet. Sci. Lett.* **41**, 395–7 by permission of G. Wollin and Elsevier Scientific; Figure 4.8 reproduced from *Biogeochemistry of amino acids* (P. E. Hare *et al.*, eds) by permission of John Wiley & Sons, © 1980 by John Wiley & Sons; Figure 4.9 reprinted by permission of G. Miller and the publisher from *Nature* **281**, 539–43, © 1979 Macmillan Journals; Figure 4.10 reproduced from R. M. Mitterer, *Geol.* **2**, 425–9 by permission of the author and the Geological Society of America; Figure 4.11 reproduced from K. L. Pierce *et al.*, *Geol Soc. Am. Bull.* **87**, 703–10 by permission of K. L. Pierce and the Geological Society of America; Regents of the University of Colorado (4.13, 4.15, 7.7, 7.10); G. H. Miller (4.14); University of Toronto Press (4.16); Figure 5.3 reproduced from S. Epstein & R. P. Sharp, *J. Geol.* **67**, 88–102 by permission of the publisher, © 1959 by the University of Chicago; Figure 5.4 reproduced from G. de Q. Robin, *Phil Trans R. Soc. B* **280** 143–68 by permission of The Royal Society and the author; Figure 5.5 reprinted by permission of E. Picciotto and publisher from *Nature* **187**, 857–9, © 1960 Macmillan Journals; Figure 5.6 reproduced from L. Aldaz & S. Deutsch, *Earth Planet. Sci. Lett.* **3**, 267–74 by permission of Elsevier Scientific; Figure 5.7 reproduced from W. Dansgaard & H. Tauber, *Science* **166**, 499–502 by permission of W. Dansgaard and the publisher, © 1969 by the American Association for the Advancement of Science; W. Dansgaard and the Kommissionen for Videnskabelige Undersogelser I Grønland (5.8); Figure 5.9 reproduced from the *Journal of Glaciology* by permission of the International Glaciological Society and F. M. Koerner; Figures 5.10 and 5.20 reprinted by permission of the publisher and S. J. Johnsen from *Nature* **235**, 429–34, © 1972 Macmillan Journals; S. J. Johnsen and the International Association of Scientific Hydrology (5.11); C. U. Hammer and the International Association of Scientific Hydrology (5.12); Figure 5.13 reproduced from the *Journal of Glaciology* by permission of the International Glaciological Society, C. H.

Hendy and A. T. Wilson; Figures 5.14–15 reprinted by permission of the publisher and C. U. Hammer from *Nature* **288**, 230–5, © 1980 Macmillan Journals; Figure 5.16 reproduced from the *Journal of Glaciology* by permission of the International Glaciological Society and N. Rech; Figure 5.17 reproduced from W. Dansgaard *et al.*, *Science* **166**, 377–81 by permission of W. Dansgaard and the publisher, © 1969 by the American Association for the Advancement of Science; Figures 5.18 & 19 reproduced from *Late Cenozoic glacial ages* (K. K. Turekian, ed.) by permission of Yale University Press and W. Dansgaard, © 1971 Yale University Press; Figure 5.21 reproduced from W. Dansgaard *et al.*, *Science* **218**, 1273–7 by permission of W. Dansgaard and the publisher, © 1982 by the American Association for the Advancement of Science; Figures 5.22–23 reprinted by permission of the publisher and W. S. B. Paterson from *Nature* **266**, 508–11, © 1977 Macmillan Journals; Figure 5.24 reprinted by permission of the publisher and W. Dansgaard from *Nature* **255**, 24–8, © 1975 Macmillan Journals; Figure 5.25 reprinted by permission of the publisher and S. J. Johnsen from *Nature* **277**, 482–3, © 1970 Macmillan Journals; Figure 5.26 reprinted by permission of the publisher and M. M. Herron from *Nature* **293**, 389–91, © 1981 Macmillan Journals; Figure 5.27 reprinted by permission of the publisher and D. Raynaud from *Nature* **281**, 289–91, © 1979 Macmillan Journals; Figure 5.27 reprinted by permission of the publisher and D. Raynaud from *Nature* **281**, 289–91, © 1979 Macmillan Journals; Figure 5.28 reprinted by permission of the publisher and R. J. Delmas from *Nature* **284**, 155–7, © 1980 Macmillan Journals; Figure 5.29 kindly supplied by L. G. Thompson, up-dated from Thompson & Mosley–Thompson, 1981; Arctic Institute of North America and J. H. Cragin (5.30); Figure 5.31 reproduced from A. J. Gow & T. Williamson, *Earth Planet. Sci. Lett.* **13**, 210–18 by permission of A. J. Gow and Elsevier Scientific; Figures 5.32–33 reproduced by permission of the American Geophysical Union; W. W. Hay (6.1); Figures 6.2–3 & 6.5 reproduced courtesy of W. F. Ruddiman, Lamont–Doherty Geological Observatory; Figure 6.6 reprinted by permission of the publisher and A. Boersma from *Introduction to marine micropaleontology* (B. U. Haq & A. Boersma, eds), © 1978 by Elsevier Science Publishing; Figure 6.7 reproduced from C. Emiliani, *Earth Planet. Sci. Lett.* **37**, 349–52 by permission of Elsevier Scientific; Figures 6.8 & 9 reprinted from *Physical oceanography*, Vol. 1 (A. Defant) by permission of Pergamon Press; Figure 6.10 reproduced from N. J. Shackleton & N. D. Opdyke, *Mem. Geol Soc. Am.* 145, by permission of N. J. Shackleton and the Geological Society of America; Figures 6.11 and 6.32 reproduced from W. F. Ruddiman, *Geol Soc. Am. Bull.* **88**, 1813–27 by permission of the author and the Geological Society of America; Figure 6.12 reproduced from L. Lidz, *Science* **154**, 1448–52 by permission of B. H. Lidz and the publisher, © 1966 by the American Association for the Advancement of Science; Figure 6.13 reproduced from W. F. Ruddiman, *Geol Soc. Am. Bull.* **82**, 283–302 by permission of the author and the Geological Society of America; Figure 6.17 reproduced from *Late Cenozoic glacial ages* (K. K. Turekian, ed.) by permission of Yale University Press and J. Imbrie, © 1971 Yale University Press; J. Imbrie and the Editor, *Quaternary Research* (6.18 & 19); Figures 6.20–21 reproduced from A. McIntyre, *Mem. Geol Soc. Am.* 145, by permission of the author and the Geological Society of America; A. McIntyre (6.22); Figures 6.23–24 reproduced from T. C. Moore, *Marine Micropaleontology* **5**, 215–47 by permission of the author and Elsevier Scientific; W. I. Prell and the Editor, *Quaternary Research* (6.25–26); Figure 6.27 reproduced from A. W. H. Bé and J. C. Duplessy, *Science* **194**, 419–22 by permission of A. W. H. Bé and the publisher, © 1976 by the American Association for the Advancement of Science; A. W. H. Bé (6.28); W. H. Berger (6.29); Figure 6.30 reprinted by permission of W. F. Ruddiman and the publisher, from W. F. Ruddiman & B. C. Heezen, *Deep Sea Research* **14**, 801–8, © 1967 Pergamon Press; Figure 6.33 reproduced from W. F. Ruddiman and A. McIntyre. *Science* **204**, 173–5 by permission of W. F. Ruddiman and the publisher, © 1979 by the American Association for the Advancement of Science; Figures 6.34–35 reproduced from W. F. Ruddiman and A. McIntyre, *Science* **212**, 617–27 by permission of W. F. Ruddiman and the publisher, © 1981 by the American Association for the Advancement of Science; L. Washburn (7.1 & 2); *Biuletyn Periglacjalry* (7.3); T. Péwé (7.4); Figure 7.6 reproduced from the *Journal of Glaciology* by permission of the International Glaciological Society and S. Hastenrath; Figure 7.8 reproduced from the *Journal of Glaciology* by permission of the International Glaciological Society; Edward Arnold (7.10); R. Cooke (7.11); S. P. Nicholson and D. Reidel Publishing (7.19 & 20); R. S. Harmon and the National Research Council of Canada (7.24); Figure 7.25 reproduced from P. Thompson *et al.*, *Geol Soc. Am. Bull.* **87**, 1730–8 by permission of P. Thompson and the Geological Society of America; C. J. Sorensen (8.1); J. C. Ritchie and the Editor, *Quaternary Research* (8.2); W. Karlén and the Editor, *Geografiska Annaler* (8.3); Figure 8.13 reproduced by permission of the Royal Society; Figure 9.1 reproduced from *History of the British flora* (H. Godwin) by permission of Cambridge University Press; Figures 9.2–3 reproduced from *Quaternary plant ecology* (H. J. B. Birks & R. G. West, eds) by permission of Blackwell Scientific; M. B. Davis and the Editor, *American Scientist* (9.4 & 9); Figures 9.6–7 reproduced from T. Webb III, *Ecology* **55**, p. 23 by permission of the publisher and author, © 1974, the Ecological Society of America; Figure 9.8 kindly supplied by W. Patterson: Figure 9.10 reproduced from T. Webb III, *Rev. Paleobotany and Palynology* **16**, 137–56 by permission of the author and Elsevier Scientific; Figures 9.11–12 reproduced from *Rev. Paleobotany and Palynology* by permission of Elsevier Scientific; J. C. Bernabo and the Editor, *Quaternary Research* (9.13 & 23); H. Nichols (9.15); Figure 9.16 reproduced by permission of the author and the Geological Society of America; Figure 9.19 reproduced by permission of the publisher from D. M. Churchill, *Aust. J. Bot.* **16**, 128; T. Webb III and the Editor, *Quaternary Research* (9.22); H. C. Fritts (10.1–11, 10.18, 10.22–24, 10.27 and see subsequent entries); the Editor, *Quaternary Research* (10.2–4, 10.7, 10.22–24); Figures 10.12–13 reproduced from V. C. Lamarche, *Science* **183**, 1043–8 by permission of the author and publisher, © 1974 by the American Association for the Advancement of Science; Figure 10.14 reprinted by permission of the publisher and E. R. Cook from *Nature* **282**, 390–2, © 1979 Macmillan Journals; Figures 10.15–16 reprinted by permission from *Long-term streamflow records reconstructed from tree rings*, Paper #5 of the Laboratory of Tree-Ring Research (Charles W. Stockton), © 1975 University of Arizona Press; Figure 10.17 reproduced from C. W. Stockton & H. C. Fritts, *Water Resources Bull.* **9**, 1006–27 by permission of the American Water Resources Association; Figures 10.19–21 reproduced from H. C. Fritts *et al.*, *J. Appl. Meter.* **10**, 845–64 by permission of H. C. Fritts and the American Meteorological Society; T. J. Blasing, H. C. Fritts and G. Weller (10.25); F. H. Schweingruber and the Editor, *Boreas* (10.26) F. W. Jones (10.27); F. H. Schweingruber and the Editor, *Tree Ring Bull.* (10.28); Figure 10.29 reprinted by permission of the publisher and S. Epstein from *Nature* **266**, 477–8, © 1977 Macmillan Journals; Figure 10.30 reprinted from S. Epstein & C. J. Yapp, *Earth Planet. Sci. Lett.* **30**, 252–66 by permission of S. Epstein and Elsevier Scientific; Figure 10.31 reprinted by permission of the publisher and J. Gray from *Nature* **262**, 481–2, © 1976 Macmillan Journals; Figure 10.32 reprinted by permission of the publisher and L. M. Libby from *Nature* **261**, 284–8, © 1976 Macmillan Journals; Figures 10.33–34 reproduced from C. J. Yapp & S. Epstein, *Earth Planet. Sci. Lett.* **34**, 333–50 by permission of C. J. Yapp and Elsevier Scientific; the Editor, *Jökull* (11.2 & 3); Figure 11.4 reprinted by permission of the publisher and T. M. L. Wigley from *Nature* **276**, 329–34, © 1978 Macmillan Journals; C. Pfister (11.5); T. Yamamoto and the Meteorological Society of Japan (11.16); Z. Beijing (11.7); H. H. Lamb (11.8); Figure 11.9 reproduced from H. Neuberger, *Weather* **25**, 46–56 by permission of the Royal Meteorological Society; Figure 11.17 reproduced from A. J. W. Catchpole *et al.*, *Can. Geographer* **20**, 279–97 by permission of the authors and the Canadian Association of Geographers; Figure 11.18 reproduced from A. A. Lindsey & J. E. Newman, *Ecology* **37**, 812–23 by permission of the Ecological Society of America; Figure 11.19 reprinted from *J. Interdisciplinary History* **X** (1980), 844–5 by permission of the *Journal of Interdisciplinary History* and the MIT Press, Cambridge, Mass., © 1982 by the Massachusetts Institute of Technology and the Editors of the *Journal of Interdisciplinary History*.

Index